Water and Climate Change

Water and Climate Change

Sustainable Development, Environmental and Policy Issues

Edited by

Trevor M. Letcher

*School of Chemistry, University of KwaZulu-Natal,
Durban, South Africa*

ELSEVIER

Elsevier
Radarweg 29, PO Box 211, 1000 AE Amsterdam, Netherlands
The Boulevard, Langford Lane, Kidlington, Oxford OX5 1GB, United Kingdom
50 Hampshire Street, 5th Floor, Cambridge, MA 02139, United States

Copyright © 2022 Elsevier Inc. All rights reserved.

No part of this publication may be reproduced or transmitted in any form or by any means, electronic or mechanical, including photocopying, recording, or any information storage and retrieval system, without permission in writing from the publisher. Details on how to seek permission, further information about the Publisher's permissions policies and our arrangements with organizations such as the Copyright Clearance Center and the Copyright Licensing Agency, can be found at our website: www.elsevier.com/permissions.

This book and the individual contributions contained in it are protected under copyright by the Publisher (other than as may be noted herein).

Notices

Knowledge and best practice in this field are constantly changing. As new research and experience broaden our understanding, changes in research methods, professional practices, or medical treatment may become necessary.

Practitioners and researchers must always rely on their own experience and knowledge in evaluating and using any information, methods, compounds, or experiments described herein. In using such information or methods they should be mindful of their own safety and the safety of others, including parties for whom they have a professional responsibility.

To the fullest extent of the law, neither the Publisher nor the authors, contributors, or editors, assume any liability for any injury and/or damage to persons or property as a matter of products liability, negligence or otherwise, or from any use or operation of any methods, products, instructions, or ideas contained in the material herein.

ISBN: 978-0-323-99875-8

For Information on all Elsevier publications
visit our website at https://www.elsevier.com/books-and-journals

Publisher: Candice Janco
Acquisitions Editor: Louisa Munro
Editorial Project Manager: Michelle Fisher
Production Project Manager: Sruthi Satheesh
Cover Designer: Miles Hitchen

Typeset by MPS Limited, Chennai, India

Contents

List of contributors .... *xvii*
Preface .... *xix*

Section A Introduction 1

Chapter 1: Introduction: water, the vital chemical 3
Trevor M. Letcher

    1.1 Introduction 3
    1.2 The unique chemical properties of water 4
          1.2.1 The polar nature of water 4
          1.2.2 The high enthalpy of vaporization of water 4
          1.2.3 The high heat capacity of water 5
          1.2.4 The anomalous density of frozen water 5
          1.2.5 Water, the "universal" solvent 6
          1.2.6 Acid−base property 7
          1.2.7 High surface tension, low viscosity, and cohesive and adhesive properties 7
    1.3 Water and climate change 8
    1.4 The origin of water on Earth 9
          1.4.1 The water cycle 9
    1.5 The scarcity of water 10
    1.6 Conclusions 10
    References 10

Chapter 2: The root causes of climate change and the role played by water 13
Trevor M. Letcher

    2.1 Introduction 13

2.2 Water and CO_2 and the greenhouse effect 14
2.3 The main greenhouse gases of anthropogenic origin 16
2.4 Feedback mechanisms and tipping points 18
2.5 Where are we in solving the problem of global heating? 19
2.6 Solutions 22
2.7 Conclusion 25
References 25

Chapter 3: Water resource planning and climate change 27

Rabee Rustum, Adebayo J. Adeloye and Quan Dau

3.1 Introduction 27
3.2 Climate change and water budget components 28
3.3 Water reservoirs 29
    3.3.1 Indices of reservoir performance 30
    3.3.2 Reservoir operation 31
    3.3.3 Climate change effect on reservoir performance in the Indus Basin 32
3.4 Conclusions 38
References 38

Chapter 4: Potential impacts of climate change on biogeochemical cycling 41

Daniel A. Vallero

4.1 Introduction 41
4.2 Thermodynamics of cycling 42
4.3 Partitioning 45
    4.3.1 Dissolution 45
    4.3.2 Sorption 47
4.4 Biochemodynamics 49
4.5 Biogeochemistry of oxygen in water 55
4.6 The carbon cycle and greenhouse effect 57
4.7 Conclusions 59
References 59
Further reading 62

Section B Sustainable development and environmental issues 63

Chapter 5: Water quality engineering: physical, chemical, and biological treatment for a sustainable future 65

Daniel A. Vallero

5.1 Introduction 65

5.2	Water quality	66
5.3	Chemical processes	69
5.4	Wastewater treatment	73
5.5	Water treatment to manage risks	79
5.6	Disposal of sludge	80
5.7	Mixed reactors	82
5.8	Aerobic reactors	83
5.9	Anaerobic reactors	86
5.10	Treating contaminated ground water	89
5.11	Conclusions	92
References		92

Chapter 6: Urban water supplies in developing countries with a focus on climate change 95

Josephine Treacy

6.1	Introduction	95
6.2	Urbanization in developing countries	97
6.3	Urban water supplies	97
	6.3.1 Source waters	99
	6.3.2 Other water sources innovative needs	100
	6.3.3 Urban water supplies as a resource	101
	6.3.4 Centralized and noncentralized water supplies	101
6.4	Climate change challenges and urban waters	102
	6.4.1 Water Supplies and demand	103
6.5	Waste management and climate change and its impact on urban water	104
6.6	Opportunities and global initiatives	104
	6.6.1 Adaptation and mitigation	105
	6.6.2 Understanding the water baseline	105
	6.6.3 Environmental impact modeling risk	105
6.7	Water security	106
	6.7.1 Policy makers	106
6.8	Conclusions	107
References		109
Further reading		112

Chapter 7: Water purity and sustainable water treatment systems for developing countries 115

Joanne Mac Mahon

7.1	Introduction: access to clean water in developing countries	115

viii Contents

    7.2 Environmental challenges to water purity in developing countries.....................118
    7.3 WHO guidelines for water purity..120
        7.3.1 Microbial guidelines 120
        7.3.2 Chemical guidelines..121
        7.3.3 Radiological guidelines...122
        7.3.4 Acceptability: taste, odor, appearance 122
        7.3.5 Other considerations 122
    7.4 Water supply sources used in developing countries 123
    7.5 Water treatment systems used in developing countries 124
        7.5.1 Centralized water treatment systems 124
        7.5.2 Decentralized water treatment systems.....................................126
        7.5.3 Water Safety Plans..126
        7.5.4 Commonly used water treatment methods................................128
        7.5.5 Water storage...132
    7.6 Sustainable water management systems...133
    7.7 Conclusions 135
    References 136

Chapter 8: Water purification techniques for the developing world 145
Aniruddha B. Pandit and Jyoti Kishen Kumar

    8.1 Introduction 145
    8.2 Water purification methodologies 147
    8.3 Solar treatment and its intensification 147
    8.4 Water disinfection by boiling...150
    8.5 Chemical treatment...152
    8.6 Filtration techniques...154
        8.6.1 Traditional filtration methods...154
        8.6.2 Recent advances/modifications in traditional filtration methods............156
        8.6.3 Hybrid filtration methods...158
    8.7 Natural treatment methods 159
    8.8 Cavitation-based water hand pumps...160
    8.9 Sustainable disinfection of harvested rainwater...163
    8.10 Recent research on emerging methods...166
        8.10.1 Filtration 166
        8.10.2 Solar disinfection 166
        8.10.3 Hybrid techniques 167
    8.11 Impact of Covid 19 pandemic on the water sector in developing countries......168
    8.12 Comparison of various purification techniques 170
    8.13 Conclusions 173

8.14 Recommendations for future work 174
References 174

Chapter 9: Plastic pollution in waterways and in the oceans 179
Lei Mai, Hui He and Eddy Y. Zeng

9.1 Introduction 179
9.2 Global marine plastic pollution 180
    9.2.1 Sources of marine plastic debris 180
    9.2.2 Distribution of microplastics in the marine environment 182
9.3 Plastic pollution in rivers 183
    9.3.1 Distribution of plastics in global rivers 183
    9.3.2 Riverine transport of microplastics to coastal oceans/seas 187
9.4 Riverine plastic outflows 188
    9.4.1 Field measured riverine microplastic outflows 188
    9.4.2 Comparison of riverine plastic outflows between measurements and model estimates 190
References 192

Chapter 10: Desalination and sustainability 197
Anju Vijayan Nair and Veera Gnaneswar Gude

10.1 Introduction 197
10.2 Description of desalination processes 198
    10.2.1 Solar stills 198
    10.2.2 Multi-stage flash desalination 199
    10.2.3 Multi-effect distillation 199
    10.2.4 Vapor compression 199
    10.2.5 Energy efficiency of thermal desalination 200
    10.2.6 Reverse osmosis 200
    10.2.7 Membrane distillation 201
10.3 Desalination and sustainability 203
    10.3.1 Environmental footprint 203
    10.3.2 Economics of desalination 205
    10.3.3 Social aspects of desalination 206
10.4 Renewable energy integration 207
    10.4.1 Selection process of desalination process 209
10.5 Future of sustainable desalination 210
    10.5.1 Desalination at household level 210
    10.5.2 Desalination at community or municipal scale 211
References 212

Chapter 11: Groundwater sustainability in a digital world 215

Ahmed S. Elshall, Ming Ye and Yongshan Wan

11.1 Introduction 215
11.2 Groundwater sustainability 217
11.3 Digital groundwater 220
11.4 Internet of Things—based data collection 221
11.5 Web-based data sharing 222
11.6 Workflow for data processing 225
11.7 Scenarios for data usage 228
11.8 Perspectives of web-based groundwater platforms 230
11.9 Disclaimer 233
Acknowledgments 233
References 234

Chapter 12: Economics of water productivity and scarcity in irrigated agriculture 241

James F. Booker

12.1 Introduction 241
12.2 Productivity concepts as indicators 243
12.3 Production and scarcity in economics 248
12.4 Constructing water productivity indicators: physical and economic considerations 249
    12.4.1 Physical considerations 250
    12.4.2 Economic considerations 251
    12.4.3 Net water productivity 252
12.5 Trends in water productivity 253
    12.5.1 Physical and economic water productivity trends 253
    12.5.2 Economic net water productivity trends: value added, optimization, and hydroeconomic models 254
    12.5.3 Other economic approaches to water productivity and efficiency 256
12.6 Conclusions 257
References 258

Chapter 13: Potential of municipal wastewater for resource recovery and reuse 263

Manzoor Qadir

13.1 Introduction 263
13.2 Wastewater as a water source 264
13.3 Wastewater as a nutrient source 265

13.4 Wastewater as an energy source 268
13.5 Conclusions and prospects 269
References 270

Chapter 14: Sustainable freshwater management—the South African approach 273
Oghenekaro Nelson Odume

14.1 Introduction 273
14.2 The classification system and classification of water resources 275
14.3 The determination of the reserve and resource quality objectives 277
14.4 Reflections on the resource directed measure components using the Vaal Barrage catchment in South Africa as a case study 278
14.5 Source directed controls 287
14.6 Linking resource directed measure and source directed control instruments for sustainable freshwater resources management 287
14.7 Conclusions 290
Acknowledgments 290
References 290

Chapter 15: Sustainable water management with a focus on climate change 293
Thomas Shahady

15.1 Introduction and background 294
    15.1.1 What is sustainable water management 294
    15.1.2 The central premise of climate change impacting water resources 294
15.2 Impacts from climate change 295
    15.2.1 Changes to the water cycle 295
    15.2.2 Changes in water storage 296
    15.2.3 Changes in precipitation patterns 297
    15.2.4 Changes to evapotranspiration rates 298
    15.2.5 Changes in hydraulic retention time 298
    15.2.6 Changes in pollutant loading and processing 298
15.3 Managing worsening concerns 299
    15.3.1 Eutrophication 300
    15.3.2 Dead zones 300
    15.3.3 Red tides 300
    15.3.4 Flooding 301
    15.3.5 Drought 302
    15.3.6 Pollutants of emerging concern 302
15.4 Sustainable water management: a need for change 304
    15.4.1 Challenges to effective management 304

	15.4.2	Stream restoration	305
	15.4.3	Green infrastructure	306
	15.4.4	Dam removal and wetland creation	307
	15.4.5	Stormwater management	308
	15.4.6	Sanitation	308
	15.4.7	Sociopolitics and economics	309
	15.4.8	Mitigation, protection, and ecological services	310
15.5	A sustainable water future		310
References			311

Chapter 16: Food-energy-water nexus and assessment models 317

Anju Vijayan Nair and Veera Gnaneswar Gude

16.1	Introduction		317
16.2	Assessment of food-energy-water nexus		318
16.3	Food-energy-water nexus model development		319
	16.3.1	WEF Nexus Tool 2.0	319
	16.3.2	CLEWS	320
	16.3.3	WEF Nexus Rapid Appraisal Tool	320
	16.3.4	MuSIASEM	320
	16.3.5	Foreseer	320
	16.3.6	WEAP-LEAP	321
16.4	Comparison of different models		322
16.5	Applications of F-E-W nexus models		322
16.6	Future considerations		328
References			329

Chapter 17: Emerging water pollutants 331

Daniel A. Vallero

17.1	Introduction	331
17.2	Dose-response relationships	332
17.3	Uncertainty	337
17.4	Conclusions	339
References		340

Chapter 18: Local representations of a changing climate 343

Juan Baztan, Scott Bremer, Charlotte da Cunha, Anne De Rudder, Lionel Jaffrès, Bethany Jorgensen, Werner Krauß, Benedikt Marschütz, Didier Peeters, Elisabeth Schøyen Jensen, Jean-Paul Vanderlinden, Arjan Wardekker and Zhiwei Zhu

18.1	Introduction	344

18.2 Local conditions of a changing climate 344
    18.2.1 Bergen, Norway 344
    18.2.2 Brest, Kerourien, France 345
    18.2.3 Dordrecht, the Netherlands 346
    18.2.4 Gulf of Morbihan, France 347
    18.2.5 Jade Bay, Germany 348
18.3 Art and science local representation processes by site and related challenges 349
    18.3.1 Bergen, Norway 349
    18.3.2 Brest, Kerourien, France 349
    18.3.3 Dordrecht, the Netherlands 351
    18.3.4 Gulf of Morbihan, France 352
    18.3.5 Jade Bay, Germany 354
18.4 Metadata and dynamic mapping perspectives for local representations 356
18.5 Lessons learned and final conclusions 358
    18.5.1 Conclusions on local representations for codeveloping climate services 359
    18.5.2 Final conclusions 360
Acknowledgments 361
References 361

Chapter 19: Agricultural water pollution 365
Thomas Shahady

19.1 Introduction and background 365
    19.1.1 Water consumption and contamination tied to agriculture 365
    19.1.2 Pollutants 366
    19.1.3 Land Use 367
19.2 Pollution problems generated from agricultural practices 368
    19.2.1 Biosolids and contamination 368
    19.2.2 Fertilizers and eutrophication 369
    19.2.3 Toxic contamination from pesticides 370
    19.2.4 Bacterial contamination 370
19.3 Shifting practices and climate change 372
    19.3.1 Globalization and trade 372
    19.3.2 Climate change 372
    19.3.3 Land management (best management practices) 373
    19.3.4 Emerging concerns 373
19.4 Water pollution control 374
    19.4.1 Water quality 374
    19.4.2 Best management practices 375

xiv Contents

    19.4.3 Land management practices 375
    19.4.4 Intensive management practices 376
 19.5 Agriculture and a sustainable future 377
 References 377

Chapter 20: Environmental impacts on global water resources and poverty, with a focus on climate change 383

Claudia Yazmín Ortega Montoya and Juan Carlos Tejeda González

 20.1 Environmental impacts on national/regional water resources 384
 20.2 Impacts of climate change on poverty 385
 20.3 Exposure and vulnerability impacts 387
 20.4 Slow-onset events 389
 20.5 Future trends 393
 References 394

Section C Policy issues 397

Chapter 21: Climate change and water justice 399

M. Mills-Novoa, R. Boelens and J. Hoogesteger

 21.1 Introduction 399
 21.2 What is justice and for whom? Examples from the water sector 401
 21.3 Defining water justice 403
 21.4 Water justice and climate change vulnerability 404
 21.5 Water justice and climate change's proposed solutions 406
    21.5.1 Mitigation 407
    21.5.2 Adaptation 408
 21.6 Naturalizing climate change 410
 21.7 Struggle(s) for water and climate justice 411
 21.8 Conclusions 412
 References 412

Chapter 22: Environmental ethics and sustainable freshwater resource management 419

Oghenekaro Nelson Odume and Chris de Wet

 22.1 Introduction 419
 22.2 Key issues and the need for environmental ethical considerations in water resource management 421
 22.3 Approaches to environmental ethics from a western perspective 423

		22.3.1 Value-oriented environmental ethics 423
		22.3.2 Relationship-based environmental ethics 427
		22.3.3 African environmental ethics 430
22.4	Relating environmental ethics to water resource management in the context of complex socio-ecological systems 432	
22.5	Conclusions 433	
Acknowledgments 434		
References 434		

Index 439

List of contributors

Adebayo J. Adeloye Heriot-Watt University, Edinburgh, United Kingdom

Juan Baztan University of Versailles Saint-Quentin-en-Yvelines, UPS-CEARC, Guyancourt, France

R. Boelens Water Resources Management Group, Department of Environmental Sciences, Wageningen University, Wageningen, The Netherlands; Centre for Latin American Research and Documentation (CEDLA), University of Amsterdam, Amsterdam, The Netherlands; Faculty of Agronomy, Central University of Ecuador, Quito, Ecuador

James F. Booker Siena College, Loudonville, NY, United States

Scott Bremer Centre for the Study of the Sciences and the Humanities, University of Bergen, Bergen Norway

Charlotte da Cunha University of Versailles Saint-Quentin-en-Yvelines, UPS-CEARC, Guyancourt, France

Quan Dau School of Climate Change and Adaptation, University of Prince Edward Island, Charlottetown, Prince Edward Island, Canada

Anne De Rudder Institut Royal d'Aéronomie Spatiale de Belgique, Bruxelles, Belgium

Chris de Wet Unilever Centre for Environmental Water Quality, Institute for Water Research, Rhodes University, Makhanda, South Africa

Ahmed S. Elshall Department of Earth, Ocean, and Atmospheric Science, Florida State University, Tallahassee, FL, United States

Veera Gnaneswar Gude Richard A Rula School of Civil and Environmental Engineering, Mississippi State University, Starkville, MS, United States

Hui He Guangdong Key Laboratory of Environmental Pollution and Health, Center for Environmental Microplastics Studies, School of Environment, Jinan University, Guangzhou, P.R. China

J. Hoogesteger Water Resources Management Group, Department of Environmental Sciences, Wageningen University, Wageningen, The Netherlands

Lionel Jaffrès Theatre du Grain, Brest, France

Elisabeth Schøyen Jensen Centre for the Study of the Sciences and the Humanities, University of Bergen, Bergen Norway

Bethany Jorgensen Civic Ecology Lab, Cornell University, Ithaca, NY, United States

Werner Krauß University of Bremen, Bremen, Germany

Jyoti Kishen Kumar Formerly Institute of Chemical Technology, Matunga, Mumbai, India

Trevor M. Letcher School of Chemistry, University of KwaZulu-Natal, Durban, South Africa

Joanne Mac Mahon   Trinity College Dublin, Dublin, Ireland

Lei Mai   Guangdong Key Laboratory of Environmental Pollution and Health, Center for Environmental Microplastics Studies, School of Environment, Jinan University, Guangzhou, P.R. China

Benedikt Marschütz   Klima und Energiefonds, Vienna, Austria

M. Mills-Novoa   Department of Environmental Science, Policy and Management, University of California: Berkeley, Berkeley, CA, United States; Energy and Resources Group, University of California: Berkeley, Berkeley, CA, United States

Anju Vijayan Nair   Richard A Rula School of Civil and Environmental Engineering, Mississippi State University, Starkville, MS, United States

Oghenekaro Nelson Odume   Unilever Centre for Environmental Water Quality, Institute for Water Research, Rhodes University, Makhanda, South Africa

Claudia Yazmín Ortega Montoya   Escuela de Humanidades y Educación, Tecnologico de Monterrey, Torreón, Mexico

Aniruddha B. Pandit   Institute of Chemical Technology, Matunga, Mumbai, India

Didier Peeters   Université Libre de Bruxelles, Bruxelles, Belgium

Manzoor Qadir   United Nations University Institute for Water, Environment and Health (UNU-INWEH), Hamilton, ON, Canada

Rabee Rustum   Heriot-Watt University, Dubai Campus, Dubai Knowledge Park, Dubai

Thomas Shahady   University of Lynchburg, Lynchburg, VA, United States

Juan Carlos Tejeda González   Facultad de Ingeniería Civil, Universidad de Colima, Colima, Mexico

Josephine Treacy   Technological University of the Shannon Midlands Midwest, TUS, Limerick City, Ireland

Daniel A. Vallero   Pratt School of Engineering, Duke University, Durham, NC, United States

Jean-Paul Vanderlinden   University of Versailles Saint-Quentin-en-Yvelines, UPS-CEARC, Guyancourt, France

Yongshan Wan   Center for Environmental Measurement and Modeling, United States Environmental Protection Agency, Gulf Breeze, FL, United States

Arjan Wardekker   Centre for the Study of the Sciences and the Humanities, University of Bergen, Bergen Norway

Ming Ye   Department of Earth, Ocean, and Atmospheric Science, Florida State University, Tallahassee, FL, United States; Department of Scientific Computing, Florida State University, Tallahassee, FL, United States

Eddy Y. Zeng   Guangdong Key Laboratory of Environmental Pollution and Health, Center for Environmental Microplastics Studies, School of Environment, Jinan University, Guangzhou, P.R. China

Zhiwei Zhu   University of Versailles Saint-Quentin-en-Yvelines, UPS-CEARC, Guyancourt, France

Preface

The most essential substance to life on Earth is water. Yet, in many parts of the world people are struggling to access the quantity and quality of water needed for growing food, cooking, washing, and even drinking. In spite of the amazing progress that has been made over the past few decades in making drinking water accessible to millions of people in developing countries, globally, billions of people still lack clean water, thus locking them in poverty for generations (https://www.unicef.org/press-releases/billions-people-will-lack-access-safe-water-sanitation-and-hygiene-2030-unless). The importance of addressing the global water crisis has been recognized by the United Nations by naming March 22 as World Water Day (https://www.un.org/en/observances/water-day).

Writing and editing books on global warming (Letcher, 2019, 2021a, 2021b) and also on waste (Letcher, 2020; Letcher and Vallero, 2019) has highlighted the problems of water availability due to our changing climate and also of water pollution due to human activity. This has prompted my desire to compile this book, Water and Climate Change.

The book contains 22 chapters and is divided into three sections:

- Introduction
- Sustainable Development and Environmental Issues
- Policy Issues

Global warming and climate change are upon us, and water resources are being compromised. An understanding of all the parameters involved in climate change is going to be necessary if we are to protect ourselves from future extremes. Water will play a major role in how we adapt to these changes.

Water quality is paramount and a conservative estimate links water pollution to 1.8 million deaths per year—many of them children (Mayor, 2017). Water is in crisis on a number of fronts:

- Global warming is changing the way rain falls or does not fall, bringing flooding and droughts;
- Ground water is being depleted as a result of population needs and creating an unsustainable situation;

- With global population increasing and more and more people demanding water, water availability is decreasing, resulting in water security and political issues which have and will certainly lead to water-wars. Another unsustainable situation in development;
- Water infrastructure is in disrepair worldwide and treatment plants are being compromised;
- Natural infrastructure is being ignored - building on flood plains, deforestation, overgrazing, all resulting in an unsustainable situation;
- Water is being wasted—it is cheaper to use clean fresh water than to use treated water; and
- The quality of water is becoming poorer due to the runoff from farms resulting in a build-up of concentrations of hormones, nitrates, and ammonia in rivers.

The audience we hope to reach with this new volume are: policy makers in local and central governments; students, teachers, researchers, professors, scientists, engineers, and managers working in fields related to climate change and water; editors and newspaper reporters responsible for informing the public; and the general public who need to be aware of the impending disasters that a warmer Earth will bring. An introduction is provided at the beginning of each chapter for those interested in a brief synopsis, and copious references are provided for those wishing to study each chapter topic in greater detail.

Many of the authors were not involved in recent assessments of the IPCC, and here they present fresh evaluations of the evidence testifying to a problem that was described as long ago as in 2008 by Sir David King as the most severe calamity our civilization is yet to face (David, 2008).

IPCC assessments have produced two basic conclusions: first, current climate changes are unequivocal, and second, this is largely due to the emission of greenhouse gases resulting from human activity. This book reinforces these two conclusions.

The International System of Quantities (SI units) has been used throughout the book, and where necessary, other units are given in parentheses. Furthermore, the authors have rigorously adhered to the IUPAC notation and spelling of physical quantities.

This book has an advantage that each chapter has been written by world-class experts working in their respective fields. As a result, this volume presents a balanced picture across the whole spectrum of climate change. Furthermore, the authors are from both the developing and developed countries thus giving a worldwide perspective of looming climatic problems. The 12 countries represented are Canada, China, Costa Rica, France, Ireland, India, Mexico, The Netherlands, South Africa, Sweden, the United Kingdom, and the United States of America.

The success of the book ultimately rests with the 39 authors and coauthors. As an editor, I would like to thank all of them for their cooperation and their highly valued, willing, and enthusiastic contributions. I would also like to thank Victoria Hume for her help, and also my wife, Valerie, for her patience and help while I wrote and edited this volume. Finally, my thanks are due to Louisa Munroe of Elsevier whose expertise steered this book to its publication.

Trevor M. Letcher

School of Chemistry, University of KwaZulu-Natal,
Durban, South Africa

References

David, K. S. (2008). In T. M. Letcher (Ed.), *Foreword to* Future Energy: Improved, Sustainable and Clean Options for our Planet (1st Edition). Oxford: Elsevier, ISBN:978-0-08–054808-1.

Letcher, T. M. (Ed.), (2021a). *Climate Change: Observed Impacts on Planet Earth* (3rd Edition). New York, USA: Elsevier, ISBN: 978-0-12–821575-3.

Letcher, T. M. (Ed.), (2021b). *Impacts of Climate Change: a Comprehensive Study of Physical, Societal and Political Issues.* Cambridge, MA: Elsevier, ISBN: 978-0-12-822373-4.

Letcher, T. M. (Ed.), (2019). *Managing Global Warming: an Interface of Technical and Human Issues.* Cambridge, MA: Elsevier, ISBN: 978-0-12-814104-5.

Letcher, T. M., & Vallero, D. A. (Eds.), (2019). *Waste: A Handbook for Management* (2nd edition). New York, NY: Elsevier, ISBN: 9780128150603.

Letcher, T. M. (Ed.), (2020). *Plastic Waste and Recycling.* Oxford: Elsevier, ISBN: 9780128178805.

Mayor, S. (2017). Research News: Pollution is linked to one in six deaths world wide, study estimates. *British Medical Journal, 357*, 4844. Available from https://doi.org/10.1136/bmj.j4844. (Published 20 October 2017).

SECTION A

Introduction

CHAPTER 1

Introduction: water, the vital chemical

Trevor M. Letcher
School of Chemistry, University of KwaZulu-Natal, Durban, South Africa

Chapter Outline
1.1 Introduction   3
1.2 The unique chemical properties of water   4
    1.2.1 The polar nature of water   4
    1.2.2 The high enthalpy of vaporization of water   4
    1.2.3 The high heat capacity of water   5
    1.2.4 The anomalous density of frozen water   5
    1.2.5 Water, the "universal" solvent   6
    1.2.6 Acid—base property   7
    1.2.7 High surface tension, low viscosity, and cohesive and adhesive properties   7
1.3 Water and climate change   8
1.4 The origin of water on Earth   9
    1.4.1 The water cycle   9
1.5 The scarcity of water   10
1.6 Conclusions   10
References   10

1.1 Introduction

This book focuses on the importance of water to life on Earth, the important role water plays in heating our planet and in our changing climate, and on the impact of climate change on water resources. We will look at the properties of water and see why it is such a vital chemical for plant, animal, and human existence. Life on Earth evolved in and around water and as a result, life in all its forms is totally dependent on water.

Water makes up 60%—75%, by mass, of the human body. From a human point of view, a loss of about 4% of body water leads to severe symptoms of dehydration, and a loss of 15% can result in death (https://rehydrate.org/dehydration/). Humans can survive a month without food but would die after 3 days without water. This dependence reflects the origins of life on Earth in a water environment 3.7 billion years ago with the evolution of microscopic microbes (https://naturalhistory.si.edu/education/teaching-resources/life-science/early-life-earth-animal-origins). This is also 0.8 billion years after the formation of planet Earth and almost 10 billion years after the big bang and the formation of the Universe. It is the unique properties of water that has made life of Earth possible. And indeed, water has rightly been called the "molecule that made us."

1.2 The unique chemical properties of water

The unique properties of water include its polarity, a high enthalpy of vaporization, a high heat capacity, an anomalous density of solid, "universal" solvating properties, buffering property, low viscosity, high surface tension, and cohesive and adhesive properties.

1.2.1 The polar nature of water

The O—H bonds that make up the water molecule, involve an unequal sharing of electrons between the oxygen atom and the hydrogen atom. This is due to the oxygen atom being more electronegative than the hydrogen atom, and the bonding electrons are attracted more to the oxygen atom. This results in an asymmetrical molecule with an angle of 104 degrees between the H—O—H, with the hydrogen ends of the H_2O molecule being slightly more positive than the oxygen end. The water molecule behaves like a magnet, but in this case, one end is positive and the other negative. This polarity allows for electrostatic attractions, called hydrogen bonding, between the water molecules and other polar molecules, and it is this hydrogen bonding that is responsible for many of the unique properties of water.

1.2.2 The high enthalpy of vaporization of water

The process of evaporation, that is, a liquid changing into a gas or vapor, requires energy and this energy is known as the enthalpy of vaporization, ΔH_{vap}. Water has an anomalously high ΔH_{vap}, as a result of hydrogen bonding between the water molecules, implying that energy has to be expended to break these bonds in order to vaporize water.

At the boiling point of water, the enthalpy of vaporization can be expressed as:

$$H_2O(liquid) = H_2O(gas) \quad \Delta H_{vap}^0(373K) = +40.66 \text{ kJ mol}^{-1}, \quad (1.1)$$

implying that 40.66 kJ must be added to one mole of water (18.02 g) to vaporize it at 373K.

Looking at this process in another way, when water evaporates such as when a breeze or wind blows over the water, some of the water evaporates and heat must be supplied, so the surroundings cool down. This is known as evaporative cooling.

In the reverse process of condensation in which a vapor or gas condenses to a liquid, energy is released. This reverse process to Eq. (1.1) involves an amount of energy equal to -40.66 kJ mol^{-1} indicating that heat is released:

$$\Delta H_{vap} = -H\Delta_{cond}$$

When humid air condenses and forms clouds of liquid water droplets, energy is released in the form of heat, which helps to create thunderstorms.

It is this high enthalpy of vaporization that has meant that the water in the oceans does not readily evaporate and for this reason that our oceans have not vaporized and disappeared into space over the past many millions of years. In humans and in other organisms it is this same property that maintains a steady body temperature. Our bodies sweat when we are physically active and this sweat, upon evaporation, takes heat from the body which is then cooled. Evaporative cooling of the skin is nature's way of keeping our bodies at a constant temperature; water acts as a thermostat.

1.2.3 The high heat capacity of water

Water has a relatively high heat capacity (specific heat), which is due to the hydrogen bonds holding the water molecules together. Heat capacity refers to the amount of heat required to raise the temperature of a mass of the substance 1°C. In the case of water, heat must first be supplied to break some of the H-bonds, before heating the water by 1°C—hence water's relatively large heat capacity.

This high heat capacity is the reason why when the sun shines, the oceans warm up more slowly than does the land. For example, the soil or sand around a pool of water or a lake may be too hot to walk on while the water feels cool. During the night, the Earth loses its heat faster than does the water in the pool or lake and the soil or sand feels cool whereas the water remains relatively warm.

It is this high heat capacity that allows water to absorb and release heat at a much slower rate than on land, and temperatures in areas near large bodies of water tend to have smaller fluctuations. The high heat capacity keeps the oceans at average temperatures well below that of land surfaces and keeps regions in coastal areas and island counties at reasonably constant temperatures. Furthermore, the daily temperature fluctuations of our planet are more moderate than they would be if the planet was devoid of water. All plant and animal life contain a high fraction of water, and it is the high heat capacity property that helps to resist changes in temperature.

1.2.4 The anomalous density of frozen water

Ice has an anomalous density which is less than that of liquid water at the same temperature, and as a result ice floats and when ice does form in fresh- or seawater, it allows the fish to swim below the floating ice. Ice can form an insulating barrier between cold air and liquid water which helps to keep the water under the ice from freezing and thus, again, allows fish to swim and live.

The reason for the anomaly is again due to hydrogen bonding between the water molecules. In solid water (ice), the molecules of water are arranged in a crystal lattice and the lattice is held in place by hydrogen bonds. This structure is less dense, less compact, and more open than the structure of liquid water which is also held together by hydrogen bonds but in a much less compact way, allowing the molecules to move. There are very few known substances that have this anomalous density at their freezing point. Silica is one and that might be the reason that the Earth has continents.

1.2.5 Water, the "universal" solvent

The water in our bodies is largely contained in our cells. The cell walls keep the water in place with the water acting as a solvent for many chemicals, such as enzymes, oxygen, nutrients and salts on which our bodies depend.

The blood in our bodies is mainly water and the solvent properties of water are vital in supporting and transporting chemicals such as oxygen gas, hormones, and toxins such as urea. Blood is also the medium for transporting drugs to targets in the body. Many body functions rely on diffusion and osmosis and both processes rely on water as a transport medium.

Water also has vital structural role in maintaining shape and structure to the cells in our bodies and indeed in all forms of life. The shape of cells is important in many biological processes. Water maintains its shape by creating pressure against the cell walls. Water in cells have an important role in creating and stabilizing the cell walls (membranes). Membranes are made up of two layers of molecules called phospholipids. These compounds are made up of a nonpolar tail and a polar head. The heads interact with the polar water molecules while the nonpolar (hydrocarbon) tails avoid the polar water and interact with other nonpolar tails. The membrane is composed of bilayers with the polar heads in the water and interacts with the nonpolar tails to create cell walls. Without water, such cell structures would be impossible. The membranes induce biological functionality to take place by allowing nutrients and salts to enter and exit cells.

The high solubility of a compound in water is largely due to the polar nature of water. Water can form hydrogen bonds with a solute; for example, sugar with its many hydroxyl groups forms hydrogen bonds with water molecules and as a result, the solubility of sugar in water is high. Also, ionic compounds such as sodium chloride readily dissolve in water as the water molecules surround each of the ions, Cl^- and Na^+ and effectively break up the salt crystal.

Water plays a crucial role in biological processes in the folding of large molecules such as amino-acid chains, protein, enzymes, in order for these macromolecules to carry out their life-giving reactions such as reaction catalysis, contraction of muscles, digestion, and decoding DNA to follow instructions. All of this is done through hydrogen bonding.

1.2.6 Acid–base property

Water molecules have another interesting property based on the ability of a water molecule to give up a hydrogen atom to become an OH^- ion, thus making the water basic. Furthermore, water can accept a hydrogen atom to become H_3O^+ thus acting as an acid. This ability allows water to combat drastic changes of pH due to the addition of harmful acidic or basic chemicals. This buffering process is very important in the equilibrium of cells and in other biological processes; see a recent report Wen et al. (2021).

1.2.7 High surface tension, low viscosity, and cohesive and adhesive properties

Hydrogen bonds are responsible for water having a high surface tension and cohesive properties that help plants take up water from their roots.

The pumping of blood (an aqueous solution of about 0.8% salt) around a body is dictated by Poiseuille's Law (Atkins & de Paula, 2002):

$$dV/dt = (\Delta p \pi r^4)/(16\, l\, \eta\, p_0), \tag{1.2}$$

where dV/dt, measured at a blood pressure of p_0, is the flow rate of a liquid of viscosity η through a pipe of length l and radius r. The pressure gradient along the pipe is Δp (Secomb, 2016). Water has a relatively low viscosity making the pumping of blood much easier. The relatively low viscosity indicates that the water molecules can slip past each other with relative ease in spite of the hydrogen bonds—a truly remarkable fluid!

Eq. (1.2) also summarizes the rising of sap up a plant or tree (Denny, 2012).

In the case of trees, the pipes or conduits (of radius of about 50 μm) are made up of special cells (e.g., xylem cells in the case of hardwoods). The sap (an aqueous solution of 10 mol m^{-3}), is sucked up by the capillaries of a negative pressure created by the evaporation of water from the leaves of the trees. The energy for the evaporative process comes from sunlight. These negative pressures can reach up to 30 atmospheres and can cause cavitation within the conduits. The cohesion of the molecules to the hydrophilic conduit walls helps to maintain the continuous column of sap from root to leaf. This can involve sap columns of over 100 m in height. Apparently, this process of cavitation can be heard as a rhythmic thumping within the tree using a specialized listening devise (https://phys.org/news/2013-04-cavitation-noise-trees.html).

Osmotic pressure at the roots of trees and plants is also a contributing factor responsible in the transport of sap in plants or trees. In the case of capillary action, the height is dictated not only by the narrowness of the capillaries but also by the surface tension which, in the case of water, is relatively large. Capillary action (Atkins & de Paula, 2002) can be summarized as:

$$h = 2\gamma/(\rho\, g\, r), \quad (1.3)$$

where h is the height of the liquid of surface tension γ, in the capillary of radius r and g is the gravitational acceleration of 9.81 M s^{-2}. The value of r for the xylem tubes is of the order of 50 μm and the capillaries, which are linked to tubes, are of the order of 5 μm (Denny, 2012).

There are other properties of water that have contributed to life on Earth such as a high boiling point, a high melting point, and a high enthalpy of melting. If water did not behave in these unusual ways, it is questionable whether life could have developed on planet Earth (Ball, 1999).

1.3 Water and climate change

On the positive side, water molecules are largely responsible for keeping our planet warm through the greenhouse effect. On the negative side, water is also largely responsible for global warming. This is due to the bond vibration between oxygen and hydrogen atoms. This will be discussed in more detail in Chapter 2.

Most disasters related to climate change involve water in way or another, such as flooding, unseasonal rain storms, washaways, rising sea levels, and contaminated water from floods and storm damage. Furthermore, the lack of water and droughts in many parts of the world is a consequence of climate change. Global warming has been slowly increasing since the start of the industrial revolution but with so many tipping points beginning to take hold, the effects of climate change are now very obvious (Letcher, 2021). Scientists are now working on ways to manage global warming (see chapters in Letcher, 2019).

Furthermore, attempts are being made to improve our lives in spite of global warming and droughts. For example, scientists are beginning to understand and perhaps introduce drought resistance into food plants by studying resurrection plants (Farrant et al., 2020).

Before leaving the unique properties of water and the relationship of water to climate change, it is worth mentioning that solid water (ice and snow) has a vital role to play in climate change. The solid ice caps of the Earth influence climate in many ways and one of these is linked to the whiteness of snow and ice. The polar caps, being white, reflect sunlight and reflection helps to maintain these massive ice sheets. With global warming, some of this ice is melting, reducing the amount of reflected soar radiation and allowing the sun to warm the melted water, causing further global warming. Pure solid ice is indeed transparent and the only reason why snow and ice are white is the result of trapped air bubbles, which cause incident light to be reflected in all directions and hence appears white in sunlight.

1.4 The origin of water on Earth

The origin of water on Earth is unknown but recent research points to water forming from the accretion of meteorites (enstatite chondrites), which contained water. These meteorites had formed in the outer solar system, where water was more abundant (Piani et al., 2020).

Other research appears to indicate that water and oceans on the Earth was a result of volcanic action and that the water was part of the original material from which the Earth was formed (Peslier, 2020).

1.4.1 The water cycle

The water vapor in the atmosphere is controlled by temperature. The relationship (Atkins & de Paula, 2002) is well defined by the Clausius Clapeyron equation:

$$d\ln p/dT = \Delta H_{vap}/RT^2 \tag{1.4}$$

where p refers to the vapor pressure, T the absolute temperature, and ΔH_{vap} the enthalpy of vaporization. This means that where there is water, such in the oceans, an increase in temperature results in an increase in the vapor pressure of water and hence an increase in evaporation and the amount of water in the air above the oceans or other water bodies.

Eq. (1.4) highlights the importance of the enthalpy of vaporization in the relationship between temperature and vapor pressure. The relatively high value of ΔH_{vap} plays an important role in determining our weather and our long-term climate patterns. Water is the major vehicle for the weather and climate on Earth. This is done through the water cycle; the sun heats the oceans, causing water to evaporate forming water vapor, which then condenses and falls on land as rain or as snow. Atmospheric circulation of the air, as a result of the Earth's rotation and differential solar radiation on land and on the oceans, moves water vapor round the Earth. Over time, the water returns to the oceans by rivers and rain and the cycle repeats itself. In the evaporative stage of the water cycle, the water is purified (it is indeed a distillation process) bringing freshwater to land making the water cycle essential to maintaining life on Earth.

Despite the relatively small amount of water vapor in the atmosphere, water vapor has a huge influence on the Earth's climate. Water is the dominant greenhouse gas as we shall see in Chapter 2 and contributes 24°C to global warming (CO_2 contributes about 6°C).

Studies of the water cycle have shown that with global warming and the subsequent increase in the amount of water vapor in the atmosphere, wetter regions on the Earth can expect increased rainfall while drier parts of the Earth can expect more drought conditions (Bengtsson, 2010).

1.5 The scarcity of water

Water covers 71% of the Earth's surface, and most of the water on Earth (96.5%) is ocean water. Only 2.5% of all water on Earth is freshwater and most of that is locked up in glaciers, icefields, and snowfields. Only 1% of all freshwater is in rivers, lakes, or streams and unfortunately some of this water is polluted in one way or another. In short, there is a vast amount of water on the Earth but most of it is saline (ocean water) or in solid form. The water cycle is vital in moving freshwater around the Earth but unfortunately, this freshwater is not evenly distributed on Earth and many areas are severely drought stricken. Furthermore, the relatively small amount of freshwater on land is easily contaminated by natural means and more importantly by human activity. Contaminated water can result in cholera and typhoid, to mention just two diseases. Furthermore, droughts as a result of a lack of water can be just as pernicious.

1.6 Conclusions

In this chapter, we have looked at the unique properties and the paramount importance of water to life on Earth. As we have seen, life could not have been established on Earth without water. In the next chapter, the focus will be on the major contribution water makes in keeping the planet relatively warm and also on its role in precipitating global warming and climate change. The rest of the book focuses on water pollution, water availability, the need for clean water, treatment techniques for purifying unclean water, water supply, the impact of climate change on water availability, water management, and policies of justice and ethics related to water management.

References

Atkins, P., & de Paula, Julio (2002). *Physical chemistry* (7th ed., p. 831)Oxford.: Oxford University Press.
Atkins, P., & de Paula, Julio (2002). *Physical chemistry* (7th ed., p. 147)Oxford.: Oxford University Press.
Atkins, P., & de Paula, Julio (2002). *Physical chemistry* (7th ed., p. 153)Oxford.: Oxford University Press.
Ball, P. (1999). *H$_2$O: A biography of water*. London: Weidenfeld & Nicolson.
Bengtsson, L. (2010). The global atmospheric water cycle. *Environmental Research Letters*, 5, 025202.
Denny, M. (2012). Tree hydraulics: How sap rises. *European Journal of Physics*, 33, 43–53.
Farrant, J. M., Moore, J. P., & Hilhorst, H. W. M. (2020). Editorial: Unifying insights into the desiccation tolerance mechanisms of resurrection plants and seeds. *Plant Science (Shannon, Ireland)*, 11, 1089. Available from https://doi.org/10.3389/fpls.2020.01089.

Letcher, Trevor M. (Ed.), (2019). *Managing global warming: An interface of technical and human issues.* Cambridge, MA: Elsevier, ISBN: 978-0-12-814104-5.

Letcher, T. M. (Ed.), (2021). *The impacts of climate change.* Oxford: Elsevier, ISBN: 978-0-12-822373-4.

Peslier, A. H. (2020). The origins of water. *Science, 369*(6507), 1058. Available from https://doi.org/10.1126/science.abc1338.

Piani, L., Marrocchi, Y., Rigandier, T., Vacher, L. G., Thomassin, D., & Marty, B. (2020). An unexpected source of water. *Science (New York, N.Y.), 369*(6507), 1110−1113. Available from https://doi.org/10.1126/science.aba1948.

Secomb, T. W. (2016). Hemodynamics. *Comprehensive Physiology, 6*(2), 975−1003. Available from https://doi.org/10.1002/cphy.c150038.

Wen, Y., Salamat-Miller, N., Jain, K., et al. (2021). Self-buffering capacity of a human sulfatase for central nervous system delivery. *Scientific Reports, 11*, 6727. Available from https://doi.org/10.1038/s41598-021-86178-2.

CHAPTER 2

The root causes of climate change and the role played by water

Trevor M. Letcher

School of Chemistry, University of KwaZulu-Natal, Durban, South Africa

Chapter Outline
2.1 Introduction   13
2.2 Water and CO_2 and the greenhouse effect   14
2.3 The main greenhouse gases of anthropogenic origin   16
2.4 Feedback mechanisms and tipping points   18
2.5 Where are we in solving the problem of global heating?   19
2.6 Solutions   22
2.7 Conclusion   25
References   25

2.1 Introduction

NASA has recently reported that the Earth's average surface temperature in 2020 tied with 2016 as the warmest year on record (https://www.nasa.gov/press-release/2020-tied-for-warmest-year-on-record-nasa-analysis-shows). The globally averaged temperature was greater than 1°C warmer than the baseline 1951–80 mean. The last seven years have been the warmest seven years on record, typifying the ongoing and dramatic warming trend. Single year records are less important than long-term trends. This particular trend is a critical indicator of the impact of human activity and a trend that looks as though it will continue.

The amplified weather conditions such as devastating floods, destructive wind blowing at record speeds, superstorms, and intense heat waves precipitating wildfires, seen around the world in recent times are just a foretaste of what can be expected if global heating continues. The basic cause is that global warming raises the temperature of the land and oceans. On land, warmer temperatures cause an increase in evaporation which can eventually result in droughts. In the oceans, warmer temperatures result in more water vapor in the atmosphere which, in turn, produces higher rainfall events. Furthermore, increasing atmospheric air temperatures changes air circulation patterns, which can result in some areas having increased rainfall while others become drier.

These extreme weather patterns will become the norm unless climate breakdown can be stopped, which likely means that the global economy must move to a low-carbon scenario within the next decade. Otherwise, it is possible that irreversible changes with multiple feedback mechanisms will result in catastrophic climate events far more serious than any we have experienced over the past few years. This is the message we take from the latest and sixth (August 9, 2021) Intergovernmental Panel on Climate Change (IPCC) report (https://www.metoffice.gov.uk/weather/climate-change/organisations-and-reports/ipcc-sixth-assessment-report). The report makes the point that world leaders must agree to a detailed and achievable plan to cut emissions now when they meet in Glasgow in November 2021. This report is an indication, the starkest warning yet, that human behavior is alarmingly accelerating global warming. We must act now before it is too late.

The heating of our planet and the changing climate is being felt in all countries, and almost every week there is a new report of extreme weather conditions or out of control fires. The impacts of climate change and the root cause of global warming has been discussed in our previous books (Letcher, 2021; Letcher, 2021; Letcher, 2019). However, it is pertinent to again discuss the root cause here with a focus on the role of water.

Global heating is affecting deaths and the recently reported US Government's natural hazard statistics for 2020 (http://www.weather.gov/hazstat/) states that heat deaths are the most prevalent of all deaths by natural disasters in the United States. Indeed, the statistics show that these deaths are higher than those caused by tornadoes, hurricanes, or even flooding.

2.2 Water and CO_2 and the greenhouse effect

The Earth is kept relatively warm by the greenhouse effect, which raises the temperature of the atmosphere due to the presence of relatively small amounts of greenhouse gases in the atmosphere, notably H_2O and CO_2. The mechanism can be described very simply as follows.

Both the sun and the Earth are black body emitters of electromagnetic radiation. This radiation is defined by the laws of Planck, Stefan, and Weiss and can be summed up as the total energy emitted per unit time integrated over all wavelengths and is proportional to the fourth power of the absolute temperature of the black body concerned, T^4. As a result, the sun emits UV/visible radiation with a peak at about 500 nm (20,000 cm^{-1} or 600×10^{12} Hz) characteristic of $T_{sun} = 5780K$. The Earth's temperature is about a factor of 20 lower and its emitted black body radiation is in the infrared region, with a peak at about 10 μm and a majority of the radiation in the range of 6–25 μm (1700–400 cm^{-1} with a frequency of 50–10 $\times 10^{12}$ Hz) (Tuckett, 2021).

All multiatomic gases in the atmosphere, such as N_2, O_2, CO_2, O_3, CH_4, H_2O, have bonds binding the atoms together, and these bonds have vibration frequencies in the IR range. For example, the O—C—O bending mode wave number for CO_2 is 526 cm^{-1} (frequency is 15×10^{12} Hz). The UV/vis radiation from the sun passes through the gases of the atmosphere without affecting the gases in any way, as the vibrating frequencies of the sun's radiation are much larger than the vibrating frequencies of the molecular bonds of the atmospheric gases. This UV/visible radiation heats the Earth. The radiation from the heated Earth, largely in the IR region, is in the range of the vibrating frequencies of the molecular bonds binding the atoms of some of the gases, notably CO_2, H_2O, but not N_2 or O_2.

The Earth's IR radiation amplifies the vibration in these molecules through a process of sympathetic mechanical vibration. This effectively heats the CO_2 and H_2O molecules and this heat then passes on by kinetic motion to the major components in the atmosphere—nitrogen and oxygen. This greenhouse effect has been responsible for keeping the Earth at an average temperature of 290K (17°C) for at least a million years. This happens in spite of the IR sensitive gases comprising only a small fraction of the gases in the atmosphere; the CO_2 concentration is 0.041% and the H_2O concentration varies from almost zero in deserts to 4% on very hot and humid days with the global average of less than 1%.

Today, conditions are different from what they were in preindustrial times; the concentration of CO_2 in 1880 was 250 ppm (or 250 μmol mol^{-1}) and in May 2021, the CO_2 concentration peaked at 419 ppm, as measured at Mauna Loa in Hawaii (https://gml.noaa.gov/ccgg/trends/weekly.html). It is this rise of over 60% in CO_2 that is considered to be the cause of the heating of the planet. However, CO_2 is not the main greenhouse gas; water being more than 25 times more concentrated, is the main culprit. The accepted mechanism is as follows. With the rise of CO_2 levels in the atmosphere over the past 140 years, there has been a corresponding rise in temperature (albeit small) of the atmosphere due to the greenhouse effect of CO_2. This has increased the ocean temperature which in turn has resulted in more water evaporating into the atmosphere which in turn has caused further global heating through the greenhouse effect. The average temperature of the Earth is now 1.2°C above preindustrial times as reported by the World Meteorological Organization in January 2021 (https://public.wmo.int/en/media/press-release/2020-was-one-of-three-warmest-years-record). This rise in temperature also causes the temperature of the oceans to rise and as a result some of the dissolved CO_2 in the oceans enters the atmosphere (the solubility of CO_2 in the water is temperature dependent), resulting in a feedback mechanism creating further global heating.

To sum up, the CO_2 in the atmosphere, albeit in a very small concentration, is the trigger for initiating global warming, with H_2O being the main greenhouse gas. It has been estimated that CO_2 is responsible for 20% and H_2O for 60%—80% of global warming (https://www.acs.org/content/acs/en/climatescience/.../its-water-vapor-not-the-co2.html; https://www.nasa.gov/topics/earth/features/vapor_warming.html).

The picture given here is a very much simplified version and more details can be found in another reference (Tuckett, 2021). In this reference, mention is made of other greenhouse gases including CH_4 and many halogenated hydrocarbons.

Most disasters related to global warming involve water in way or another. For example, flooding, unseasonal rain storms, washaways, rising sea levels, and contaminated water from floods can result in causing cholera and typhoid. Furthermore, droughts as a result of lack of water can be just as pernicious.

2.3 The main greenhouse gases of anthropogenic origin

The rise of CO_2 in the atmosphere over the past 140 years is largely due to the burning of fossil fuel (https://www.c2es.org/content/international-emissions/).

The amount of atmospheric CO_2 produced annually by natural processes, is about 750 Gt (https://welcome.arcadia.com/energy-101/environmental-impact/greenhouse-gas-emissions-natural-vs-man-made) and is derived from multiple sources such as volcanic outgassing, the combustion of organic matter, and the respiration processes of animal life and living aerobic organisms. The annual amount of CO_2 (2020) produced by humans is significantly less at 33 Gt (https://www.epa.gov/ghgemissions/global-greenhouse-gas-emissions-data).

Apart from water the main greenhouse gases are CO_2 (72%) and CH_4 (19%) (Olivier & Peters, 2018). It has also been estimated that about 90% of the CO_2 produced from human activity is a result of burning fossil fuel. The estimated breakdown is given in Table 2.1.

Another way of looking at where the anthropogenic CO_2 pollution comes from is to look at the source of CO_2 as reported in Table 2.2.

The main sources of CH_4 as a result of human activity have also been estimated by Olivier and Peters and their results are summarized in Table 2.3 (Olivier & Peters, 2018).

The relatively large amount of CO_2 produced naturally (750 Gt) has been responsible for keeping the Earth at its equitable temperature for thousands of years. Much of this CO_2 is adsorbed by the oceans and is taken up by marine life for their growth in much the same way as plant life on land absorbs CO_2 to grow the carbon structures that make up life. This equilibrium of this process has been disturbed by the anthrophomorphic production of CO_2 (33 Gt produced annually). Some of this additional CO_2 is also taken up by the oceans and

Table 2.1: Estimated breakdown of CO_2 produced from human activity (Olivier & Peters, 2018).

Coal (largely for electricity production)	39%
Oil (largely used for transport)	31%
Natural gas (CH_4) (largely used for heating)	18%
Cement manufacture	4%

Table 2.2: Worldwide source of CO_2 (mostly fossil fuel) emissions, 2018 (Letcher, 2021; https://www.epa.gov/ghgemissions/sources-greenhouse-gas-emissions).

Source	CO_2 emissions (%)
Electricity	27
Transport	28
Industrial (including cement manufacture)	22
Residential (heating, wood fires)	12
Agriculture and other sources	11

Table 2.3: The main sources of methane (Olivier & Peters, 2018).

Source	Percentage
Cattle	21
Rice production	10
Gas/oil production	23
Coal mining	10
Landfill	10

Table 2.4: Greenhouse gas properties.

Gas	Atmospheric concentration (%)	Greenhouse potential
CO_2	0.0413	1
CH_4	0.00019	28
N_2O	0.00003	265

has caused an increase in the acidity of the seas. The remainder is responsible for triggering global warming. The amount of CO_2 produced by human activities is relatively small when compared to the natural output, but for a world that had been in equilibrium with its naturally produced CO_2 for thousands of years, this extra 33 Gt does apparently make a significant change to weather patterns.

Other serious greenhouse gases include CH_4, N_2O, and fluorinated hydrocarbons. The greenhouse effect is not identical in each gas and the greenhouse potentials of many such gases have been well documented by Tuckett (Tuckett, 2021). Table 2.4 details the greenhouse potential and atmospheric concentration of three gases.

Methane gas levels have more than doubled since the industrial revolution. Methane is produced from: burning biomass; leakage from gas fields; leakage from ancient peat permafrost areas and from undersea methane clathrates; and from anaerobic processes as found in landfills, decaying organic matter; paddy fields and livestock farming. Although its concentration in the atmosphere is low, it is 28 times more effective than CO_2 as a greenhouse gas (https://research.noaa.gov/article/ArtMID/587/ArticleID/2742/Despite-pandemic-shutdowns-carbon-dioxide-and-methane-surged-in-2020).

The rate of increase of CO_2 in the atmosphere has been accelerating at an alarming rate of 2.4% per annum excluding the slight reduction seen in 2020 as a result of the COVID epidemic (https://www.esrl.noaa.gov/gmd/ccgg/trends/). The increase is a stark warning that something must be done soon to avert major problems. To make matters worse, the world's population is increasing at a rate of 1.05% per year at the moment and this means that more energy for electricity and transport will be required. The expected increase in global electricity generation is 2% per year (https://www.worldometers.info/world-population/world-population-by-year/).

One slight glimmer of hope on the horizon is the fact that natural gas, methane, (including shale gas) is better for the planet than burning coal or even oil. The reason why natural gas is better than coal is that the amount of CO_2 produced from burning CH_4 per unit of energy (50 g MJ^{-1}) is less than it is for coal (92 g MJ^{-1}) and moreover, coal burning produces particulates. Of course, the burning of CH_4 still produces CO_2:

$$CH_4 + 2O_2 = CO_2 + 2H_2O$$

However, once the world is rid of burning coal, the next step must be to stop burning oil and natural gas.

It is of interest to note that five countries and the European Union emit 63% of global greenhouse gas emissions. China accounts for 27%, the United States 13%, the EU 9%, India 7%, Russia 5%, and Japan only 3% (Olivier & Peters, 2018).

2.4 Feedback mechanisms and tipping points

The major role played by water in creating global warming involves feedback which is described in Section 2 above and has been discussed before (Letcher, 2021).

However, there are a number of other feedback mechanisms that have accelerated global warming. One of these is the melting of glacial and ice sheets. When ice melts, land or open water is exposed and both land and open water are very much less reflective than ice. As a result, both land and water absorb solar radiation and heat up. In turn, this causes more melting, and so the cycle continues.

The oceans contain vast amounts of dissolved CO_2. The amount is governed by the solubility of CO_2 in seawater which is dependent on temperature. With global warming, the oceans become warmer resulting in a lowering of the CO_2 solubility and CO_2 leaving the oceans and entering the atmosphere, which then increases global warming in an ever-increasing cycle.

Another feedback mechanism is at play in the peat bogs and permafrost regions of the world, such as Siberia and Greenland. Rising global temperatures are melting the permafrost and in time will release vast quantities of methane gas. This poses a real threat for our future.

Yet another feedback mechanism involves methane clathrates, a form of water ice that contains methane within its crystalline structure. Extremely large deposits have been found under the sediments of ocean floors. An increase in temperature breaks the crystal structure releasing the caged methane. Rising sea temperatures could cause a sudden release of vast amounts of methane from such clathrates and result in runaway global warming.

2.5 Where are we in solving the problem of global heating?

We are hopelessly unprepared to deal with the changes necessary to reduce climate change although scientists have been warning society about it for decades. However, we do have the tools available to deliver rapidly with speed and scale the action needed to avoid further catastrophic weather patterns.

The obvious solution to global warming would be to reduce our dependence on fossil fuel and move to renewable forms of energy. However, even if we stopped using fossil fuel today, global warming would be with us for many decades as the half-life of CO_2 in the atmosphere is on the order of 100 years. One solution would be to use CO_2 as a chemical feedstock. That is not a simple option as it is not easily transformed into other chemicals; a large amount of energy has to be expended to facilitate reactions with CO_2. It is the most oxidized form of carbon and as a result is thermodynamically very stable and hence any chemical reaction involving CO_2 will require a significant input of energy. In short, in the language of thermodynamics, it has a large and negative Gibbs energy of formation. Of course, we could use photosynthesis to chemically convert atmospheric CO_2 into useful products (using sunlight as the energy source) but this is not going to take up enough CO_2 to make a major difference. Perhaps, the only viable way to get rid of it is to collect it and store it using carbon capture and storage processes which are now being investigate and deployed in a small way. Thirty new carbon capture, utilization, and storage (CCUS) (https://www.iea.org/fuels-and-technologies/carbon-capture-utilisation-and-storage) facilities have been announced over the past four years. These are mainly in the United States and in Europe. Others have also been proposed in Australia, China, Korea, Middle East and New Zealand. Once these projects are up and working would mean capturing 130 Mt per year. This however is only 0.25% of the CO_2 annually produced by human activity—but it is a start!

It is worth looking at how far we have got in solving the global warming crisis. Globally, wind and solar power produces only 10% of all electricity. For some countries, the figures are much higher, for example, Germany (32%) Spain (29%), United Kingdom

(28%). Renewable energy including hydro (18%) produces 30% of the world's electricity with fossil fuels (coal 34% and natural gas 25%) still the major producers of electricity (https://www.iea.org/reports/global-energy-review-2020/electricity).

At present, about 40% of all energy sources used to produce electricity are either renewable (wind, solar, hydropower, biomass, tidal and geothermal) or nuclear (10%) (https://www.iea.org/reports/global-energy-review-2020/electricity). The changeover from fossil fuel to renewables is slow and the prediction for 2040 is that renewables will produce nearly 50% of the world's electricity, with coal still a significant supplier of energy. Replacing fossil fuel is going to be a mammoth task (BP Energy outlook, 2020).

If there is the necessary political will to do so, we can replace the fossil fuel-derived electricity with renewable forms of energy, such as nuclear energy or hydropower. However, we do have a problem with replacing transport fuel and piped natural gas.

We could one day have electric cars replace gas-based vehicles and possibly even diesel vehicles, once all our electricity is produced by renewables, but replacing fossil fuel for air travel and sea travel is going to be difficult, if not impossible. Attempts at replacing gas in transport with renewable fuel derived from biomass (sugar cane as done in Brazil or corn as done in the United States for petrol, and palm oil in Malaysia for biodiesel) has had some success but the overall contribution has been relatively small. In 2018 biofuels contributed 3% to the world's transport fuels. The United States, Brazil, and Malaysia are the world leaders in biofuels (https://www.iea.org/reports/transport-biofuels).

Attempts at using hydrogen gas for transport purposes are ongoing, as is the aim of replacing piped natural gas with hydrogen gas. Hydrogen is perhaps the safest gas known (the diffusion rate for CO_2 is very high), and it burns in air to form water. Using renewables to make hydrogen by electrolysis and piping the hydrogen to householders and to industry, as is presently done with natural gas, should be our aim.

All of this does indicate that the world is not on top of solving the global warming problem, in spite of the steady increase in the deployment of renewable forms of energy. The changeover from fossil fuel to renewables is just too slow.

Much of this chapter relates to the need for the world to replace fossil fuels with renewable forms of energy in order to reduce the concentration of CO_2 in the atmosphere. However, there is another, typically overlooked reason for moving to a world powered by renewable energy, which is the fact that fossil fuel reserves are finite. It has been estimated (https://www.iea.org/reports/transport-biofuels) that globally, we currently consume the equivalent of over 11 billion tons of oil from fossil fuels every year. Crude oil reserves are vanishing at a rate of more than 4 billion tons a year—so if we carry on as we are, our known oil deposits could run out in just over 50 years. If we increase gas production to fill the energy gap left by oil, our known gas reserves can give us just over another 50 years. Although it is often claimed that we have enough coal to last hundreds of years, this does not consider the need for increased production. If we step up production to make up for depleted oil and gas reserves, our known coal deposits could be gone in 150 years. Another set of estimates have been given by BP in 2018. The figures were a little less optimistic. Their estimation of the time left for fossil fuel as a result of present-day usage was predicted to be: 30 years for oil, 40 years for gas, and 70 years for coal (https://www.ecotricity.co.uk/our-green-energy/energy-independence/the-end-of-fossil-fuels).

Our future mindset must however not be seduced by the convenient properties of fossil fuel, but for the sake of the planet, the reserves must stay forever below ground and non-fossil fuel sources of energy should be embraced.

The coronavirus pandemic has been linked to climate change issues. There seems to be little doubt that there is a link between population density, human encroachment on natural areas, and zoonotic disease transmission (Christine et al., 2020).

Climate change, through drought, flooding, rising sea levels, unpredictable weather conditions, is slowly reducing the arable land in many parts of the world, forcing people to move into areas close to wildlife populations that humans had not previously been in close contact with. The disruption of pristine forests driven by logging, mining, the need to find new places to live, the spread of urban development and population growth are bringing people into closer contact with animal species. In the case of COVID-19, the contact was most likely with bats. As Jane Goodall says, "COVID-19 is a product of our unhealthy relationship with animals and the environment and that our exploitation of animals and the environment has contributed to pandemics, including the COVID-19 crisis. Wildlife trafficking, factory farming, and the destruction of habitats are drivers of zoonotic diseases" (Goodall, 2020). With global warming on the increase, we can expect that climate change will further impact on the spread of infectious diseases through animals as it has in the past with rabies, the plague, Ebola, SARS, MERS, and ZIKA to mention but a few recent zoonotic diseases.

The present COVID-19 pandemic has shown that all countries of the world can work together to fight a common enemy. This is exactly what is needed to reduce the onset of global warming. The carry-over to climate change when the pandemic has run its course could well have been made that much easier.

Another spin-off from the COVID-19 pandemic is that during lockdown, people have had a chance to look at their futures and compare them to the less polluting, less frantic, and calmer times. Life may never return to what it was and the fight against climate change might just improve.

Yet another spin-off is that the world population has got used to listening to scientists for direction and perhaps this will help overcome the barriers hindering an international effort to reduce the emission of CO_2 and replace fossil fuels with renewable energy.

Once the pandemic is over and even with the CO_2 levels still rising, the people of the world are a little better prepared to tackle the next major world catastrophe—global heating.

2.6 Solutions

There are a number of things we can do:

1. Money talks, as the saying goes. In discussing the replacement of fossil fuel by renewables, it is encouraging to note that the commonly held belief that renewable energy is expensive is simply not true! Solar power and onshore wind are the cheapest renewable ways of electricity, which means the energy they produce is cheaper than using nuclear, gas and fossil fuels. The cost of renewables has fallen faster than predicted.

   The relative costs of producing electricity have been analyzed by the International Renewable Energy Agency (IRENA) (https://www.irena.org/-/media/Files/IRENA/Agency/Publication/2019/May/IRENA_Renewable-Power-Generations-Costs-in-2018.pdf?la = en&hash = 99683CDDBC40A729A5F51C20DA7B6C297F794C5D) in 2018. The figures are given in Table 2.5.

   According to new cost analysis from the International Renewable Energy Agency (IRENA) (https://www.irena.org/newsroom/pressreleases/2018/Jan/Onshore-Wind-Power-Now-as-Affordable-as-Any-Other-Source), the cost of generating power from onshore wind has fallen by around a quarter since 2010, and solar photovoltaic (PV) electricity costs have fallen by over 70% in that time. The report also highlights that solar PV could be delivering electricity for an equivalent of USD 3 cents per kilowatt hour, or less within the next two years.

2. When discussing fuel costs, the environment should be counted in. Today, when buying fossil fuels of any kind, or electricity produced from fossil fuels, we do not pay for

Table 2.5: The cost of producing electricity (https://www.irena.org/-/media/Files/IRENA/Agency/Publication/2019/May/IRENA_Renewable-Power-Generations-Costs-in-2018.pdf?la = en&hash = 99683CDDBC40A729A5F51C20DA7B6C297F794C5D).

Energy type	Cost (US$ (kWh)$^{-1}$)
Fossil fuel	0.05–0.17
Bioenergy	0.05–0.24
Geothermal	0.06–0.14
Hydro	0.03–0.14
Solar photovoltaics	0.06–0.22
Solar concentrated solar power SP	0.11–0.27
Offshore wind	0.10–0.20
Onshore wind	0.04–0.01
Nuclear	0.1–0.15

cleaning up the pollution that burning the fossil fuels generate. The environment should be considered in the costing and accounting. It should be part of every transaction, be it plastic or fuel or electricity. Adding clean-up costs to fossil fuels will make the argument for renewables even more attractive. The environment and the atmosphere, in particular, are limited resources and clean-up costs are expensive. The atmosphere will absorb only so much GHGs and black carbon before it is deemed unclean. Logically, there is no reason why people buying electricity made from fossil fuels should not be paying for the clean-up. Furthermore, the longer we leave it the costlier will the clean-up get. As described by the United Nations (https://www.un.org/en/sections/issues-depth/climate-change/):

"Climate change is the defining issue of our time and we are at a defining moment. From shifting weather patterns that threaten food production, to rising sea levels that increase the risk of catastrophic flooding, the impacts of climate change are global in scope and unprecedented in scale. Without drastic action today, adapting to these impacts in the future will be more difficult and costly."

3. When dealing with hard-nosed climate change deniers, one must not keep plugging away with facts and figures and theories of carbon dioxide and global warming. Instead, one should build a bridge between the facts and the science behind climate change and people's cultural convictions and values and beliefs. One should appeal to more diverse values, particularly those values held by the deniers themselves. Perhaps the focus should be on investing in more profitable energy options as laid out in solution 1 above, or the need for a cleaner environment as laid out in solution 2 above. Furthermore, fossil fuels have a relatively short life as discussed earlier and that too should help. It has been suggested that in the US context, appealing to conservative values of patriotism, obeying authority and defending the purity of nature can encourage conservatives to support pro-environmental actions. Brow-beating deniers with further climate science are unlikely to succeed: their faculty of reason is motivated to defend itself from revising its beliefs.

4. People power is having an effect and organizations are disinvesting their fossil fuel shares. The recent protests led by Greta Thunberg have had a major impact and support for doing something to replace fossil fuels and clean up the environment is growing. We must support grassroots movements such as Fridays for Future, Youth for Climate, Climate Strike or Youth Strike for Climate, Greenpeace, Friends of the Earth, Global Justice Now, and Extinction Rebellion. Collective action is required. The present COVID-19 pandemic has shown the world that it is possible for everyone to work toward a common goal and this bodes well for the next major catastrophe—global heating.

5. Most of the solutions are in the hands of governments and big oil companies. What we can do is to target governments (by voting perhaps?) to reduce subsidies given to coal, oil, and natural gas companies and target these companies to embrace renewable

energy. Companies and organizations (such as pension funds, for example) should be encouraged to divest themselves of their shares in fossil fuels. Governments should tax imports from countries that still use coal to make electricity.

6. We need to ensure that our governments are working toward climate change solutions at a global level. A single country does not hold the answers to solving greenhouse gas emissions. We need leaders who will support global initiatives. As stated by the United Nations: "Climate change is a global challenge and requires a global solution. Greenhouse gas emissions have the same impact on the atmosphere whether they originate in Washington, London, or Beijing. Consequently, action by one country to reduce emissions will do little to slow global warming unless other countries act as well. Ultimately, an effective strategy will require commitments and action by all the major emitting countries" (https://www.un.org/en/climatechange/un-climate-summit-2019.shtml).

7. The global effort to manage climate change has been organized through what is called the United Nations Framework Convention on Climate Change (UNFCCC). The UNFCCC was launched at the 1992 Rio Earth Summit to reduce GHG concentrations. With the failure of the Rio initiatives, the then 191 signatories to the UNFCCC agreed to meet in Kyoto in 1997 to establish a more stringent regime. The resulting Kyoto Protocol created a global trading system for carbon credits and binding GHG reductions for ratifying countries. (The United States did not sign; China and India were exempt as developing countries.) The UNFCCC set up annual meetings called Conferences of the Parties (COPs) in places such as The Hague, Cancun, Durban, and Doha to assess progress in dealing with climate change. The meetings began the mid-1990s, to negotiate the Kyoto Protocol to establish legally binding obligations for developed countries to reduce their greenhouse gas emissions. The most recent meeting of COP (the 26th session) took place early November 2021, in Glasgow, United Kingdom. As with previous COP meetings, commitments by the major players were weak. More than 40 countries have pledged to shift away from coal. These include major coal using countries, Poland, Vietnam and Chile but not China, Australia, or the United States.

    Perhaps leading by example will bring the selfish nations to the negotiating table.

8. We need to remember the advice of Riccardo Mastini in reference (http://unevenearth.org/2020/02/a-post-growth-green-new-deal/):

    "To summarize from a post-growth perspective, a Green New Deal must pursue three distinct but interrelated goals: decreasing energy and material use, decommodifying the basic necessities of life, and democratizing economic production. Any Green New Deal proposal that does not directly address the drivers of economic growth is doomed to fall short of the challenge of steering away from the worst scenarios of ecological breakdown." This is also the sentiment of Jason Hickel who wrote in reference (Hickel, 2020):

    "The world has finally awoken to the reality of climate breakdown and ecological collapse. Now we must face up to its primary cause. Capitalism demands perpetual expansion, which is devastating the living world. There is only one solution that will lead to meaningful and immediate change: degrowth."

9. And lastly, we should also consider reducing our own use of fossil fuel by supporting renewable energy utility companies and reducing our carbon footprint on unnecessary travel.

2.7 Conclusion

We are already seeing the devastating effects of climate change on global food supplies, increasing migration, conflict, disease and global instability, and this will only get worse if we do not act now. The evidence is irrefutable. Man-made climate change is the biggest environmental crisis of our time. As Sir David King stated 14 years ago in his foreword to the first edition of *Future Energy: Improved, Sustainable and Clean Options for our Planet, 1st Edition*, "climate change is the most severe calamity our civilization has yet to face" (King, 2008).

Global warming threatens the future of the planet that we depend on for our survival, and we are the last generation that can do something about it.

References

BP Energy Economics. (2020). *BP Energy outlook*. Edinburgh: Centre for Energy Economics Research and Policy, Heriot-Watt University.

Christine, K., Johnson, P. L., Hitchens, P. S., Pandit, J., Rushmore, T., Smiley Evans, C. C. W., Young., & Megan, M. D. (2020). Global shifts in mammalian population trends reveal key predictors of virus spillover risk. *Proceedings of the Royal Society. B (Biological Sciences)*. Available from https://doi.org/10.1098/rspb.2019.2736.

Goodall, J. (2020). Mongabay, News: Covid-19 is a product of our unhealthy relationship with animals and the environment. Available from https://news.mongabay.com/2020/05/jane-goodall-covid-19-is-a-product-of-our-unhealthy-relationship-with-animals-and-the-environment/.

Jason, H. (2020). *Less is more: How degrowth will save the world*. Milton Keynes: William Heinemann. Available from https://www.penguin.co.uk/books/1119823/less-is-more/9781785152498.html.

King, D. (2008). Foreword, ISBN:978-0-08-054808-1 In T. M. Letcher (Ed.), *Future energy: Improved, sustainable and clean options for our planet* (1st ed.). Oxford: Elsevier.

Why do we have global warming? In T. M. Letcher (Ed.), Managing global warming. Oxford: Elsevier, chapter 1.

Climate change – A complex situation In T. M. Letcher (Ed.), *Climate change* (3rd ed., pp. 3–18). Oxford: Elsevier, Chapter 1.

Why discus the impacts of climate change. In T. M. Letcher (Ed.), *The Impacts of climate Change*. Oxford: Elsevier, Chapter 1.

Olivier, J. G. J., & Peters, J. A. H. W. (2018). *Trends in global CO_2 and total GHG emissions, 2018 report*. The Hague: PBL Netherlands Environmental Assessment Agency. https://www.pbl.nl/en/publications/trends-in-global-co2-and-total-greenhouse-gas-emissions-2018-report.

Tuckett, R. P. (2021). The role of atmospheric gases in global warming. In T. M. Letcher (Ed.), *Climate change, observed impacts on planet earth* (3rd ed., pp. 19–45). Oxford: Elsevier, Chapter 2.

CHAPTER 3

Water resource planning and climate change

Rabee Rustum[1], Adebayo J. Adeloye[2] and Quan Dau[3]
[1]*Heriot-Watt University, Dubai Campus, Dubai Knowledge Park, Dubai,* [2]*Heriot-Watt University, Edinburgh, United Kingdom,* [3]*School of Climate Change and Adaptation, University of Prince Edward Island, Charlottetown, Prince Edward Island, Canada*

Chapter Outline
3.1 Introduction 27
3.2 Climate change and water budget components 28
3.3 Water reservoirs 29
  3.3.1 Indices of reservoir performance 30
  3.3.2 Reservoir operation 31
  3.3.3 Climate change effect on reservoir performance in the Indus Basin 32
3.4 Conclusions 38
References 38

3.1 Introduction

Water is the most valuable resource and the most crucial issue of our time (Strang, 2004). However, climate change impacts water resource systems leading to water stress in many parts of the world, especially in developing countries, where the problem is compounded by rapid population growth, poverty, and lack of basic infrastructure (Pradhanang & Jahan, 2021). Increased water stress caused by scarcity will not only hinder the provision of water for drinking and industry, but it will also limit crop production and the ability to generate electricity from hydroelectric projects. Mitigating and adapting to these impacts will require improvements in the planning and operation of reservoirs and other large water facilities.

Therefore, water resources should be managed to support human well-being and ecosystem integrity in a robust economy as they are the core to sustainable development (Connor, 2015). The UN envisages that sustainable development goals should envision a more holistic and integrated agenda for advancing human well-being (Dickens et al., 2019).

Thus, managing water resource sustainability in the face of climate change impacts and challenges should be guided by the principles of IWRM (Integrated Water Resources Management). In that, water, land, and related resources should be managed to maximize the resultant economic and social welfare equitably without compromising the sustainability of vital ecosystems (Janardhanan, 2021).

In addition, climate change and the increase in population and urban development worldwide pose a significant risk to freshwater, both surface and groundwater. For example, they brought a significant impact on sediment transport, threats to ecohydrological function (Ashraf, 2013), and reduction in surface runoff, causing deterioration in reservoir vulnerability and reliability (Adeloye & Dau, 2019). However, recent studies have suggested that the socioeconomic impacts on water resources are much more significant than climate change (Dau et al., 2021; Momblanch et al., 2020). Therefore, managing water storage is a crucial strategy for tackling the challenges mentioned above, such as controlling the release and consumption of the stored water to further optimize these resources.

This chapter focuses on the impact of climate change on the performance of water reservoirs. This is because reservoirs are a major component of most water resources systems in various parts of the world. The way they are operated can go a long way in achieving adaptation plans for climate change impacts. In general, reservoirs are planned or designed using historical records of runoff at the reservoir site, but where future runoff during operation differs significantly from the historical, performance problems will result. Climate change will intensify the discord between historical and operational hydrology and will require innovative ideas to overcome these issues. This is because climate change will affect precipitation patterns and soil moisture, affecting the radiative forcing. The World Water Development Report (WWDR, 2020) suggests that much of the extreme impacts of climate change and water stress will manifest in the tropical zones where most developing countries are situated (UNESCO and UN-Water, 2020). The chapter also provides insights on understanding reservoir management for developing effective adaptation responses under climate change.

3.2 Climate change and water budget components

Increases in temperatures and solar radiation have been proven to increase both evaporation and evapotranspiration from open water and vegetation (Abtew, 2013). The risk of droughts is increasing because of drying and rising temperatures caused by climate change. Rainfall deficits and low humidity experienced during a drought create a favorable environment for accelerated water losses from lakes, reservoirs, and soil due to evaporation and evapotranspiration.

For example, Pan et al. (2015) found that terrestrial evapotranspiration has direct correlations with temperature and precipitation, with climate change effects accounting for 91.3% of inter-annual variations in evapotranspiration. The lack of available water supply at high temperatures suggests precipitation deficits may limit evapotranspiration. Decreases in soil moisture may also be attributable to playing a dominant role in the decline of evapotranspiration in the high-temperature scenario in the event precipitation does not change. The study utilized future scenario projections that arose from the work of the early IPCC (Intergovernmental Panel on Climate Change) forecasting up to 2099 for various scenarios.

The increase in potential evapotranspiration suggests crop water requirements are on the rise, as discussed by Dinpashoh et al. (2019) and Hasenmueller (2013). The soil hydrology or soil water content available for the plants is also affected by climate change as it is a function of the ratio between precipitation and evapotranspiration. Thus, the higher the evapotranspiration ratio, the higher the soil water content. The land surface also plays a vital role within the climate system. This feature is degraded through deforestation, agriculture, and other degrading uses with retrospective impacts on the regional atmospheric circulation. The increase in temperature causes alterations in the surface parameters and land cover characteristics. Altered patterns in precipitation influence the production of terrestrial vegetation and the availability of soil moisture with changes in evapotranspiration (Siva Kumar, 2007).

Zhongming et al. (2021) stress the need to move toward Sustainable Land Management (SLM) which is aptly defined as "the stewardship and use of land resources, including soils, water, animals and plants, to meet changing human needs, while simultaneously ensuring the long term productive potential of these resources and the maintenance of their environmental functions". In the context of extreme events and climate change challenges, SLM is viewed as a risk reduction tool to protect the vulnerable from declining socioeconomic impacts. Of these resources, water reservoirs are discussed in the next section.

3.3 Water reservoirs

As noted earlier, reservoirs are a major component of large scale water supply systems that rely on surface water resources. Reservoirs store excess water during high river flows, which is then released during low flows to meet demands. The planning of reservoirs using any of the available techniques (see Adeloye, 2012; McMahon & Adeloye, 2005) relies on the available historic runoff data at the reservoir site to determine the storage capacity of the reservoir that will meet the demand during the operation of the reservoir. Where the data are available, the effects of secondary processes such as net evaporation and sediment accumulation can be accommodated in the planning analysis, although this increases the computational effort (Pinarlik et al., 2021). The assumption implicit in planning analysis is that the historic runoff and other inputs used for the planning will be repeated exactly in the future, in which case the reservoir will continue to perform as intended. However, because the historic record used for planning is just a realization of several possible realizations at the site, deviations from the intended performance can occur, including situations where the reservoir system fails catastrophically to perform. Climate change is an additional stressor that impacts the variability of the climate and hence surface runoff with consequences for the performance of reservoirs.

Mitigating the impact of climate change will require that significant reductions in emitted greenhouse gases occur and remains a long term objective. Much more immediate is how water resources systems can adapt to the effect of climate change through improved operational practices of the reservoirs in such a way that severe water shortages are averted. It would be better to have a series of small water shortages spread over a long period than a large crippling shortage concentrated over a few short periods. Therefore, the starting point is to assess the performance of the reservoirs with and without climate change, evaluate any deterioration in performance, and develop improvements in operational practices that will temper the deterioration. In this chapter, examples will be presented of recent studies that have investigated climate change impacts of reservoir performance and improvements in operating policy that have largely eliminated the problem. However, before presenting the examples, the indices of reservoir performance will be reviewed briefly.

3.3.1 Indices of reservoir performance

The available indices of reservoir performance are as follows (Adeloye, 2012):

3.3.1.1 Reliability

Reliability measures either the proportion of time in which full demand is met (time-reliability, Eq. 3.1) or the proportion of total demand met during operation (volume-based, Eq. 3.2):

$$R_t = \frac{n - N_s}{n} \tag{3.1}$$

$$R_v = \frac{v - V_s}{v} \tag{3.2}$$

where R_t is the time-reliability, R_v is the volumetric reliability, N_s is the number of periods of unsatisfied demand, n is the total periods in operation, V_s is the total volume of shortage over the operation period and v is the total volume of water demanded over the operation period. In general, the R_v and R_t will give different indications of the performance, with $R_v \geq R_t$, resulting in the need for tradeoffs when using the two indices simultaneously for performance evaluation. Adeloye et al. (2017) harmonized the two indices that have removed the need for the tradeoffs.

3.3.1.2 Resilience

Resilience measures the reservoir's ability to recover from failure (Adeloye & Soundharajan, 2018).

$$\Phi = \frac{f_s}{f_d} (f_d \neq 0) \tag{3.3}$$

where Φ is the reservoir resilience, f_s is the number of continuous failure sequences over the planning period, f_d is the total number of failure periods. Resilience is measured in a fraction between 0 and 1:

$$0 < \Phi \leq 1 \tag{3.4}$$

3.3.1.3 Vulnerability

Vulnerability is a measure of the impact of water shortage and is usually estimated as the mean of the maximum water shortage occurring across the series of water shortage sequences in the simulation:

$$\eta = \frac{\sum_{t=1}^{f_d} \left[(D_t - Y_t)/D_t\right]}{f_d} ; \ t \in f_d \tag{3.5}$$

where η is the vulnerability is dimensionless, D_t is the demand in period t, Y_t is the actual supply (less any spills) in the same period and the other variables are as defined previously.

3.3.2 Reservoir operation

To evaluate the above indices will require simulation of the behavior of the reservoirs based on an operating policy. The operating policy determines how the stored water in the reservoir is released following consideration of the demand and the available water in the reservoir (McMahon & Adeloye, 2005). Fig. 3.1 is an example of zone-based rule curves used for the purpose. When the available water is between the upper and lower rules curves, attempts are made to supply the full water demand without going below the lower rule curve (LRC). If the supply will cause the water to fall below the LRC, the supply is reduced to prevent this. A situation where the supply is lesser than the demand is called a failure, and the water shortage can grow if the inflow situation does not increase to the point that the shortage becomes unbearable, that is, the vulnerability is maximized. Fig. 3.1 also includes a critical rule curve which is a way of managing the shortage so

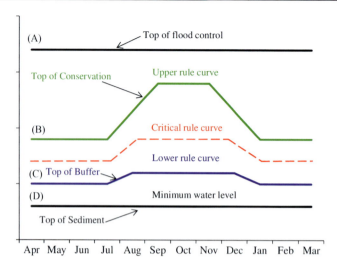

Figure 3.1
A schematic showing example of water reservoir rule curves (Adeloye & Soundharajan, 2018). A: Top of flood control. B: Upper Rule Curve. C: Lower Rule curve. D: Minimum water level.

that the situation of excessive shortage is prevented. This is called hedging, that is, reducing the water supply by small amounts so that users can adapt instead of potentially large shortages that can occur without hedging and the critical rule curve.

Adeloye et al. (2017) have investigated the effectiveness of hedging in adapting to the impacts of climate change and have shown that significant reductions in vulnerability can be achieved by hedging. An example in the Indus Basin in India will now be presented that demonstrates the development of the enhanced operating rules and their use in driving reservoir simulation to produce reductions in system vulnerability.

3.3.3 Climate change effect on reservoir performance in the Indus Basin

3.3.3.1 System description

The Beas-Sutlej river basin, part of the Indus Basin in northern India, was selected for the study (Fig. 3.2). The Beas basin lies entirely in India with a total area of 12,569 km^2, of which 780 km^2 is under permanent snow; whereas the Sutlej basin covers 56,860 km^2 with 22,275 km^2 of this lying in India. There are two main reservoirs, that is, the Pong on the Beas River and the Bhakra on the Sutlej River, which serve irrigation water supply, hydropower generation, and flood control (see Table 3.1). The average annual runoff of the Sutlej River is estimated at around 16,000 million m^3 (MCM), with around 59% of this runoff deriving from snowmelt (Singh & Jain, 2002). In the Beas River, the average annual runoff is approximately 14,800 MCM, and about 35% of this is also from snowmelt (Kumar et al., 2007). Mean annual

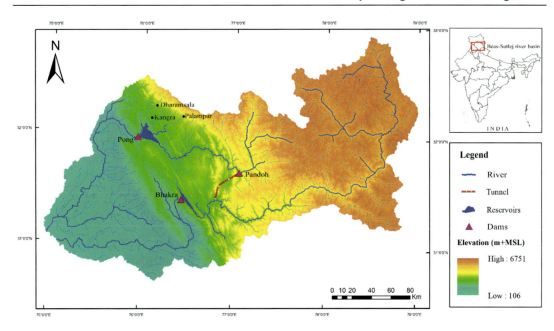

Figure 3.2
The Beas-Sutlej river basin, India (Dau & Adeloye, 2021a).

Table 3.1: Salient features for the Pong and Bhakra reservoirs.

Elements	Pong	Bhakra
Crest length (m)	1950.7	518.16
Width at crest (m)	13.72	9.14
Catchment area (km^2)	12,560	56,980
Length of reservoir (km)	41.8	96.56
Dead storage elevation (m + MSL)	384.05	445.62
Normal reservoir elevation (m + MSL)	423.67	512.06
Gross storage capacity (MCM)	8570	9621
Live storage capacity (MCM)	7290	7191
Dead storage capacity (MCM)	1280	2430
Hydropower capacity (megawatt, MW)	396	1325

rainfall in the Beas and Sutlej catchments is estimated at 1800 and 1260 mm, respectively. Except during the monsoon season (July to August), evaporation usually exceeds the amount of rainfall all year round, which leads to serious impacts on water security in the basin.

3.3.3.2 Design of rule curves and hedging policy

The ordinates of the upper rule curve (URC) and LRC were optimized using the genetic algorithm (GA) optimizer. In the GA, a process of fitness-based selection and recombination is derived from random chromosomes to produce a successor population. In this stage, parent chromosomes are selected, and their genes are recombined to generate new offspring (children) through selection, crossover, and mutation. This process is repeated until the specified number of iterations is reached, or the stopping criterion has been attained. Since the initial solutions are randomly generated, it means they are not unique as they will change with every implementation of the generation routine. Repeating the GA optimization over several realizations of the initial solution will help capture the variability caused by the random generation of the initial solution. A trial and error approach is usually applied to select the appropriate population size. As shown by Adeloye et al. (2016), the genetic operation carried out over 100 generations using 100 population sizes is often enough.

For a diffused prior situation in which a uniform density function is assumed for randomly generating the initial solutions, both the upper and lower bounds of these must be specified before optimizing with the GA. These were derived from the implementation of the Sequent Peak Algorithm (SPA) (Loucks et al., 1981) and its modification by Adeloye et al. (2001) to accommodate evaporation loss. These bounds were then used as boundaries for randomly generating the initial solutions in GA optimization (see Fig. 3.3).

The critical rule curves that initiate hedging were derived using the same GA optimization for the URC and LRC. Associated with the hedging curve are the rationing factors which determine the maximum proportion of the target demand that can be met during hedging.

The resulting rules curves with integrated-hedging curves are shown in Fig. 3.4. The rationing factors during hedging, that is, 0.82 for the Pong reservoir and 0.78 for the

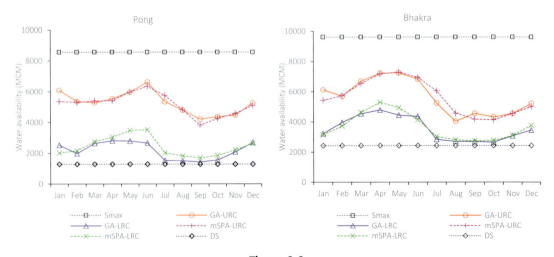

Figure 3.3
Optimal zone-based rule curves using a coupled mSPA-GA optimizer (Dau & Adeloye, 2021a).

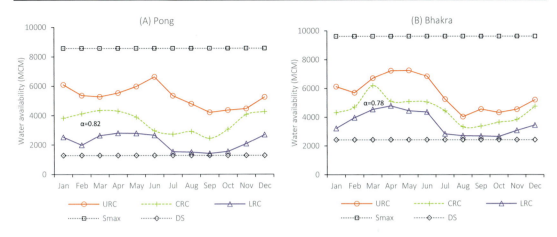

Figure 3.4
Integrated-hedging rule curve for (A) Pong and (B) Bhakra reservoirs (Dau & Adeloye, 2021a).

Table 3.2: Performance measure for irrigation water supply in standalone operation.

	No hedging		With hedging	
Index (%)	Pong	Bhakra	Pong	Bhakra
R_t	92	75	78	70
R_v	96	88	95	92
φ	85	66	38	57
η	40	54	20	50

Bhakra, remain constant throughout the year. Adeloye and Dau (2019) investigated the effect of dynamically varying these factors seasonally, but the effect was minimal.

3.3.3.3 Reservoir performance indices

The evaluated performance indices—reliabilities, resilience, and vulnerability—both for hedging and no hedging are shown in Table 3.2. As seen in the Table, the reliabilities for no hedging are high, with the volume-based reliability higher than its time-based equivalent as expected. As hedging is introduced, the time-based reliability plunged significantly due to the deliberate action to not meet the full demand. The resilience indices were also high for no hedging, implying reservoir systems that are capable of immediate recovery from a failed situation; though, it dropped during hedging because the total failure duration was deliberately prolonged as a consequence of hedging. For vulnerability, hedging proved a significant effect for the Pong reservoir, which has improved from a high vulnerability of 40% (no hedging) to 20% with hedging. Similarly, the improvement situation at Bhakra is less remarkable, only dropping from 58% (no hedging) to 54% (with hedging).

The system was then tested with the projections of climate and land use changes at three irrigated areas (Punjab, Haryana, and Rajasthan) located downstream of the Beas-Sutlej Basin, as given in Fig. 3.5. Based on the results of ensemble climate models (Dau & Adeloye, 2021b), mean annual precipitation in the Beas-Sutlej basin will increase from 15% to 40% (mid-century) and 10% to 30% (end-century) relative to the baseline, respectively. The temperature is also expected to increase considerably between 3°C and 6°C in the basin, with the highest rise projected by the GDDL-CM3 model. It suggests that more rainfall would occur in the Punjab State, especially during the end of the century, while a moderate change is observed in the Haryana. The study also found that the relative dryness of Rajasthan State in the baseline is likely to change insignificantly in the future, causing it to become more challenging for future adaptations. Simultaneous with climate change, land use also shows a rapid shift in the coming decades. For example, irrigation land will decrease by 15%−30% in the Punjab State and 5%−10% in the Haryana State. The situation is inverted in the Rajasthan irrigation land, which is expected to expand from 12% to 18% in the future relative to the baseline period. Changes in climate and land use have significant impacts on the hydrological regime and reservoir performances. This has been demonstrated from the previous studies conducted by Dau and Adeloye (2021b) and Dau et al. (2021).

Figure 3.5
Performance indices for irrigation water demands (Dau et al., 2021).

Regarding hydropower generation, the performance indices reveal a very marked reliability improvement relative to the baseline at the Bhakra reservoir. The vulnerability trend is also desirable, with this projected to decline in the future (see Fig. 3.6).

For flood control, the optimized storage levels in Fig. 3.7 represent a reduction relative to the nonoptimized level of about 12 m (~1300 MCM) at the Bhakra reservoir and 3 m (~580 MCM) at the Pong reservoir during high inflow periods. The financial and other benefits of these extra flood storage spaces during the flood season will be huge, although the actual quantification of these is beyond the purview of this study.

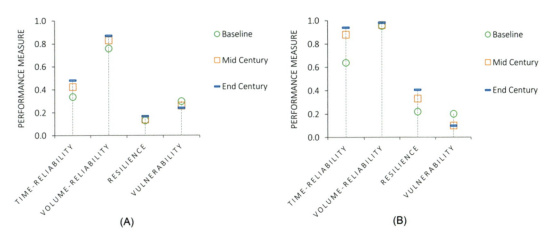

Figure 3.6
Performance indices for hydropower generation at (A) Pong and (B) Bhakra reservoirs (Dau et al., 2021).

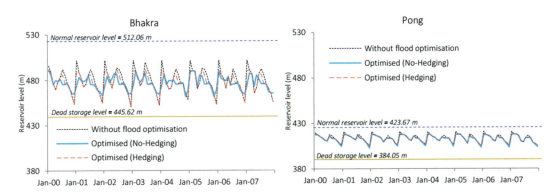

Figure 3.7
Optimized flood control reservoir levels at Bhakra and Pong reservoirs (Dau & Adeloye, 2021a).

3.4 Conclusions

In this chapter, a potential valuation of the effects of global climate changes on water resource security was carried out. The chapter shows the importance of accounting for climate changes on future water security risks. Only in this way is it possible to develop appropriate adaptation measures that can ensure reliable access to water resources in the future and support sustainable development.

In terms of the performance implications for the two reservoirs used as examples, the Pong and Bhakra, there were general improvements because of the enhanced runoff in the upstream part and reduced irrigation water demands in the downstream. Also, due to high inflows expected to arrive at the Pong and Bhakra reservoirs in the future and the reduced allocation to irrigation needs, hydropower generation at the two reservoirs is projected to be enhanced in the future leading to improved performance.

The results in the case studies indicate that integrated operation did enhance irrigation performance relative to the standalone operation (Dau & Adeloye, 2021a). For example, although hedging with standalone operation did help to reduce system vulnerability, the same outcome was achieved without hedging if the reservoirs were operated jointly.

The methodology used for performance evaluation in this chapter can be useful for planning adaptation strategies to cope with the drivers (climate and socioeconomic changes) of water insecurity in many river basins. For example, adaptation is necessary for sustaining water resources, mainly to minimize water shortages and to anticipate water supply challenges in the longer term. Changing crops such as rice to wheat appears to be an effective option to reduce water requirements while ensuring food security. Further detailed studies on reservoir operation rules could shed more light on potential supply management solutions to enhance water supply for irrigation and hydropower production.

To sum up, joint reservoir operation enhances the system yield and improves the system performance, especially in terms of the vulnerability that quantifies the intensity of water shortage.

References

Abtew, W. (2013). In A. Melesse (Ed.), Evaporation and evapotranspiration: Measurements and estimations *by Wossenu Abtew, Assefa Melesse* (1st ed.). Dordrecht: Springer Netherlands: Imprint: Springer.

Adeloye, A. J. (2012). *Hydrological sizing of water supply reservoirs. Encyclopedia of lakes and reservoirs* (pp. 346–355). Springer.

Adeloye, A. J., & Dau, Q. V. (2019). Hedging as an adaptive measure for climate change induced water shortage at the Pong reservoir in the Indus Basin Beas River, India. *Science of the Total Environment, 687*, 554–566. Available from https://doi.org/10.1016/j.scitotenv.2019.06.021.

Adeloye, A. J., & Soundharajan, B. S. (2018). Effect of reservoir zones and hedging factor dynamism on reservoir adaptive capacity for climate change impacts. *Proceedings of the International Association of Hydrological Sciences*, *379*, 21−29.

Adeloye, A. J., Montaseri, M., & Garmann, C. (2001). Curing the misbehavior of reservoir capacity statistics by controlling shortfall during failures using the modified Sequent Peak Algorithm. *Water Resources Research*, *37*(1), 73−82. Available from https://doi.org/10.1029/2000WR900237.

Adeloye, A. J., Soundharajan, B. S., & Mohammed, S. (2017). Harmonization of reliability performance indices for planning and operational evaluation of water supply reservoirs. *Water Resources Management*, *31*(3), 1013−1029.

Adeloye, A. J., Soundharajan, B.-S., Ojha, C. S. P., & Remesan, R. (2016). Effect of Hedging-integrated rule curves on the performance of the Pong reservoir (India) during scenario-neutral climate change Perturbations. *Water Resources Management*, *30*(2), 445−470. Available from https://doi.org/10.1007/s11269-015-1171-z.

Ashraf, A. (2013). *Changing hydrology of the Himalayan watershed*. London: IntechOpen. Available from https://doi.org/10.5772/54492.

Connor, R. (2015). *The United Nations world water development report 2015: Water for a sustainable world* (Vol. 1). UNESCO publishing.

Dau, Q. V., & Adeloye, A. J. (2021a). Influence of reservoir joint operation on performance of the Pong-Bhakra multipurpose, multi-reservoir system in northern India. *Journal of Water Resources Planning and Management*, *147*(11), 04021076. Available from https://doi.org/10.1061/(asce)wr.1943-5452.0001462.

Dau, Q. V., & Adeloye, A. J. (2021b). Water security implications of climate and socioeconomic stressors for river basin management. *Hydrological Sciences Journal*, *66*(7), 1097−1112. Available from https://www.doi.org/10.1080/02626667.2021.1909032.

Dau, Q. V., Momblanch, A., & Adeloye, A. J. (2021). Adaptation in a Himalayan water resources system under a sustainable socioeconomic pathway in a high-emission context. *Journal of Hydrologic Engineering*, *26*(3), 1−21, 04021003. Available from https://www.doi.org/10.1061/(ASCE)HE.1943-5584.0002064.

Dickens, C., Nhlengethwa, S., & Ndhlovu, B. (2019). *Mainstreaming the sustainable development goals in developing countries*. IWMI.

Dinpashoh, Y., Singh, V. P., Biazar, S. M., & Kavehkar, S. (2019). Impact of climate change on streamflow timing (case study: Guilan Province). *Theoretical and Applied Climatology*, *138*(1), 65−76.

Hasenmueller, E. A. (2013). *Water balance estimates of evapotranspiration rates in areas with varying land use*. IntechOpen.

Janardhanan, R. (2021). Water management: A key to sustainable development. *Handbook of research on future opportunities for technology management education* (pp. 387−400). IGI Global.

Kumar, V., Singh, P., & Singh, V. (2007). Snow and glacier melt contribution in the Beas River at Pandoh Dam, Himachal Pradesh, India. *Hydrological Sciences Journal*, *52*(2), 376−388. Available from https://doi.org/10.1623/hysj.52.2.376.

Loucks, D. P., Stedinger, J. R., & Haith, D. A. (1981). *Water resource systems planning and analysis*. Englewood Cliffs, NJ: Prentice-Hall.

McMahon, T. A., & Adeloye, A. J. (2005). *Water resources yield*. Water Resources Publication.

Momblanch, A., Beevers, L., Srinivasalu, P., Kulkarni, A., & Holman, I. P. (2020). Enhancing production and flow of freshwater ecosystem services in a managed Himalayan river system under uncertain future climate. *Climatic Change*, *162*(2), 343−361. Available from https://doi.org/10.1007/s10584-020-02795-2.

Pan, S., Tian, H., Dangal, S. R. S., Yang, Q., Yang, J., Lu, C., Tao, B., Ren, W., & Ouyang, Z. (2015). Responses of global terrestrial evapotranspiration to climate change and increasing atmospheric CO_2 in the 21st century. *Earth's Future*, *3*(1), 15−35.

Pinarlik, M., Adeloye, A. J., & Selek, Z. (2021). Impacts of ignored evaporation and sedimentation fluxes at planning on reservoir performance in operation. *Water Resources Management*, *35*(11), 3539−3570. Available from https://doi.org/10.1007/s11269-021-02904-5.

Pradhanang, S. M., & Jahan, K. (2021). Urban water security for sustainable cities in the context of climate change. *Water, Climate Change, and Sustainability*, 213−224.

Singh, P., & Jain, S. K. (2002). Snow and glacier melt in the Satluj River at Bhakra Dam in the western Himalayan region. *Hydrological Sciences Journal*, *47*(1), 93–106. Available from https://doi.org/10.1080/02626660209492910.

Siva Kumar, M. V. K. (2007). In M. V. K. Sivakumar, & N. Ndiang'ui (Eds.), *Climate and land degradation* (1st ed.). Berlin, Heidelberg: Springer.

Strang, V. (2004). *The meaning of water* (1st ed.). Routledge. Available from https://doi.org/10.4324/9781003087090.

UNESCO and UN-Water. (2020). *United Nations world water development report 2020: Water and climate change*. Paris.

WWDR. (2020). United Nations World Water Development Report. *United Nations World Water Assessment and UN-Water, UNESCO*. Routledge. Available from https://unesdoc.unesco.org/ark:/48223/pf0000372985. locale = en.

CHAPTER 4

Potential impacts of climate change on biogeochemical cycling

Daniel A. Vallero

Pratt School of Engineering, Duke University, Durham, NC, United States

Chapter Outline

4.1 Introduction 41
4.2 Thermodynamics of cycling 42
4.3 Partitioning 45
    4.3.1 Dissolution 45
    4.3.2 Sorption 47
4.4 Biochemodynamics 49
4.5 Biogeochemistry of oxygen in water 55
4.6 The carbon cycle and greenhouse effect 57
4.7 Conclusions 59
References 59
Further reading 62

4.1 Introduction

The earth is a reactor in which matter and energy continuously move within and among cycles at scales ranging from molecular to planetary. One of the most obvious examples of energy cycling commences with the release of plumes from stationary and mobile combustion sources to the atmosphere wherein the released heat and pollutants move between and among the various phases of water and other carriers within the hydrosphere, atmosphere, and lithosphere. For instance, the atmosphere's aerosols both contain global greenhouse gases (GHGs) and heat energy. The GHGs may return to the lithosphere through natural processes of precipitation and deposition of the aerosols. However, the return to the earth can be enhanced through engineering projects, for example, carbon sequestration within rock structures (Resnik & Vallero, 2011; Shepherd, 2009; Vallero & Resnik, 2013).

4.2 Thermodynamics of cycling

The first law of thermodynamics requires that all mass and energy are balanced within all the earth's reactors, including the planetary-scale reactor. Chemical, biological, and physical agents continuously cycle within and between systems. Chemical agents include myriad classes and species, including toxic pollutants and the GHGs. Biological agents include organisms, for example, viruses, bacteria, and other microbes, as well as components of organisms like pollen, cysts, and spores. Physical agents include energy at various wavelengths, especially incoming solar radiation at the visible and ultraviolet wavelengths and black body releases of infrared energy. The three types intermingle, such as when organisms and their components store and use energy and transform chemicals at various trophic levels.

The structure and function of an environmental system determine the rates and types of biogeochemical cycling. The growth of plants and microbes depends on the amount and wavelengths of light energy, that is, the light budget. However, the light budget is one of many budgets that determine the rates of growth and survival of organisms and the condition of ecosystems. For example, the available nutrients, for example, the concentrations of phosphorus and nitrogen compounds are often rate limiting in water systems. The system needs light, water, carbon, and other inputs, but these are often available in surplus, so the growth of microbes like algae are limited by the presence of P or N. Thus, in the winter, light energy may indeed by the rate-limiting factor, but in the summer, algal growth in a lake can be managed and fish kills can be prevented by reducing inflows of N and P compounds (Chorus & Spijkerman, 2021; Schreckenbach et al., 2001).

Chemical cycling is directly affected by energy cycling, such as the direct proportionality of water temperature and a water body's dissolved oxygen (DO) content. Thus, temperature limits the abundance and type of fish communities. A stream may not be able to sustain a trout or salmon populations with increasing atmospheres, if the heat exchange between the air and water leads to mean water temperatures increases lead to drops in DO levels below game fish survival thresholds. Rough fish populations, like carp, may thrive at much lower ambient DO concentrations. In this case, the water ecosystem can be dramatically altered with concomitant increases in atmospheric temperatures (Vallero, 2015a).

Decreasing DO is certainly not the only indirect impact on a water ecosystem and, for that matter, public health. The amount of molecular oxygen (O_2) and the bioavailability of metals are affected by a waterbody's heat budget. Increasing water temperature decreases the O_2 and enhances the solubility of metals and other toxic substances (see Fig. 4.1). Concentrations of mercury, cadmium, and lead, for example, increase not only with increasing water temperature, with the toxic chemical species also increasing in concentration. The resulting decrease in O_2 concentrations creates a chemically reduced

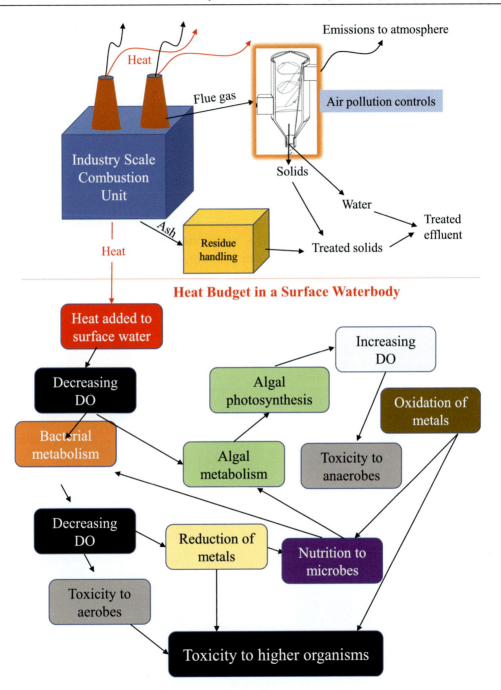

Figure 4.1

The fate of matter leaving the combustion chamber. Biota respond to the additional heat, with a concomitant abiotic response, that is, decreased dissolved oxygen (DO) concentrations in the water. Next, biotic processes generate an either increase or decrease in DO and increase adverse environmental and public health outcomes, for example, low DO and increase in metal bioavailability.

Modified from: Vallero, D. A. (2014). Fundamentals of air pollution (5th ed.). Elsevier Academic Press.

environment in which the metals form reduced inorganic species like sulfides, but rates of alkylation of the metals also increase, for example, the formation of highly toxic methylmercury. Thus, the synergistic impact of combining hypoxic water and reduced metal compounds forms a threatening cascade of harm to the stream's ecosystems, but also to the species up the food chain, including humans (Aberg et al., 1969; Dong et al., 2010).

The behavior of pollutants and their effects on biological systems are highly complex. For example, both bacteria and algae undergo population decreases with declining DO levels. However, algae also directly convert light energy to cell building via photosynthesis, which oxygenates the water. Thus, bacterial growth is absolutely decreasing DO levels, but algal growth is a net DO increase/decrease, depending on the difference in the rates of algal respiration and photosynthesis (Vernier Corporation, 2018). Large, complex molecules in the waste are degraded into simpler molecules by decomposers, a process known as mineralization. As shown in Fig. 4.2, the cycling of larger to smaller molecules enriches

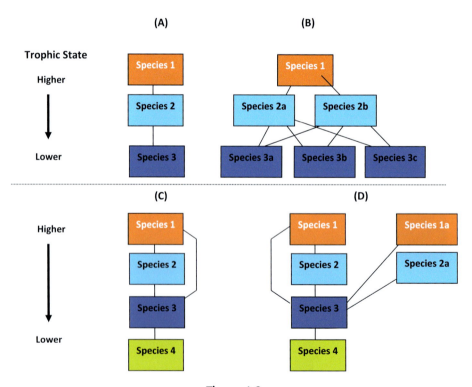

Figure 4.2
Mass and energy flows in food chains: (A) Linear; (B) Multi-level; (C) Omnivorous; and (D) Multi-level with predation and omnivorousness. *From: Vallero, D. A., & Letcher, T. M. (2012). Unraveling environmental disasters. Newnes. Data from Graedel, T. E. (1996). On the concept of industrial ecology. Annual Review of Energy and the Environment, 21(1), 69–98.*

soil, which supports plant life that is available to be consumed by higher order organisms, including human populations (Graedel, 1996; Vallero & Letcher, 2012).

Human activities change the rates of mass and energy cycling. For example, the extraction of natural resources, transportation, manufacturing trade and disposal generate wastes, including hazardous wastes. The fate of a pollutant is determined by various pathways, beginning with the release of a parent compound into the environment, followed by various physical, chemical, and biological transformation processes, until the ultimate fate in the environment or human tissues. Chemical agents are transported before residing in various environmental media along the pathway. The residence time, often expressed as a media-specific half-life can range from microseconds to decades. These media half-lives are part of predictive models on the likelihood and potential extent of the time that a human or other receptor encounters the pollutant. Thus, it is a key component of the exposure estimation steps in a risk assessment (Vallero et al., 2010, 2013). The exposure estimate dictates the dose received by a receptor (Dellarco et al., 1998; Lioy & Smith, 2013; Pleil & Sheldon, 2011).

4.3 Partitioning

Chemical compounds find their way into water systems predominantly by two mechanisms, that is, dissolution and sorption to a particle.

4.3.1 Dissolution

Polarity is an important physicochemical characteristic of a that determines a compound's solubility. A molecule's polarity is an expression of the unevenness in charge. Water's single oxygen and two hydrogen atoms are configured to yield a slight negative charge at the oxygen and a slight positive charge at the hydrogen ends, so that polar substances tend toward dissolution in water, whereas nonpolar substances resist being dissolved in water. The increasing kinetic energy with increasing temperature in a system increases the velocity of the molecules, which weakens intermolecular forces. Intermolecular forces may be relatively weak or strong. The weak forces in liquids and gases are known as van der Waals forces.

Compounds readily dissolved in water are hydrophilic and those that resist aqueous dissolution are hydrophobic. Most environmental contaminants are organic, consisting of molecules held together by carbon-to-carbon bonds and/or carbon-to-hydrogen bonds. Furthermore, the solubility of organic compounds in organic solvents is high, that is, lipophilic. A compound not easily dissolved in organic solvents is lipophobic. Often, a hydrophilic compound is also lipophobic.

According to the first law of thermodynamics, that is, the conservation law, every molecule moving into and out of the control volume must be accounted for, including all chemical changes within the control volume. Control volume is often envisioned as a simple cube reactor (Fig. 4.3A) to make the math simpler. More realistic environmental control volumes can consist of a cell within an organism, the entire organism, or the organism's habitat (Fig. 4.3B).

Any change in storage of a substance's mass in a control volume must equal the difference between the mass of the chemical transported into the system less the mass of the chemical

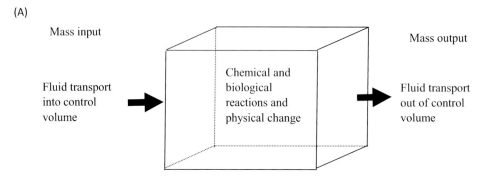

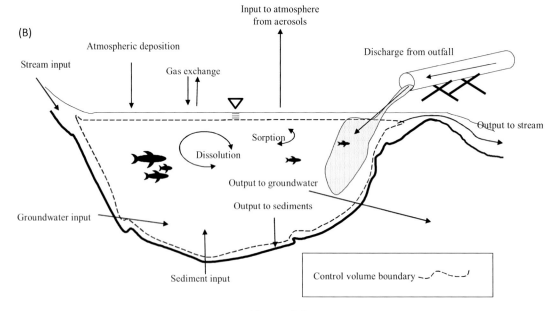

Figure 4.3
(A) Control volume of an environmental matrix (e.g., soil, sediment, or other unconsolidated material) or fluid (e.g., water, air, or blood). (B) Lake system. Both volumes have equal masses entering and exiting, with abiotic and biotic transformations and physical changes occurring within the control volume. From: Vallero, D. A. (2014). Fundamentals of air pollution (5th ed.). Elsevier Academic Press.

transported out of the system. ("Nonequilibrium Condition—An Overview | ScienceDirect Topics") The entering mass transported equals the inflow to the system that includes pollutant discharges, transfer from other control volumes and other media, and formation of degradation products by abiotic chemistry and biological transformation. Conversely, the outflow is the mass transported out of the control volume, which includes uptake by biota, physical transfer out of the control volume, and abiotic and biological degradation of the parent and degradation products. ("Nonequilibrium Condition—An Overview | ScienceDirect Topics") Stated as a differential equation, the rate of change a parent compound is:

$$\frac{d[A]}{dt} = -v \cdot \frac{d[A]}{dx} + \frac{d}{dx}\left(D \cdot \frac{d[A]}{dx}\right) + r \qquad (4.1)$$

where v is the fluid velocity, $d[A]/dx$ is the concentration gradient of chemical A, r is the internal sinks and sources within the control volume.

To determine a mass balance within an environmental system, partitioning relationships that control the "leaving" and "gaining" compartments must be known. Chemical compounds exit and enter the atmosphere from clouds, water bodies, soil, and biota, as well as move within soil and sediment matrices, in and on particles, within the atmosphere, and within organic tissues. ("Nonequilibrium Condition—An Overview | ScienceDirect Topics") Environmental systems can be envisioned as a cascade of control volumes. Within each control volume, an individual compartment may gain or lose mass of a chemical species, but the overall mass must always balance.

Some of the mass will remain unmoved and unchanged. Some of the unchanged mass fraction will be transported out of the control volume by advection and diffusion. Some of the mass fraction is chemically transformed with all products remaining products staying in the compartment where they were generated. ("Bioengineering—An Overview | ScienceDirect Topics") Finally, a fraction of the original parent mass is transformed and then moved to another compartment.

Engineers often refer to the mass that is dissolved in water as the total dissolved solids (TDS), which they differentiate to the mass that is suspended in water, that is, total suspended solids (TSS). This distinction is important since the engineered systems designed to remove TDS differ substantially from those designed to remove TSS.

4.3.2 Sorption

Sorption determines how easily transported, bioavailable, or toxic a compound will be in surface water, ground water, or the atmosphere (Lyman, 1995; Vallero, 2014). Sorption occurs at the microscopic scale in water.

The process whereby a dissolved compounds attaches to a solid surface is known as adsorption This is a common sorption process of organic compounds sorbing to clayey soils (Johnson et al., 1989). Absorption occurs in porous materials so that the solute can diffuse into the particle and be sorbed onto the inside surfaces of the particle. ("Chemical Phenomena—An Overview | ScienceDirect Topics") The main process for absorption electrostatic interactions between the surface and the contaminant. Chemisorption integrates a chemical into a porous materials surface via chemical reactions, such as a covalent reaction between a mineral surface and a compound in soil.

Ion exchange is the process by which positively charged ions (cations) are attracted to negatively charged particle surfaces or negatively charged ions (anions) are attracted to positively charged particle surfaces, causing ions on the particle surfaces to be displaced. Particles undergoing ion exchange can include soils, sediment, airborne particulate matter, or even biota, such as pollen particles. Cation exchange has been characterized as being the second most important chemical process on earth, after photosynthesis. This is because the cation exchange capacity, and to a lesser degree anion exchange capacity in tropical soils, is how nutrients are made available to plant roots. Without this process, the atmospheric nutrients and the minerals in the soil would not come together to provide for the abundant plant life on planet earth (Richter, 1996). ("Chemical Phenomena—An Overview | ScienceDirect Topics").

These four types of sorption combine physical and chemical phenomena at surfaces. Adsorption and absorption are chiefly controlled by physical factors, whereas chemisorption and ion exchanges are combinations of chemical reactions and physical processes. Generally, sorption reactions affect partitioning in three ways (Westfall, 1987):

1. A compound's transport potential changes because of distributions between the aqueous phase and particles.
2. The aggregation and transport of the compound is affected by the electrostatic properties of suspended solids to which the compound is sorbed.
3. Dissociation, surface-catalysis, and precipitation are changed.

When a compound reaches a solid matrix like soil or sediment, a portion of it mass remains in solution in the interstitial water, whereas the rest is adsorbed onto the surfaces of the soil particles. Residence time in the matrix is strongly driven by sorption. The movement of strongly-sorbed compounds is restricted between a particle and the surrounding water, air, or other fluids in the matrix.

Chemical compounds ultimately reach a balance between the mass on solid surfaces and the mass in solution. Molecules migrate among physical phases to maintain this balance. The properties of both the chemical and the matrix determine the partitioning rates. These physicochemical relationships, known as sorption isotherms, are found experimentally and are reported in handbooks and journal articles (Burant et al., 2013; Finizio et al., 1997; Li et al., 2021; Mackay et al., 2018; Mackay & Wania, 1995; McKone et al., 2006; Miller & Weber, 1984; Muñoz et al., 2007; Oste et al., 2002; Sauvé et al., 2000; Storey et al., 1995; Tourinho et al., 2019; Weschler et al., 2008).

4.4 Biochemodynamics

The natural and anthropogenic sources of GHGs and other pollutants is driven by both the previously discussed thermodynamics, as well as by fluid dynamics, Thus, the interactions and independent, parallel processes of the movement and change of substances can be expressed by motion, chemical reactions, and biological processes, that is, biochemodynamics. At every level of biological organization, these processes are at work, but vary in space and time.

The continuous exchange of mass and energy within and between the abiotic and biotic components of ecosystems and human populations cycle through, between and among organisms and their surroundings. For example, if a persistent pesticide enters the food chain, it may alter the health and survival of organisms to varying extents, killing some, reducing the ability of others to reproduce, and simply being stored by others that are eaten by predators, which may build up pesticide concentrations. The elevated concentrations may affect endogenous processes that compromise the predator's ability to withstand increased temperatures or to survive under drought conditions. Thus, exposure to a toxic substance released to the environment, in this case, led to a decreased survival rate when combined with previously survival climate change effects. As result, the change affects diversity and species abundance, meaning that both the ecological function and structure were harmed due to concomitant changes in the growth and survival of a certain species. Structure can also be altered by ongoing adaptation and species replacement. Finally, as shown in Fig. 4.4, the pesticide and its degradation products, which may be more toxic than the parent pesticide, can enter the food supply of human populations via bioaccumulation (Hale et al., 2020; Mackay & Fraser, 2000).

Perturbations and responses occur at every level of biological organization and trophic state. An adverse outcome pathway (AOP) depicts the mechanisms that follow an organism's exposure to a substance that ultimately lead to harm (Versteeg & Naciff, 2015). The source-to-outcome pathway may be direct, indirect, or stepwise. For example, if containment of a potentially dangerous genetically modified microbe is breached and a microbe is released into the environment, the microbe would be the primary stressor of concern; it is behaving directly as a pathogen. When climatic factors change the environment, for example, flooding or drought, the already toxic situation is exacerbated. In an instance where the microbe is not directly pathogenic, the microbe may release

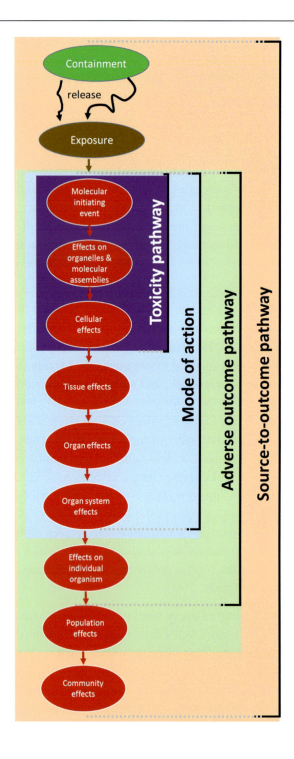

(*Continued*)

biological or chemical substances that are inherently toxic, for example, endotoxins and allergens. This indirect pathway requires an additional step between containment and exposure (see Fig. 4.5).

Thus, the AOP begins with exposure to an agent, uptake, and absorption within an organism, leading to a molecular initiating event (MIE) that leads to an adverse outcome (see Fig. 4.6). The pathway requires additional arrows and boxes between the initiating event and the outcome, that is, a sequence of key events (KEs) following the MIE and with other KEs, each linked by key event relationships (KERs). Ultimately, an adverse outcome occur in an organism, and if exposure is sufficiently distributed, to a larger population (Groh et al., 2015) (see Fig. 4.7).

Each response in an AOP culminates from many biological processes, including interactions between cellular receptors[1] and ligands, DNA binding, and protein oxidation, for example. The environmental assessment drives the type of AOP to be investigated. An organism-level response may be sufficient for an animal study; but a regulatory decision on the acceptable level of ecological risk posed by a chemical entering the marketplace may need to estimate a community or population outcome, for example, loss of diversity (Vallero, 2017). It is important to note that up to this point the discussion has addressed single agents, but there are often simultaneous pathways that are not be part of the AOP that can lead to environmental damage. For example, bioactivation may occur as the lower trophic state organism accumulates the parent compound and converts it to toxic forms (see Fig. 4.8). The resulting toxic metabolite is transferred to the predator via the food chain where the MIE in the predator may be affected by the exposure to the chemical and the decreased species abundance of its prey. Therefore, thought must be given to the steps before the MIE, that is, endogenous AOPs must be combined with exogenous AOPs to

◀

Figure 4.4
Stages among pathways that lead to an adverse environmental outcome. *From: Vallero, D. (2015a). Environmental biotechnology: A biosystems approach. Elsevier Science. Data from: Croft, D., O'Kelly, G., Wu, G., Haw, R., Gillespie, M., Matthews, L., . . . Jassal, B. (2010). Reactome: A database of reactions, pathways and biological processes. Nucleic Acids Research, gkq1018; Organisation for Economic Co-operation and Development. (2011). Report of the workshop on using mechanistic information in forming chemical categories (ENV/JM/MONO(2011)8). Crystal City VA, USA Environment Directorate Joint Meeting of The Chemicals Committee and The Working Party on Chemicals, Pesticides and Biotechnology.*

---

[1] In this discussion "receptor" denotes the location on the cell membrane that links to a biological ligand. The other connotation of receptor in this chapter refers to the entity receiving the stress, for example, a human, ecosystem, or material.

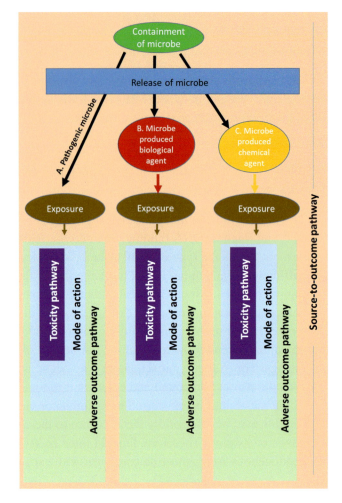

Figure 4.5
After containment of a microbe is breached, the microbe may be (A) pathogenic and lead directly to an adverse outcome; (B) not pathogenic but releases a toxic biological agent (e.g., spore or cyst) that initiates a toxicity pathway; or (C) not pathogenic but releases a toxic chemical agent (e.g., a volatile organic compound to the air or an endotoxin endogenously to the blood). "Each of these has its own pathways and mechanisms of action." ("Waste Constituent Pathways—ScienceDirect"). *From: Vallero, D. (2015a). Environmental biotechnology: A biosystems approach. Elsevier Science.*

predict the entire AOP of ecosystem damage. Even within a single organism, numerous AOPs can occur simultaneously.

Certain aspects of water-energy relationships during cycling are obvious and direct, such as when thermal inversions in the lower atmosphere introduce vertical differences in air

Figure 4.6
The three steps of the adverse outcome pathway, which is a conceptual framework of the logical sequence of events that lead to diseases, ecotoxic effects, and other adverse outcomes. ("Waste Constituent Pathways—ScienceDirect"). *From: Vallero, D. (2015a). Environmental biotechnology: A biosystems approach. Elsevier Science. Data from Allen, T. E., Goodman, J. M., Gutsell, S., & Russell, P. J. (2014). Defining molecular initiating events in the adverse outcome pathway framework for risk assessment. Chemical Research in Toxicology, 27(12), 2100–2112.*

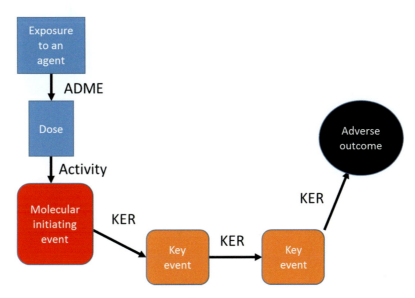

Figure 4.7
Adverse outcome pathway (AOP), consisting of a sequence of key events and key event relationships (KERs). Following absorption, distribution, and metabolism within an organism, enough of the chemical agent that is not eliminated becomes available to begin the AOP process. *From: Vallero, D. (2015a). Environmental biotechnology: A biosystems approach. Elsevier Science. Data from Groh, K. J., Carvalho, R. N., Chipman, J. K., Denslow, N. D., Halder, M., Murphy, C. A., ... Watanabe, K. H. (2015). Development and application of the adverse outcome pathway framework for understanding and predicting chronic toxicity: I. Challenges and research needs in ecotoxicology. Chemosphere, 120, 764–777.*

density and lead to exchanges of heat energy in water vapor between various layers in the troposphere. The transfer of energy between trophic states in an ecosystem depend partly on energy and water exchanges among living tissue and nonliving substrates. During photosynthesis, phosphorous compounds transfer energy from sunlight to plant tissue.

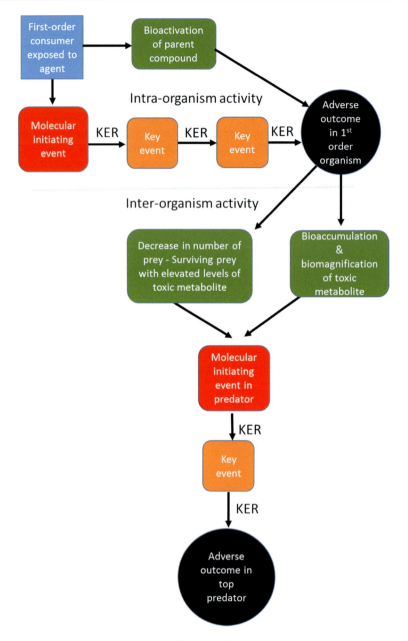

Figure 4.8
Linked adverse outcome pathways (AOPs), in two levels of biological organization. The AOP plus other endogenous processes of the first-order producer will lead to an AOP in the top predator. From: Vallero, D. (2015a). *Environmental biotechnology: A biosystems approach*. Elsevier Science.

The cycle continues with the generation and emission of molecular oxygen to the atmosphere. These are but a few of the many examples of water-energy cycling within the hydrosphere, biosphere, lithosphere, and atmosphere.

4.5 Biogeochemistry of oxygen in water

Without proper management of water systems at the facility-level, undesirable releases of pollutants can result, for example, allowing heated water to be released to surface waters may lower DO levels, harming aquatic life. Recall from Fig. 4.1 that the heat from boilers and other industry-scale operations is going to be exchanged. This is a consideration of any pollution management decision. The DO is a limiting factor of the diversity and dominance of fish communities that can be supported by a water body (see Tables 4.1 and 4.2). The net increase in heat stresses the biotic integrity of an aquatic ecosystem. Less tolerant game fish would be inordinately threatened, and undesirable fish would thrive. From an ecological standpoint, the biodiversity of the aquatic ecosystem is threatened. Less diverse

Table 4.1: Relationship between water temperature and maximum dissolved oxygen concentration in water (at 1 atm).

Temperature (°C)	Dissolved oxygen (mg L^{-1})	Temperature (°C)	Dissolved oxygen (mg L^{-1})
0	14.60	23	8.56
1	14.19	24	8.40
2	13.81	25	8.24
3	13.44	26	8.09
4	13.09	27	7.95
5	12.75	28	7.81
6	12.43	29	7.67
7	12.12	30	7.54
8	11.83	31	7.41
9	11.55	32	7.28
10	11.27	33	7.16
11	11.01	34	7.16
12	10.76	35	6.93
13	10.52	36	6.82
14	10.29	37	6.71
15	10.07	38	6.61
16	9.85	39	6.51
17	9.65	40	6.41
18	9.45	41	6.41
19	9.26	42	6.22
20	9.07	43	6.13
21	8.90	44	6.04
22	8.72	45	5.95

From: Vallero, D. (2015a). Environmental biotechnology: A biosystems approach. *Elsevier Science. Data from Dohner, E., Markowitz, A., Barbour, M., Simpson, J., Byrne, J., & Dates, G. (1997). 5.2. Monitoring and assessing water quality. In* Volunteer stream monitoring: A methods manual; *Vallero, D. (2010). Environmental biotechnology: A biosystems approach. Academic Press.*

Table 4.2: Normal temperature tolerances of aquatic organisms.

Organism	Taxonomy	Range in temperature tolerance (°C)	Minimum dissolved oxygen (mg L^{-1})
Trout	Salma, Oncorhynchus and Salvelinus spp.	5–20	6.5
Smallmouth bass	Micopterus dolomieu	5–28	6.5
Caddisfly larvae	Brachvcentrus spp.	10–25	4.0
Mayfly larvae	Ephemerella invaria	10–25	4.0
Stonefly larvae	Pterongrcys spp.	10–25	4.0
Catfish	Order Siluriformes	20–25	2.5
Carp	Cyprinus spp.	10–25	2.0
Water boatmen	Notonecta spp.	10–25	2.0
Mosquito larvae	Family Culicidae	10–25	1.0

Data from Vallero, D. (2010). Environmental biotechnology: A biosystems approach. *Academic Press; Vernier Corporation.* (2018). Computer 19: Dissolved oxygen in water. Retrieved from http://www2.vernier.com/sample_labs/BWV-19-COMP-dissolved_oxygen.pdf.

systems are more vulnerable to disease, sensitive species may be threatened, and the overall resilience of the ecosystem is impaired.

Physics, chemistry, and biology combine to create an adverse outcome, such as when increasing water temperature raises the aqueous solubility of otherwise less soluble compounds that are toxic to organisms. The solubility of mercury, cadmium, lead, and other toxic metals increases with elevated temperatures. Meanwhile, decreasing DO concentrations leads to changes in the aquatic system's redox, that is, the increased likelihood of a reduced environment where the metal sulfides and metallic organic compounds can more readily form. These can be highly toxic to fish and other organisms, including humans. One thing leads to another. The harm chaos and cascade begin with elevation of water temperature, followed by the resultant decrease in DO levels and increasing metal compound concentrations, the synergistic impact of the combining the hypoxic water and reduced metal compounds, and ultimately leads to a degraded stream ecosystem, accumulation in the food chain, and a threat to public health.

The relationships between abiotic and biotic systems affect the rates of growth and metabolism among different species of microbes. When respiring, growing, and reproducing, aerobic and facultative bacteria use O_2, thereby decreasing DO levels. Algae both consume DO for metabolism and produce DO by photosynthesis.

Certain chemical reactions are mediated by the anaerobic and facultative bacteria that reduce the metal compounds as the oxygen levels drop. The reduced conditions in the water and sediment can stress aquatic ecosystems, resulting in diminished biodiversity and even fish kills. Yet, in oxygen-rich regions of the aquatic system, the opposite is true, that is, the metals form oxides, which may also be toxic, but may less readily accumulate and may be more easily eliminated from the organism. For example, reduced forms of mercury bioaccumulate in fish tissue, concentrate in the food chain, and are neurotoxic to humans who consume the fate organisms, for example, seafood (Aberg et al., 1969; Dong et al., 2010; Lin & Pehkonen, 1999; Lindqvist et al., 1991; Munthe et al., 2001; Keating et al., 1997).

Even forms of metals in relatively low toxicity oxidations can be metabolized within the environment and within an organism to form alkylated metal compounds. Any form of mercury is toxic, but speciation can lead to highly toxic and environmentally mobile forms. As evidence, elemental, zero-valence mercury (Hg^0) can be released from mining operations, but when deposited to the surface the Hg speciation changes according to the ionic strength of the receiving waters (Stoddard et al., 2005; U.S. Environmental Protection Agency, 2016, 2017, 2018). Depending on the total Hg content of coal being burned, divalent mercury (Hg^{2+}) is released from coal-fired powerplant stacks. Both are less mobile and far less toxic than alkylated forms of mercury. However, Hg^{2+} emitted from a stack and Hg^0 released from a mine can be converted to the highly toxic dimethylmercury during transport.

4.6 The carbon cycle and greenhouse effect

Modeling climate change must characterize and quantify many drivers and constraints (see Fig. 4.9). The effects are seldom linear but can be synergistic or antagonistic as result of the interactions of these drivers and constraints. When wetland structure is altered by a change in microbial population from predominantly aerobic to anaerobic, releases of methane (CH_4) increase. An overly simplistic model would predict a direct increase in global temperatures simply from the increasing CH_4. However, this does not factor in biological activity and increased photosynthesis triggered by the increase in CO_2 when wetlands are lost to agricultural production or when the upper canopy cutting leads to greater amounts of insolation that will decrease forest floor detritus mass, converting the forest floor to predominantly aerobic microbial populations that will lead to lower releases of CH_4. There will be increases and decreases at various scales, so the net effects on a complex, planetary system are highly uncertain (Vallero, 2021).

The concentration of CH_4 in the atmosphere has been steady at about 0.75 for over a thousand years, and then increased to 0.85 ppm in 1900. Since then, CH_4 concentrations have doubled to about 1.7 ppm (Vallero, 2021).

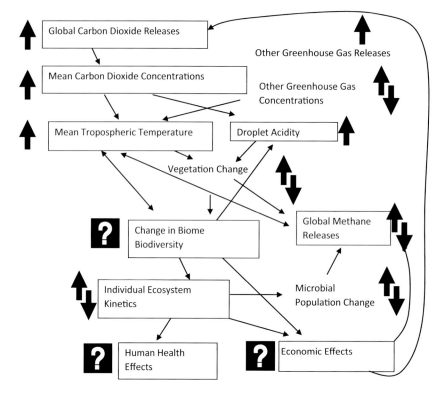

Figure 4.9
Changes in tropospheric carbon dioxide. Thick arrows indicate whether this factor will increase (up arrow), decrease (down arrow), or will vary depending on the specifics, for example, the chlorofluorocarbons and some gases can cool the atmosphere, for example, sulfate aerosols. Question mark indicates that the type and/or direction of change are unknown or mixed. Thin arrows connect the factors as drivers toward downstream effects. *From Vallero, D. (2015a). Environmental biotechnology: A biosystems approach. Elsevier Science.*

The principal mechanism for removing CH_4 from the atmosphere is by reaction with free radicals, an atom, molecule, or ion that has an unpaired electron. The most important is with the reaction with the hydroxyl radical (•OH):

$$CH_4 + \cdot OH + 9O_2 \rightarrow CO_2 + 0.5H_2 + 2H_2O + 5O_3 \tag{4.2}$$

This indicates that the reaction creates carbon dioxide, water vapor, and ozone, which are all GHGs. Thus, CH_4 molecule, itself a GHG, is the source of others.

Gas concentrations and exchange coefficients between the atmosphere and surface waters determine rate of a molecule of gas moving across the ocean-atmosphere boundary. The CO_2 in surface ocean takes about a year to equilibrate with atmospheric CO_2, thus large atmosphere-ocean differences in CO_2 concentrations are common. Oceans contain enormous C reservoirs, wherein the CO_2 reacts with water to form carbonic acid and its dissociation products. With the increased atmospheric CO_2 concentrations, the interaction with the ocean surface alters the chemistry of the seawater and leads to ocean acidification (National Oceanic and Atmospheric Administration, 2018; Vallero, 2015b).

Ocean uptake of anthropogenic CO_2 is primarily physical in response to increasing atmospheric CO_2 concentrations. Increasing the partial pressure of a gas in the atmosphere directly above the body of water makes the gas diffuse into that water until the partial pressures across the air-water interface are equilibrated. The effects are complex, for example, increasing CO_2 also modifies the climate which in turn may change ocean circulation, which changes the rate of ocean CO_2 uptake. Marine ecosystem changes also alter the uptake (National Oceanic and Atmospheric Administration, 2018).

Halocarbons, that is, those that contain halogen and carbon atoms, are the same classes of compounds involved in the destruction of atmospheric ozone are also at work in promoting global warming. The most effective global warming gases are chlorofluorocarbons (CFCs), for example, CFC-11 and CFC-12, both of which are no longer manufactured, and the banning of these substances has shown a leveling off in the stratosphere.

Many, but not all, GHGs are C-based. Notably, nitrous oxide (N_2O; also known as dinitrogen monoxide) is the third most emitted to the atmosphere predominantly from human activities that also release C-based GHGs; especially the cutting and clearing of tropical forests, which are vital sources of atmospheric oxygen. The greatest problem with nitrous oxide is that there appear to be no natural removal processes for this gas and so its residence time in the stratosphere is quite long.

4.7 Conclusions

Chemical, physical, and biological processes drive the amounts and rates of myriad planetary phenomena, including climate change and pollution. Substances cycle through the lithosphere, atmosphere, biosphere, and hydrosphere, all the while changing characteristics. Some of the changes result in GHGs and toxic substances. Some result in pollution. An appreciation for biogeochemistry allows for ways to characterize the incremental and cumulative effects of human and natural activities. These must be considered in a systematic way, given that the processes are ongoing and interactive.

References

Aberg, B., Ekman, L., Falk, R., Greitz, U., Persson, G., & Snihs, J. O. (1969). Metabolism of methyl mercury (203Hg) compounds in man: Excretion and distribution. *Archives of Environmental Health: An International Journal*, *19*(4), 478–484.

Allen, T. E., Goodman, J. M., Gutsell, S., & Russell, P. J. (2014). Defining molecular initiating events in the adverse outcome pathway framework for risk assessment. *Chemical Research in Toxicology*, *27*(12), 2100–2112.

Burant, A., Lowry, G. V., & Karamalidis, A. K. (2013). Partitioning behavior of organic contaminants in carbon storage environments: A critical review. *Environmental Science & Technology*, *47*(1), 37–54.

Chorus, I., & Spijkerman, E. (2021). What Colin Reynolds could tell us about nutrient limitation, N: P ratios and eutrophication control. *Hydrobiologia*, *848*(1), 95–111.

Croft, D., O'Kelly, G., Wu, G., Haw, R., Gillespie, M., Matthews, L., . . . Jassal, B. (2010). Reactome: A database of reactions, pathways and biological processes. *Nucleic Acids Research*, gkq1018.

Dellarco, V., Gibb, H., McMaster, S., Orme-Zavaleta, J., & Pahl, D. (1998). Human health risk assessment research strategy. *External Review Draft*.

Dong, W., Liang, L., Brooks, S., Southworth, G., & Gu, B. (2010). Roles of dissolved organic matter in the speciation of mercury and methylmercury in a contaminated ecosystem in Oak Ridge, Tennessee. *Environmental Chemistry*, *7*(1), 94–102.

Finizio, A., Mackay, D., Bidleman, T., & Harner, T. (1997). Octanol-air partition coefficient as a predictor of partitioning of semi-volatile organic chemicals to aerosols. *Atmospheric Environment*, *31*(15), 2289–2296.

Graedel, T. E. (1996). On the concept of industrial ecology. *Annual Review of Energy and the Environment*, *21*(1), 69–98.

Groh, K. J., Carvalho, R. N., Chipman, J. K., Denslow, N. D., Halder, M., Murphy, C. A., . . . Watanabe, K. H. (2015). Development and application of the adverse outcome pathway framework for understanding and predicting chronic toxicity: I. Challenges and research needs in ecotoxicology. *Chemosphere*, *120*, 764–777.

Hale, S. E., Arp, H. P. H., Schliebner, I., & Neumann, M. (2020). Persistent, mobile and toxic (PMT) and very persistent and very mobile (vPvM) substances pose an equivalent level of concern to persistent, bioaccumulative and toxic (PBT) and very persistent and very bioaccumulative (vPvB) substances under REACH. *Environmental Sciences Europe*, *32*(1), 1–15.

Johnson, R. L., Cherry, J. A., & Pankow, J. F. (1989). Diffusive contaminant transport in natural clay: A field example and implications for clay-lined waste disposal sites. *Environmental Science & Technology*, *23*(3), 340–349.

Keating, M.H., Mahaffey, K.R., Schoeny, R., Rice, G.E., & Bullock, O.R. (1997). *Mercury study report to congress*. U.S. Environmental Protection Agency, Office of Air Quality Planning and Standards and Office of Research and Development. Report No. EPA-452/R-97-003. https://www.osti.gov/servlets/purl/575110; accessed on March 7, 2022.

Li, Y.-F., Qin, M., Yang, P.-F., Liu, L.-Y., Zhou, L.-J., Liu, J.-N., . . . Tian, C.-G. (2021). Treatment of particle/gas partitioning using level III fugacity models in a six-compartment system. *Chemosphere*, *271*, 129580.

Lin, C.-J., & Pehkonen, S. O. (1999). The chemistry of atmospheric mercury: A review. *Atmospheric Environment*, *33*(13), 2067–2079.

Lindqvist, O., Johansson, K., Bringmark, L., Timm, B., Aastrup, M., Andersson, A., . . . Meili, M. (1991). Mercury in the Swedish environment—Recent research on causes, consequences and corrective methods. *Water, Air, and Soil Pollution*, *55*(1–2), xi–261.

Lioy, P. J., & Smith, K. R. (2013). A discussion of exposure science in the 21st century:Avision and a strategy. *Environmental Health Perspectives*, *121*(4), 405.

Lyman, W. (1995). Transport and transformation processes. *Rand GM fundamentals of aquatic toxicology. Effects, environmental fate, and risk assessment* (2nd ed., pp. 449–492). Washington, DC: Taylor and Francis.

Mackay, D., & Fraser, A. (2000). Bioaccumulation of persistent organic chemicals: Mechanisms and models. *Environmental Pollution*, *110*(3), 375–391.

Mackay, D., & Wania, F. (1995). Transport of contaminants to the Arctic: Partitioning, processes and models. *Science of the Total Environment*, *160*, 25–38.

Mackay, D., Celsie, A. K., & Parnis, J. M. (2018). Kinetic delay in partitioning and parallel particle pathways: Underappreciated aspects of environmental transport. *Environmental Science & Technology*, *53*(1), 234–241.

McKone, T., Riley, W., Maddalena, R., Rosenbaum, R., & Vallero, D. (2006). Common issues in human and ecosystem exposure assessment: The significance of partitioning, kinetics, and uptake at biological exchange surfaces. *Epidemiology, 17*(6), S134.

Miller, C. T., & Weber, W. J., Jr (1984). Modeling organic contaminant partitioning in ground-water systems. *Groundwater, 22*(5), 584–592.

Muñoz, R., Villaverde, S., Guieysse, B., & Revah, S. (2007). Two-phase partitioning bioreactors for treatment of volatile organic compounds. *Biotechnology Advances, 25*(4), 410–422.

Munthe, J., Kindbom, K., Kruger, O., Petersen, G., Pacyna, J., & Iverfeldt, Å. (2001). Emission, deposition, and atmospheric pathways of mercury in Sweden. *Water, Air and Soil Pollution*.

National Oceanic and Atmospheric Administration. (2018). *Carbon education tools*. Retrieved from https://www.pmel.noaa.gov/co2/file/Carbon + Cycle + Graphics.

Organisation for Economic Co-operation and Development. (2011). *Report of the workshop on using mechanistic information in forming chemical categories* (ENV/JM/MONO(2011)8). Crystal City VA, USA Environment Directorate Joint Meeting of The Chemicals Committee and The Working Party on Chemicals, Pesticides and Biotechnology.

Oste, L. A., Temminghoff, E. J., & Riemsdijk, W. V. (2002). Solid-solution partitioning of organic matter in soils as influenced by an increase in pH or Ca concentration. *Environmental Science & Technology, 36*(2), 208–214.

Pleil, J. D., & Sheldon, L. S. (2011). Adapting concepts from systems biology to develop systems exposure event networks for exposure science research. *Biomarkers, 16*(2), 99–105.

Resnik, D. B., & Vallero, D. A. (2011). Geoengineering: An idea whose time has come? *Journal of Earth Science & Climatic Change*.

Richter, D. (1996). *Ion exchange*.

Sauvé, S., Hendershot, W., & Allen, H. E. (2000). Solid-solution partitioning of metals in contaminated soils: Dependence on pH, total metal burden, and organic matter. *Environmental Science & Technology, 34*(7), 1125–1131.

Schreckenbach, K., Knosche, R., & Ebert, K. (2001). Nutrient and energy content of freshwater fishes. *Journal of Applied Ichthyology, 17*(3), 142–144.

Shepherd, J. (2009). *Geoengineering the climate: Science, governance and uncertainty*. Royal Society.

Stoddard, J., Peck, D., Paulsen, S., Van Sickle, J., Hawkins, C., Herlihy, A., . . . Lomnicky, G. (2005). *An ecological assessment of western streams and rivers*. Washington, DC: US Environmental Protection Agency. Available at http://www.epa.gov/nheerl/arm/documents/EMAP.W.Assessment.final.pdf (accessed 3 December 2010).

Storey, J. M., Luo, W., Isabelle, L. M., & Pankow, J. F. (1995). Gas/solid partitioning of semivolatile organic compounds to model atmospheric solid surfaces as a function of relative humidity. 1. Clean quartz. *Environmental Science & Technology, 29*(9), 2420–2428.

Tourinho, P. S., Kočí, V., Loureiro, S., & van Gestel, C. A. (2019). Partitioning of chemical contaminants to microplastics: Sorption mechanisms, environmental distribution and effects on toxicity and bioaccumulation. *Environmental Pollution, 252*, 1246–1256.

U.S. Environmental Protection Agency (1997). Office of Water. Chapter 5: Water Quality Conditions. In: Volunteer Stream Monitoring: A Methods Manual. Report No. EPA 841-B-97-003. https://www.epa.gov/sites/default/files/2015-04/documents/volunteer_stream_monitoring_a_methods_manual.pdf; accessed on March 7, 2022.

U.S. Environmental Protection Agency. (2016, July 19). *Gold king mine watershed fact sheet*. How Did the August 2015 Release from the Gold King Mine Happen? https://www.epa.gov/goldkingmine/how-did-august-2015-release-gold-king-mine-happen; accessed on March 7, 2022.

U.S. Environmental Protection Agency. (2017). *Analysis of the Transport and Fate of Metals Released from the Gold King Mine in the Animas and San Juan Rivers: Executive Summary*. (EPA/600/R-16/296). Washington, DC.

U.S. Environmental Protection Agency. (2018, January 4). *Emergency response to August 2015 release from gold king mine*. Retrieved from https://www.epa.gov/goldkingmine.

Vallero, D. (2010). *Environmental biotechnology: A biosystems approach*. Academic Press.
Vallero, D. (2015a). *Environmental biotechnology: A biosystems approach*. Elsevier Science.
Vallero, D. (2017). *Translating diverse environmental data into reliable information: How to coordinate evidence from different sources*. Academic Press.
Vallero, D. A. (2013). Measurements in environmental engineering. In Handbook of measurement in science and engineering.
Vallero, D. A. (2014). In M. A. Waltham (Ed.), *Fundamentals of air pollution* (5th ed.). Elsevier Academic Press.
Vallero, D. A. (2015b). Engineering aspects of climate change. *Climate change* (2nd ed., pp. 547–568). Elsevier.
Vallero, D. A. (2021). *Environmental systems science: Theory and practical applications* (1st ed.). Elsevier Academic Press.
Vallero, D. A., & Letcher, T. M. (2012). *Unraveling environmental disasters*. Newnes.
Vallero, D. A., & Resnik, D. B. (2013). Geoengineering: Enhancing cloud albedo. *Access Science*.
Vallero, D. A., S.S.I., Zartarian, V. G., Mccurdy, T. R., Mckone, T., Georgopoulos, P., & Dary, C. C. (2010). Modeling and predicting pesticide exposures. In R. Krieger (Ed.), *Hayes handbook of pesticide toxicology* (pp. 995–1020). New York: Elsevier Science.
Vernier Corporation. (2018). *Computer 19: Dissolved oxygen in water*. Retrieved from http://www2.vernier.com/sample_labs/BWV-19-COMP-dissolved_oxygen.pdf.
Versteeg, D. J., & Naciff, J. M. (2015). In response: Ecotoxicogenomics addressing future needs: An industry perspective. *Environmental Toxicology and Chemistry*, *34*(4), 704–706.
Weschler, C. J., Salthammer, T., & Fromme, H. (2008). Partitioning of phthalates among the gas phase, airborne particles and settled dust in indoor environments. *Atmospheric Environment*, *42*(7), 1449–1460.
Westfall, J. (1987). Adsorption mechanisms in aquatic surface chemistry. In W. Stumm (Ed.), *Aquatic surface chemistry: Chemical processes at the particle-water interface* (Vol. 87). John Wiley & Sons.

Further reading

Emergency Response to Gold King Mine Release. Retrieved from https://www.epa.gov/goldkingmine/gold-king-mine-watershed-fact-sheet.

SECTION B

Sustainable development and environmental issues

CHAPTER 5

Water quality engineering: physical, chemical, and biological treatment for a sustainable future

Daniel A. Vallero

Pratt School of Engineering, Duke University, Durham, NC, United States

Chapter Outline
5.1 Introduction 65
5.2 Water quality 66
5.3 Chemical processes 69
5.4 Wastewater treatment 73
5.5 Water treatment to manage risks 79
5.6 Disposal of sludge 80
5.7 Mixed reactors 82
5.8 Aerobic reactors 83
5.9 Anaerobic reactors 86
5.10 Treating contaminated ground water 89
5.11 Conclusions 92
References 92

5.1 Introduction

From an environmental engineering perspective, water is a resource that is optimized according to what it does not contain. The closer water is to being pure, the closer we are to a desired condition. Generally, the difference between acceptable and unacceptable water quality is the concentration of substances other than molecular H_2O. Indeed, even the concentratons of hydrogen and hydroxyl ions of natural water, pH and pOH respectively, can render water polluted. Polluting substances fall into two general categories, dissolved and suspended. Therefore, the first determination of water quality is some threshold of the concentration of substances in aqueous solution and substances that are suspended.

5.2 Water quality

Regulators define water quality criteria based on a list substances matched against maximum concentrations. Threshold concentrations apply to effluents, that is, water that is discharged to a receiving waterbody. They also apply to water supply, such as maximum contaminant levels (MCLs) leaving a water treatment plant prior to entering the water distribution system that carries the treated water to users. Threshold criteria also apply to point-of-use scenarios, such as water at the tap in a residence or bottled water sold by retailers. Various agencies at different levels of government regulate water quality. For example, among the many federal agencies with roles in protecting the waters of the US, the Environmental Protection Agency sets national limits on effluents and MCLs, the Food and Drug Administration has oversight of bottled water and other consumables, the US Fish and Wildlife and US Army play major roles in protecting wetlands, and the US Geological Survey tests water quality in rivers and streams. The States in the US each have unique regulatory structures that vary even more than those of the federal government.

Wastewater treatment is distinguished from water treatment. The domain of wastewater treatment is water that has already been used for some purpose and is now in a polluted state, that is, sewage. This domain requires treatment to a point where the treated water is allowed to re-enter the environment, for example, from an outfall structure. The domain of water treatment is to provide potable water to the user. The sources of the water entering a water treatment plant include ground water from wells and surface waterbodies like lakes and even estuaries and ocean water. Saline water is increasingly treated by advanced technologies, like distillation, osmosis, and electrodialysis, and nanotubes (Subramani & Jacangelo, 2015).

Both wastewater and water treatment involve decreasing the concentrations of undesired substances produced by applying sound engineering controls to reduce or eliminate exposures to these substances during and after removal, for example, sludge management. From an engineering perspective, most undesired substances in water are there by accident. Facilities that generate the pollutants are attempting to produce substances demanded by the marketplace, but in the process and under the wrong circumstances these substances or their components find their way to the surface waters, ground water, soil, and air. Ideally, engineers and plant managers can design and operate facilities in ways to prevent, contain, decontaminate, and otherwise treat these wastes before they are released into the environment. This is the heart of pollution prevention and waste reduction. However, wastewater treatment engineering is most often associated with interventions after the pollutants have left the source, so engineers are called upon to address these wastes at some point downstream from the source.

The threshold concentrations are designed to protect humans and ecosystems. The level of protection and water quality criterion for a pollutant is based on its hazard, that is, the potential of an unacceptable outcome. Often, there are two approaches for addressing a hazardous pollutant, that is, effluent limits and water quality limits. Effluent limits apply to any source that discharges substances to a waterbody, for example, 100 ppb of Chemical X or 10 ng L^{-1} for Chemical Y. Often, general indirect indicators of water quality are used, especially biochemical oxygen demand (BOD), the amount of oxygen that bacteria and other microbes will consume in the process of decomposing organic matter under aerobic conditions. Thus, the larger the BOD, the more potential harm the effluent will cause to a receiving waterbody (Ashley & Word, 1991). BOD_5 is an almost universally applied metric, which is merely the measured dissolved oxygen (DO) at time zero, that is, the initial measurement of DO (D_1), which is measured immediately after it is taken from the effluent, minus the DO of the same effluent measured exactly 5 days later, that is, D_5:

$$\text{BOD} = \frac{D_1 - D_5}{P} \qquad (5.1)$$

where P is the decimal volumetric fraction of water utilized. D units are in mass per volume, usually mg L^{-1}. If the dilution water is seeded, the calculation becomes:

$$\text{BOD} = \frac{(D_1 - D_5) - (B_1 - B_5)f}{P} \qquad (5.2)$$

where B_1 is the initial DO of seed control; B_5 is the final DO of seed control; and f is the ratio of seed in sample to seed in control $= \frac{\% \text{ seed in } D_1}{\% \text{ seed in } B_1}$. B units are in mg L^{-1}.

For example, to find the BOD_5 value for a 10 mL water sample added to 300 mL of dilution water with a measured DO of 7 mg L^{-1} and a measured DO of 4 mg L^{-1} 5 days later:

$$P = \frac{10}{300} = 0.03$$

$$BOD_5 = \frac{7-4}{0.03} = 100 \text{ mg } L^{-1}$$

Thus, the microbial population in this water is demanding 100 mg L^{-1} dissolved oxygen over the 5-day period. So, if a conventional municipal wastewater treatment system is achieving 95% treatment efficiency, the effluent discharged from this plant would be 5 mg L^{-1}.

Chemical oxygen demand (COD) does not differentiate between biologically available and inert organic matter, and it is a measure of the total quantity of oxygen required to oxidize all organic material completely to carbon dioxide and water. COD values always exceed BOD values for the same sample. COD (mg L^{-1}) is measured by oxidation using potassium dichromate ($K_2Cr_2O_7$) in the presence of sulfuric acid (H_2SO_4) and silver. By convention, 1 g of carbohydrate or 1 g of protein accounts for about 1 g of COD. On average, the ratio

BOD:COD is 0.5. If the ratio is <0.3, the water sample likely contains elevated concentrations of *recalcitrant* organic compounds, that is, compounds that resist biodegradation (Gerba & Pepper, 2009). That is, there are numerous carbon-based compounds in the sample, but the microbial populations are not efficiently using them for carbon and energy sources. This is the advantage of having both BOD and COD measurements. Sometimes, however, COD measurements are conducted simply because they require only a few hours compared to the five days for BOD.

Since available carbon is a limiting factor, the carbonaceous BOD reaches a plateau, that is, *the ultimate carbonaceous BOD* (see Fig. 5.1). However, carbonaceous compounds are the only substances demanding oxygen. Microbial populations will continue to demand O_2 from the water to degrade other compounds, especially nitrogenous compounds, which account for the bump in the BOD curve. Thus, in addition to serving as an indication of the amount of molecular oxygen needed for biological treatment of the organic matter, BOD also provides a guide to sizing a treatment process, assigns its efficiency, and gives operators and regulators information about whether the facility is meeting its design criteria and is complying with pollution control permits.

If effluent with high BOD concentrations reaches surface waters, it may diminish DO to levels lethal some fish and many aquatic insects. As the water body re-aerates due to mixing with the atmosphere and by algal photosynthesis, O_2 is added to the water, the

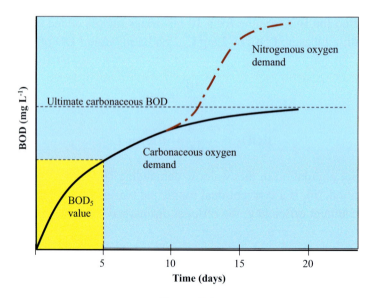

Figure 5.1
Biochemical oxygen demand (BOD) curve. Showing ultimate carbonaceous BOD and nitrogenous BOD. Source: *Used with permission from Gerba, C. P., & Pepper, I. L. (2009). Wastewater treatment and biosolids reuse. In* Environmental microbiology *(2nd ed.) (pp. 503–530). United States: Elsevier Inc.*

oxygen levels will slowly increase downstream. The drop and rise in DO concentrations downstream from a source of BOD is known as the DO sag curve, because the concentration of dissolved oxygen "sags" as the microbes deplete it. So, the falling O_2 concentrations fall with both time and distance from the point where the high BOD substances enter the water (see Fig. 5.2).

The stress on aquatic life due to decreasing DO is indirectly indicated by the BOD. Water bodies that receive pollutant loads require that microbial populations adapt. Sometimes they deploy mechanisms to outlast the low DO conditions through dormancy, for example, production of cysts. The DO concentrations are usually a net result of inputs and outputs of oxygen due to increased nutrient loads. Given that the various genera of biota vary in their tolerance ranges and metabolic processes to product cells and to obtain energy, that is, anabolism and catabolism respectively, the net amount of oxygen in a waterbody varies in time and space. For example, both algae and bacteria demand O_2 for growth and respiration, drawing DO from the water, algae also undergo photosynthesis, adding O_2.

5.3 Chemical processes

The treatment of water pollutants usually occurs by a handful of chemical processes, that is, synthesis, decomposition, single replacement, and double replacement (Vallero, 2010).

Synthesis. Here, two or more substances react to form a single substance:

$$A + B \rightarrow AB \tag{5.3}$$

Two types of synthesis reactions are at work in wastewater treatment, that is, formation and hydration. Formation occurs when elements combine to form a compound. An example is the formation reaction of ferric oxide:

$$Fe(s) + 3\,O_2(g) \rightarrow 2\,Fe_2O_3(s) \tag{5.4}$$

Hydration results form the addition of water to synthesize a new compound, for example, in reaction when calcium oxide is hydrated to form calcium hydroxide:

$$CaO(s) + H_2O(l) \rightarrow Ca(OH)_2(s) \tag{5.5}$$

Hydration also occurs when phosphate is hydrated to form phosphoric acid:

$$P_2O_5(s) + 3H_2O(l) \rightarrow 2H_3PO_4(aq) \tag{5.6}$$

Decomposition. This is an important process for degrading organic compounds into simpler compounds:

$$AB \rightarrow A + B \tag{5.7}$$

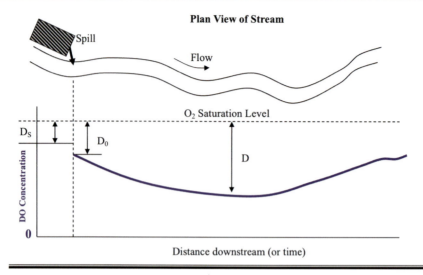

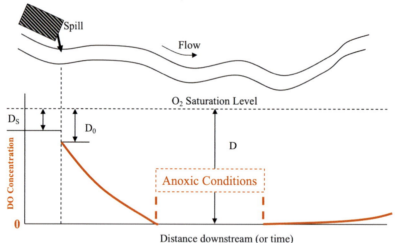

Figure 5.2
Dissolved oxygen sag curve downstream from an O_2-depleting contaminant release to a streams. The concentration of dissolved oxygen in Curve A remains above 0. Although the available oxygen is reduced, the system remains aerobic. Curve B sags until dissolved oxygen falls to 0, and anaerobic conditions result and continue until the DO concentrations begin to increase. D_S is the background oxygen deficit before the pollutants enter the stream. D_0 is the oxygen deficit after the pollutant is mixed. D is the deficit for contaminant A which may be measured at any point downstream. This indicates both distance and time of microbial exposure to the source. For example, if the stream's average velocity is 5 km h^{-1}, D measured 10 km downstream also represents 2 h of microbial activity to degrade the pollutant.

Here, one substance, that is, the parent compound, breaks down in two or more new substances, that is, the degradation products. For example, in reaction calcium carbonate breaks down into calcium oxide and carbon dioxide:

$$CaCO_3(s) \rightarrow CaO(s) + CO_2(g) \tag{5.8}$$

Of course, the degradation of larger molecules is much more complex than this simple decomposition. For example, polymers break down with a series of reactions, beginning with a biochemical reaction, for example, an aerobic bacteria population, in which the polymer is oxidized, producing a residue, biomass, CO_2, and H_2O. If oxygen is deficient, the reaction will be anaerobic, producing residue, biomass, CO_2, CH_4, and H_2O. Thus, the degradation of large, organic molecules will result in more than two degradation products from a series of reactions (Abramowicz, 1995; Alexander, 1999; Boonyaroj et al., 2017; Ikada & Tsuji, 2000).

Single replacement (single displacement). In this reaction, one reactant replaces another reactant:

$$A + BC \rightarrow AC + B \tag{5.9}$$

For example, a less toxic ion can be used to replace a more toxic ion, or if the cationic replacement makes collection, disposal or additional treatment easier. For example, when silver nitrate reactions with solid chromium, it produces chromium nitrate and solid silver:

$$3AgNO_3(aq) + Cr(s) \rightarrow Cr(NO_3)_3(aq) + 3Ag(s) \tag{5.10}$$

Double replacement (metathesis or double displacement). In metathesis, molecules exchange their cations and anions:

$$AB + CD \rightarrow AD + CB \tag{5.11}$$

These reactions are commonly used to precipitate metals out of solution, for example, when lead precipitates during a reaction of a lead salt with an acid such as potassium chloride:

$$Pb(ClO_3)_2(aq) + 2KCl(aq) \rightarrow PbCl_2(s) + 2KClO_3(aq) \tag{5.12}$$

A few commonly applied chemical treatment processes include neutralization, precipitation, ion exchange, dehalogenation, and oxidation-reduction.

All four types of chemical reactions discussed above can take place within an organism (Vallero & Peirce, 2003). Contaminants, if completely organic in structure are, in theory, fully destructible using microorganisms with the engineering inputs and outputs summarized as:

$$\text{Hydrocarbons} + O_2 + \text{microbes} + \text{energy} \rightarrow CO_2 + H_2O + \text{microbes} \tag{5.13}$$

Biodegradation occurs in the biofilm around the microbes, as well as within the microbes. In time, the byproducts of gaseous carbon dioxide and water generated will exit the top of the reaction vessel, while a solid mass of microorganisms is produced to exit the bottom of the reaction vessel.

If the waste contains other chemical constituents, chlorine and/or heavy metals, and if the microorganisms can withstand such an environment, the simple input and output relationship in reaction (5.13) becomes:

Hydrocarbons + O_2 + microbes + energy + Cl or heavy metal(s) + H_2O + inorganic salts
+ nitrogen compounds + sulfur compounds + phosphorus compounds →
CO_2 + H_2O + chlorinated hydrocarbons or heavy metal(s) inorganic salts
+ nitrogen compounds + sulfur compounds + phosphorus compounds

(5.14)

During reaction (5.14), waste products may be produced. These may be toxic and sometimes are of greater toxicity and more mobile than the original wastes, for example, containing new chlorinated hydrocarbons, higher heavy-metal concentrations.

Biodegradation efficiency depends on the growth and survival of microbial populations and the ability of these organisms to come into contact with the substances that need to be degraded into less toxic compounds. Sufficient numbers of microorganisms must be present and active to make bioremediation successful and the preferred cleanup choice. The WWTP design must modify the environment (e.g., dilution, change in pH, pumped oxygen, adding organic matter, etc.) to make it habitable for the microbial populations. Microbial optimization can include artificially adding microbial populations known to break down the compounds of concern. Sludge treatment efficiency decreases with the molecular size and complexity, for example, number of halogen substitutions and the type of particles. For example, the methanogenic granular sludge structure has been found to play a key role in high-rate anaerobic processes (Cheremisinoff, 1997). Granular sludge is an aggregation of several metabolic groups of bacteria living synergistically.

The benefits of biological treatment systems include (1) the potential for energy recovery; (2) volume reduction of the hazardous waste; (3) detoxification as selected molecules are reformulated; (4) the basic scientific principles, engineering designs, and technologies are well understood from a wide range of other applications including municipal wastewater treatment at facilities; (5) application to most organic contaminants which as a group compose a large percentage of the total hazardous waste generated nationwide; and (6) the possibility to scale the technologies to handle a single liter or kilogram or millions of liters or tons of waste per day.

The drawbacks of the biotreatment systems include (1) the operation of the equipment requires very skilled operators and is costlier as input contaminant characteristics change over time and correctional controls become necessary; and (2) the ultimate disposal of the waste microorganisms is necessary and particularly troublesome and costly if heavy metals and/or chlorinated compounds are found during the monitoring activities. In general, hazardous waste containing heavy metals should not be bio-processed. Variations on three different types of bioprocessors are available to treat hazardous wastes: trickling filter, activated sludge, and aeration lagoons.

5.4 Wastewater treatment

Various combinations of physical, chemical and biological processes are applied in wastewater treatment. In the first steps when untreated water reaches a treatment facility, physical capture by size separation, for example, grit chambers, remove solid materials. Grinding equipment reduces the size of suspended matter, and settling tanks simultaneously remove low-density material near the tank surface and heavier materials are collected at the bottom. Advanced physical processes also come into play toward the end of the treatment process. Notably filtration is used to remove particles, such as micro-, ultrafiltration. Nanofiltration is deployed for extremely small-diameter particles. Diffusion is used for large molecules that are at least in part dissolved in the water, for example, reverse osmosis removes solutes by passing the water through a semipermeable membranes. Chemical processes include precipitation, in which coagulants are added to the water to induce flocculation, that is, colloidal matter binds together and leaves suspension to form aggregates that are easier to separate and collect than the unaggregated matter and also can become sufficiently heavy and sink (Bridle et al., 2021). Other commonly applied chemical processes include ion exchange, biodegradation, and biosorption.

Wastewater treatment designs vary considerably, but some of the most common are trickling filters, digesters, and lagoons. After the design and construction, the operation and maintenance (O&M) of wastewater treatment facilities largely determine removal efficiencies. Often, as systems become more complex, O&M becomes more difficult and critical. A simpler system may involve passive processes that require less frequent and easier upkeep than complex systems. For example, treatment processes fall on a continuum between "natural" or entirely "electromechanical," depending on the amounts of electrical energy needed to keep them operating within an optimal range. A natural system requires little O&M, such as a bioretention system (see Fig. 5.3). Other natural systems include waste stabilization ponds, storage and treatment reservoirs, septic tanks, UASB reactors, high-rate anaerobic ponds, Imhoff tanks, and constructed wetlands. Energy-intensive systems include aerated lagoons.

Electromechanical systems have numerous components depending on the type of pollutant being removed or detoxified. At present, the removal of microbial pathogens and highly toxic chemical compounds is most commonly achieved through electromechanical systems.

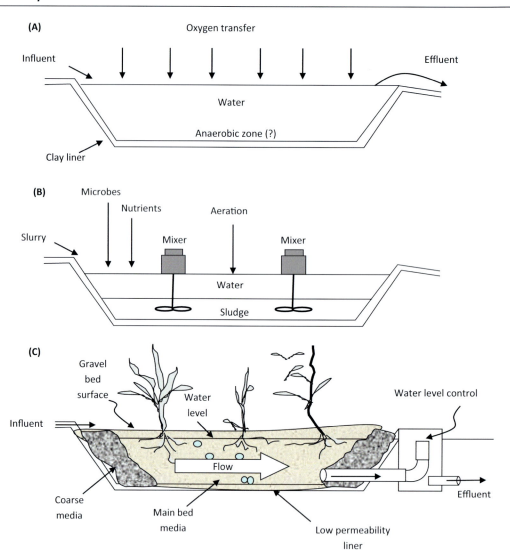

Figure 5.3
(A) Passive system: aeration pond. (B) Active system: in situ slurry-phase lagoon. (C) Combined active and passive system: engineered and constructed wetland. All three systems make use of microbes, nutrients and oxygen, but the active system increases the contact between the microbes and organic matter by mechanically mixing the sludge layer (which contains the organic contaminants to be degraded) with the water in the lagoon. The engineered wetland also incorporates plant processes in the degradation process. Source: *Adapted from EPA/540/A5-91/009. (1993). Pilot-scale demonstration of slurry-phase biological reactor for creosote-contaminated soil: Applications analysis report; Wallace, S. (2009). Engineered wetlands lead the way. Land and Water, 48(5). [Online]. Available: http://www.landandwater.com/features/vol48no5/vol48no5_1.html.*

This can be problematic for developing countries and low income regions, where electrical services are unavailable or less reliable and where personnel training and retention can be challenging (Jiménez et al., 2009).

Wastewater treatment systems are classified as whether the water is to be treated onsite (e.g., septic tank), or whether it is to be collected and distributed to a central wastewater treatment plant (WWTP). In developing nations, large amounts of wastewater may not be treated at all, rather it is directly released into ditches or onto soil, where it may leach into ground water and/or allowed to drain into surface waters. Even in developed nations, a large fraction of residences treat their wastes onsite; for example, about 25% of US population daily dispose of 1.5×10^7 m^3 wastewater by means of onsite disposal and treatment systems. An estimated 10% of these systems do not function properly, so that a fraction of the contaminants are not treated according to the design specifications (EPA No. 832-B-05-001, 2005).

The efficiency and effectiveness of any system is expressed as the rate and amount of destruction of harmful agents. For onsite wastewater treatment systems (OWTSs) the principal contaminants needed to be treated are those produced by residential activities and product usage, for example, pathogenic microbes, nutrients, and toxic substances (see Fig. 5.4). An efficient system keeps pollutants from reaching drinking water supplies, for example, aquifers and surface waters. For example, a system loses efficiency with decreasing retention time, since beneficial bacteria do not have sufficient time to

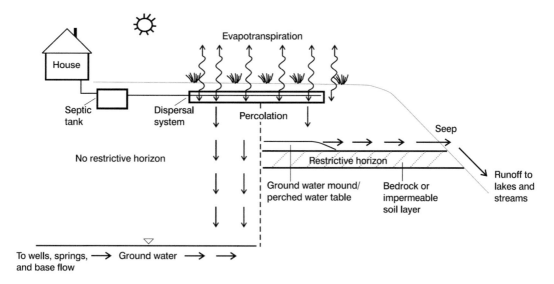

Figure 5.4
Flow of wastewater and the constituents that it carries in an onsite wastewater treatment system (OWTS). *From: EPA/625/R-00/008. (2002). Onsite wastewater treatment design manual.*

metabolize their "food," that is, the organic compounds in the wastewater. For example, if the soil is too permeable, biosorption is occurring at a suboptimal rate. The operation of a successful onsite system depends on the application of knowledge about the degradation processes.

Onsite treatment usually involves a single unit for a residence. However, the processes can be expanded to include several units for entire neighborhoods. At the municipal scale, the WWTP employs the same physical and biological processes as those used in an onsite system, but for much larger volumes of wastewater. After receiving the wastewater from a sanitary sewer or other collection system, the WWTP uses of gravity to remove solids. The larger solids are removed first by grit chambers and, next, the smaller suspended solids removed by sedimentation tanks, that is, primary treatment.

Secondary treatment mechanisms that follow are designed to remove up to 90% of organic matter is removed using biotechnologies (Rittmann, 2009; Rittmann & McCarty, 2012; Scragg, 2005; Vallero, 2015; Vallero & Gunsch, 2020). Trickling filters and other "attached growth" or "fixed film" processes are commonly used by municipalities for secondary. Here, microbial populations adhere to various types of media which are coated with biofilm (see Fig. 5.5). Secondary treatment also includes "suspended growth processes," especially activated sludge systems (see Fig. 5.6). Aerated lagoons, slow rate methods like land application, and constructed wetlands and other bioretention systems are used to remove organic matter.

In an activated sludge system, a tank of liquid hazardous waste is injected with a mass of microorganisms. Oxygen is supplied through pipes in the aeration basin as the microorganisms come in contact, become adsorbed, and metabolize the waste. For complete

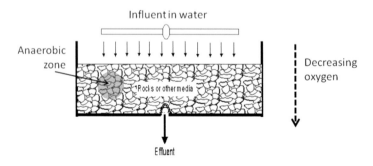

Figure 5.5
Trickling filter system used for secondary treatment of wastewater. Oxygen concentrations available to the microbial populations generally decrease with depth in the porous media, but there are also zones of varying concentrations of dissolved oxygen, rending a trickling filter to be a mixed reactor. Source: *Adapted from Vallero, D. A., & Peirce, J. J. (2003). Engineering the risks of hazardous wastes (pp. xxi, 306 p). Burlington, MA: Butterworth-Heinemann.*

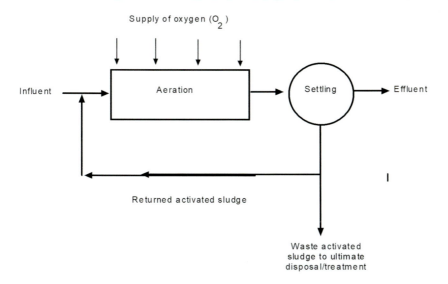

Figure 5.6
Activated sludge treatment system. Incoming wastewater is combined with recycled biomass and aerated to maintain a target dissolved oxygen (DO) content. Organisms use the organic components of waste as food, decreasing the organic levels in the wastewater. Oxygen concentrations must be controlled to maintain optimal treatment efficiencies. One means of achieving optimal DO content is tapered aeration, which provides highest concentrations of oxygen near the influent to accommodate the large oxygen demand from microbes as waste is introduced to the aeration tank (photo). Source: *Adapted from Vallero, D. A., & Peirce, J. J. (2003). Engineering the risks of hazardous wastes (pp. xxi, 306 p). Burlington, MA: Butterworth-Heinemann; photo courtesy D.J. Vallero.*

and ideal conditions, this would convert all organic matter into
$CO_2 + H_2O$ + microorganisms + energy. The heavy, that is, well-fed, microorganisms then flow into a quiescent tank where they settle by gravity and are collected disposed. Depending on the operating conditions of the facility, some or many of the settled (now hungry) and active microorganisms are returned to the aeration basin where they feed again. Liquid effluent from the activated sludge system may require additional microbiological and/or chemical processing prior to release into a receiving stream or sewer system.

In many cases, secondary treatment is insufficient to meet water quality standards and regulations. If so, advanced treatment or tertiary treatment is needed. After most of the solids and organic matter is removed, specific treatment technologies are deployed. Often, for example, excessive concentrations of nutrients, especially nitrogen and phosphorus remain after secondary treatment. If released into receiving waters, these elevated nutrient levels could induce eutrophication and associated problems. Above certain concentrations, nutrients encourage the growth of algae, ultimately leading to algal blooms. When the algae die, they fall to the bottom of the waterbody. Decomposition leads to greater demand for dissolved oxygen (DO) by bacteria and by abiotic processes, further depressing DO concentrations. As a result, organisms within the food chain are stressed, especially game fish, resulting alterations in in aquatic habitats and even fish kills (Chorus & Spijkerman, 2021; Vallero et al., 2007; Zhao et al., 2022).

Advanced treatment can consist of both biotic and abiotic processes, often combinations of each. For example, biotic processes can remove nitrogen and phosphorus compounds. Tertiary treatment systems include intermittently decanted extended aeration lagoon systems and biologically enhanced phosphorus removal systems. one modification of the activated sludge process which provides biological treatment for the removal of biodegradable organic wastes under aerobic conditions (EPA 832-F-00-016, 2000).

Oxygen is supplied either by mechanical mixing or by diffused aeration to induce and maintain the aerobic conditions. The aeration is simultaneously provided to keep the microbes in contact with the dissolved organic compounds. Microbial populations also require an optimal pH range, which must be maintained in throughout the wastewater flow, from immediate screening upon entry to remove settleable and floating solids that may interrupt the processes and damage downstream equipment (see Fig. 5.7). A grinder reduces large-diameter particles not captured in the screening process. If needed, the effluent will next flow into equalization basins to regulate peak wastewater flow. Wastewater then enters the aeration chamber, where it is mixed with air. The mixed liquor next reaches a clarifier or settling chamber. There, most of the microbial mass settles to the bottom, with a portion pumped back to the incoming wastewater at the entry to the plant, that is, return activated sludge (RAS). The fraction that is not returned, that is, waste activated sludge (WAS), is removed from the WWTP and treated (EPA 832-F-00-016, 2000).

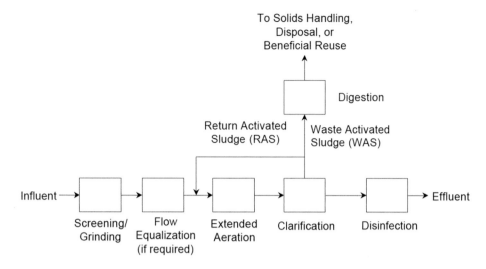

Figure 5.7
Process flow diagram for a typical extended aeration plant. *From: EPA 832-F-00-016. (2000). Wastewater technology fact sheet: Package plants.*

The biological removal of nitrogen requires two steps. First, ammonia nitrogen is oxidized into nitrate-nitrogen by nitrifying bacteria. This step is known as nitrification. Second, the nitrate is reduced to nitrogen gas, that is, denitrification. Abiotic, physicochemical processes include deep-bed filtration, floating media filtration, and membrane filtration.

Other advanced wastewater treatment technologies include coagulation-sedimentation, granulated activated carbon adsorption, sand filtration, ozone treatment, ultraviolet (UV) treatment, membrane bioreactors (MBR), advanced oxidation processes (AOP), nanotechnology, automatic variable filtration, and combinations of these techniques. These can treat and/or remove a wide variety contaminants from wastewater (EPA 832-F-00-016, 2000).

Following all the processes within the WWTP, the final step is the removal and handling of biosolids and other wastewater residuals, that is, sludge. This can consist of variations among land application, incineration, and marketable products, such as dried biosolid pellets and fertilizers for landscapes.

5.5 Water treatment to manage risks

Hazards come in many forms, but the principal form related to water quality is toxicity, both human hazards and ecological hazards. A substance may be hazardous to public welfare if it damages property values or physical materials; for example, expressed as its corrosiveness or acidity. The hazard may also relate to environmental quality, such as an ecosystem stress indicated by diminished productivity and biodiversity, loss of important habitats, and decreases in the size of the population of sensitive species. The hazard may be inherent to the substance. But more than likely, the hazard depends upon the situation and conditions where the exposure may occur. The substance is most hazardous when a number of conditions exist simultaneously, such as the hazard to firefighters using water in the presence of oxidizers.

Since 1980, specific methods for testing for hazards is delineated in the EPA publication SW-846, *Test Methods for Evaluating Solid Waste, Physical/Chemical Methods*, which is a compendium of analytical and sampling methods that comply with the RCRA regulations. The SW-846 provides guidance, although not necessarily a legal requirement, when conducting RCRA-related sampling and analysis requirements. This guidance is updated and revised as new information become available. It is currently in its third edition as a fully integrated 3500 page manual (U.S. Environmental Protection Agency, 1986).

5.6 Disposal of sludge

Sludge treatment technologies include physical treatment, chemical treatment, biological treatment, incineration, and solidification or stabilization treatment. These processes recycle and reuse waste materials, reduce the volume and toxicity of a waste stream, or produce a final residual material that is suitable for disposal.

The selection of the most effective technology depends upon the characteristics of the wastes being treated (U.S. Environmental Protection Agency, 1994). Likewise, the characteristics of the media in need of treatment determine the performance of any contaminant treatment or control. For example, if the waste is found in sediment, sludge, slurries, and soil, these media's characteristics will influence the efficacy of treatment technologies include particle size, solids content, and high contaminant concentration (see Table 5.1). The US Environmental Protection Agency's Remediation Guidance Document (U.S. Environmental Protection Agency, 1994) provides details on several of the treatment technologies discussed in this chapter.

Particle size is an important limiting characteristic when applying treatment technologies to porous media. Most treatment technologies work well on sandy soils and sediments. The presence of fine-grained material reduces the effectiveness of treatment system emission controls because it increases particulate generation during thermal drying; it is more difficult to de-water, and it has a greater attraction for the contaminants; especially fine-grained clays. Clayey sediments that are cohesive also present materials-handling problems in most processing systems.

Solids content generally ranges from high (30%−60% solids by weight) to low (10%−30% solids by weight). Treatment of slurries is better at lower solids contents, but this can be achieved for high solids contents by adding water at the time of processing. It is more difficult to change from a lower to a higher solids content, but evaporative and dewatering

Table 5.1: Effect of particle size, solids content and extent of contamination on decontamination efficiencies.

Treatment technology	Predominant practical size			Solids content		High contaminant concentration	
	Sand	Silt	Clay	High (slurry)	Low (in situ)	Organic compounds	Metals
Conventional incineration	N	X	X	F	X	F	X
Innovative incineration	N	X	X	F	X	F	F
Pyrolysis	N	N	N	F	X	F	F
Vitrification	F	X	X	F	X	F	F
Supercritical water oxidation	X	F	F	X	F	F	X
Wet air oxidation	X	F	F	X	F	F	X
Thermal desorption	F	X	X	F	X	F	N
Immobilization	F	X	X	F	X	X	N
Solvent extraction	F	F	X	F	X	X	N
Soil washing	F	F	X	N	F	N	N
Dechlorination	U	U	U	F	X	X	N
Oxidation	F	X	X	N	F	X	X
Bioslurry process	N	F	N	N	F	X	X
Composting	F	N	X	F	X	F	X
Contained treatment facility	F	N	X	F	X	X	X

Data from: Averett, D. (1994). Chapter 7: Treatment technologies. In Assessment and remediation of contaminated sediments (ARCS) program: Remediation guidance document. *Chicago, IL: G. L. N. P. O. U.S. Environmental Protection Agency (Ed.).*

approaches, such as those used for municipal sludges, may be used. Also, thermal and dehalogenation processes are decreasingly efficient as the solids content is reduced. More water means increased chemical costs and an increased need for wastewater treatment.

Chemical composition of the wastes must also be considered in the treatment of wastewater treatments. Elevated levels of organic compounds or heavy metals in high concentrations may require special safeguards. Organic matter can be a bonus, given that higher total organic carbon (TOC) content favors incineration and oxidation processes. The TOC can be the contaminant of concern or any organic, since they are combustibles with caloric value. Conversely, higher metal concentrations may make a technology less favorable by increasing contaminant mobility of certain metal species following processing.

Regulatory compliance and community perception are always a part of decisions regarding an incineration system. Land use considerations, such as the area needed, are commonly confronted in solidification and solid-phase bioremediation projects, as well as in sludge farming and land application. Disposing of ash and other residues following treatment must be part of any process. Treating water effluent and air emissions must be part of the decontamination decision-making process.

After treatment is completed, an inorganic residue remains that must be disposed of safely. Most often, this residue has no economic value. There are five options for disposing of hazardous waste. (1) Underground injection wells are steel- and concrete-encased shafts placed deep below the surface of the earth into which hazardous wastes are deposited by force and under pressure. Some liquid waste streams are commonly disposed of in underground injection wells. (2) Surface impoundment involves natural or engineered depressions or diked areas that can be used to treat, store, or dispose of hazardous waste. Surface impoundments are often referred to as pits, ponds, lagoons, and basins. (3) Landfills are disposal facilities where hazardous waste is placed in or on land. Properly designed and operated landfills are lined to prevent leakage and contain systems to collect potentially contaminated surface water run-off. Most landfills isolate wastes in discrete cells or trenches, thereby preventing potential contact of incompatible wastes. (4) Land treatment is a disposal process in which hazardous waste is applied onto or incorporated into the soil surface. Natural microbes in the soil break down or immobilize the hazardous constituents. Land treatment facilities are also known as land application or land-farming facilities. (5) Waste piles are noncontainerized accumulations of solid, nonflowing hazardous waste. While some are used for final disposal, many waste piles are used for temporary storage until the waste is transferred to its final disposal site.

5.7 Mixed reactors

In nature, degradation varies in space and time according to available oxygen, moisture, nutrients, and biota. Thus, natural systems are usually a mixture of aerobic and anaerobic systems. These processes may be mimicked in engineered systems, for example, mixed reactors. The most notable mixed reactor is the trickling filter (see Fig. 5.5), consisting of a bed of media, often rocks averaging about 10 cm in diameter, but in fact ranging in size and shape. The media is enclosed in a rectangular or cylindrical structure, upon which the wastewater is sprayed, and through which wastes flow. Biofilms grow on the media and provide the habitat for the bacteria to degrade organic wastes. As the liquid waste moves with gravity downward through the bed, the microorganisms contact the organic contaminant/food source and ideally metabolize the waste into
$CO_2 + H_2O$ + microorganisms + energy. Oxygen is supplied by blowing air into the bottom of the reactor which forces air upward through the bed. The treated waste moves downward through the bed and subsequently enters a quiescent tank where the microorganisms that have sloughed off the rocks can settle and can be collected and disposed. Generally, the higher-oxygen layers of the media are near the surface. This is where most of the aerobic activity occurs, whereas the anaerobes grow in the regions lower in the system. However, there are pockets of low-O_2 regions in the upper areas and high-O_2 regions in the lower areas of the media, depending on the air movement within the media.

Another mixed reactor system is the landfill. This is not a wastewater treatment technology, but can be a source of water pollutants, since landfills often have highly persistent and recalcitrant chemical compounds that resist biodegradation. This often requires enhancements to the landfill environment, notably bioaugmentation and biostimulation. Bioaugmentation is a method of helping indigenous microbial populations degrade pollutants by adding new microbes shown to metabolize the compounds of concern, for example, oil-degrading bacteria (Leahy & Colwell, 1990). Biostimulation is the process of adding nutrients and injecting air and water into the cells of the landfill. For example, adding enriched nitrifying membrane bioreactor sludge can enhance the destruction of aromatic compounds (Boonyaroj et al., 2017). Another type of biostimulation is Fenton oxidation, that is, using hydrogen peroxide (H_2O_2) as an oxidation agent and divalent iron (Fe^{2+}) as a catalyst, to increase degradation rates of water pollutants (Singa et al., 2018). Increasing the amount of water in the land layers and leachate re-circulation can also enhance the growth and efficiency of land systems (Pohland, 1973).

To be sustainable and prevent water and air pollution, landfills and land application systems must include a run-on control system, a run-off management system, and control of the wind dispersal of particulate matter (Meegoda et al., 2016). The run-on control system's capacity must prevent flow onto the active portion of the landfill during peak discharge of a storm event, for example, 25-year. The run-off management system must be designed to collect and control water from a 24-h or a 25-year storm and the run-off's chemical constituents must be tested. Other retention methods can include holding tanks and basins emptied expeditiously after storm events.

The aerobic/anaerobic processes can be a function of space and/or time. The trickling filter is an example of space, whereas landfill reactions are predominantly aerobic in the early stages and become progressively anaerobic with time. The majority of landfill decomposition by volume occurs under these later anaerobic conditions, but toward the end of the landfill life cycle, there is a nearly dichotomous amount of anaerobic and aerobic activity (EPA/600/R-07/060, 2007).

5.8 Aerobic reactors

The quantity of oxygen is always a key factor in designing treatment systems. Numerous water pollutants can be most efficiently degraded under aerobic, that is, excess oxygen conditions. In an activated sludge system, microorganisms are recycled within the system (see Fig. 5.6). Here, microbial populations can evolve over time and adapt to the changing characteristics of the wastewater.

In addition to activated sludge systems, other aerobic reactors include aeration lagoons or ponds treat liquid and dissolved contaminants for long time periods, for example, years. Persistent organic molecules, those not readily degraded in trickling filter or activated sludge systems, may be degraded by specific microbes into

CO_2 + H_2O + microorganisms + energy, if given sufficient amounts of time (Zhao et al., 2022). Usually, rain and other precipitation provide water and winds and diffusion from the air provide oxygen to these systems. They vary in shape and size, but their general dimensions are:

- Pond size: 1–8 ha (0.45–20 acres).
- Pond depth: 0.3–9 m (1–30 ft.).
- Operation: In series with other treatment systems, other ponds, or stand-alone.
- Operation: The flow to the pond is either continuous or intermittent; and,
- Operation: Additional oxygen to the system may be needed, for example, active systems like blowers and diffusers.

A potentially sustainable aerobic reactor is the engineered wetland (see Fig. 5.8), which functions as a biofilter. Wetlands naturally contain diverse and abundant microbial populations. These features are combined with growth, nutrient extraction, photosynthesis, and ion exchange processes of plants. For example, a radial-flow constructed wetland system can consist of subsurface beds of crushed concrete insulated with layers of mulch. Above these layers, species native to the region can be planted in separate sections of the wetland system. Large volumes of water can be passively treated each day, degrading even large organic molecules. The rock/mulch insulated layers allow biological activity to

Figure 5.8
Engineered and constructed wetland system in San Diego, California, utilizing macrophytes, that is, water hyacinths (*Eichhornia spp.*). The system takes advantage of both microbial and larger plant species (bioremediation and phyto-remediation, respectively). Source: *Gerba, C. P., & Pepper, I. L. (2009). Wastewater treatment and biosolids reuse. In* Environmental microbiology *(2nd ed., pp. 503–530). United States: Elsevier Inc.*

continue throughout the year, even in the cold winters. Since this is a passive treatment system, it is cost effective compared to most active systems. As evidence, estimated construction costs are about 20% of a pump-and-treat system that meet similar criteria, for example, air stripping and catalytic oxidation (Wallace, 2009).

The crucial engineering concerns in the design and operation of ponds and other environmental biotechnologies are the identification and maintenance of microbial populations that metabolize the specific contaminant of concern. Once selected, the conditions most favorable to the microbial populations can be determined and controlled.

The fluid dynamics and biological principles of the slurry-phase lagoon (see Fig. 5.9) and the engineered wetland can be used to treat air pollutants in a simple biofilter. However, polluted air rather than water is pumped into the system and mixed with water and pushed into the bottom of a 1-m-deep trench covered with unconsolidated material (soil or compost). The saturated air and water containing the organic contaminants move into the gravel and contact the microbial biofilm. The vapor pressure allows the more volatile compounds to move first through the media, but with time less volatile compound vapors will percolate through the unconsolidated material. During percolation, the air contacts the biofilm of microbes sorbed to the particles, which degrade the organic contaminants. A more intricate version is the packed bed biological control system to treat volatile compounds anaerobically.

Highly hazardous waste will not often be allowed to be treated in these open systems, without very strong containment systems surrounding them in all directions, that is, including the need for covers and collectors for any volatilized or aerosolized material

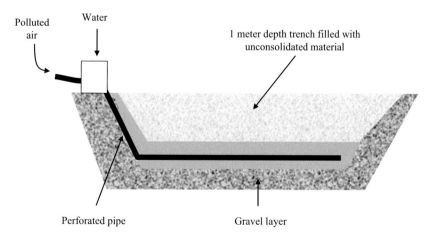

Figure 5.9
Biofilter system used to treat air contaminated with organic pollutants. Source: *Based on: Scragg, A. H. (2005). Environmental biotechnology. New York: Oxford University Press.*

emitted to the air. In addition to scientific and engineering constraints, legal restrictions and public perception may prevent in situ treatment.

5.9 Anaerobic reactors

Anaerobic and aerobic processes frequently occur in different parts of the same system, as mentioned in the trickling filter discussions. One of the challenges of an aerobic system is maintaining the aerobes' contact with molecular oxygen. If the reactor's oxygen level drops or if the tank is not completely mixed, then anoxic and reduced regions can induce the growth of anaerobes. Bioreactions vary according to the concentration of O_2 as the electron acceptor. An indication of low-O_2 conditions in a WWTP is that one of the reactors is "going anaerobic," often because reduced forms of sulfur are released, for example, hydrogen sulfide and mercaptans, which are malodourous. However, these unintentional processes in an aerobic reactor can be simulated in an aerobic reactor to treat otherwise recalcitrant organic compounds (see Fig. 5.10).

Figure 5.10
Anaerobic reactor. The lighter substances that have migrated to the surface are fats that have been separated physically during the treatment process, or bubbles from gases, such as methane produced when the anaerobes degrade the wastes. Note that although this is an anoxic chamber, a thin film layer at the surface will be aerobic given its contact with the atmosphere. Source: *Photo courtesy D.J. Vallero.*

During anaerobic digestion, polysaccharides, proteins, fats and other complex molecules are hydrolyzed. These anaerobic reactions are catalyzed by enzymes, which generates more water-soluble compounds. The new compounds enter the cellular biofilm where they move by advection and diffusion and bind to receptors on cellular membranes. Advection is the transport with the flow of the mass, that is, in this instance, the water flowing into and through the cell membrane. Diffusion is the movement of a molecule by concentration gradient, that is, a chemical compound in the water has a greater concentration than the biofilm concentration, so the compound enters the biofilm and passes through the cell membrane. The chemical, that is, ligand is usually aided by an enzyme, that is, a biological catalyst, that is specific to a class of chemicals. The enzyme-ligand complex binds with a receptor on the cell membrane and is transported into the cell. This process can be likened to a key (ligand) fit to a specific lock (receptor), which allows the chemical to enter the cell.

The rate of production and ensuing escape of CH_4 indicates that the rate at which the organic material has degraded. The retention time of the solids suspended in the wastewater must be sufficient for contact and reaction by the microbes with the substrate (Bal & Dhagat, 2001; Zhao et al., 2022). For example, if the input of rapidly degradable organics is too rapid, acidic and toxic conditions will occur, including the buildup of organic acids, which can foul the reactor by inhibiting methanogenesis. Instead of the desired methane-water products, the system will continue to produce acids. If the wastewater contains substantial amounts of nitrogen, ammonia (NH_3), can also be inhibitory and cause anaerobic digestion to fail (Jiang et al., 2019).

Anaerobes' acclimation to a substrate can take several weeks. Conventional anaerobic treatment processes combine waste and sludge that contains large microbial populations. As in the aerobic activated sludge systems, these biosolids can be added continuously or in semi-batches and mixed in the reactor. Theoretically, aerobic digestion is a once-through, completely mixed system. Thus, the hydraulic retention time (HRT) would be equal to the solids retention time (SRT). The digestion efficiency depends directly on the contact with the biosolids, that is, the SRT. In the real world, however, HRT and SRT can be decoupled. For example, installing an anaerobic upflow filter can substantially improve anaerobic degradation volumes and rates, because the filter can catch and sustain high concentrations of biosolids. By retaining the solids, attenuated SRTs allow can increase throughput needed to degrade greater amounts of low-strength organic wastes (Abramowicz, 1995; Bal & Dhagat, 2001; Christensen et al., 1984; Dagnew et al., 2010; Elefsiniotis, 1993; Li et al., 2020; Vallero, 2019).

A bed upflow reactor is an anaerobic system that depends on the sorption of biomass onto media surfaces. Wastewater passes upward through a bed of sand-sized media at a velocity necessary to fluidize and partially expand the sand bed (Rittmann & McCarty, 2012; Vallero, 2015). The upflow anaerobic sludge blanket process (UASB) induces flocculation and settling of anaerobic sludge, allowing for much higher HRT loadings and partitioning of gases (e.g., H_2, CH_4) from the sludge solids (see Fig. 5.11). The UASB separates solids and gases from the liquid, followed by degradation of organic matter. Biogas production enhances continuous contact between substrate and anaerobes. The UASB reactor, under optimal conditions, would be a completely mixed reactor. The contact of the biosolids with the incoming organic liquids is enhanced by the agitation caused by the release of gases from the biodegradation, along with an inlet that evenly distributes incoming materials in the lower level of the reactor.

For any biological treatment system, additional chemical or physical treatment may be required to extract inorganic materials from a waste stream prior to the biological treatment. If not, toxic substances may destroy beneficial microbial populations and/or may be present in the effluent or sludge, leading to the need for hazardous waste handling (Vallero, 2019; Vallero & Peirce, 2003).

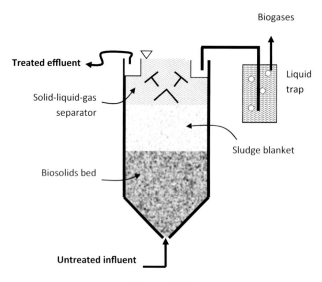

Figure 5.11
Schematic of upflow anaerobic sludge blanket system. Source: *Adapted from Christensen, D., Gerick, J., & Eblen, J. (1984). Design and operation of an upflow anaerobic sludge blanket reactor.* Journal (Water Pollution Control Federation), *1059–1062.*

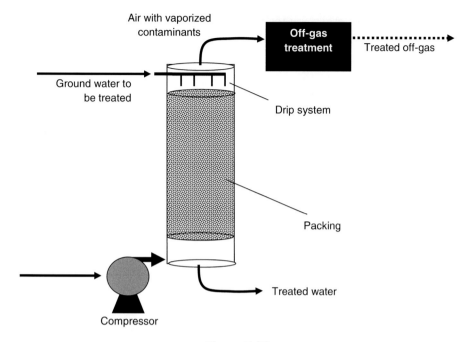

Figure 5.12
Air stripping system to treat volatile compounds in water. Source: *Vallero, D. (2015). Environmental biotechnology: A biosystems approach. Elsevier Science.*

5.10 Treating contaminated ground water

Ground water can be treated by drilling recovery wells to pump contaminated groundwater to the surface. Commonly used groundwater treatment approaches include air stripping, filtering with granulated activated carbon (GAC), and air sparging.

Air stripping moves volatile compounds from water into the air (Fig. 5.12). In a tower filled with a permeable material, ground water moves by gravity downward while a stream of air flows upward. Another method bubbles pressurized air through contaminated water in a tank. The air leaving the tank is treated by removing the pollutants in their vapor phase. Contaminated ground water can also be filtered by pumping the water through the GAC to trap the contaminants. Air sparging consists of pumping air into the ground water in situ. A soil venting system is often combined with an air sparging system for vapor extraction (Vallero, 2015).

The previously discussed microbial activity can be added as an enhancement for vapor phase compounds. The microbes are grown on porous media. The air containing volatile compounds is passed through the biologically active media, where the microbes break down

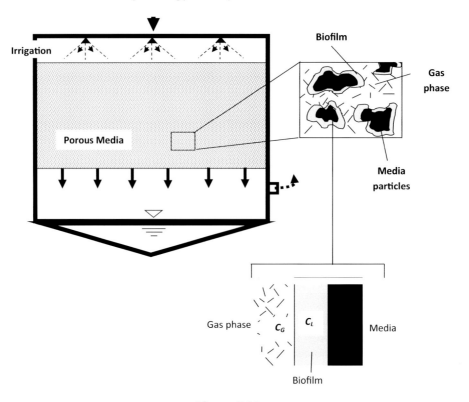

Figure 5.13
Packed bed biological control system to treat volatile compounds. Air containing gas phase pollutants (C_G) moves through porous media. The soluble portion of the volatilized compounds in the air stream partition into the biofilm (C_L) according to Henry's Law: $C_L = C_G/H$; where H is the Henry's Law constant. *From: Vallero, D. A. (2014). Fundamentals of air pollution (5th ed., 999 p). Waltham, MA: Elsevier Academic Press; Ergas, S. J., & Kinney, K. A. (2000). Biological control systems. In Air pollution engineering manual (p. 55).*

the compounds to simpler compounds. The major difference between biofiltration and trickling systems mainly is in the liquid interfaces with the microbes. The liquid phase is stationary in a biofilter (see Fig. 5.13), whereas liquids move through the porous media of the system, that is, the liquid "trickles" as it does in Fig. 5.14. Compost can be used as the porous media. Compost contains numerous species of beneficial microbes that are already acclimated to organic wastes. Efficiency in treatment depends on the air stream rates, the pollutants' fugacity and solubility needed to enter the microbes' biofilm, and pollutant loading rates (Rittmann & McCarty, 2012).

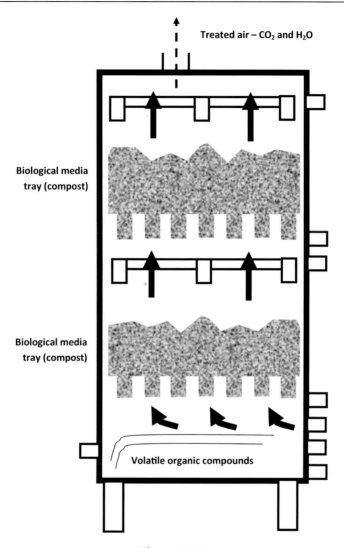

Figure 5.14

Biofiltration without a liquid phase used to treat vapor phase pollutants. Air carrying the volatilized contaminants upward through porous media (e.g., compost) containing microbes acclimated to break down the particular contaminants. The wastes at the bottom of the system can be heated to increase the partitioning to the gas phase. Microbes in the biofilm surrounding each individual compost particle metabolize the contaminants into simpler compounds, eventually converting them into carbon dioxide and water vapor. *From: Vallero, D. A. (2014). Fundamentals of air pollution (5th ed., 999 p). Waltham, MA: Elsevier Academic Press.*

5.11 Conclusions

The amounts and quality of water are crucial to the survival of human populations. Surface and ground water systems and habitats are threatened by myriad natural processes and human activities. This chapter has addressed physical, chemical, and biological processes that are at work in water pollution and in the technologies needed to address this pollution. Finding ways to prevent the pollution in the first place is most desirable. When this is not possible, engineers play a key role in selecting, designing, operating, and maintaining the most sustainable systems to treat wastewater, which is crucial to proper water governance and future needs and demands.

References

Abramowicz, D. A. (1995). Aerobic and anaerobic PCB biodegradation in the environment. *Environmental Health Perspectives*, *103*(Suppl. 5), 97.

Alexander, M. (1999). *Biodegradation and bioremediation*. Gulf Professional Publishing.

Ashley, D. M., & Word, D. M. (1991). *Developing watershed protection criteria for Georgia*. Georgia Institute of Technology.

Averett, D. (1994). Chapter 7: Treatment technologies. *Assessment and remediation of contaminated sediments (ARCS) program: Remediation guidance document*. Chicago, IL: G. L. N. P. O. U.S. Environmental Protection Agency (Ed.).

Bal, A., & Dhagat, N. (2001). Upflow anaerobic sludge blanket reactor: A review. *Indian Journal of Environmental Health*, *43*(2), 1–82.

Boonyaroj, V., Chiemchaisri, C., Chiemchaisri, W., & Yamamoto, K. (2017). Enhanced biodegradation of phenolic compounds in landfill leachate by enriched nitrifying membrane bioreactor sludge. *Journal of Hazardous Materials*, *323*, 311–318.

Bridle, H., Jacobsson, K., & Schultz, A. C. (2021). Sample processing. *Waterborne pathogens* (pp. 63–109). Elsevier.

Cheremisinoff, N. P. (1997). *Biotechnology for waste and wastewater treatment*. Elsevier.

Chorus, I., & Spijkerman, E. (2021). What Colin Reynolds could tell us about nutrient limitation, N: P ratios and eutrophication control. *Hydrobiologia*, *848*(1), 95–111.

Christensen, D., Gerick, J., & Eblen, J. (1984). Design and operation of an upflow anaerobic sludge blanket reactor. *Journal (Water Pollution Control Federation)*, 1059–1062.

Dagnew, M., Parker, W. J., & Seto, P. (2010). A pilot study of anaerobic membrane digesters for concurrent thickening and digestion of waste activated sludge (WAS). *Water Science and Technology*, *61*(6), 1451–1458.

Elefsiniotis, P. (1993). *The effect of operational and environmental parameters on the acid-phase anaerobic digestion of primary sludge*. University of British Columbia.

EPA/540/A5-91/009. (1993). *Pilot-scale demonstration of slurry-phase biological reactor for creosote-contaminated soil: Applications analysis report*.

EPA/600/R-07/060. (2007). *Landfill bioreactor performance: Second interim report: Outer loop recycling & disposal facility*. Louisville, Kentucky.

EPA/625/R-00/008. (2002). *Onsite wastewater treatment design manual*.

EPA 832-F-00-016. (2000). *Wastewater technology fact sheet: Package plants*.

EPA No. 832-B-05-001. (2005). *Handbook for managing onsite and clustered (Decentralized) wastewater treatment systems: An introduction to management tools and information for implementing EPA's management guidelines*.

Ergas, S. J., & Kinney, K. A. (2000). Biological control systems. In *Air pollution engineering manual* (p. 55).

Gerba, C. P., & Pepper, I. L. (2009). Wastewater treatment and biosolids reuse. *Environmental microbiology* (2nd ed., pp. 503–530). United States: Elsevier Inc.

Ikada, Y., & Tsuji, H. (2000). Biodegradable polyesters for medical and ecological applications. *Macromolecular Rapid Communications*, *21*(3), 117–132.

Jiang, Y., McAdam, E., Zhang, Y., Heaven, S., Banks, C., & Longhurst, P. (2019). Ammonia inhibition and toxicity in anaerobic digestion: A critical review. *Journal of Water Process Engineering*, *32*, 100899.

Jiménez, B., Mara, D., Carr, R., & Brissaud, F. (2009). Wastewater treatment for pathogen removal and nutrient conservation: Suitable systems for use in developing countries. *Wastewater irrigation and health* (pp. 175–196). Routledge.

Leahy, J. G., & Colwell, R. R. (1990). Microbial degradation of hydrocarbons in the environment. *Microbiological Reviews*, *54*(3), 305–315.

Li, B., et al. (2020). Recuperative thickening for sludge retention time and throughput management in anaerobic digestion with thermal hydrolysis pretreatment. *Water Environment Research*, *92*(3), 465–477.

Meegoda, J. N., Hettiarachchi, H., & Hettiaratchi, P. (2016). Landfill design and operation. *Sustainable Solid Waste Management*, 577–604.

Pohland, F. G. (1973). *Sanitary landfill stabilization with leachate recycle and residual treatment*. Georgia Institute of Technology.

Rittmann, B. E. (2009). Environmental biotechnology in water and wastewater treatment. *Journal of Environmental Engineering*, *136*(4), 348–353.

Rittmann, B. E., & McCarty, P. L. (2012). *Environmental biotechnology: Principles and applications*. Tata McGraw-Hill Education.

Scragg, A. H. (2005). *Environmental biotechnology*. New York: Oxford University Press.

Singa, P. K., Isa, M. H., Ho, Y.-C., & Lim, J.-W. (2018). Treatment of hazardous waste landfill leachate using Fenton oxidation process, *E3S Web of conferences* (Vol. 34, p. 02034). EDP Sciences.

Subramani, A., & Jacangelo, J. G. (2015). Emerging desalination technologies for water treatment: A critical review. *Water Research*, *75*, 164–187.

U.S. Environmental Protection Agency. (1986). *Test methods for evaluating solid waste, physical/chemical methods*. Cincinnati, OH: US EPA.

U.S. Environmental Protection Agency. (1994). 905-R94-003 *Assessment and remediation of contaminated sediments (ARCS) program: Remediation guidance document*. Chicago, IL: EPA.

Vallero, D. (2010). *Environmental contaminants: Assessment and control*. Academic Press.

Vallero, D. (2015). *Environmental biotechnology: A biosystems approach*. Elsevier Science.

Vallero, D. A. (2014). *Fundamentals of air pollution* (5th ed.). Waltham, MA: Elsevier Academic Press, 999 p.

Vallero, D. A. (2019). Chapter 31 - Hazardous wastes. In T. M. Letcher, & D. A. Vallero (Eds.), *Waste* (2nd ed., pp. 585–630). Academic Press.

Vallero, D. A., & Gunsch, C. K. (2020). Applications and implications of emerging biotechnologies in environmental engineering. *Journal of Environmental Engineering*, *146*(6), 03120005.

Vallero, D. A., & Peirce, J. J. (2003). *Engineering the risks of hazardous wastes* (p. xxi) Burlington, MA: Butterworth-Heinemann, 306 p.

Vallero, D. A., Reckhow, K. H., & Gronewold, A. D. (2007). Application of multimedia models for human and ecological exposure analysis. *International conference on environmental epidemiology and exposure*. Durham, NC: International Society of Exposure Analysis.

Wallace, S. (2009). Engineered wetlands lead the way. *Land and Water*, *48*(5). [Online]. Available: http://www.landandwater.com/features/vol48no5/vol48no5_1.html.

Zhao, F., et al. (2022). New insights into eutrophication management: Importance of temperature and water residence time. *Journal of Environmental Sciences*, *111*, 229–239.

CHAPTER 6

Urban water supplies in developing countries with a focus on climate change

Josephine Treacy

Technological University of the Shannon Midlands Midwest, TUS, Limerick City, Ireland

Chapter Outline

6.1 Introduction   95
6.2 Urbanization in developing countries   97
6.3 Urban water supplies   97
    6.3.1 Source waters   99
    6.3.2 Other water sources innovative needs   100
    6.3.3 Urban water supplies as a resource   101
    6.3.4 Centralized and noncentralized water supplies   101
6.4 Climate change challenges and urban waters   102
    6.4.1 Water Supplies and demand   103
6.5 Waste management and climate change and its impact on urban water   104
6.6 Opportunities and global initiatives   104
    6.6.1 Adaptation and mitigation   105
    6.6.2 Understanding the water baseline   105
    6.6.3 Environmental impact modeling risk   105
6.7 Water security   106
    6.7.1 Policy makers   106
6.8 Conclusions   107
References   109
Further reading   112

6.1 Introduction

Water is important for human existence. One of the greatest impacts of climate change is disturbances of the water cycle. In line with these disturbances are severe climate and weather events that can influence water treatment infrastructure and also quality and availability of source water. Global water resources are coming under increasing pressure due to human impacts on climate change. Water demand is beginning to outpace supply.

This is particularly relevant in developing countries. The water supply itself is being impacted in terms of both quantity and quality by a range of issues such as hydrologic imbalances, climate change, environmental degradation, water demand and inefficient use of water. An important reason why urban water supplies in developing countries should be studied is the fact that half of humanity lives in developing countries. Society, in general, has seen a large surge of people moving to live in cities and this trend also applies to developing countries. The majority of people on Earth live in developing countries comprising Africa, Asia, Eastern Europe, Latin America, the Caribbean and the Middle East. This chapter will focus on urban water supplies in developing countries with a focus on climate change. Three variables are important to consider when examining water supplies: land use and planning, climate change, and population (Cohen, 2006). The United Nations in their Millennium Declaration draws attention to the importance of water and water related activities in supporting development and eradicating poverty (Vairavamoorthy et al., 2008). One of the targets of goal 7 of the Millennium Development Goals (MDGs signed in 2000) is to provide access to safe drinking water and basic sanitation. Further policies building from this include the Paris Convention 2015 and Agenda 2030 and its 17 Sustainable Development Goals, SDGs, in which this chapter is particularly relevant to Sustainable Development Goal SDG 7. Sustainable Development Goal SDG 7 is related to water and sanitation and highlights the urgent challenges of the influences of climate change on water and the right for clean water (WHO/UNICEF, 2017).

The World Bank has emphasized, "Water availability and efficient management impacts on whether poor girls are educated, whether cities are healthy places to live, and whether growing industries or poor villages can withstand the impacts of floods or droughts" (World Bank, 2017).

Water security is in line with seasonal water demand influenced by climate change. The focus of this chapter support the regime of both sustainable cities and smart cities. This chapter will discuss influences of climate change on water supplies, surface and ground water sources and related water networks and treatments. Challenges of droughts and floods, water security, water infrastructure resilience and robustness, nature solutions and water treatment needs, all influenced by climate change will be discussed. Other areas such as water conservation, water harvesting, and circular water initiatives including water stewardship and water productivity strategies will be outlined. Mitigation and adaptation to climate change and related challenges on urban water supplies will be discussed. As stated by the Intergovernmental Panel on Climate Change (IPCC, 2013), the main causes of climate change are anthropogenic activities giving rise to excess of carbon dioxide CO_2 methane CH_4 and nitrous oxide N_2O in the atmosphere. These anthropogenic activities contribute up to 5% of global green houses gases (GHG) emissions, namely carbon dioxide (CO_2) from energy consumption, as well as emissions of methane (CH_4) and nitrous oxide (N_2O) from wastewater treatment plants.

6.2 Urbanization in developing countries

Urbanization is associated with human development but can also be associated with human inequalities and health problems (Kuddus et al., 2020). It is estimated that by the year 2041 over half of the world's population will live in urban areas (United Nations, 2019). Water plays a central role in the human development index of a country which is benched marked under economic growth, health, education and well being. The terminologies surrounding urbanization are important to outline to fully understand the scope of the challenges.

There is no universal definition that can be applied to a city. Different countries, municipalities, and scientists use different definitions. Definitions surrounding mega cities vary from cities having over 5 million to over 20 million inhabitants. In the State of the World's Cities 2010/2011 report, the UN-HABITAT defines a mega city as a city with 20 million or more inhabitants (UN-Habitat, 2010). Unfortunately, urbanization can also give rise to slums. A slum is considered households where groups of individuals are living under the same roof lacking one or more of the following basic needs; access to improved water, access to improved sanitation facilities, sufficient living areas including not more than three people sharing the same room, structural quality and durability of dwellings and security of tenure.

One of the biggest challenges with urbanization is unplanned settlements and unfortunately bad planning. One of the important criteria within a planning protocol entails proper water management and proper water resource sustainability strategies, which should be integrated with waste management (Hadi et al., 2021). Climate change is also a present and future concern in the planning for urban living. From the United Nations 2018 report urban living perspective, one can see from Fig. 6.1 the density of urban living in developing countries. The highest density of urban living globally occurs in developing countries.

Most urban communities are in coastal areas and have extra risks due to coastal erosion and extreme weather events, such as tropical storms (Sinisi & Aertgeerts, 2011). These risks are greatly influenced and enhanced by climate change. Water supplies and infrastructure can be challenged by disaster risks. Urban water supplies and climate challenges are very complex and should be examined in terms of an integrated management lens approach (Esmail & Suleiman, 2020) (Fig. 6.2).

6.3 Urban water supplies

Water has a central role to achieving the three pillars of sustainability. For example, both economic and social development can have large impacts on water demand. Water availability has an important role in ensuring good health, food and nutrition (Agarwal et al., 2010). Water also has the role through its ecological services of sustaining ecosystems and habitats. Water sustainability and water security are very much influenced

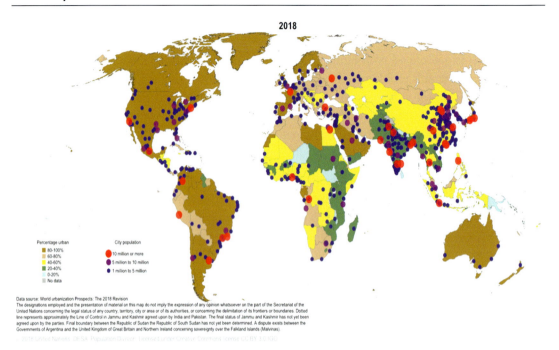

Figure 6.1
The extent of urbanization and extent in developing countries (UN Report, 2018).

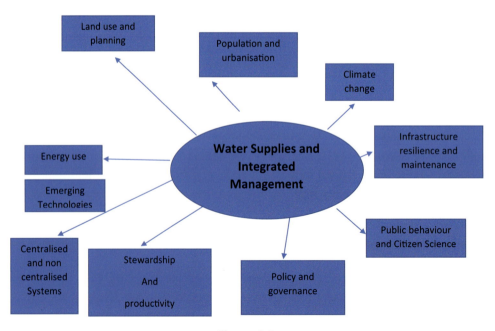

Figure 6.2
Integrated approaches to urban water supplies and climate change challenges.

by climate change. Despite water's importance disappointingly 663 million people in the world still have no access to improved drinking water sources. It is also important to state access and appropriate cost of water can be further challenged by climate change. The Sustainable Development Goals, SDGs, concerning water and sustainability proposes the Agenda 2030 which encourages universal and equitable access to safe and affordable drinking water for all. The Agenda 2030 also encourages access to adequate and equitable sanitation and hygiene for all which includes no open sewers. Unfortunately, open sewers can be problematic in slums. The Agenda 2030 reflects the growing importance of water and sanitation as a human right (World Bank, 2017). Urban water supplies quality standard and availability are a major concern for developing countries but an extra layer within this challenge is climate change.

6.3.1 Source waters

Source water, also known as abstraction water, can be characterized as conventional and nonconventional. Conventional urban water supplies, which are considered natural renewable water sources, come from surface waters such as rivers and lakes and ground water aquifers. Ground waters include aquifers namely confined and nonconfined. Aquifer shallow wells and pore holes are readily used in developing counties (Dagdeviren and Robertson, 2012) Shallow wells are often individual house owned and are sometimes partially treated or untreated (Durmade, 2017). Water from this facility is usually manually drawn, but there are some which are covered and fitted with pumps. An interesting review of Nigerian peri-urban dwellings show that of the water facilities, 70% are shallow wells water and 30% are bore holes (Durmade, 2017). Bore holes are managed in most cases by local authorities and private contractors. The quality and quantity of the source water is influenced by climate change and can have a bearing on the treatment required and the toxicology of the water. Biological treatment should receive priority over chemical treatment. Ground waters can be less polluted than surface water and thus are more readily used. In Burkina, a country in West Africa, their large cities are investing in drilling three new boreholes (into sedimentary rocks) and building three new water towers to store 4000 m^3 in total for network distribution (Newborne, 2016). In some instances, because of demand for water in urban areas the source water must be extracted from a distant rural area in proximity to the urban area and this again is impacted by climate change. This approach can also affect the aquatic ecology of rural areas. With regard conventional water sources stresses can be due to rainwater which is seasonal, which can have increase in intensity and shortness of duration due to climate change (Danilenko et al., 2010; Delpla et al., 2009). Weak land resource management, impact of increased submergence of coastal regions, loss of wetlands, sea level raises, and increase of pollution can all effect the quality of source waters. Risk of sea water intrusion due to sea level raises can affect both aquifers and surface water sources. In general, evaporation is more prevalent to precipitation when it comes to climate change. The dry season gives rise to lack of recharge of aquifers. Aquifers can have problems with inclusion of saline water due to climate change. The management of the catchment is important in terms of where the water is drawn from and the extra challenges due to climate change such as run off and pollution. Strategies to prevent pollution runoff both diffusion and point sources is important to manage as storm water runoff can affect the quality of surface waters and ground waters, both shallow wells and bore holes (Clifton et al., 2010; Cook & Smith, 2005). Turbidity and bacteriological contamination are the largest concerns and treatment challenges. Surface waters, for example lake thermocline problems, and algae bloom accelerated due to climate change, can give rise to more expensive treatments such as agitation of water and activated carbon treatment (Ali, 2010).

6.3.2 Other water sources innovative needs

Other water initiatives that can be used for drinking water include rainwater harvesting, desalination technologies, and recycling waste water such as grey waters. These are known as nonconventional water sources. Raised sea water affects intakes of desalination plants. This leads to higher costs on nonconventional water desalination treatment. Increases in salinity caused by increased evaporation of water again leads to more costly treatment needed for the desalination process. Wherever possible, more awareness surrounding eco circular water approaches should be demonstrated. Climate change can give rise to more rainwater and thus more water harvesting potentials. Collection of rainwater through harvesting can give rise to a source of water and can have a second role as a form of mitigation against flooding.

6.3.2.1 Water desalination

Given that about 97.5% of global water is saline water, global desalination has become increasingly important in all parts of the world. Ocean water contains a high percentage of salt, bacteria and suspended and dissolved solids particulates. For human consumption, ocean water must be desalinated and purified. However, desalination remains an energy-intensive process and is prohibitively expensive. Technologies include thermal systems and membrane systems. Corrosion, scaling and fouling problems are more serious in thermal processes compare to the membrane processes. Membrane systems have challenges due to clogging of membranes by particulates. Usually, the water undergoes a pretreatment step to prolong the life of the membranes. Membrane technologies are the preferred choice as this process is the cheapest (Saadat et al., 2018). One of the biggest challenges of desalination plants and climate change is the increases in evaporation of water given rise to a higher salinity of the water thus higher treatment costs (Elimelech & Phillip, 2011). Innovations in solar energy for desalination can reduce the carbon footprint of the process (Caldera et al., 2016; Elimelech & Phillip, 2011).

6.3.2.2 Atmospheric water harvesting

Atmospheric water harvesting has emerged as a potential extra source of water. The two approaches to collect water from the atmosphere that are being researched include obtaining water from fog and clouds and also dew through the condensation of vapor on surfaces with a temperature below the dew point. Micro- and nanostructured surfaces with different surface wetting properties and having hydrophilic or superhydrophobic functionalities are efficient for fog and dew trapping of water. Drop mobility surface superhydrophobicity and micronanotopography are the most important factors for water capture from fog. Interestingly, superhydrophobic surfaces show 40%−65% higher collection rates than flat hydrophilic designs. Dew harvesting is based on a combination of drop mobility and nucleation rate. Superhydrophobic surfaces exhibiting 40% higher water collection rate compared to the flat hydrophilic or hydrophobic surfaces due to a high surface area available for nucleation (Nioras et al., 2021). Trapping water from the atmosphere in dew, fog, and clouds using different material surface interaction approaches can be considered a low carbon-based technology (Jarimi et al., 2020;(Bhushan and Feng, 2020).

6.3.2.3 Water recycling

This can reduce the pressure on the treatment of abstraction water for drinking water purposes. Portable water reuse or treated drinking water at the level of gray water is the most realistic choice. Regulation of water recycling is very important. Policies and regulations would need to include education and community engagement (Macpherson & Snyder, 2013).

6.3.3 Urban water supplies as a resource

Urban water supplies as a resource can be utilized for residential, commercial, and industrial activities including agricultural. An interesting NASA study studied night-time light data sets as a proxy for industrial activity in urban areas. The NASA approach helps to gather more knowledge surrounding industrial activities which can be difficult to measure (NASA, 2017). Productivity management and water stewardship as well as an appreciation of the source water thus aiding controlling the demand for water are important (Treacy, 2021).

6.3.4 Centralized and noncentralized water supplies

Delivery to the consumer can entail centralized or noncentralized systems. Centralized systems are more regulated. Centralized systems pump water from treatment plants through a network of piping systems to the point of use. One of the challenges is the network can be too expensive or too difficult to fix (Boin & Hart, 2003).

City and mega city designs can be supplied through centralized networks in most cases. Noncentralized systems are the least regulated and can be challenged by both volume and quality. The human development index of a country can be impacted on the water volume used. Leakage of network pipes in the case of centralized systems and infrastructure resilience influenced by weather and climate are important to manage. Run off, floodwater, and droughts can all give rise to changes in water table levels and levels of treatment required.

Vendors with water tankers and illegal connections are the least regulated and relate to noncentralized approaches. Private contractors are also used when the demand for water exceeds the centralized supply. The cost of water is greatly nonregulated in noncentralized approaches. Slums have the greatest problems in terms of water availability, nonregulation, and in-house storage contamination problems are very challenging to control. The biggest problems are biological contamination such as microbes or microbiological contaminants; examples of biological or microbial contaminants include bacteria, viruses, protozoa, and parasites (Treacy, 2019).

An interesting approach in Istanbul involved installing water meters in buildings, so consumption can be measured and attributed to individual households. In Istanbul, public wells were taken out of service. When repairs to water lines networks were introduced security of supply improved and water leakage in network piping systems decreased. The city refinanced these investments by raising the price of drinking water. Today, the residents pay a larger share of their income to receive municipal water. Poorer people have fewer opportunities to obtain drinking water free from public wells or from water pipe systems that are not metered (Huy, et al., 2015).

Urban planning, education, treatment of wastewater, drinking water storage, infrastructure design and treatments, network systems and water delivery can all involve maintenance and financial problems. As stated earlier as the water demand in urban areas increases so the need to abstract water from far distances rural areas increases, which again is challenged by climate change. The challenges of pumping water from a distance can be difficult in developing countries. Infrastructure resilience due to storms and malfunction of operations are important to manage inline with financial supports (Ferreira et al., 2021).

6.4 Climate change challenges and urban waters

Due to the global positioning of developing countries in the subtropical low to mid latitudes there is a larger tendency of rapid drying and extreme weather events in general. Floods and droughts are the major impacts of climate change on society and urban areas. Other challenges including water security, quality of water determined by the physicochemical chemical and biological parameters. Temperature variation in water can influence dissolved oxygen levels; higher temperatures gives rise to less dissolved oxygen in water. Stream water flow and transboundary pollution effects due to transportation of water with increased flow can be caused by excessive rain. Thermocline development can impact abstraction water and the quality of abstraction water, especially due to oxygen depletion, and the extra challenges of having to agitate the water in order to increase oxygen levels. Rainwater interlinking with storm water flow and sewage overflow are all major concerns which are again enhanced by climate change. Drought can lower water levels thus increasing the risk of pollution which are due to input flows from point and diffuse sources of pollution.

An interesting example relates to East Africa where the climate will continue to warm with rainfall events becoming more intense resulting in flash flooding. These flash floods will impact the urban sanitation service delivery in East Africa and contribute to negative health outcomes. As urban areas grow, the problem will get worse especially with the low priority placed on investment in urban drainage construction and management (Karikari, 1996; Muller, 2015). Furthermore, the high turbidity of watercourses increases the cost and complexity of water treatment. Infrastructure resilience and the unpredictability of the consumer to obtaining water is also a concern in relation to climate change (Hoffmann, 2020).

6.4.1 Water Supplies and demand

A further problem, is when the demand for water outweighs the supply, this can give rise to conflicts and potential wars over water (Ford, 2002). Environmental impact assessment with catchment auditing to improve the source water quality can prevent pollution and consequently reduce the cost and minimize energy for treatment. In terms of the challenges of water treatment, the management and improvement of water consumption rather than increasing the cost and demand on energy for treatment is salient. Both public and private partnership can give financial support to aid modernization of infrastructure and support maintenance requirements (Mathur, 2017; Patel & Bhagat, 2019). Furthermore, climate change introduces an extra inequality in the human right for water (Fig. 6.3).

The service gap highlights the gap between provisions of piped water and the growth of urban populations. The availability of piped water is a service challenge (Fig. 6.3).

To reach the water demand for example, Istanbul ensures a strong relationship between the city and its catchment region. Municipal and private water supplies are needed to sustain the water demands of the city. Seven rivers and lakes, located primarily outside of Istanbul, are where most people's drinking water comes from. The communities water services are obtained from either municipal water lines or from private water suppliers.

The same fate hits agriculture in the areas where the water is sourced. Urban water supply hinges on municipal and private water suppliers from Istanbul involving also local and regional authorities outside the city limits.

104   Chapter 6

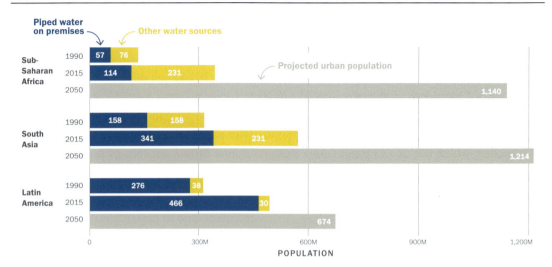

Figure 6.3
The service gap (Mitlin et al., 2019).

6.5 Waste management and climate change and its impact on urban water

It is interesting to note clean water is only as clean as the waste management strategies in the vicinity of the clean water. Waste management is particularly important with climate change due to issues of overflows storm overflows and potential mixing of clean water with wastewater. Unregulated waste disposal is a major concern. Waste management in developing countries is more challenging than in developed countries because rubbish dumps and landfills are the general choice for disposing of solid waste because these are the cheapest options (Waste Advantage Magazine, 2021). Greenhouse gases mitigation through food waste reduction and organic waste diversion is needed in developing countries. Adoption of treatment and disposal technologies that capture biogas and landfill gas are needed. Education and citizen science to support resilience by reducing waste disposal in waterways, addressing debris management, and safeguarding waste treatment infrastructure against flooding, which can affect the quality of abstraction waters, is salient (ECSSR, 2019; Grant et al., 2012).

6.6 Opportunities and global initiatives

Global initiatives and research to aid the sustainability of urban water supplies include areas such as, risk management controlling risk, modeling disaster management, water conservation, training, public private partnership, understanding the network, mitigation for climate change and ensuring the human right to clean water. National level planning, scenario planning, water storage management and water risk monitoring at local level. Trans-regional, and transboundary catchment management can give rise to confidence in the subject of water resource (UN, 2014).

6.6.1 Adaptation and mitigation

When addressing climate change, it is important to include both mitigation and adaptation. Unfortunately, adaptation is important as some of the consequences of climate change are very slow to reverse (Adger et al., 2003; AsLenzholzer et al., 2020; Tjandraatmadja et al., 2012). If possible mitigation should be the first line of defense.

Water resource management including permaculture design to protect river systems and groundwater systems such as buffer zones will aid both mitigation and adaptation. Utilizing nature to aid mitigation and adaptation from disasters such as coastal erosion and storm surge preventions is important. Other adaptation initiatives include water productivity and water stewardship including community involvement. Conservation and preservation initiatives during the drought and flood seasons can minimize the risk of not meeting the water demand. Urban trapping of rainwater in large holding tanks during the wet season, education on mitigation needs, are all important. In harmony with nature, as Pope Francis related to the COP 26 Pope's address highlights the words of the cry of nature and the cry of the poor. Multibarrier approach to catchment management is important in line with water stewardship (Treacy, 2021).

6.6.2 Understanding the water baseline

Understanding the baseline is important in terms of total annual withdrawal and seasonal water withdrawal. This needs to be evaluated in terms of the available water resources and how it can be impacted by climate change. Sub-basin management is related to where water is coming from in line with catchment activities and water quality levels are also important for baseline assessment (United Nations, 2019; US Agency, 2009).

6.6.3 Environmental impact modeling risk

Managing risk in the wet and dry season are important to understand and research to ensure the basic right to clean water. Risk is defined in the International Standard ISO, 2009:(section 2.1) as the "effect of uncertainty on objectives." Vulnerability is a result of the risk. Once the risks and vulnerabilities are known mitigation and adaptation variables can be decided on.

$$\text{Risk} = \text{hazard} \times \text{exposure}$$

The strategies to increase resilience from climate chnage and water availalbility require the identification of the relevant risks and vulnerabilities (Cullis et al., 2019). In urban dwellings it is important to measure the risk in the wet and dry season. With respect to the network in the case of a piped water network, there is an obvious risk that a main pipe will break at some point interrupting the water supply (Biswas, 1979). Mitigation would involve more auditing and more routine maintenance.

Risk management can be based on ensuring the freshwater reserve, rainwater harvesting, education on conservation, water storage during the wet season, clouds and fog water trapping technologies. Other risk management strategies include miminising the water challenges related to the cost of water , pollution reduction, and from our experiences of COVID 19 lessons learned and building a better further. Education awareness to manage water especially in relation to noncentralized water supplies in areas such as protecting the water from contamination once collected is also embedded in risk management. Others bench markers include the human development index and the multibarrier approach design and deployment.

6.7 Water security

By developing grassroots strategies with grassroots knowledge involving younger generations and innovators, this action can ensure water security. Knowledge of water demand and an appreciation of reserves of resource water will enable this. Furthermore, inclusion of local people and citizen science in decision making enhances water security strategies (He et al., 2021).

6.7.1 Policy makers

Policy and community involved in policy design and deployment are salient when it comes to climate change and water (Foucault, 1991). Energy and water are interlinked when it comes to understanding water security. For policy makers, the technologies and processes supplying energy and water services are of concern because part of the SDGs goals target clean water and energy for all. As 2.1 billion people still lack access to an improved water source and 1.1 billion lack access to electricity, thus higglighting water, energy and climate change policies, are interlinked. Moreover, achieving the other SDGs, such as those related to health, ecosystems, and poverty, will be contingent on meeting water and energy sustainability objectives. A large amount of energy is required to pump and treat water resources. Identifying long-term infrastructure strategies that effectively balance water demand with availability, tackle water quality, energy and human development objectives in an integrated manner can assist in achieving the SDGs (Fig. 6.4).

Due to the need to interlink water supplies with climate change, the literature has shown a new paradigm approach to governance, management, and infrastructure needs. Traditional systems are considered rigid and not resilience in terms of environmental instabilities due to

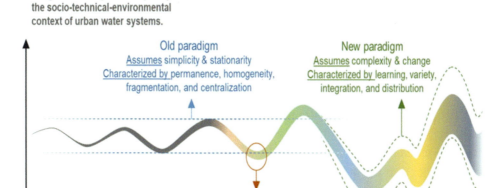

Figure 6.4
The paradigms static and dynamic paradigms (Franco-Torres et al., 2020).

climate change, growing populations and innovations in technologies. The new paradigm approach focuses on these (Fig. 6.4).

Citizen science involvement in monitoring the use of catchment and water quality linking with policies is important. This can also give communities an awareness and appreciation of their own role in sustaining and increasing security of water locations.

Major key players in water management include World Meteorological Organization (WMO), United Nations Children's Fund (UNICEF), the Food and Agriculture Organization (FAO); the UN Educational, Scientific and Cultural Organization (UNESCO); the World Health Organization (WHO); the International Fund for Agricultural Development (IFAD); the UN Environment Program (UNEP); the UN University (UNU); the UN Economic Commission for Europe (UNECE), and the Global Water Partnership (GWP).

These key players highlight the importance of a proactive approach to integrating water and climate. Having adaptation and resilience protocols at governmental and local levels can ensure strong sustainable policies. The importance of monitoring systems to provide timely information about current and future water availability and sources and climate change consequences utilising realtime monitoring and early warning systems are important.

6.8 Conclusions

The extent and speed of which climate change is occurring is creating a need for fast strategies including innovation in water management and governance (Adger et al., 2003;

Lennart & Head, 2015). Development agencies and governments must engage in adaptation and mitigation plans. Local water resources must be understood. The aim of international agreements and actions is to promote adaptive capacity in the context of competing sustainable development objectives such as UN-habitats and Agenda 2030 SDGs and most recently, COP 26 (COP 26, 2021). The water crisis is not an isolated problem but is linked with food, energy, hygiene and health problems, water and biodiversity sustainability.

The concept of urbanization, a high human development index, understanding water and climate change; floods and droughts, security of source waters, centralized and noncentralized technologies are all a global concern (OECD, 2014). The human development index can aid in bench marking and highlighting the importance of supporting underdeveloped cities with climate challenges related to water supplies (Human Development UN, 2021).

Integrated approaches and links with root cause understanding to minimise climate change must be part of the solution. Inequality, loss of natural systems unplanned urban planning, frequent and intense droughts, and floods undermine water security (Koncagül et al., 2019). Integration between stakeholders including public and private partnerships can enable solutions to climate change benefiting challenges in other areas.

Loss of natural soils, wetlands, and forests impact water replenishment and drainage making it difficult to cope with water security these areas should be taken into consideration under planning laws in line with enhanced risks of flooding and landslides. Urban living is important and urban areas in developing countries can be considered urban engines of economic growth.

Modern innovations surrounding provisions of water services, water recycling initiatives, and water trapping from the atmosphere, are important for water sustainable needs into the future (Ali, 2010). Creating sponges to minimize flooding problems involving trapping and storage of water in cities are all important for further urban city living.

It is interesting to note that even when aquifers and dams are full, there is still the fact that 63% of Saharan African lacks access to basic water and sanitation (Jenson & Khalis, 2020).

Around the globe as people tackle COVID 19, there is a real opportunity to build back better to build cleaner, more resilient sustainable societies and a fairer world.

The fears of wars about water are again threatened by climate change. Climate change unfortunately creates more human inequalities in relation to water availability. On researching the literature, broader ideas surrounding water management water supplies in developing countries have emerged including collaborating horizontally and vertically and encouraging both bottom up and top down approaches. The importance of working with existing infrastructure and making cities smarter and more resilient should be nurtured into the future

(Younos, 2017). The empowerment for all levels of society in relation to climate challenges is in line with education and training can ensure more water for all. Conserving our resources especially water, helps with water security. Precautionary land use can help to conserve ground waters and surface waters. Urban shift research strategies for urban living are based on the 5Ts namely: templates, technologies, transfer of knowledge tools and training which demonstrates the complexity of urban living (UrbanShift, 2021).

Interesting thoughts from the UN state that the 2020 UN World Water Development Report further emphasized that water is the "climate connector" that allows for greater collaboration across the majority of global targets for climate response, sustainable development, and disaster risk reduction.

Prior to the commencement of COP 26, the water cycle occurred with droughts, floods, and water stress which was considered an important topic surrounding climate change. The WMO emphasized the call for urgent integrated water and climate action policies, as climate change is disturbing the water cycle.

The final text from COP 26 Glasgow states the "important role" that "indigenous peoples, local communities, youth, children, local and regional governments and other stakeholders" play in tackling the climate crisis (COP 26, 2021).

> This again highlights citizen science's role and every level of societies in water which applies to developing counties and climate change.

The worry of the challenge of scarce, naturally renewable freshwater resources is a global problem that many countries face, especially those located in arid and hyper-arid zones within developing counties (Zhang, 2016). Education, innovation, logical thinking, and shared thinking is so important for the future development of water security in developing urban living needs in line with climate change.

We are all together when it comes to water and climate challenges. The most important sustainable development goal is goal number 17, a global partnership which involves learning and working together to solve our world crisis related to water supplies and climate.

References

Adger, W. N., Huq, S., Brown, K., Conway, D., & Hulme, M. (2003). Adaptation to climate change in the developing world. *Progress in Development Studies*, *3*, 179–195. Available from https://doi.org/10.1191/1464993403ps060oa.

Agarwal, S., Srivastava, A., & Kumar, S. (2010). Urban health in developing countries (2010). In M. C. Gibbons, R. Bali, & N. Wickramasinghe (Eds.), *Perspectives of knowledge management in Urban health* (p. 6194). Springer. Available at SSRN: https://ssrn.com/abstract = 2781296.

Ali, S. I. (2010). Alternatives for safe water provision in urban and peri-urban slums. *Journal of Water and Health*, *8*, 720–734.

Ali, S. I. (2010). Alternatives for safe water provision in urban and peri-urban slums. *Journal of Water and Health*, 8(4), 720−734. Available from https://doi.org/10.2166/wh.2010.1418.

AsLenzholzer, S., Carsjens, G.-J., Brown, R. D., Tavares, S., Vanos, J., Kim, Y., & Lee, K. (2020). Awareness of urban climate adaptation strateegies—An international overview. *Urban Climate*, 34100705.

Biswas, A. (1979). Water development in developing countries: Problems and prospects. *GeoJournal*, 3(5), 445−456. Retrieved June 29, 2021, from http://www.jstor.org/stable/41142293.

Boin, A., & Hart, P. (2003). Public Leadership in Times of Crisis. *Public Administration Review*, 63(5), 544−553.

Caldera, U., Bogdanov, D., & Breyer, C. (2016). Local cost of seawater RO desalination based on solar PV and wind energy: A global estimate. *Desalination*, 385, 207−216.

Clifton, C., Evans, R., Hayes, S., Hirji, R., Puz, G., & Pizarro, C. (2010). *Water and climate change: Impacts on groundwater resources and adaptation options*. Water Working Notes; No. 25. World Bank, Washington, DC. © World Bank. Online https://openknowledge.worldbank.org/handle/10986/27857 License: CC BY 3.0 IGO.

Cohen, B. (2006). Urbanization in developing countries current trends future projects and challenges for sustainability. *Technology in Society* (28), 63−80.

Cook, F., & Smith, L. (2005). *Catchment management relevant in developed and developing countries Waterlines 24 2−3 ment*.

COP 26. (2021). *Online HOME—UN Climate Change Conference (COP26) at the SEC − Glasgow 2021*. ukcop26.org. Accessed November 10, 2021.

Cullis, J. D. S., Horn, A., Rossouw, N., Fisher-Jeffes, L., Kunneke, M. M., & Hoffman, W. (2019). Urbanisation, climate change and its impact on water quality and economic risks in a water scarce and rapidly urbanising catchment: Case study of the Berg River Catchment. $H_2Open\ Journal$, 2(1), 146−167. Available from https://doi.org/10.2166/h2oj.2019.027.

Dagdeviren, H., & Robertson, S. (2012). Access to water in the slums of sub-Saharan Africa. *Development Policy Review*, 29(4), 485−505.

Danilenko, A., Dickson, E., & Jacobsen, M. (2010). *Climate change and urban water utilities: challenges and opportunities (English)*. Water working notes; note no. 24 Washington, DC: World Bank Group. http://documents.worldbank.org/curated/en/628561468174918089/Climate-change-and-urban-water-utilities-challenges-and-opportunities.

Delpla, A. V., Jung, E., Baures, M., & Clement, O. T. (2009). Impacts of climate change on surface water quality in relation to drinking water production. *Environmental International*, 35, 1225−1233.

Durmade, I. (2017). Sustainable water supply: An overview of water supply systems in some Nigerian peri-urban communities. *Journal of Economics and Sustainable Development*, 8, 10.

ECSSR. (2019). *Climate change and the future of water*. The Emirates Centre for Strategic Studies and Research, 2019. ISBN 9789948245414. https://search.ebscohost.com/login.aspx?direct = true&AuthType = sso&db = nlebk&AN = 2441098&scope = site. Accessed November 5, 2021.

Elimelech, M., & Phillip, W. A. (2011). The future of seawater desalination: Energy, technology, and the environment. *Science (New York, N.Y.)*, 333, 712−717. Available from https://doi.org/10.1126/science.1200488.

Esmail, B. A., & Suleiman, L. (2020). Analyzing evidence of sustainable urban water management systems: A review through the lenses of sociotechnical transitions. *Sustainability*, 12, 4481. Available from https://doi.org/10.3390/su12114481.

Ferreira, D. C., Graziele, I., Marques, R. C., & Gonçalves, J. (2021). Investment in drinking water and sanitation infrastructure and its impact on waterborne diseases dissemination: The Brazilian case. *Science of the Total Environment*, 779146279.

Ford, N. (2002). Battle over water privatisation. *African Business*, 272, 31−32.

Foucault, M. (1991). Governmentality. In G. Burchell, C. Gordon, & P. Miller (Eds.), *The foucault effect* (pp. 87−104). Chicago, IL: University of Chicago Press.

Franco-Torres, M., Briony, C., Rogers, B. C., & Harder, R. (2020). Articulating the new urban water paradigm. *Critical Reviews in Environmental Science and Technology*. Available from https://doi.org/10.1080/10643389.2020.1803686.

Grant, S. B., et al. (2012). Taking the waste out of wastewater for human water security and ecosystem sustainability. *Science (New York, N.Y.)*, *337*, 681–686.

Hadi, H., Mazdak, A., Travis, W., & Sybil, S. (2021). Effects of urban development patterns on municipal water shortage. *Frontiers in Water*, *3*, 77. https://www.frontiersin.org/article/10.3389/frwa.2021.694817. Accessed November 2, 2021.

He, C., Liu, Z., Wu, J., Pan, X., Fang, Z., Li, J., & Bryan, B. A. (2021). Future global urban water scarcity and potential solutions. *Nature Communications*, *12*, 4667.

Hoffmann, B. (2020). Resilient infrastructure is crucial when planning for an uncertain future. In *Inter-American Development Bank Blog*. https://cutt.ly/Qvdu9oW.

Human Development UN. (2021). *Online Human Development Index (HDI). Human Development Reports*. undp.org. Accessed November 18, 2021.

Huy, N.N., Văn Giải Phóng, T., Tyler, S., La Sen, S., & Hồng, L. (2015). *Rationalising peri-urban water supply in Can Tho*. Vietnam Published by IIED, December 2015 Asian Cities Climate Resilience Policy brief 2015.

IPCC. (2013). Summary for policymakers. In *Climate change 2013: The physical science basis*. Contribution of Working Group I to the Fifth Assessment Report of the Intergovernmental Panel on Climate Change [Stocker, T.F., D. Qin, G.-K. Plattner, M. Tignor, S.K. Allen, J. Boschung, A. Nauels, Y. Xia, V. Bex and P.M. Midgley (eds.)]. Cambridge University Press, Cambridge, United Kingdom and New York.

Jarimi, H., Powell, R., & Riffat, S. (2020). Review of sustainable methods for atmospheric water harvesting. *International Journal of Low-Carbon Technologies*, *15*(2), 253–276. Available from https://doi.org/10.1093/ijlct/ctz072.

Jenson, O., & Khalis, A. (2020). Urban water system: Development of micro level indications to support integrated policy. *PLoS One*, *15*(2)e0228295.

Karikari, K. (1996). *Water supply management in rural Ghana: Overview and case studies*. http://www.idrc.ca. Accessed September 12, 2008.

Koncagül, E., Tran, M., Connor, R., & Uhlenbrook, S. (2019). *The United Nations world water development report 2019: Leaving no one behind, facts and figures UNESCO World Water Assessment Programme*.

Kuddus, M. A., Tynan, E., & McBryde, E. (2020). Urbanization: A problem for the rich and the poor? *Public Health Reviews*, *41*(1). Available from https://doi.org/10.1186/s40985-019-0116-0.

Lennart, O., & Head, B. H. (2015). Urban water governance in times of multiple stressors: An editorial. *Ecology and Society*, *20*(1). http://www.jstor.org/stable/26269710. Accessed July 31, 2021. JSTOR.

Muller, M. (2015). Urban water security in Africa: The face of climate and development challenges. *Development Southern Africa*, *33*(1), 67–80. Available from https://doi.org/10.1080/0376835X.2015.1113121. 2016. [s. l.]. Disponível em: https://search.ebscohost.com/login.aspx?direct = true&AuthType = sso&db = bth&AN = 112508667&scope = site. Acesso em: 5 nov. 2021.

Macpherson, L., & Snyder, S. (2013). *Downstream—Context, understanding, acceptance: Effect of prior knowledge of unplanned potable reuse on the acceptance of planned potable reuse*. Alexandria, VA: Water Reuse Research Foundation.

Mathur, S. (2017). Public-private partnership for municipal water supply in developing countries: Lessons from Karnataka, India, Urban Water Supply Improvement Project. *Cities (London, England)*, *68*, 56–62.

Mitlin, D. Beard, V. A. Satterthwaite, J., & Du, J. (2019). Unaffordable and undrinkable Rethinking Urban water access in the Global South; working paper Washington D.C. World Resources Institute. Available from http://www.citiesforall.org [Accessed 19 November 2019].

NASA. (2017). *New night lights maps open up possible real-time applications online new night lights maps open up possible real-time applications*. NASA. Accessed October 15, 2021.

Newborne, P. (2016). Water for cities and rural areas in contexts of climate variability: Assessing paths to shared prosperity – The example of Burkina Faso. *Journal of Field Action Field Action Science Report Special Edition 14*. openedition.org. Accessed November 18, 2021.

Nioras, D., Kosmas Ellinas, K., Constantoudis, V., & Gogolides, E. (2021). How different are fog collection and dew water harvesting on surfaces with different wetting behaviors? *ACS Applied Materials & Interfaces*, *13*(40), 48322–48332. Available from https://doi.org/10.1021/acsami.1c16609.

OECD. (2014). *Managing water for future cities policy perspectives*. Available from https://www.oecd.org/environment/resources/Policy-Perspectives-Managing-Water-For-Future-Cities.pdf. Accessed June 29, 2021.

Patel, B., & Bhagat, S. (2019). Significance of public-private partnerships in India. *International Research Journal of Engineering and Technology, 6*(12), 1523–1528.

Saadat, A. H. M., Islam, M. S., Fahmida, P., & Sultana, A. (2018). Desalination technologies for developing countries: A review. *Journal of Scientific Research, 10*(1), 77–97.

Sinisi, L., & Aertgeerts, R. (2011). *Guidance on water supply and sanitation in extreme weather events WHO publication 2021*.

Tjandraatmadja, G., Stone-Jovicich, S., Muryanto, I., Suspawati, E., Gunasekara, C., Iman, M. N., & Talebe, A. (2012). *Tools for urban water management and adaptation to climate change*. CSIRO-AusAID Research for Development Alliance and CSIRO Climate Adaptation flagship, CSIRO, Australia.

Treacy, J. (2019). Drinking water treatment and challenges in developing countries. In N. Potgieter & A. N. T. Hoffman (Eds.), *The relevance of hygiene to health in developing countries*. IntechOpen. doi: 10.5772/intechopen.80780. Available from: https://www.intechopen.com/chapters/63707.

Treacy, J. (2021). Water stewardship as an aid to water productivity. *Water Productivity Journal, 1*(4), 53–66. Available from https://doi.org/10.22034/wpj.2021.280056.1037.

UN. (2014). *International decade for action 'water for life' 2005–2015*. https://www.un.org/waterforlifedecade/water_cities.shtml. Accessed June 2021.

United Nations. (2019). *World urbanization prospects*. New York.

UN-Habitat. (2010/2011). *State of the World's Cities 2010/2011- Cities for All: Bridging the Urban Divide*. UN-Habitat (unhabitat.org). unhabitat.org/state-of-theworlds-cities-for-all-bridging-the-urban-divide. Accessed October 5, 2021.

UN Report. (2018). 2018 United Nations DESA Population Division Licensed under creative common license by 3.0 IGO.

UrbanShift. (2021). UrbanShift. World Resources Institute (wri.org). www.wri.org/initatives/urbanshift. Accessed September 2021.

US Agency. (2009). *Addressing water challenges in the developing world: A framework for action*. Bureau of Economic Growth, Agriculture, and Trade U.S. Agency for International Development.

Vairavamoorthy, K., Gorantiwar, S. D., & Pathirana, A. (2008). Managing urban water supplies in developing countries – Climate change and water scarcity scenarios. *Physics and Chemistry of the Earth, Parts A/B/C, 33*(5), 330–339.

Waste Advantage Magazine. (2021). *Waste management practices in developing countries 2021*. Waste Management Practices in Developing Countries—Waste Advantage Magazine.

WHO/UNICEF. (2017). *Joint monitoring program for water supply, sanitation and hygiene (JMP)—2017 Update and SDG baselines*.

World Bank. (2017). Water supplies. https://www.worldbank.org/en/topic/watersupply.

Younos, T. (2017). *Challenges of urban water infrastructure*. https://www.waterworld.com/home/article/14070334/challenges-of-urban-water-infrastructure. Accessed July 29, 2021.

Zhang, X. Q. (2016). The trends, promises and challenges of urbanization in the world. *Habitat International, 54*(13), 241–252.

Further reading

Bhushan, B., & Feng, W. (2020). Water collection and transport in bioinspired nested triangular patterns. *Philosophical Transactions of the Royal Society A, 378*, 2019044120190441. Available from http://doi.org/10.1098/rsta.2019.0441.

Mc Granahan, G., Walnycki, A., Dominick, F., Kombe, W., Kyessi, A., Limbumba, T., Magambo, H., Mkanga, M., & Ndezi, T. (2016). *Universalising water and sanitation coverage in urban areas: From global targets to local realities in Dar es Salaam, and back*. IIED Working Paper. IIED, London. http://pubs.iied.org/10812IIED.

Huy, N., Phong, T., Tyler, S., La Sen, S., & Hong, L. (2015). Rationalising peri-urban water supply in Chan Tho Vietnam. *IIED Asian Cities Resilience Policy brief.*

Muller, M. (2007). Adapting to climate change: Water management for urban resilience. *Environment and Urbanization, 19*, 99–113.

Muller, M. (2009). *Towards the regulation of the competences of South Africa's water services managers.* WRC Report No TT 401/09, Water Research Commission, Pretoria. DEVELOPMENT SOUTHERN AFRICA 79.

Muller, M. (2012). Water management institutions for more resilient societies. *Proceedings of the ICE – Civil Engineering, 165*(6), 33–39.

Muller, M. (2016). Urban water security in Africa: The face of climate and development challenges. *Development Southern Africa, 33*(1), 67–80. Available from https://doi.org/10.1080/0376835X.2015.1113121. [s. l.]. https://search.ebscohost.com/login.aspx?direct = true&AuthType = sso&db = bth&AN = 112508667&scope = site. Acessed 5 nov. 2021.

Parkinson, S., et al. (2019). *Environmental Research Letters, 14* (1). © 2019 The Author(s). Published by IOP Publishing Ltd.

Parkinson, S., Krey, V., Huppmann, D., Kahil, T., McCollum, D., Fricko, O., Byers, E., Gidden, M. J., Mayor, B., & Khan, Z. (2019). Balancing clean water-climate change mitigation trade-offs. *Environmental Research Letters, 14*014009.

Pirouz, B., Palermo, S. A., & Turco, Michele (2021). Improving the efficiency of green roofs using atmospheric water harvesting systems (an innovative design). *Water, 13*(4), 546. Available from https://doi.org/10.3390/w13040546.

Rahaman, A. S., Everett, J., & Neu, D. (2013). Trust, morality, and the privatization of water services in developing countries. *Business & Society Review, 118*(4), 539–575. Available from https://doi.org/10.1111/basr.12021. (00453609), [s. l.]. Disponível em: https://search.ebscohost.com/login.aspx?direct = true&AuthType = sso&db = ofm&AN = 92692853&scope = site. Acesso em: 5 nov. 2021.

Sari, S. Y. I., Sunjaya, D. K., Shimizu-Furusawa, H., Watanabe, C., & Raksanagara, A. S. (2018). Water sources quality in urban slum settlement along the contaminated river Basin in Indonesia: Application of quantitative microbial risk assessment. *Journal of Environmental and Public Health, 2018*, 7, 3806537. Available from https://doi.org/10.1155/2018/3806537.

Satterthwaite, D., & Dodman, D. (2013). Towards resilience and transformation for cities within a finite planet. *Environment and Urbanization, 25*(2), 291–298.

Sen, L. S. (2013). *Technical Report No. 2: Additional Study on Water Pollution and Water Supply.* Report prepared for Rockefeller Foundation under project 2011 CAC311, Enhancing Can Tho City Resilience to Saline Intrusion caused by Climate Change, Can Tho Department of Natural Resources and Environment.

Syed, I. A. (2010a). Alternatives for safe water provision in peri-urban slums. *Journal of Water and Health.*

Syed, I. A. (2010b). Alternatives for safe water provision in urban and peri-urban slums. *Journal of Water and Health.*

UN. (2015). *UN World Water Development Report 2015: 'Water for a Sustainable World' UN-Habitat: New Urban Agenda.*

UN. (2021). *Take action for the sustainable development goals – United Nations sustainable development.* www.un.org/sustainabledevelopment/sustainable-development-goals. Accessed November 17, 2021.

UN-Water. (2021). *Summary progress update 2021: SDG 6—Water and sanitation for all.*

WHO/UNICEF. (2021). *JMP progress on household drinking water, sanitation and hygiene 2000–2020.*

WWAP (UNESCO World Water Assessment Programme). (2019). *The United Nations world water development report 2019: Leaving no one behind.* Paris: UNESCO.

CHAPTER 7

Water purity and sustainable water treatment systems for developing countries

Joanne Mac Mahon
Trinity College Dublin, Dublin, Ireland

Chapter Outline
7.1 Introduction: access to clean water in developing countries   115
7.2 Environmental challenges to water purity in developing countries   118
7.3 WHO guidelines for water purity   120
    7.3.1 Microbial guidelines   120
    7.3.2 Chemical guidelines   121
    7.3.3 Radiological guidelines   122
    7.3.4 Acceptability: taste, odor, appearance   122
    7.3.5 Other considerations   122
7.4 Water supply sources used in developing countries   123
7.5 Water treatment systems used in developing countries   124
    7.5.1 Centralized water treatment systems   124
    7.5.2 Decentralized water treatment systems   126
    7.5.3 Water Safety Plans   126
    7.5.4 Commonly used water treatment methods   128
    7.5.5 Water storage   132
7.6 Sustainable water management systems   133
7.7 Conclusions   135
References   136

7.1 Introduction: access to clean water in developing countries

A developing country is one with a relatively low standard of living, undeveloped industrial base, and moderate to low Human Development Index (Baumann, 2021), and these countries generally have poor access to safe drinking water supplies, causing detrimental damage to public health and economies (World Health Organization, 2017). To reduce the risk of death and disease through exposure to contaminated water, public health strategies in developing countries focus on improving access to safe water, sanitation and hygiene

(Baumann, 2021; World Health Organization, 2017). Safe drinking water of sufficient quality and quantity is required for drinking, food preparation, and personal hygiene and is defined as water with microbial, chemical, and physical characteristics that meet WHO guidelines, and which "does not represent any significant risk to health over a lifetime of consumption" (World Health Organization, 2017). The international community has committed to improvements in water quality and access to safe drinking water and sanitation globally through various sustainability initiatives including the Millennium Development Goals in 2000 (United Nations, 2015); declarations by the United Nations (UN) General Assembly in 2010 that safe and clean drinking water and sanitation is a human right, essential to the full enjoyment of life and all other human rights (United Nations General Assembly, 2010); declarations of the periods 2005−15 and 2018−28 as International Decades for Action, "Water for Life" (United Nations Office to support the International Decade for Action, 2015) and "Water for Sustainable Development" (United Nations, 2018) respectively; and the adoption of the Sustainable Development Goals (SDGs) by countries in 2015 (United Nations, 2021). SDG 6 in particular aims to achieve universal access to safe drinking water by 2030 and includes targets and indicators for sanitation, water quality, water resource use and scarcity, and water resource management (UN Water, 2018). In the context of these commitments, WHO and UNICEF's Joint Monitoring Program (JMP), which tracks global access to safe drinking water, reports that 2 billion people gained access to safely managed water services between 2000 and 2020 and 2.4 billion people gained access to safely managed sanitation services [World Health Organization (WHO)/ United Nations Children's Fund (UNICEF), 2021], while the global population increased by 1.7 billion people during the same period.

Despite this significant progress, huge inequalities persist between and within countries and almost half of people drinking water from unsafe sources live in sub-Saharan Africa, eight in ten live in rural areas, and there are large gaps between wealthy and poorer populations [United Nations Children's Fund (UNICEF) and World Health Organization (WHO), 2020; World Health Organization (WHO), 2019; World Health Organization (WHO)/ United Nations Children's Fund (UNICEF), 2017a]. The JMP reports that the proportion of the global population using safely managed water services in 2020 was 74%, with urban coverage at 86% and rural coverage at 60% [World Health Organization (WHO)/ United Nations Children's Fund (UNICEF), 2021]. However, 2 billion people still lacked access to safely managed water services, including 771 million without even a basic level of service, 122 million of whom collected drinking water directly from surface water sources [World Health Organization (WHO)/ United Nations Children's Fund (UNICEF), 2021]. 3.6 billion people lacked access to safely managed sanitation services in 2020 [World Health Organization (WHO)/ United Nations Children's Fund (UNICEF), 2021], which in turn significantly impacts local water quality and the risk of exposure to pathogens. Sanitation in this context is defined as facilities that ensure hygienic separation

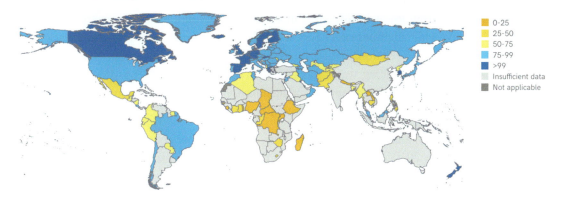

Figure 7.1
Proportion of population using safely managed drinking water services in 2020 (shown as percentage). Source: *Image reproduced from WHO/UNICEF Joint Monitoring Programme (JMP) for Water and Sanitation. Progress on household drinking water, sanitation and hygiene 2000–2020: Five years into the SDGs. 2021. Page 8. Copyright WHO/UNICEF 2021. Reprinted with permission.*

of human excreta from human contact [World Health Organization (WHO)/ United Nations Children's Fund (UNICEF), 2016]. Fig. 7.1 shows the percentage of population by country with access to a safely managed drinking water supply in 2020. Achieving universal access to safely managed water services by 2030 (SDG 6) will require a fourfold increase in current rates of progress (ten-fold in least developed countries) [World Health Organization (WHO)/ United Nations Children's Fund (UNICEF), 2021].

The impacts of lack of access to safe drinking water in developing countries are stark. Contaminated water is a major public health risk, transmitting diseases such diarrhea, cholera, dysentery, typhoid, polio, and other waterborne diseases. An estimated 485,000 diarrheal deaths occur each year as a direct result of contaminated drinking water and the number increases to 829,000 deaths (297,000 under the age of five years) if the impacts of inadequate sanitation and hand hygiene are also accounted for (World Health Organization (WHO), 2019). In addition, millions of people suffer from high intensity intestinal helminth infections (e.g., Ascariasis, Trichuriasis, and Hookworm disease), which often leads to cognitive impairment, severe dysentery, or anemia (Sanctuary et al., 2007). This enormous disease burden causes huge economic disadvantages for people in developing countries through lost working hours (Malik et al., 2012) and access to safe drinking water is a critical component of sustainable development at national, regional and local levels. It has been estimated that universal access to drinking water in developing countries could be equivalent to economic gains of on average 15% in terms of GDP (Frontier Economics. Exploring the links between water & economic growth, 2012) but more for countries significantly behind in water access. Successful water supply interventions can therefore be an effective part of poverty alleviation strategies (World Health Organization, 2017), reducing the long-term need for economic assistance. Every $1 invested in water and sanitation provides a

$4 economic return from lower health costs, more productivity and fewer premature deaths (Hutton, 2012) and while approximately $260 billion is lost globally each year due to lack of basic water and sanitation (Hutton, 2012), the World Bank estimates the annual capital costs of meeting SDG targets for drinking water and sanitation (SDGs 6.1 and 6.2) as US$114 billion per year, excluding costs for operation and maintenance, monitoring, institutional support, sector strengthening and human resources (Hutton & Varughese, 2016).

In addition to the health and economic implications for developing countries, the onset of the Covid-19 pandemic in 2019/2020 has highlighted the urgency of achieving universal access to safe drinking water, sanitation and hygiene as essential elements of any global pandemic prevention and mitigation strategy, and therefore has significant public health implications for all countries (Organisation for Economic Co-operation & Development, 2020).

7.2 Environmental challenges to water purity in developing countries

The challenges of improving access to water of sufficient quality and quantity in developing countries is greatly exacerbated by the global problem of water stress or water scarcity. Water scarcity can refer to the lack of available water in nature to meet requirements or to the lack of access to water despite a relative abundance of water resources, due to institutional, economic and technological constraints (Dell'Angelo et al., 2018). Water availability for human abstraction relies on the dynamics of the hydrological cycle but these are disrupted through human activities such as large scale extractions and dam construction and increasingly by the environmental impacts of climate change (Huber & Gulledge, 2011), land use changes, deforestation, urban sprawl, and area waterproofing (Howard et al., 2010, 2016). Water extractions increased sixfold during the 20th century (Vörösmarty et al., 2005), leading to water scarcity and a deterioration in the quality of accessible water in many regions (Pichel et al., 2019; Wada et al., 2011). 2.3 billion people now live in water-stressed countries (UN Water, 2021), with 4 billion people experiencing severe water scarcity during at least one month of the year (Mekonnen & Hoekstra, 2016). Fig. 7.2 shows global baseline water stress by country in 2019, with 17 countries (equivalent to one-quarter of the world's population) facing "extremely high" levels of baseline water stress, where irrigated agriculture, industries and municipalities withdraw more than 80% of their available supply on average every year (Hofste et al., 2019). Forty-four countries (equivalent to one-third of the global population) face "high" levels of stress, where on average more than 40% of available supply is withdrawn every year (Hofste et al., 2019).

Climate change is now a major source of disruption to the water cycle and is expected to further affect water supply and demand (Hagemann et al., 2013), with some areas globally seeing increased rainfall and others increased drought (Haddeland et al., 2014). The past four decades have been hotter than any others previously recorded (Lindsey & Dahlman, 2021) and very dry areas across the globe have doubled in extent since the 1970s

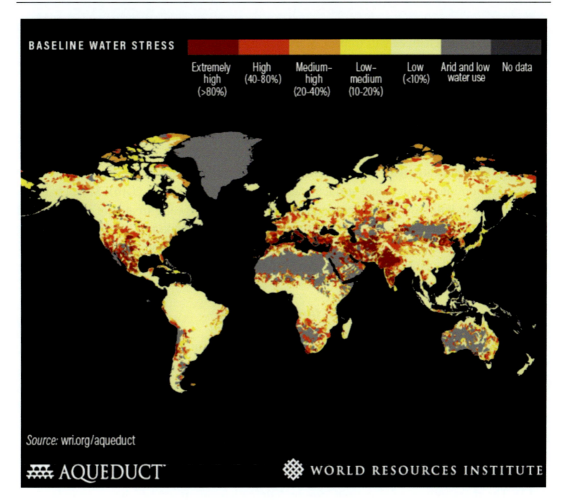

Figure 7.2
Baseline global water stress by country in 2019. Source: *Image reproduced from, Hofste RW, Reig P, Schleifer L. 17 Countries, Home to One-Quarter of the World's Population, Face Extremely High Water Stress. World Resources Institute. 2019, Aug 6. Copyright World Resources Institute 2019. Reprinted with permission.*

(Dai, 2011a,b), with persistent drying trends evident in many developing countries, particularly in Africa, East and South Asia, and northern South America (Dai, 2011a,b). Climate related disasters including flash floods, prolonged droughts, heatwaves, wildfires and hurricanes have also increased dramatically from 3656 between 1980 and 1999 to 6681 between 2000 and 2019 (Mizutori & Guha-Sapir, 2020). Floods and storms were the most frequent events (Mizutori & Guha-Sapir, 2020), causing widespread damage and loss of access to water infrastructure on a regular basis in developing and developed countries (Howard et al., 2010, 2016), thus increasing the risks of waterborne disease.

In addition to the increasing extreme weather events caused by climate change, other long-term ambient changes will impact water quality including elevated temperatures, changes in precipitation patterns, changes in soil drying-rewetting cycles, increased evaporation, water flow and recharge changes, warming oceans and increased solar radiation intensity (Howard et al., 2010, 2016). For example, drought can concentrate nutrient loads and pollutants in reduced surface waters (Howard et al., 2010, 2016; Vörösmarty et al., 2005), since the dilution effect is inversely related to the amount of water in circulation (Vörösmarty et al., 2005). In the case of heavy rainfall, increased loads of suspended solids are likely in surface waters and changing sea water levels could lead to saline intrusion into aquifers and surface waters (Howard et al., 2010, 2016). Such changes in water quality will have serious consequences for traditional water treatment systems used in developing countries and will make the production of safe drinking water more expensive and difficult, requiring more sophisticated technologies and management systems (García-Ávila et al., 2021; Howard et al., 2010, 2016; Verlicchi & Grillini, 2020).

7.3 WHO guidelines for water purity

Drinking water standards vary among countries and regions and no single approach is universally applicable (World Health Organization, 2017). Developing countries in particular may have very limited resources to allocate to water projects and management. Current WHO guidelines for drinking water quality support efforts to ensure safe collection, treatment, and storage of drinking water. Water sources used for drinking purposes can be contaminated microbiologically and chemically, either through natural processes or through human and agricultural activities and exposure can cause acute and chronic health impacts, while also impacting the taste, odor and appearance of water (World Health Organization, 2017). The main sources of contamination include pathogenic microbes, heavy metals, organic substances and inorganic chemicals (Nzung'a et al., 2021). Fecal contamination is of particular concern, often occurring due to inadequate protection of the water source from human or animal feces (including birds), damaged water distribution systems, unhygienic practices of the community at the source, and poor household handling and storage practices (Gizachew et al., 2020).

The areas of contamination covered by WHO guidelines include microbial, chemical, radiological and issues of consumer acceptability (World Health Organization, 2017) and will be outlined briefly here.

7.3.1 Microbial guidelines

Infectious diseases caused by pathogenic bacteria, viruses and parasites (such as protozoa and helminths) are the most common health risk associated with drinking water in developing countries (World Health Organization, 2017) and are generally encountered through ingestion of water contaminated with feces. Fecal contamination of drinking water is monitored using fecal indicator bacteria (World Health Organization, 2017), with *Escherichia coli* the preferred indicator, followed by thermotolerant coliforms (TTC) (World Health Organization, 2017). In the majority of cases, monitoring for *E. coli* or TTC provides a high degree of assurance because of their large numbers in polluted waters and the aim of water treatment and supply systems should be to provide drinking water that contains no fecal indicator organisms. WHO provides quality guidelines indicating that *E. coli* or TTC should not be detectable in any 100 mL sample for all water directly intended for drinking, treated water entering the distribution system and treated water in the distribution system. However, in many developing countries, a high proportion of household and small community drinking water systems, in particular, fail to meet requirements for water safety, including the absence of *E. coli*. In these situations, WHO recommends that realistic goals for improvement are identified and implemented, including grading schemes for water safety linked to priority for action, which are useful for community and household supplies where the frequency of testing is low and therefore reliance on analytical results alone is inappropriate (World Health Organization, 2017).

E. coli and TTC are not ideal indicators for the presence of enteric viruses and protozoa, which are more resistant to conventional treatment technologies, including filtration and disinfection, and may be present in treated drinking water in the absence of *E. coli* (World Health Organization, 2017*)*. Therefore the inclusion of more resistant microorganisms, such as bacteriophages and/or bacterial spores, as indicators of persistent microbial hazards may be appropriate in some cases (World Health Organization, 2017). Other microbial hazards, such as guinea worm (*Dracunculus medinensis*), toxic cyanobacteria and Legionella, may also be of public health importance (World Health Organization, 2017).

It is important to note that microbial contamination of source water may not be constant throughout the year and water quality parameters in surface waters (Ouyang et al., 2006), unimproved drinking water sources (Wright, 1986), and improved drinking water sources (Dongzagla et al., 2021; Kostyla et al., 2015) have been shown to follow seasonal patterns. (See section 7.4 for definition of improved drinking water source). Although fecal contamination in improved drinking water sources tends to be greater during the wet season, drinking water surveys to track quality are often conducted in the dry season for logistical reasons (Kostyla et al., 2015) and so the true extent of contamination may not be reported. These issues are of increasing importance as the patterns and severity of wet and dry seasons are affected by climate change.

7.3.2 Chemical guidelines

The health risks associated with chemical constituents are mostly concerned with their ability to cause adverse health effects after prolonged periods of exposure. Less important in the context of water management in developing countries is the potential for accidental contamination of a drinking water supply which would cause problems with a single exposure. World Health Organization (2017) outlines acceptable levels of chemical constituents of concern, which could include harmful chemicals from industrial and human activities, including pesticides, fertilizers and heavy metals and naturally occurring chemicals and minerals such as arsenic, salts and fluorides. Arsenic in particular is present in high levels naturally in the groundwater of many South East Asian countries, notably India and Bangladesh, as well as regions of South America (Kabir & Chowdhury, 2017; Ng et al., 2003) and must be removed to prevent chronic arsenic poisoning (arsenicosis). Large scale arsenic poisoning of populations has occurred in Bangladesh and some regions of India and continues to be a huge challenge with many people still consuming considerably more than the 10 μg L^{-1} advised by WHO (Kabir & Chowdhury, 2017; Ng et al., 2003). Reducing human exposure to heavy metals can be a particular challenge in developing countries, due to their limited economic capacities to use advanced technologies for heavy metal removal (Chowdhury et al., 2016).

7.3.3 Radiological guidelines

WHO advises that the health risks associated with the presence of naturally occurring radionuclides in drinking water be considered but the contribution of drinking water to total exposure to radionuclides is very small under normal circumstances (World Health Organization, 2017).

7.3.4 Acceptability: taste, odor, appearance

The appearance, taste and odor of drinking water should be acceptable to the majority of consumers (World Health Organization, 2017). Water that is esthetically unacceptable can lead to the use of water from sources that are esthetically more acceptable, but potentially less safe (World Health Organization, 2017). Naturally occurring contaminants which impact color, odor and taste but which are not harmful at low concentrations include particulate matter, humic material, zinc or iron (Treacy, 2019). Turbidity is measured in standardized nephalometric turbidity units (NTUs) and high turbidity interferes with the effectiveness of disinfection by chlorine, ozonation, and ultra-violet (UV) light. The mean turbidity of water being treated with these disinfection methods should be below 1 NTU, with no single sample having turbidity exceeding 5 NTU (World Health Organization, 2017).

7.3.5 Other considerations

In addition to the quality of drinking water in terms of safety for human use, supply must also be reliable, of sufficient quantity, accessible and available at a cost which allows families to meet basic domestic needs. Thus issues of quantity, continuity of supply, accessibility and affordability are very important (World Health Organization, 2017). Of particular importance is the distance to a safe drinking water source, as significant time can be spent fetching and carrying water long distances. In particular, this takes a large toll on women and children as they are often responsible for collecting water, impacting time spent in work, education, and caring for family [United Nations Children's Fund (UNICEF), 2016].

In terms of overall water quality indicators, WHO advocates incremental improvement towards long-term water quality targets, which is an important concept in the allocation of resources to improving drinking water safety. This involves prioritizing the most urgent problems (e.g., protection from pathogens), while also setting long-term targets of further water quality improvements (e.g., improvements in taste, odor and appearance) (World Health Organization, 2017).

7.4 Water supply sources used in developing countries

Water supply solutions used in developing countries are different to those of economically advanced countries, where reliable and safe piped water supply is the norm in urban and most rural households. Water sources may include boreholes, open wells, surface water such as rivers and lakes, brackish waters, and rainwater, and supply options are often restricted by insufficient resources to access and treat the water. In urban settings, piped and bottled water may form part of the water supply used by households. Point sources such as wells, boreholes, and springs are vulnerable to pollution and usually have to be shared between many households, which makes their management and maintenance difficult (Thomas, 2014).

Global access to safe drinking water is tracked by the JMP as the proportion of the population using an improved drinking water source, which is one that, "by nature of its construction, adequately protects the source from outside contamination, particularly fecal matter" [World Health Organization WHO/ United Nations Children's Fund UNICEF, 2015]. This indicator is a proxy for safety and does not account for actual contamination, as defined by the WHO guidelines (Kostyla et al., 2015) and what qualifies as "improved" drinking water sources under the international monitoring criteria do not necessarily correspond to potable water sources (Sorlini et al., 2013); hence, reported estimates for the percentage of populations with access to safe drinking water may be overstated (UNESCO, 2015). Until 2017, JMP data reflected access to an improved technology but did not provide details on the level of service in terms of quality and quantities supplied or the accessibility and reliability of the supplies, which are critical features of a safe and secure water service (Moriarty et al., 2013). Since 2017, however, the JMP has introduced a more ambitious indicator for SDG monitoring that takes account of accessibility, availability and quality of drinking water [World Health Organization (WHO)/ United Nations Children's Fund (UNICEF), 2017b].

Improved drinking water sources defined by the JMP include piped water into a dwelling, yard or plot and other sources of drinking water likely to be protected from outside contamination, particularly fecal matter, including a public tap or stand-pipe, a tube-well or borehole, a protected dug well, a protected spring, or rainwater collection [World Health Organization (WHO)/ United Nations Children's Fund (UNICEF), 2021]. Unimproved drinking water sources include an unprotected dug well, an unprotected spring, a cart with small tank or drum, a tanker truck, a surface water source (river, dam, lake, pond, stream, canal, irrigation channel), or bottled water [World Health Organization (WHO)/ United Nations Children's Fund (UNICEF), 2021]. In practice, most households in developing countries will rely on a number of water sources to meet their needs, the combination of which may change seasonally. Bottled water, in both bottles and plastic sachets, has become a significant source of drinking water in some developing countries and is considered an improved drinking water source only when another improved source is also used for cooking and personal hygiene [World Health Organization (WHO)/ United Nations Children's Fund (UNICEF), 2008]. Bottled water can present significant environmental hazards in terms of plastic pollution.

7.5 Water treatment systems used in developing countries

Water is treated to remove pathogenic microorganisms, to decrease turbidity, to eliminate taste and odor and to reduce or eliminate hazardous or undesirable chemicals such as arsenic. The type of treatment required will depend on the water source and generally surface waters such as lakes and rivers will require more treatment than groundwater, which may require little or no treatment (Howard et al., 2006). The goal of water treatment in developing countries is to reduce the occurrence of waterborne disease but it is often difficult to distinguish the direct cause of diarrheal infection due to the complex interrelationships between water, sanitation and hygiene. For example, if water is not readily available, people may decide handwashing is not a priority, thus increasing the risk of diarrhea and other diseases (World Health Organization (WHO), 2019). Thus focusing on drinking water quality alone may not be enough to reduce disease prevalence and improve outcomes for developing countries (Clasen et al., 2007, 2015). It is important to emphasize therefore that water, sanitation and hygiene measures need to be addressed together in order to have sustainable public health results and prevention strategies such as water treatment, education in safe storage of drinking water, and education in improved sanitation techniques, when used together can significantly reduce risk of exposure (Omarova et al., 2018).

7.5.1 Centralized water treatment systems

Water treatment systems can be classified as *centralized or decentralized systems*. Centralized water treatment systems, typically used in urban areas, involve abstraction of water from a large (often distant) water source such as a river, lake, or groundwater and treatment at a central location, followed by distribution to the consumer via dedicated distribution networks. Treatment capacity tends to be medium to large scale, serving the needs of towns and cities and they generally use conventional treatment processes such as a combined process of coagulation, flocculation, sedimentation, filtration, and disinfection as follows:

1. Coagulation and flocculation help remove particles through the addition of alum (or other metal salts) to form coagulated masses called floc that attract other particles.
2. Sedimentation of the coagulated, heavy particles occurs through gravity.
3. Filtration of the water fraction (following sedimentation) is carried out through sand, gravel, coal, activated carbon, or membranes to remove smaller solid particles not already removed.
4. Disinfection by the addition of chlorine destroys or inactivates microorganisms and an adequate residual is generally ensured to maintain disinfection during storage and distribution (Water Quality & Health Council, 2017). (See schematic diagram of water treatment process in Fig. 7.3).

The centralized water treatment approach is well established and can effectively remove raw water turbidity along with harmful pathogens, including bacteria, virus, and protozoa. Challenges of centralized systems which are particularly relevant for developing countries include high capital and operating costs, the significant maintenance requirements of the large infrastructure, and prevention of network contamination (Holler, 2003; Pain). Centralized systems are generally managed by public and private companies that are

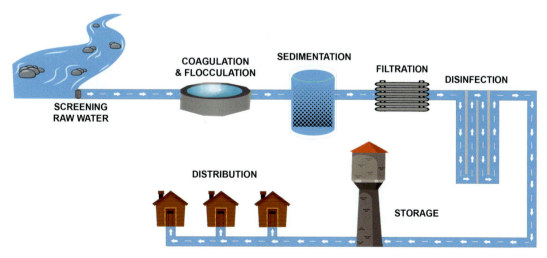

Figure 7.3
Conventional centralized drinking water treatment system.

subjected to some regulatory control and are more common in urban areas. However, urban populations in developing countries are becoming increasingly reliant on decentralized water treatment options, due to inadequate and deteriorating centralized water supply infrastructure (Pain).

7.5.2 Decentralized water treatment systems

Decentralized water supply refers to the small-scale purification and distribution of water. Decentralized treatment systems fall into three main categories: point-of-use systems (POU), point-of-entry systems (POE), and small-scale systems (SSS). POU and POE systems are designed for individual households, while SSS can provide for community water supply, for emergency water supply in camps, or can be used to purify water for sale in water kiosks [World Health Organization (WHO), 2021]. Decentralized water treatment technologies must be effective in terms of water quality and quantity, easy to use, have relatively low maintenance and operational requirements, be economically viable, environmentally sustainable and socioculturally acceptable (Nzung'a et al., 2021) and the choice of decentralized supply system depends on the combination of these factors in the local context [World Health Organization (WHO), 2021]. Decentralized systems for small communities require some form of community management to operate and maintain the system, while household treatment systems are owned and operated by individual families.

Household water treatment and safe storage (HWTS) is an important public health intervention among those who rely on water from unimproved sources or from unreliable / unsafe piped water supplies (Clasen, 2015). These systems provide affordable, convenient ways for households to obtain safe drinking water and can also be an effective emergency response intervention [Clasen, 2015; Dangol & Spuhle; World Health Organization (WHO)/ United Nations Children's Fund (UNICEF), 2011]. Several HWTS methods, including filtration, chemical disinfection, disinfection with heat (boiling, pasteurization) and flocculants / disinfectants, have been proven to significantly improve drinking water quality in the laboratory and in field trials in developing countries (Clasen et al., 2015; World Health Organization (WHO), 2012). These HWTS methods are illustrated in Fig. 7.4 [World Health Organization (WHO), 2012] and a combination of methods may be used to increase the efficacy of treatment. Education and training are essential to ensure household methods are employed correctly (World Health Organization, 2017). Although there are many advanced small-scale water treatment technologies available, the technologies implemented in low cost decentralized treatment systems are basic and focus mainly on the removal of waterborne pathogens (Pooi & Ng, 2018).

7.5.3 Water Safety Plans

WHO sets guidelines for the performance of centralized and decentralized water treatment systems and ensuring the microbial safety of drinking water supplies is based on the use of

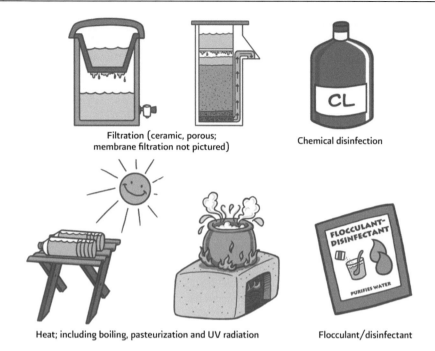

Figure 7.4
Household water treatment system (HWTS) methods proven to improve water quality. Source: Image reproduced from World Health Organization (WHO). A toolkit for monitoring and evaluating household water treatment and safe storage programs. WHO. Geneva. 2012. Page 8. Original images from Centre for Affordable Water and Sanitation Technology (CAWST). Copyright WHO 2012. Reprinted with permission.

multiple barriers, from catchment to consumer, including protection of water resources; proper selection and operation of a series of treatment steps; and management of distribution systems (piped or otherwise) to maintain and protect treated water quality (World Health Organization, 2017). WHO advocates using a risk management approach that places the primary emphasis on preventing or reducing the entry of pathogens into water sources and reducing reliance on treatment processes for removal of pathogens (World Health Organization, 2017). Central to this is the design of a water safety plan [Bartram, 2009; World Health Organization (WHO), 2012] as shown in Fig. 7.5 (String & Lantagne, 2016) and WHO provides guidelines on systematic assessment of risks throughout the drinking water supply chain and identification of the ways in which these risks can be managed on a day-to-day basis and in the event of upsets and failures (World Health Organization, 2017). Monitoring of water quality is a challenge in developing countries due to limited ability to carry out quality checks and is particularly difficult for household systems, which rely on individual households following safety guidelines to prevent contamination (World Health Organization, 2017).

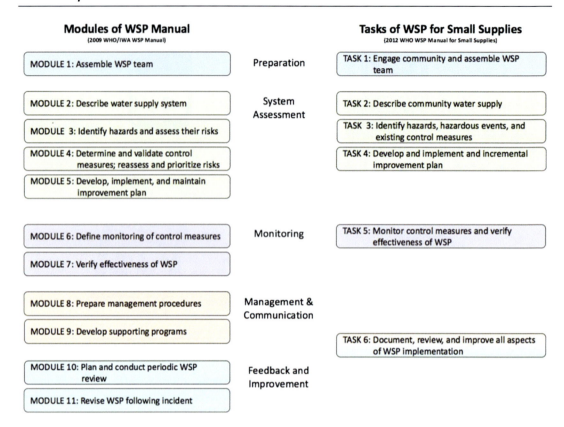

Figure 7.5
Steps of the Water Safety Plan process and associated modules and tasks. Source: *Image reproduced from String G, Lantagne D. A systematic review of outcomes and lessons learned from general, rural, and country-specific Water Safety Plan implementations. Water Science and Technology: Water Supply. 2016 Dec;16(6): Page 1581. Copyright IWA 2016. Reprinted with permission.*

7.5.4 Commonly used water treatment methods

Details of water treatment methods commonly used in developing countries are outlined briefly here, with an emphasis on sustainability.

1. *Traditional coagulation and flocculation* systems, used in both centralized and decentralized treatment systems, have low energy requirements and are reliable. They are mainly used for highly turbid water but cannot be used as a standalone water treatment solution and must incorporate some level of disinfection such as chlorination (Pooi and Ng, 2018). Natural coagulants used for household sedimentation include *Moringa oleifera* seeds, and protein from red bean, sugar maize and red maize (Gunaratna et al., 2007).

2. *Disinfection* is the inactivation and destruction of microorganisms to a safe level, traditionally achieved through chlorination in both centralized and decentralized systems (Pooi & Ng, 2018). For household treatment, chlorination requires the addition of sodium hypochlorite solution to clear water, which is mixed thoroughly by agitating, and water is ready for consumption in about 30 minutes (Gallo & Lantagne, 2008). It inactivates most bacteria and viruses that cause diarrheal disease but does not remove protozoa, such as cryptosporidium (Mengistie et al., 2013; Sobsey et al., 2003). Chlorination is inexpensive, and easy to use but it may cause potential long-term health issues, such as colorectal cancers, it changes to the taste and color of water, and is not effective in turbid water (Botlagunta et al., 2015; Zinn et al., 2018).
3. *Solar Water Disinfection (SODIS)* uses UV solar radiation to kill pathogens in water and involves filling a clear standard 2-liter PET bottle (or suitable PET bag) and exposing it to sunlight for at least 6 hours (Luzi et al., 2016), as shown in Fig. 7.6 (longer exposure is required in cloudy conditions). SODIS is an inexpensive and easy method to improve the quality of drinking water in a household and has proven to be effective in reducing the occurrence of diarrheal disease in numerous studies (Bitew et al., 2018; Conroy et al., 1999, 2001; McGuigan et al., 2012). Placing the bottles on reflective backings, such as corrugated rooftops, can improve SODIS performance

Figure 7.6
SODIS steps and SODIS bottles on a corrugated iron sheet. Source: *Images reproduced from Luzi S, Tobler M, Suter F, Meierhofer R. SODIS manual: Guidance on solar water disinfection. Eawag, Switzerland. 2016. Pages 8 & 12. Reprinted with permission from Eawag.*

(Amin & Han, 2009). Regrowth of bacteria can occur following solar disinfection but incorporation of a photocatalyst such as titanium dioxide in the process can prevent this (Gelover et al., 2006; Mac Mahon et al., 2017; Misstear & Gill, 2012). Continuous flow solar disinfection systems have been tested and could potentially be used as community water treatment systems (Chauque et al., 2021; Gill & Price, 2010; Mac Mahon & Gill, 2018) and combining solar disinfection with filtration can improve the outcome of both methods (Nzung'a et al., 2021).

4. *Pasteurization* involves heating water to 63°C for at least 30 minutes or to over 72°C for at least 15 seconds. It can be achieved by boiling but this is very energy intensive and has potential for particulate emissions depending on the heat source used. It can also be carried out using solar radiation with a simple and inexpensive solar pasteurization box (Tratschin et al.; Pejack, 2011). Issues of sustainability for solar pasteurization in terms of life cycle costs of the pasteurization box (Burch & Thomas, 1998) are improving with newer technology (Rossi et al., 2019).

5. *Filtration* removes colloids, suspended solids, and microorganisms from drinking water by size exclusion, whereby components larger than the pore size of the filter will be retained within the system (Pooi & Ng, 2018). Commonly used filtration systems in developing countries include ceramic, slow sand, biosand, rapid filtration and membrane filters. *Ceramic filters* are affordable, low energy, low maintenance, easy to use and produce acceptable water in terms of taste, appearance and temperature (Clasen, 2015; Clasen et al., 2015; Sobsey et al., 2008; Yang et al., 2020; Zinn et al., 2018). They are not effective in removing viruses but the use of chitosan coagulation as a pretreatment has been shown to increase virus and bacteria reductions (Abebe et al., 2016). Silver-modified ceramic filters can be effective at removing protozoan (Abebe et al., 2014, 2015). *Rapid and slow filters* reduce turbidity and can achieve large reductions in bacterial and viral contamination, as well as removing larger biological contaminants such as cryptosporidium, giardia, amoebae, and parasite eggs (Gadgil, 1998). Any given design of a filter will have inherent limits on the level of turbidity and total suspended solids it can treat (Gadgil, 1998; Zinn et al., 2018). *Biosand filtration* is a slow sand filter adapted for use in the home and is one of the simplest filtration systems to use (Ngai et al., 2014). The products needed for biosand filtration can be made locally, at a low cost and they have a long life span (Zinn et al., 2018). Biosand filters can remove turbidity and pathogens but require constant maintenance to ensure effectiveness (Ngai et al., 2014). *Membrane filtration* involves the use of a single layer of material (i.e., the membrane), usually made of woven fibers, ceramics, polymeric or metallic materials (Hoslett et al., 2018). Membrane filters can be produced and adapted to filter out almost any substance, including pathogens, arsenic, and other harmful chemical pollutants, but high costs are currently a barrier for use in developing countries (Hoslett et al., 2018; Zinn et al., 2018).

6. *Arsenic removal* methods used in developed countries are generally too expensive to be employed in developing countries but low cost solutions include flocculation and sedimentation, filtration, precipitation, adsorption (Kabir & Chowdhury, 2017) and solar oxidation (Mukherjee et al., 2007; O'Farrell et al., 2016; Wegelin et al., 2000). *Heavy metals* and metalloids are commonly removed from wastewater and surface water using granular activated carbon, forward osmosis, reverse osmosis, ultra- and nanofiltration but high costs prevent these treatment options being used widely in developing countries (Chowdhury et al., 2016; Hoslett et al., 2018).

7. *Rainwater* has been used throughout history to fulfill water demands (Sivanappan, 2006) and while some locations receive more useful quantities of rainfall than others, access to this resource is generally unrestricted and its high quality makes it suitable for most household uses (UNEP, 2009). Rainwater can be harvested in both rural and urban settings to provide a drinking water source (using rooftop run off stored in storage tanks) or to augment groundwater sources, which subsequently can be used for drinking water or agricultural purposes. In areas where rainfall is seasonal, rainwater can provide an important water source during dry periods if harvested effectively, and rainwater can provide a safe local alternative in areas where local supply is contaminated or unavailable. There are many options for augmenting groundwater with rain in rural settings including percolation tanks, check dams, recharge shafts, dug well recharge, and groundwater dams (Sivanappan, 2006; UNEP, 2009). In urban settings, rooftop rainwater and run off can be used to augment groundwater via recharge pits and trenches, tube-wells and recharge wells, and can also serve to manage stormwater effectively (Sivanappan, 2006; UNEP, 2009). Rooftop rainwater harvesting is increasingly practiced in areas with very different rainfall patterns and needs such as Kenya, India, and Brazil (UNEP, 2009). Problems with rooftop rainwater harvesting for use as drinking water which must be addressed to ensure sustainability include insufficient or unreliable rainfall volumes; excessive installation costs; difficult water management; uncertain water quality; and poor installation, maintenance, and longevity (as shown in Fig. 7.7). In some cases rainwater may be more useful as a supplemental rather than main drinking water supply source (Thomas, 2014).

8. *Reuse of wastewater* to recover water, nutrients, or energy, is becoming an important water supply strategy, with many potential benefits for water provision and management [World Health Organization (WHO), 2019]. Greywater may be defined as household wastewater which does not have any input from toilets, that is, without urine, feces and toilet paper, such as wastewater from bathing or kitchen use (Allen et al., 2010). Greywater may also be defined as low polluted wastewater since the concentration of microorganisms and some nutrients (e.g., nitrogen and potassium) is lower than in combined wastewater (Warner, 2006). Greywater can be collected at source in the household, before it is mixed with other pollutants, and may be reused on-site with the installation of relatively simple technologies, leading to a more efficient

Figure 7.7
Rooftop rainwater harvesting system in a state of disrepair at a school in rural Kenya (author's image).

use of available water. Greywater reuse systems range from simple installations at the household level to more complex installations that collect water from a cluster of buildings, a neighborhood or a municipality. In developing countries, untreated greywater may be reused in urban and rural agriculture as a means for improving food and water security and ultimately, alleviating poverty (Domènech, 2011; Oteng-Peprah et al., 2018; Warner, 2006). Currently wastewater is used for irrigation in 7% of irrigated land in developing countries and while this practice may pose health risks if not managed appropriately, safe management of wastewater can yield multiple benefits, including increased food production [World Health Organization (WHO), 2019].

New water supply and treatment technologies such as removal of water from humid air (Salehi et al., 2020) and biological drinking water treatment (Hasan & Muhammad, 2020) are emerging but are not yet an economical or practical choice for developing countries. Advancing new affordable water treatment technologies for use in both developing and developed countries will be an essential part of achieving global resilience to climate change.

7.5.5 Water storage

In addition to adequate water treatment, effective water storage is also essential, as one of the most common forms of water contamination in developing countries comes from water that has been stored in poor conditions [World Health Organization (WHO), 2000]. Water storage tanks and reservoirs are critical components of distribution systems, particularly where supply is intermittent, ensuring that there is sufficient water for short- or long-term usage but they can impact water quality if not used and maintained correctly. Common problems with storage tanks and reservoirs include leaking valve seals, poor connections, inadequate covering, exposed pipes, damaged tanks, bacterial regrowth, and excessive storage times (Chalchisa et al., 2018). Regrowth of bacteria in water during storage has been reported over both short and long time periods (Akuffo et al., 2013; Chalchisa et al., 2018; Zaqoot et al., 2016; Nnaji et al., 2019; Baguma et al., 2012) and storage tank material and size has a significance effect on the quality of stored water (Manga et al., 2021; Schafer & Mihelcic, 2012). Temperatures above 15°C can favor the regrowth of bacteria (Chalchisa et al., 2018) and so choosing tank materials and colors which reduce heating effects and placing tanks in shaded areas is advisable (Schafer & Mihelcic, 2012). Ensuring water storage tanks are kept clean can prevent deterioration of the microbiological quality of stored water (Manga et al., 2021; Nnaji et al., 2019).

7.6 Sustainable water management systems

Centralized water management systems, which are the standard in urban areas of developed and some developing countries, were first introduced in the mid-19th century to combat waterborne diseases. The benefits include reliable safe water supplies, flood control, food production and hydroelectricity generation (Gleick, 2000). Increasingly, however, the costs of large scale projects which bring water from new, more distant sources, and which are usually subsidized by governments and international organizations are becoming unsustainable (Gleick, 2000; Leflaive, 2009). In addition, the environmental and social costs of dams and water transfers face growing local opposition in the affected areas of both developed and developing countries (Domènech, 2011; Shah et al., 2021). The long-term costs of maintaining large water distribution networks have also become a burden in both developed and developing countries, as infrastructure ages and components fail (Cunha et al., 2020) and the energy costs associated with running these systems is a further sustainability issue. In developing countries centralized systems face additional challenges such as poor resources, weak regulation, population growth, water contamination, political uncertainty (Pallavi et al., 2021), as well as inequality in access to services (Adams, 2018, 2019; Cetrulo et al., 2020; Dos Santos et al., 2017) particularly in informal settlements (Adams, 2018; Adams et al., 2019).

Prior to the introduction of centralized drinking water and sanitation systems to urban areas in the 19th century, the main sources of water supply were local and included surface water, groundwater and rainwater but these were progressively abandoned in favor of distant sources as demand in cities grew (Domènech, 2011). Now, in the 21st century, a return to decentralized water management is emerging (Domènech, 2011; Leigh & Lee, 2019; Moglia et al., 2011; Sharma et al., 2010) and importance is increasingly placed on how best water can be allocated to meet human needs, with part or all of the water demand supplied by a combination of local water sources including stormwater, rainwater, wastewater and greywater (Domènech, 2011; Poustie et al., 2015; Quezada et al., 2016; Sharma et al., 2010). These local water sources, traditionally treated as "nuisances" (Domènech, 2011), are increasingly appreciated as valuable and sustainable resources and differentiated treatment of wastewater which has been separated from the source, enables the recovery of water for new uses (Gómez-Román et al., 2020).

In conjunction with decentralized options, community-based water management models have emerged recently in urban areas of developing countries in response to the failure of public and private water provision (Adams et al., 2019) and while relatively new in these settings, they have a long history in rural water supply. In urban areas they tend to operate either as community self-help schemes or as partnerships between public water utilities and diverse actors such as community groups, private operators, individual water vendors, faith based organizations, community-based organizations, and NGOs (Adams et al., 2019). Partnership-based water management models involving communities have great potential to fill water supply gaps in urban areas, especially underserved locations and informal settlements (Adams et al., 2019) but the particular challenges of urban settings such as more complex treatment and distribution needs, expert maintenance requirements, and lack of social cohesion and community participation need to be accounted for to ensure sustainability of these management approaches (Sally, 2011; Sally et al., 2014).

Centralized water treatment and distribution systems are generally prohibitively expensive for rural communities in developing countries, and decentralized household or small-scale community systems are a more affordable and effective way to ensure safe water supply in these areas (Peter-Varbanets et al., 2009). However, rural water supply systems are often characterized by their infrastructure and technology deficit and a lack of knowledge or experience among those operating or managing the systems (Domínguez et al., 2019). Premature failure of rural water supply systems is well documented (RWSN Rural Water Supply Network, 2010; Starkl et al., 2013), with a 30%–40% failure rate typically reported (Moriarty et al., 2013) (see example in Fig. 7.8). Failure is often a result of inadequate water management planning and lack of community engagement (Domínguez et al., 2019; RWSN Rural Water Supply Network, 2010).

Community management is based on a set of principles (both explicit and implicit) that include: (1) community participation in the system design, (2) community ownership, and (3) willingness and ability of the community to carry out operation and maintenance (Moriarty et al., 2013). However, these principles often assume community cohesion and a willingness and ability to manage the technical aspects of water supply systems in rural communities, which may not necessarily be the case (Moriarty et al., 2013; RWSN Rural Water Supply Network, 2010). In addition, donor agencies and governments tend to put

Figure 7.8
Failed water pump in rural Kenya in 2011 (author's image).

more emphasis on water supply (i.e., the number of constructed water systems) than on water management, system usage instructions and their translation into local languages (RWSN Rural Water Supply Network, 2010; Baguma et al., 2012). Therefore to achieve sustainability in the long-term, improved capacity-building in water management needs to be addressed in communities alongside new projects to improve water supply and quality (Baguma et al., 2012). The community management model could be further developed towards a wider diversity of models for different contexts, including a stronger degree of professionalization or external support for community-based service providers, while other management models, such as delegated management or self-supply, may also be sustainable options (Moriarty et al., 2013). Financial management plans are an essential aspect of sustainability and should cover all costs of the service over the whole life cycle, with a particular focus on financing maintenance and support costs (Moriarty et al., 2013).

7.7 Conclusions

The impacts of climate change are already being felt across the globe, and poor communities in developing countries (who contributed least to the problem) will feel the effects first and most intensely because of vulnerable geography and poor capacity to cope with deteriorating water supplies and damage from severe weather events and rising sea levels (Akanwa & Joe-Ikechebelu, 2019; Koubi, 2019; Mattoo & Subramanian, 2013).

WHO provides guidelines for water quality and management in developing countries, emphasizing incremental improvements which suit the particular challenges of decentralized water systems that are widely used in urban and rural communities. While significant gains have been made in access to safe drinking water over recent decades as a result of international commitments such as the SDGs, the challenges of climate change and the continued depletion of accessible, quality water resources pose a serious threat to the sustainability of existing and future interventions (Howard et al., 2016; MacAllister et al., 2020). To ensure that improvements in access to safe drinking water are sustainable for communities in developing countries in the long term, water treatment systems must account for the existing and predicted changes in climate patterns, as well as other challenges for water quality such as changes in land use and urbanization.

New water management approaches are emerging in both developed and developing countries in order to meet the challenges of the 21st century, including a more multidisciplinary approach to supply and demand, integrating social, economic, and environmental aspects with practices such as rainwater management, water conservation, wastewater reuse, energy management, nutrient recovery, and separation of wastewater at source (Gómez-Román et al., 2020). Particularly in the area of decentralized water treatment and managements systems, collaboration between developed and developing countries could identify appropriate sustainable solutions for a diverse range of situations and needs (Quezada et al., 2016), including traditional and more advanced treatment technologies. Building resilience to climate change requires water and sanitation policies, technologies and management systems capable of adapting to a wide range of potential climate scenarios (Howard et al., 2010, 2016), and efforts in this regard must guarantee adequate universal water supply, sanitation, and hygiene, all of which impact global public health, food security, poverty reduction, and equality (Pichel et al., 2019).

References

Abebe, L. S., Chen, X., & Sobsey, M. D. (2016). Chitosan coagulation to improve microbial and turbidity removal by ceramic water filtration for household drinking water treatment. *International Journal of Environmental Research and Public Health, 13*(3), 269.

Abebe, L. S., Smith, J. A., Narkiewicz, S., Oyanedel-Craver, V., Conaway, M., Singo, A., Amidou, S., Mojapelo, P., Brant, J., & Dillingham, R. (2014). Ceramic water filters impregnated with silver nanoparticles as a point-of-use water-treatment intervention for HIV-positive individuals in Limpopo Province, South Africa: A pilot study of technological performance and human health benefits. *Journal of Water and Health, 12*(2), 288–300, Jun.

Abebe, L. S., Su, Y. H., Guerrant, R. L., Swami, N. S., & Smith, J. A. (2015). Point-of-use removal of cryptosporidium parvum from water: Independent effects of disinfection by silver nanoparticles and silver ions and by physical filtration in ceramic porous media. *Environmental Science & Technology, 49*(21), 12958–12967, Nov 3.

Adams, E. A. (2018). Thirsty slums in African cities: Household water insecurity in urban informal settlements of Lilongwe, Malawi. *International Journal of Water Resources Development, 34*(6), 869–887, Nov 2.

Adams, E. A., Sambu, D., & Smiley, S. L. (2019). Urban water supply in Sub-Saharan Africa: Historical and emerging policies and institutional arrangements. *International Journal of Water Resources Development*, *35*(2), 240−263, Mar 4.

Akanwa, A. O., & Joe-Ikechebelu, N. (2019). The developing world's contribution to global warming and the resulting consequences of climate change in these regions: A Nigerian case study. *Global warming and climate change*. IntechOpen, Dec 13.

Akuffo, I., Cobbina, S. J., Alhassan, E. H., & Nkoom, M. (2013). Assessment of the quality of water before and after storage in the Nyankpala community of the Tolon-Kumbungu District, Ghan. *International Journal of Science Technology Research*, *2*(2), 2277−8616.

Allen, L., Christian-Smith, J., & Palaniappan, M. (2010). Overview of greywater reuse: The potential of greywater systems to aid sustainable water management. *Pacific Institute*, *654*(1), 19−21, Nov.

Amin, M. T., & Han, M. (2009). Roof-harvested rainwater for potable purposes: Application of solar disinfection (SODIS) and limitations. *Water Science and Technology*, *60*(2), 419−431, Jul.

Baguma, D., Aljunid, S., Hashim, J., Jung, H., & Willibald, L. (2012). Rain water and health in developing countries: a case study on Uganda. United Nations University Publication. < https://unu.edu/publications/articles/rainwater-and-health-in-developing-countries-a-case-study-on-uganda.html > Accessed 17.09.21.

Bartram, J. (2009). *Water safety plan manual: Step-by-step risk management for drinking-water suppliers.* World Health Organization.

Baumann, F. (2021). The next frontier—Human development and the anthropocene: UNDP human development report 2020. *Environment: Science and Policy for Sustainable Development*, *63*(3), 34−40. Mar 24 Accessed at https://report.hdr.undp.org on 17.09.2021.

Bitew, B. D., Gete, Y. K., Biks, G. A., & Adafrie, T. T. (2018). The effect of SODIS water treatment intervention at the household level in reducing diarrheal incidence among children under 5 years of age: A cluster randomized controlled trial in Dabat district, northwest Ethiopia. *Trials*, *19*(1), 1−5, Dec.

Botlagunta, M. A., Bondili, J. S., & Mathi, P. (2015). Water chlorination and its relevance to human health. *Asian Journal of Pharmaceutical and Clinical Research*, *8*(1), 20−24.

Burch, J. D., & Thomas, K. E. (1998). Water disinfection for developing countries and potential for solar thermal pasteurization. *Solar Energy*, *64*(1−3), 87−97, Sep 1.

Cetrulo, T. B., Ferreira, D. F., Marques, R. C., & Malheiros, T. F. (2020). Water utilities performance analysis in developing countries: On an adequate model for universal access. *Journal of Environmental Management*, *268*, 110662, Aug 15.

Chalchisa, D., Megersa, M., & Beyene, A. (2018). Assessment of the quality of drinking water in storage tanks and its implication on the safety of urban water supply in developing countries. *Environmental Systems Research*, *6*(1), 1−6, Jan.

Chauque, B. J., Benetti, A. D., Corcao, G., Silva, C. E., Goncalves, R. F., & Rott, M. B. (2021). A new continuous-flow solar water disinfection system inactivating cysts of *Acanthamoeba castellanii*, and bacteria. *Photochemical & Photobiological Sciences*, *20*(1), 123−137, Jan.

Chowdhury, S., Mazumder, M. J., Al-Attas, O., & Husain, T. (2016). Heavy metals in drinking water: occurrences, implications, and future needs in developing countries. *Science of the Total Environment*, *569*, 476−488, Nov 1.

Clasen, T. (2015). Household water treatment and safe storage to prevent diarrheal disease in developing countries. *Current Environmental Health Reports*, *2*(1), 69−74, Mar.

Clasen, T., Schmidt, W. P., Rabie, T., Roberts, I., & Cairncross, S. (2007). Interventions to improve water quality for preventing diarrhoea: Systematic review and *meta*-analysis. *BMJ (Clinical Research ed.)*, *334* (7597), 782, Apr 12.

Clasen, T. F., Alexander, K. T., Sinclair, D., Boisson, S., Peletz, R., Chang, H. H., Majorin, F., & Cairncross, S. (2015). Interventions to improve water quality for preventing diarrhoea. *Cochrane Database of Systematic Reviews* (10).

Conroy, R. M., Meegan, M. E., Joyce, T., McGuigan, K., & Barnes, J. (1999). Solar disinfection of water reduces diarrhoeal disease: An update. *Archives of Disease in Childhood*, *81*(4), 337−338, Oct 1.

Conroy, R. M., Meegan, M. E., Joyce, T., McGuigan, K., & Barnes, J. (2001). Solar disinfection of drinking water protects against cholera in children under 6 years of age. *Archives of Disease in Childhood*, *85*(4), 293–295, Oct 1.

Cunha, M., Marques, J., & Savić, D. (2020). A flexible approach for the reinforcement of water networks using multi-criteria decision analysis. *Water Resources Management*, *34*(14), 4469–4490.

Dai, A. (2011a). Characteristics and trends in various forms of the Palmer Drought Severity Index during 1900–2008. *Journal of Geophysical Research: Atmospheres*, *116*(D12), Jun 27.

Dai, A. (2011b). Drought under global warming: A review. *Wiley Interdisciplinary Reviews: Climate Change*, *2*(1), 45–65, Jan.

Dangol, B., & Spuhler, D. Household water treatment and safe storage (HWTS). SSWM. <https://sswm.info/sswm-solutions-bop-markets/affordable-wash-services-and-products/affordable-water-supply/household-water-treatment-and-safe-storage-%28hwts%29> Accessed 17.09.21.

Dell'Angelo, J., Rulli, M. C., & D'Odorico, P. (2018). The global water grabbing syndrome. *Ecological Economics*, *143*(Jan 1), 276–285.

Domènech, L. (2011). Rethinking water management: From centralised to decentralised water supply and sanitation models. *Documents d'anàlisi Geogràfica*, *57*(2), 293–310, Jul 26.

Domínguez, I., Oviedo-Ocaña, E. R., Hurtado, K., Barón, A., & Hall, R. P. (2019). Assessing sustainability in rural water supply systems in developing countries using a novel tool based on multi-criteria analysis. *Sustainability*, *11*(19), 5363, Jan.

Dongzagla, A., Jewitt, S., & O'Hara, S. (2021). Seasonality in faecal contamination of drinking water sources in the Jirapa and Kassena-Nankana Municipalities of Ghana. *Science of the Total Environment*, *752*, 141846, Jan 15.

Dos Santos, S., Adams, E. A., Neville, G., Wada, Y., De Sherbinin, A., Bernhardt, E. M., & Adamo, S. B. (2017). Urban growth and water access in sub-Saharan Africa: Progress, challenges, and emerging research directions. *Science of the Total Environment*, *607*, 497–508, Dec 31.

Frontier Economics. Exploring the links between water and economic growth. *A report prepared for HSBC by Frontier Economics*. 2012. <http://www.circleofblue.org/wp-content/uploads/2012/06/HSBC_June2012_Exploring-the-links-between-water-and-economic-growth.pdf> Accessed 17.09.21.

Gadgil, A. (1998). Drinking water in developing countries. *Annual Review of Energy and the Environment*, *23*(1), 253–286, Nov.

Gallo W., & Lantagne, D. S. (2008). Safe water for the community; a guide for establishing a community-based safe water system program. CDC.gov. <https://stacks.cdc.gov/view/cdc/5232> Accessed 17.09.21.

García-Ávila, F., Avilés-Añazco, A., Sánchez-Cordero, E., Valdiviezo-Gonzáles, L., & Ordoñez, M. D. (2021). The challenge of improving the efficiency of drinking water treatment systems in rural areas facing changes in the raw water quality. *South African Journal of Chemical Engineering*, *37*, 141–149.

Gelover, S., Gómez, L. A., Reyes, K., & Leal, M. T. (2006). A practical demonstration of water disinfection using TiO2 films and sunlight. *Water Research*, *40*(17), 3274–3280, Oct 1.

Gill, L. W., & Price, C. (2010). Preliminary observations of a continuous flow solar disinfection system for a rural community in Kenya. *Energy*, *35*(12), 4607–4611, Dec 1.

Gizachew, M., Admasie, A., Wegi, C., & Assefa, E. (2020). Bacteriological contamination of drinking water supply from protected water sources to point of use and water handling practices among beneficiary households of Boloso Sore Woreda, Wolaita Zone, Ethiopia. *International Journal of Microbiology*, *2020*, Apr 12.

Gleick, P. H. (2000). A look at twenty-first century water resources development. *Water International*, *25*(1), 127–138, Mar 1.

Gómez-Román, C., Lima, L., Vila-Tojo, S., Correa-Chica, A., Lema, J., & Sabucedo, J. M. (2020). "Who cares?": The acceptance of decentralized wastewater systems in regions without water problems. *International Journal of Environmental Research and Public Health*, *17*(23), 9060, Jan.

Gunaratna, K. R., Garcia, B., Andersson, S., & Dalhammar, G. (2007). Screening and evaluation of natural coagulants for water treatment. *Water Science and Technology: Water Supply*, *7*(5–6), 19–25, Dec.

Haddeland, I., Heinke, J., Biemans, H., Eisner, S., Flörke, M., Hanasaki, N., Konzmann, M., Ludwig, F., Masaki, Y., Schewe, J., & Stacke, T. (2014). Global water resources affected by human interventions and climate change. *Proceedings of the National Academy of Sciences*, *111*(9), 3251–3256, Mar 4.

Hagemann, S., Chen, C., Clark, D. B., Folwell, S., Gosling, S. N., Haddeland, I., Hanasaki, N., Heinke, J., Ludwig, F., Voss, F., & Wiltshire, A. J. (2013). Climate change impact on available water resources obtained using multiple global climate and hydrology models. *Earth System Dynamics*, *4*(1), 129–144, May 7.

Hasan, H. A., & Muhammad, M. H. (2020). A review of biological drinking water treatment technologies for contaminants removal from polluted water resources. *Journal of Water Process Engineering*, *33*, 101035, Feb 1.

Hofste, R. W., Reig, P., & Schleifer, L. (2019). 17 Countries, Home to One-Quarter of the World's Population, Face Extremely High Water Stress. World Resources Institute, Aug 6. <https://www.wri.org/insights/17-countries-home-one-quarter-worlds-population-face-extremely-high-water-stress> Accessed 17.09.21.

Holler, S. (2003). Decentralised infrastructure saves system costs, produces biogas energy, May 1. *Waterworld Magazine*. <https://www.waterworld.com/international/wastewater/article/16200433/decentralised-infrastructure-saves-system-costs-produces-biogas-energy> Accessed 17.09.21.

Hoslett, J., Massara, T. M., Malamis, S., Ahmad, D., van den Boogaert, I., Katsou, E., Ahmad, B., Ghazal, H., Simons, S., Wrobel, L., & Jouhara, H. (2018). Surface water filtration using granular media and membranes: A review. *Science of the Total Environment*, *639*, 1268–1282, Oct 15.

Howard, G., Bartram, J., Pedley, S., Schmoll, O., Chorus, I., & Berger, P. (2006). Groundwater and public health. *Protecting groundwater for health: Managing the quality of drinking-water sources* (pp. 3–19). London: International Water Association Publishing.

Howard, G., Calow, R., Macdonald, A., & Bartram, J. (2016). Climate change and water and sanitation: likely impacts and emerging trends for action. *Annual Review of Environment And Resources*, *41*(Oct 17), 253–276.

Howard, G., Charles, K., Pond, K., Brookshaw, A., Hossain, R., & Bartram, J. (2010). Securing 2020 vision for 2030: Climate change and ensuring resilience in water and sanitation services. *Journal of Water and Climate Change*, *1*(1), 2–16, Mar.

Huber, D. G., & Gulledge, J. (2011). Extreme weather and climate change: Understanding the link, managing the risk. Arlington: Pew Center on Global Climate Change.

Hutton, G., & Varughese, M. (2016). The costs of meeting the 2030 sustainable development goal targets on drinking water, sanitation, and hygiene. World Bank, Washington, DC. <https://openknowledge.worldbank.org/handle/10986/23681> Accessed 17.09.21.

Hutton, G., World Health Organization. (2012). Global costs and benefits of drinking-water supply and sanitation interventions to reach the MDG target and universal coverage. No. WHO/HSE/WSH/12.01. World Health Organization. <https://www.who.int/water_sanitation_health/publications/2012/globalcosts.pdf> Accessed 17.09.21.

Kabir, F., & Chowdhury, S. (2017). Arsenic removal methods for drinking water in the developing countries: Technological developments and research needs. *Environmental Science and Pollution Research*, *24*(31), 24102–24120, Nov.

Kostyla, C., Bain, R., Cronk, R., & Bartram, J. (2015). Seasonal variation of fecal contamination in drinking water sources in developing countries: A systematic review. *Science of the Total Environment*, *514*, 333–343, May 1.

Koubi, V. (2019). Sustainable development impacts of climate change and natural disaster. Background Paper Prepared for Sustainable Development Outlook.

Leflaive, X. (2009). Alternative ways of providing water—emerging options and their policy implications. *Advance copy for 5th world water forum*. OECD.

Leigh, N. G., & Lee, H. (2019). Sustainable and resilient urban water systems: The role of decentralization and planning. *Sustainability*, *11*(3), 918, Jan.

Lindsey, R., & Dahlman, L. (2021). Climate change: Global temperature. Mar 21. NOAA. Climate.gov. <https://www.climate.gov/news-features/understanding-climate/climate-change-global-temperature> Accessed 17.09.21.

Luzi, S., Tobler, M., Suter, F., & Meierhofer, R. (2016). SODIS manual: Guidance on solar water disinfection. Eawag, Switzerland. <https://www.sodis.ch/methode/anwendung/ausbildungsmaterial/dokumente_material/sodismanual_2016.pdf> Accessed 17.09.21.

Mac Mahon, J., & Gill, L. W. (2018). Sustainability of novel water treatment technologies in developing countries: Lessons learned from research trials on a pilot continuous flow solar water disinfection system in rural Kenya. *Development Engineering*, *3*, 47–59, Jan 1.

Mac Mahon, J., Pillai, S. C., Kelly, J. M., & Gill, L. W. (2017). Solar photocatalytic disinfection of E. coli and bacteriophages MS2, ΦX174 and PR772 using TiO2, ZnO and ruthenium based complexes in a continuous flow system. *Journal of Photochemistry and Photobiology B: Biology*, *170*, 79–90, May 1.

MacAllister, D. J., MacDonald, A. M., Kebede, S., Godfrey, S., & Calow, R. (2020). Comparative performance of rural water supplies during drought. *Nature Communications*, *11*(1), 1–3, Mar 4.

Malik, A., Yasar, A., Tabinda, A. B., & Abubakar, M. (2012). Water-borne diseases, cost of illness and willingness to pay for diseases interventions in rural communities of developing countries. *Iranian Journal of Public Health*, *41*(6), 39.

Manga, M., Ngobi, T. G., Okeny, L., Acheng, P., Namakula, H., Kyaterekera, E., Nansubuga, I., & Kibwami, N. (2021). The effect of household storage tanks/vessels and user practices on the quality of water: A systematic review of literature. *Environmental Systems Research*, *10*(1), 1–26, Dec.

Mattoo, A., & Subramanian, A. (2013). *Greenprint: A new approach to cooperation on climate change*. Brookings Institution Press on 17.09.2021 Mar 22. Brookings Institution Press. Accessed at. Available from https://www.cgdev.org/publication/9781933286679-greenprint-new-approach-cooperation-climate-change.

McGuigan, K. G., Conroy, R. M., Mosler, H. J., du Preez, M., Ubomba-Jaswa, E., & Fernandez-Ibanez, P. (2012). Solar water disinfection (SODIS): A review from bench-top to roof-top. *Journal of Hazardous Materials*, *235*, 29–46, Oct 15.

Mekonnen, M. M., & Hoekstra, A. Y. (2016). Four billion people facing severe water scarcity. *Science Advances*, *2*(2), e1500323, Feb 1.

Mengistie, B., Berhane, Y., & Worku, A. (2013). Household water chlorination reduces incidence of diarrhoea among under-five children in rural Ethiopia: A cluster randomized controlled trial. *PLoS One*, *8*(10), e77887, Oct 23.

Misstear, D. B., & Gill, L. W. (2012). The inactivation of phages MS2, ΦX174 and PR772 using UV and solar photocatalysis. *Journal of Photochemistry and Photobiology B: Biology*, *107*, 1–8, Feb 6.

Mizutori, M., & Guha-Sapir, D. (2020). Human cost of disasters: An overview of the last 20 years (2000–2019). Centre for research on the epidemiology of disasters (CRED) and United Nations Office for Disaster Risk Reduction (UNDRR), Belgium and Switzerland. <https://www.undrr.org/sites/default/files/inline-files/Human%20Cost%20of%20Disasters%202000-2019%20FINAL.pdf> Accessed 17.09.21.

Moglia, M., Cook, S., Sharma, A. K., & Burn, S. (2011). Assessing decentralised water solutions: Towards a framework for adaptive learning. *Water Resources Management*, *25*(1), 217–238, Jan.

Moriarty, P., Smits, S., Butterworth, J., & Franceys, R. (2013). Trends in rural water supply: Towards a service delivery approach. *Water Alternatives*, *6*(3), Oct 1.

Mukherjee, P., Chatterjee, D., Jana, J., Maity, P. B., Goswami, A., Saha, H., Sen, M., Nath, B., Shome, D., Sarkar, M. J., & Bagchi, D. (2007). Household water treatment option: Removal of arsenic in presence of natural Fe-containing groundwater by solar oxidation. *Trace Metals and other Contaminants in the Environment*, *9*, 603–622, Jan 1.

Ng, J. C., Wang, J., & Shraim, A. (2003). A global health problem caused by arsenic from natural sources. *Chemosphere*, *52*(9), 1353–1359, Sep 1.

Ngai, T. K., Coff, B., Baker, D., Lentz, R., Nakamoto, N., Graham, N., Collins, M. R., & Gimbel, R. (2014). Global review of the adoption, use, and performance of the biosand filter. *Progess in Slow Sand and Alternative Biofiltration Processes*, *14*, 309–317, May 14.

Nnaji, C. C., Nnaji, I. V., & Ekwule, R. O. (2019). Storage-induced deterioration of domestic water quality. *Journal of Water, Sanitation and Hygiene for Development*, *9*(2), 329−337.

Nzung'a, S. O., Kiplagat, K., & Paul, O. (2021). Techniques for potable water treatment using appropriate low cost natural materials in the tropics. *Journal of Microbiology, Biotechnology and Food Sciences*, *2021*, 2294−2300, Jan 6.

O'Farrell, C., Mac Mahon, J., & Gill, L. W. (2016). Development of a continuous flow solar oxidation process for the removal of arsenic for sustainable rural water supply. *Journal of Environmental Chemical Engineering*, *4*(1), 1181−1190, Mar 1.

Omarova, A., Tussupova, K., Berndtsson, R., Kalishev, M., & Sharapatova, K. (2018). Protozoan parasites in drinking water: A system approach for improved water, sanitation and hygiene in developing countries. *International Journal of Environmental Research and Public Health*, *15*(3), 495, Mar.

Organisation for Economic Co-operation and Development. (2020). Environmental health and strengthening resilience to pandemics. OECD Publishing. <https://www.oecd.org/coronavirus/policy-responses/environmental-health-and-strengthening-resilience-to-pandemics-73784e04/> Accessed 17.09.21.

Oteng-Peprah, M., Acheampong, M. A., & DeVries, N. K. (2018). Greywater characteristics, treatment systems, reuse strategies and user perception—A review. *Water, Air, & Soil Pollution*, *229*(8), 1−6, Aug.

Ouyang, Y., Nkedi-Kizza, P., Wu, Q. T., Shinde, D., & Huang, C. H. (2006). Assessment of seasonal variations in surface water quality. *Water Research*, *40*(20), 3800−3810, Dec 1.

Pain, A. Review: Decentralized systems for potable water and the potential of membrane technology. SSWM. <https://sswm.info/ar/water-nutrient-cycle/water-distribution/hardwares/water-network-distribution/decentralised-supply> Accessed 17.09.21.

Pallavi, S., Yashas, S. R., Anilkumar, K. M., Shahmoradi, B., & Shivaraju, H. P. (2021). Comprehensive understanding of urban water supply management: Towards sustainable water-socio-economic-health-environment Nexus. *Water Resources Management*, *35*(1), 315−336, Jan.

Pejack, E. (2011). Solar pasteurization. In *Drinking water treatment* (pp. 37−54). Springer, Dordrecht.

Peter-Varbanets, M., Zurbrügg, C., Swartz, C., & Pronk, W. (2009). Decentralized systems for potable water and the potential of membrane technology. *Water Research*, *43*(2), 245−265, Feb 1.

Pichel, N., Vivar, M., & Fuentes, M. (2019). The problem of drinking water access: A review of disinfection technologies with an emphasis on solar treatment methods. *Chemosphere*, *218*, 1014−1030, Mar 1.

Pooi, C. K., & Ng, H. Y. (2018). Review of low-cost point-of-use water treatment systems for developing communities. *NPJ Clean Water*, *1*(1), 1−8, Aug 6.

Poustie, M. S., Deletic, A., Brown, R. R., Wong, T., de Haan, F. J., & Skinner, R. (2015). Sustainable urban water futures in developing countries: The centralised, decentralised or hybrid dilemma. *Urban Water Journal*, *12*(7), 543−558, Oct 3.

Quezada, G., Walton, A., & Sharma, A. (2016). Risks and tensions in water industry innovation: Understanding adoption of decentralised water systems from a socio-technical transitions perspective. *Journal of Cleaner Production*, *113*, 263−273, Feb 1.

Rossi, F., Parisi, M. L., Maranghi, S., Manfrida, G., Basosi, R., & Sinicropi, A. (2019). Environmental impact analysis applied to solar pasteurization systems. *Journal of Cleaner Production*, *212*, 1368−1380, Mar 1.

RWSN (Rural Water Supply Network). (2010). Myths of the rural water supply sector. Perspectives Paper No. 4. RWSN Executive Steering Committee. St Gallen: RWSN.

Salehi, A. A., Ghannadi-Maragheh, M., Torab-Mostaedi, M., Torkaman, R., & Asadollahzadeh, M. (2020). A review on the water-energy nexus for drinking water production from humid air. *Renewable and Sustainable Energy Reviews*, *120*, 109627, Mar 1.

Sally, M. Z. (2011). Stakeholder participation and sustainability challenges confronting a small urban community-managed water supply project: Case study of Buea, Cameroon. McGill University (Canada).

Sally, Z., Gaskin, S. J., Folifac, F., & Kometa, S. S. (2014). The effect of urbanization on community-managed water supply: Case study of Buea, Cameroon. *Community Development Journal*, *49*(4), 524−540, Oct 1.

Sanctuary, M., Tropp, H., & Haller, L. (2007). Making water a part of economic development: The economic benefits of improved water management and services. Stockholm International Water Institute (SIWI):

Stockholm, Sweden. <https://www.siwi.org/wp-content/uploads/2015/09/waterandmacroecon.pdf> Accessed 17.09.21.

Schafer, C. A., & Mihelcic, J. R. (2012). Effect of storage tank material and maintenance on household water quality. *Journal-American Water Works Association*, *104*(9), E521−E529, Sep.

Shah, E., Vos, J., Veldwisch, G. J., Boelens, R., & Duarte-Abadía, B. (2021). Environmental justice movements in globalising networks: A critical discussion on social resistance against large dams. *The Journal of Peasant Studies*, *48*(5), 1008−1032, Jul 29.

Sharma, A., Burn, S., Gardner, T., & Gregory, A. (2010). Role of decentralised systems in the transition of urban water systems. *Water Science and Technology: Water Supply*, *10*(4), 577−583, Sep.

Sivanappan, R. K. (2006). Rain water harvesting, conservation and management strategies for urban and rural sectors. *International Seminar on Rainwater Harvesting and Water Management*, *11*(12), 1, Nov 11.

Sobsey, M. D., Handzel, T., & Venczel, L. (2003). Chlorination and safe storage of household drinking water in developing countries to reduce waterborne disease. *Water Science and Technology*, *47*(3), 221−228, Feb.

Sobsey, M. D., Stauber, C. E., Casanova, L. M., Brown, J. M., & Elliott, M. A. (2008). Point of use household drinking water filtration: A practical, effective solution for providing sustained access to safe drinking water in the developing world. *Environmental Science & Technology*, *42*(12), 4261−4267, Jun 15.

Sorlini, S., Palazzini, D., Mbawala, A., Ngassoum, M. B., & Collivignarelli, M. C. (2013). Is drinking water from 'improved sources' really safe? A case study in the Logone valley (Chad-Cameroon). *Journal of Water and Health*, *11*(4), 748−761, Dec.

Starkl, M., Brunner, N., & Stenstrom, T. A. (2013). Why do water and sanitation systems for the poor still fail? Policy analysis in economically advanced developing countries. *Environmental Science & Technology*, *47*(12), 6102−6110, Jun 18.

String, G., & Lantagne, D. (2016). A systematic review of outcomes and lessons learned from general, rural, and country-specific Water Safety Plan implementations. *Water Science and Technology: Water Supply*, *16*(6), 1580−1594, Dec.

Thomas, T. H. (2014). The limitations of roofwater harvesting in developing countries. *Waterlines*, 139−145, Apr 1.

Tratschin, R., & Spuhler, D. Solar Pasteurisation.S.S.W.M. <https://sswm.info/sswm-solutions-bop-markets/affordable-wash-services-and-products/affordable-water-supply/solar-pasteurisation> Accessed 17.09.21.

Treacy, J. (2019). *Drinking water treatment and challenges in developing countries. The relevance of hygiene to health in developing countries*. Intech Open on 17.09.2021 Apr 3. Available from https://www.intechopen.com/chapters/63707.

UN Water. (2018). Sustainable development Goal 6 synthesis report on water and sanitation. United Nations. New York. <https://www.unwater.org/publications/sdg-6-synthesis-report-2018-on-water-and-sanitation/> Accessed 17.09.21.

UN Water. (2021). Summary progress update 2021: SDG 6—Water and sanitation for all. 24 Feb. United Nations. <https://www.unwater.org/publications/summary-progress-update-2021-sdg-6-water-and-sanitation-for-all/> Accessed 17.09.21.

UNEP. (2009). Rainwater harvesting: A lifeline for human well being. United Nations Environment Programme. <https://www.unep.org/resources/report/rainwater-harvesting-lifeline-human-well-being> Accessed 17.09.21.

UNESCO. (2015). The United Nations World Water Development Report: Water for a Sustainable World. United Nations. <https://en.unesco.org/themes/water-security/wwap/wwdr/series> Accessed 17.09.21.

United Nations Children's Fund (UNICEF) and World Health Organization (WHO). (2020). Joint monitoring programme (JMP) for water and sanitation. Progress on household drinking water, sanitation and hygiene 2000−2017: Special focus on inequalities. United Nations.

United Nations Children's Fund (UNICEF). (2016). Press release: Collecting water is often a colossal waste of time for women and girls. <https://www.unicef.org/press-releases/unicef-collecting-water-often-colossal-waste-time-women-and-girls> Accessed 17.09.21.

United Nations General Assembly. (2010). Resolution A/RES/64/292: The human right to water and sanitation. Adopted by the General Assembly on 28 July. <https://undocs.org/A/RES/64/292> Accessed 17.09.21.

United Nations Office to support the International Decade for Action. (2015). A 10 year story. The water for life decade 2005-2015 and beyond. United Nations. <https://www.un.org/waterforlifedecade/pdf/WaterforLifeENG.pdf> Accessed 17.09.21.

United Nations. (2021). Department of economic and social affairs: Sustainable development. Sustainable development goals. <https://sdgs.un.org/goals> Accessed 17.09.21.

United Nations. (2015). The millennium development goals report. United Nations, New York. <https://www.un.org/millenniumgoals/2015_MDG_Report/pdf/MDG%202015%20rev%20(July%201).pdf> Accessed 17.09.21.

United Nations. (2018). United Nations Secretary-General's Plan: Water action decade 2018-2028. <https://wateractiondecade.org/wp-content/uploads/2018/03/UN-SG-Action-Plan_Water-Action-Decade-web.pdf> Accessed 17.09.21.

Verlicchi, P., & Grillini, V. (2020). Surface water and groundwater quality in South Africa and Mozambique—Analysis of the most critical pollutants for drinking purposes and challenges in water treatment selection. *Water*, *12*(1), 305, Jan.

Vörösmarty, C. J., Léveque, C., Revenga, C., Bos, R., Caudill, C., Chilton, J., Douglas, E. M., Meybeck, M., Prager, D., Balvanera, P., & Barker, S. (2005). Fresh water. *Millennium ecosystem assessment*, *1*, pp. 165-207, Vörösmarty CJ, Lévêque C, Revenga C. Chapter 7: Fresh Water *Millennium Ecosystem Assessment*, *2005*(1), 165–207.

Wada, Y., Van Beek, L. P., & Bierkens, M. F. (2011). Modelling global water stress of the recent past: on the relative importance of trends in water demand and climate variability. *Hydrology and Earth System Sciences*, *15*(12), 3785–3808, Dec 20.

Warner, W. (2006). Understanding greywater treatment. *Water Demand Management*, 62–81.

Water Quality & Health Council. (2017). Finished drinking water and treatment fundamentals. <https://waterandhealth.org/safe-drinking-water/finished-drinking-water-treatment-fundamentals/> Accessed 17.09.21.

Wegelin, M., Gechter, D., Hug, S., Mahmud, A., & Motaleb, A. (2000). SORAS-a simple arsenic removal process. In Proceedings of the twenty-sixth WEDC conference, Dhaka, Bangladesh.

World Health Organization (WHO). (2012). A toolkit for monitoring and evaluating household water treatment and safe storage programmes. United Nations. <https://www.who.int/household_water/WHO_UNICEF_HWTS_MonitoringToolkit_2012.pdf> Accessed 17.09.21.

World Health Organization (WHO). (2019). Drinking water factsheet. United Nations. <https://www.who.int/news-room/fact-sheets/detail/drinking-water> Accessed 17.09.21.

World Health Organization (WHO). (2000). *Global water supply and sanitation assessment*. Geneva: World Health Organization.

World Health Organization (WHO). (2021). Household water treatment and safe storage. <https://www.who.int/teams/environment-climate-change-and-health/water-sanitation-and-health/water-safety-and-quality/household-water-treatment-and-safe-storage> Accessed 17.09.21.

World Health Organization (WHO)/ United Nations Children's Fund (UNICEF). (2008). Joint Monitoring Programme (JMP) for water and sanitation. Progress on sanitation and drinking water: 2008. Special focus on sanitation. United Nations. <https://www.who.int/water_sanitation_health/monitoring/jmp_report_7_10_lores.pdf> Accessed 17.09.21.

World Health Organization (WHO)/ United Nations Children's Fund (UNICEF). (2011). Joint Monitoring Programme (JMP) for water and sanitation. Rapid assessment of drinking water quality: Pilot country reports. Geneva, Switzerland. World Health Organization/United Nations Children's Fund Joint Monitoring Program for Water Supply and Sanitation.

World Health Organization (WHO)/ United Nations Children's Fund (UNICEF). (2016). Joint Monitoring Programme (JMP) for water and sanitation. Improved and unimproved water sources and sanitation facilities. United Nations.

World Health Organization (WHO)/ United Nations Children's Fund (UNICEF). (2017a). Joint Monitoring Programme (JMP) for water and sanitation. Progress on drinking water, sanitation and hygiene. United Nations.

World Health Organization (WHO)/ United Nations Children's Fund (UNICEF). (2017b). Joint Monitoring Programme (JMP) for water and sanitation. Safely managed drinking water - thematic report on drinking water 2017. United Nations.

World Health Organization (WHO)/ United Nations Children's Fund (UNICEF). (2021). Joint Monitoring Programme (JMP) for water and sanitation. Progress on household drinking water, sanitation and hygiene 2000-2020: Five years into the SDGs. United Nations. <https://www.who.int/publications/i/item/9789240030848> Accessed 17.09.21.

World Health Organization (WHO)/ United Nations Children's Fund (UNICEF). (2015) Joint Monitoring Programme (JMP) for Water and sanitation. Progress on sanitation and drinking water: 2015 Update and MDG Assessment. <https://www.unicef.org/reports/progress-sanitation-and-drinking-water> Accessed 17.09.21.

World Health Organization. (2017). Guidelines for drinking-water quality. Fourth Addition including First Addendum. United Nations. <https://apps.who.int/iris/bitstream/handle/10665/254637/9789241549950-eng.pdf> Accessed 17.09.21.

Wright, R. C. (1986). The seasonality of bacterial quality of water in a tropical developing country (Sierra Leone). *Epidemiology & Infection*, 96(1), 75–82, Feb.

Yang, H., Xu, S., Chitwood, D. E., & Wang, Y. (2020). Ceramic water filter for point-of-use water treatment in developing countries: Principles, challenges and opportunities. *Frontiers of Environmental Science and Engineering*, 14(5), 79.

Zaqoot, H. A., Hamada, M., & Mohammed, A. (2016). Investigation of drinking water quality in the kindergartens of Gaza Strip Governorates. *Journal of Tethys*, 4(2), 088–099.

Zinn, C., Bailey, R., Barkley, N., Walsh, M. R., Hynes, A., Coleman, T., Savic, G., Soltis, K., Primm, S., & Haque, U. (2018). How are water treatment technologies used in developing countries and which are the most effective? An implication to improve global health. *Journal of Public Health and Emergency*, 2(25), 1–4.

CHAPTER 8

Water purification techniques for the developing world

Aniruddha B. Pandit[1] and Jyoti Kishen Kumar[2]

[1]Institute of Chemical Technology, Matunga, Mumbai, India, [2]Formerly Institute of Chemical Technology, Matunga, Mumbai, India

Chapter Outline
- 8.1 Introduction 145
- 8.2 Water purification methodologies 147
- 8.3 Solar treatment and its intensification 147
- 8.4 Water disinfection by boiling 150
- 8.5 Chemical treatment 152
- 8.6 Filtration techniques 154
    - 8.6.1 Traditional filtration methods 154
    - 8.6.2 Recent advances/modifications in traditional filtration methods 156
    - 8.6.3 Hybrid filtration methods 158
- 8.7 Natural treatment methods 159
- 8.8 Cavitation-based water hand pumps 160
- 8.9 Sustainable disinfection of harvested rainwater 163
- 8.10 Recent research on emerging methods 166
    - 8.10.1 Filtration 166
    - 8.10.2 Solar disinfection 166
    - 8.10.3 Hybrid techniques 167
- 8.11 Impact of Covid 19 pandemic on the water sector in developing countries 168
- 8.12 Comparison of various purification techniques 170
- 8.13 Conclusions 173
- 8.14 Recommendations for future work 174
- References 174

8.1 Introduction

Water is Life! This says it all as the importance of water for the survival of living beings cannot be over emphasized. Water is essential for a healthy life, and this is true across the globe without any exception. However, the quantity and quality of water available to humanity is a challenge faced worldwide and more pronounced in developing countries where there is a basic lack of access to safe and clean drinking water. According to the WHO, around 785 million people lack basic drinking water and around 144 million people rely on surface water for their requirements [1] (http://www.who.int/news-room/fact-sheets/detail/drinking-water).

The main reason for this is the rising population on one hand and an intensified industrial activity on the other that has led to a major pollution of the water bodies. These pollutants that enter water bodies can be either physical, chemical, and/or biological in nature and if consumed untreated, can lead to innumerable diseases in humans (Table 8.1). It is indeed alarming to note that there are over 500 thousand diarrheal deaths caused by waterborne viruses and bacteria [2] (Pichel et al., 2019). Thus this has mandated the need for various types of water treatment to be performed on the source water before it can be provided to the end user in a city, community, or individual household. On the other hand, scarcity, and shortage of water due to lack of rainfall, climatic changes and human activities like deforestation has aggravated the situation. Although these challenges exist for all globally, the developing countries of the world bear the brunt.

It is a known fact that the economic condition of a developing country is not on par with the developed counterparts as the per capita income of the developing countries is extremely low. Thus water purification technologies that are usually used in the developed nations cannot be afforded by the developing countries. This has led to myriad developments in the arena of inexpensive and sustainable water purification methods that can be easily deployed in the developing sectors of the worlds. It is interesting to note that although decades of research have gone into the augmentation of different water purification approaches, the quest for efficient and economically feasible methods continues.

Table 8.1: Some common chemical and microbial pollutants encountered in water and their health hazards in humans (Pandit & Kumar, 2019).

S. no.	Chemical/microbial contaminant	Health hazard caused	Limits as per WHO/EPA guidelines
1.	Fluoride	Affects bones and teeth and results in dental fluorosis	1.5 mg L^{-1}
2.	Arsenic	Skin, bladder and lung cancers	0.01 mg L^{-1}
3.	Uranium	Nephritis	0.03 mg L^{-1}
4.	Enteroviruses	Myocarditis, poliomyelitis	0 (MCLG)[a] mg L^{-1}
5.	Total coliforms	Enterogastritis	0 (MCLG)[a] mg L^{-1}
6.	Cryptosporidium	Diarrhea, nausea, vomiting and fever	0 (MCLG)[a] mg L^{-1}
7.	Legionella	Legionellosis	0 (MCLG)[a] mg L^{-1}

[a]MCLG, Maximum contaminant level goal.

Among a plethora of water treatment methodologies available to mankind, only those that can be successfully implemented in the developing nations and which can be scaled up, will be discussed in this chapter. These include traditional and conventional methods such as boiling, use of chemicals like chlorine, solar disinfection, filtration methods and above all innovative developments such as the cavitation-based hand pump, Lifestraw, the drinkable book, to name some. Here, it is important to mention that the type of treatment to be used for water purification depends on a multitude of factors such as the principle of the technique, the pollutants targeted, scale of operation, cost of treatment and above all its feasibility.

Thus the objective of this chapter is to highlight the various methodologies available, compare them and recommend the possible choices that can be opted for water treatment in a developing country. The relevance of water purification in the developing world cannot be overemphasized given the current global Covid-19 pandemic situation. This only pushes mankind to consistently evolve and bring out better technologies to grapple with the onslaught of global crisis such as this pandemic. A brief discussion on the impact of the pandemic on the water sector especially with respect to the developing countries is also highlighted.

8.2 Water purification methodologies

Water purification involves the removal of physical, chemical, and/or biological pollutants from the source water by means of different techniques. However, the removal of biological pollutants such as bacteria, protozoa, and viruses are termed disinfection. It is important to note that disinfection does not imply complete removal of these microorganisms as it is not practically feasible. However, the water authorities such as the WHO and the EPA have laid down standard guidelines that state the minimum allowed microbial limit of the pollutant that is permissible. Often it is expressed in the form of colony forming Units (CFU mL^{-1}) and these guidelines must be strictly adhered to for ensuring clean water after a particular treatment protocol.

Water purification methods are vast and encompass a myriad of methodologies depending on the type of pollutant being treated, the source water quality and its intended end use such as cooking, domestic uses, or drinking. Among these, the following types of treatment techniques have been proved useful in developing countries as they are generally easy to operate, economical and provide treated water that fall within the guidelines stipulated by the national or local water authorities. One such interesting technique is the use of solar radiation commonly termed Solar Disinfection (SODIS).

8.3 Solar treatment and its intensification

Harnessing the sun's energy for various uses has been practiced since ancient times. Sun's rays especially the UV rays are known to destroy microorganisms and use of this technique is often referred as solar disinfection or SODIS. It is interesting to note that the WHO has recommended the use of SODIS for water disinfection in rural areas of the developing world where conventional water purification cannot be employed and where centralized municipally treated water systems are not available (Byrne et al., 2011). In such areas, SODIS can be easily implemented.

SODIS essentially involves exposing untreated water in PET bottles to sunlight for a period of approximately 6 hours on a clear sunny day or for 48 hours when it is cloudy. Thus this method does not require skilled labor, is relatively easy to practice, and is inexpensive as it does not need expensive chemicals or high-end treatment technologies. Moreover, it is typically suited for the developing world that receive abundance of sunlight almost round the year.

Decades of research on SODIS has proved that it is amazingly effective in reducing the microbial load of water and successfully inactivates several bacteria, viruses, and removes inorganic contaminant by settling or mineralizing the same. The main mechanism of action involved is the absorption of UV rays by water which results in damage of cellular DNA of the microbes and due to the formation of reactive oxygen species (ROS) such as hydrogen peroxide, hydroxyl radicals, etc., which in turn damage the microbial cells. Additional, thermal disinfection is also believed to be a contributory factor. SODIS like any other water treatment method has its disadvantages too. For instance, the waiting times (6–48 hours) are too long, and this may not be appropriate when there may be an urgent need for drinking water. This has prompted a lot of research in enhancing the efficacy of the conventional SODIS. Fig. 8.1 explains a few of these techniques in a nutshell (Fig. 8.1).

Some of these improvements include thermal enhancements which involves partially painting the outer surface of the PET bottle black. This increases the absorption of heat as compared to the plain PET bottle. Augmented thermal energy results in enhanced SODIS effect thereby improving the disinfection efficacy. However, on cloudy days, the rate of absorption maybe be low requiring long hours of treatment.

Yet another way of enhancing SODIS is using photocatalysis. In this SODIS method, semiconductor photocatalyst such as TiO_2 (anatase form, which is photoactive) is used which absorbs the UV-A radiation of the solar rays and produce ROS such as hydroxyl radicals and hydrogen peroxide which in turn destroy the microorganisms present in water. In a very recent study on the role of photocatalytic water treatment (PWT) in SODIS has revealed that apart from TiO_2, host of other catalysts such as MoS_2, ZnO, Bi, and magnetic iron oxides have been developed with a view to using them to exploit the visible light which constitute almost 44% of the solar spectrum as compared to the UV rays (4%–5%). The authors emphasize the need to harvest visible light energy by using newer and better photocatalysts that can destroy a spectrum of microbes ranging from *Escherichia coli* to pathogenic bacteria, viruses, and protozoa. Above all, they highlight the need for real time

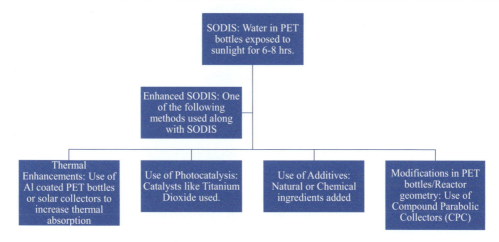

Figure 8.1
SODIS and few methods to enhance SODIS.

field studies to obtain accurate illustration of their efficacy and the feasibility of the technology in the developing countries (Bradley et al., 2020).

Interestingly, many natural and chemical additives have also been used to enhance SODIS. These additives include dyes such as methylene blue and chemicals like ascorbate, vinegar to several natural ingredient such as lime, lemon juice, plant extracts such as *Cassia alata* and *Phyllanthus niruri*. It has been observed that these additives when added to the PET bottle kept for SODIS, essentially acted as photosensitizers and absorbed solar energy. This resulted in the formation of free radicals and/or ROS which in turn were responsible for the inactivation of the microbes in the sample water. In one investigation, the researchers observed that addition of plant extracts to SODIS destroyed all pathogenic microorganisms within 2 hours of treatment. Thus treatment with additives reduced the time required for SODIS and resulted in higher number of microbes being inactivated (4 log reduction) thereby enhancing its efficacy (Makuba et al., 2016).

Finally, SODIS can also be augmented by modifications in the PET bottle or the reactor geometry. Solar Collector Disinfection (SOCODIS) system is one such case where the solar disinfection is improved by using solar collectors. This system has proved to be amazingly effective in inactivating *Cryptosporidium parvum* oocysts as the SOCODIS was able to provide extremely high intensity (>600 W m^{-2}) of sunlight that is usually required to destroy these oocysts. Aluminum foils, mirrors, lenses, reflective solar boxes and use of Compound Parabolic Collector (CPC) mirrors are some more ways of enhancing conventional SODIS (Pandit & Kumar, 2019).

Although the last decade has witnessed several research studies on accelerating SODIS and its scaling up to meet the requirements of a larger community, there is still a huge potential to develop varied solutions to meet these challenges. Recently, some interesting research has been carried out on the enhancement of SODIS apparatus that was equipped with Wood's glass and a Fresnel lens. The Wood's glass behaves as a band pass filter which not only allows UV rays to pass through and hence causes disinfection but also prevents the infrared rays which is supposed to promote growth of bacteria. The researchers conducted experiments both in a batch-mode (18 L plastic pails) and in a novel instantaneous mode which consisted of a small—scale continuous flow system (plug flow). Both these systems, when operated with Wood's glass and Fresnel lens, resulted in almost 20%–93% (batch) and 55%–85% (continuous) pathogen removal, respectively (Alkhalidi et al., 2021).

In yet another study conducted in Bangladesh in the year 2021, the investigators studied SODIS during different seasons. They used both PET and plastic bags and conducted experiments for both transmissive and reflective reactors. They observed a log reduction of more than 5 for transmissive PET reactors (6–8 hour exposure to sunlight) in summer months. During monsoon and winters, log reduction of greater than four could be achieved after 16 and 8 hours of solar radiations, respectively, employing reflective reactors. The researchers found that the plastic bags were more effective than the PET bottles. This study also revealed that regrowth of bacteria occurred which highlights the need for further research on the storage time before use. Moreover, increase in turbidity resulted in longer time for SODIS. The authors conclude by recommending SODIS as a household water treatment (HWT) for developing countries like Bangladesh especially in summers and point out the need to accelerate SODIS in the monsoon and winter months (Karim et al., 2021).

From the preceding discussion it can be concluded that SODIS appears to be a method of choice for water disinfection in developing countries due to its simplicity, efficiency and relatively inexpensive materials needed. Typically, the cost of conventional SODIS is reported as US\$ 0.5 m^{-3} assuming 100 uses of PET bottles before they cannot be used further (Plappally et al., 2013). However, further research on scaling up SODIS and methods to augment the process without compromising the economics are the immediate need.

8.4 Water disinfection by boiling

Just as utilizing solar energy is an age-old methodology of water purification, boiling of water is also a traditional and conventional method of purifying water since eons. Boiling water has been a practice adopted in almost all households worldwide for ensuring safe, clean and sometimes softened drinking water. It is a proven method of choice to treat water at the point of use during outbreaks of waterborne disease, emergencies like natural disasters and certain situations like traveling when there is uncertainty about the water source quality.

The WHO states the scientific basis for the efficacy of boiling, based on the research carried out by scientific experts spanning almost two decades. Interestingly, results reveal that bacteria are sensitive to heat and rapid kills of less than one minute per log (90%) reduction are obtained at temperatures above 65°C. Most viruses are killed at temperatures between 60°C and 65°C and in the case of few viruses such as poliovirus and hepatitis A, almost 99.99% reduction is obtained in less than a minute as temperatures increase to above 70°C. Protozoa such as Giardia cysts are inactivated at temperatures between 50°C to 70°C. (Table 8.2) (http://www.who.int/water_sanitation_health/dwq/Boiling_water_01_15.pdf).

The US Centers for Disease Control and Prevention (CDC) state that boiling is one of the methods of choice for treating drinking water during emergencies and have provided a protocol for boiling water. If the water is clear, it must be brought to a rolling boil for 1 minute (at elevations above 6500 feet, boil for 3 minutes), allow the boiled water to cool and store it in clean sanitized containers that have tight covers. If the water is cloudy, it has to be filtered either with a plain cloth or a paper filter before following the same protocol as in the case of clear water (https://www.cdc.gov/healthywater/emergency/making-water-safe.html).

Some noteworthy investigations cited in the literature bring out the effectiveness of boiling as a method of choice in treating water. In one such study conducted in Nigeria, the authors compared the economics of boiling, SODIS and granulated activated carbon (GAC) for water disinfection in developing countries. Fascinatingly, they report boiling to be a remarkably effective method compared to SODIS and GAC, as it could eliminate coliforms from the examined sample (River water and borehole water) (Pandit & Kumar, 2019).

Recently, in the year 2020, Heitzinger et al. conducted household trials to assess the feasibility and acceptability of water pasteurization indicators (WAPIs) in the Peruvian Amazon. WAPI is an indicator that shows water has reached pasteurization temperature (65°C). This saves energy by removing the necessity to boil water (100°C). Generally, WAPIs are made of polycarbonate tubes containing wax that melts when water reaches a temperature of 65°C (http://www.solarcooking.fandom.com). The authors found that ease of use, short treatment time and the taste of treated water were some key factors that influenced the acceptability of the treatment. Such studies indicate the potential of such

Table 8.2: Few examples of microbes and their temperature kills (WHO, 2015).

Type of organism	Example	Inactivation time (s)	Temperature (°C)	Log_{10} reduction
Protozoa	Cryptosporidium parvum	300	60	3.4 log
Viruses	Adenovirus 5	1260	70	>8 log
	Echovirus 6	1800	60	4.3 log
Bacteria	Enterococcus faecalis	7–19	65	Per log
	Salmonella typhimurium	<2	65	Per log

interventions to boost the access to safe drinking water in resource constrained settings (Heitzinger et al., 2020).

Thus, boiling is essentially considered as an efficacious method especially in emergency settings and can inactivate most human pathogens even in turbid waters. However, it is an energy intensive process and involves the use of expensive carbon-based fuel sources such as wood, charcoal, or electricity. In an interesting investigation, a five-month study was carried out among 218 boilers located in semiurban communities in Virar District close to Mumbai, in the state of Maharashtra, India. The approximate monthly fuel cost for boiling was stated as INR 43.8 (US$0.88) for families using liquid petroleum gas and INR 34.7 (US$0.69) for those using wood (Clasen et al., 2008). Boiling often imparts an unpleasant taste to the treated water, does not provide residual protection and is onerous when it must be done every day for long periods. Despite these shortcomings, it can be still regarded as a conducive method of water disinfection especially in developing countries where expensive water treatment methods cannot be used.

8.5 Chemical treatment

Water treatment with chemical disinfectants can be considered one of the oldest and most widely adopted practice worldwide. Although a spectrum of chemicals is available, chlorine has always been the choice for water disinfection globally. This is especially true in the low-income nations due to its low cost and high disinfection efficacy. Fascinatingly, chlorine when used for water disinfection also improves the esthetic appeal of treated water due to its odor- and color-removing (bleaching) capability. In developing countries, chlorine is generally used either as sodium hypochlorite solution or calcium hypochlorite. These are relatively safer to handle and transport as compared to chlorine gas.

An incredible aspect of chlorine in water disinfection is its ability to provide residual effect during long storage. This means that the treated water remains safe and pathogen free for long periods as the residual chlorine in the treated water ensures that microbes do not regrow in the distribution systems such as pipes. A residual free chlorine of 0.25 mg L^{-1} is generally considered sufficient for water at 20°C with a total organic carbon content of less than 0.25 mg L^{-1} (Gadgil, 1998).

The mechanism of chlorine disinfection has been widely studied over decades and several hypotheses have been put forth by various experts. Most commonly, it is stated that chlorine affects the cell membrane of the microorganism and results in the release of cellular contents. Moreover, it also attacks the nuclease of the microbial cell and inhibits its biochemical reactions related to oxygen uptake and oxidative phosphorylation. These events result in the inactivation of the pathogens. Although chlorine has been found to kill a wide range of bacteria and viruses, it is generally not found as effective in the case of some protozoa like *Giardia* and *Cryptosporidium* (Pandit & Kumar, 2019).

Like any water treatment process, chlorination has several pros and cons some of which are stated in Table 8.3.

Although chlorination is an effective and widely used technique of water disinfection particularly in developing countries, an important concern is the formation of disinfection byproducts (DBPs). These are formed due to the reaction of chlorine with the naturally present organic compounds in water that is to be treated. Some of the DBPs include trihalomethanes (THMs), haloacetic acids (HAAs), etc. Many DBPs can pose health risks to humans and several studies have been undertaken to establish the link between the presence of DBPs and their implied health hazards. For instance, the International Agency for Research on Cancer (IARC) has classified chloroform (a trihalomethane) as carcinogenic to humans and is also known to cause liver damage. Hence, the guideline value for chloroform as laid down by the WHO is 0.3 mg L^{-1} (Pandit & Kumar, 2019).

Among the numerous ongoing studies in the arena of chemical treatment for water disinfection especially for the developing nations, recently a remarkably interesting article has described the effectiveness of chlorine against *Schistosoma mansoni cercariae*. This organism causes a water borne infection called Schistosomiasis. Alarmingly, many communities live in endemic regions and rely heavily on water bodies that are polluted with these parasites. The authors conducted trials both in lab scale and field conditions. The latter was conducted at Lake Victoria, Tanzania. Both studies showed that a C_t value (residual chlorine concentration multiplied by the chlorine contact time) of 26 ± 4 mg min L^{-1} was required to achieve 2 $\log_{10}$ reduction of *S. mansoni* cercariae at pH 7.5 and temperature of 20°C. Finally, the authors suggest a C_t value of 30 mg min L^{-1} to disinfect *S. mansoni* cercaria infested water. This would be useful in providing safe drinking water for communities and individual families that lack alternative water supplies (Braun et al., 2020).

The concern regarding DBP and their health hazards has been discussed in the previous section. Although these DBPs pose health risks, there has always been a debate among various stakeholders between the use of chemicals for water treatment versus the perils of consuming treated water with DBPs. In a riveting research conducted in the year 2020, Mazhar et al. (2020) reviewed fifty-two studies about DBPs health risks and epidemiology

Table 8.3: Some merits and demerits of chlorine for water disinfection (Kausley et al., 2018).

S. no.	Advantages of chlorine disinfection	Disadvantages of chlorine disinfection
1.	Extensively used and general choice in developing countries	Forms disinfection byproducts, for example, trihalomethanes and haloacetic acids that can cause health risks
2.	Inactivates most bacteria and viruses	Some protozoan cysts are more resistant
3.	Easy to operate and useful as emergency disinfection process	When added in excess, can alter taste and odor of water
4.	Ability to leave residual concentration in water distribution systems	Effective dose of chlorine depends on the pH and turbidity of water

to understand this issue faced globally. Essentially, they studied the DBPs formed in treated municipal water during chlorination. They state that several innovations and techniques such as coagulation with alum, lime, adsorption on activated carbons, membrane techniques and ion exchange processes can be employed to control or eliminate the DBPs after chlorine treatment. Thus the authors highlight that chlorine is the main disinfectant used globally and that the microbial contamination risk is of a greater magnitude as compared to the risk presented by the DBPs (Mazhar et al., 2020).

Thus chlorination can be considered as an effective and feasible method to disinfect water in developing countries. Although there are some disadvantages in its use such as the potential health risks due to DBP formation, it is an economical method. Therefore chlorination can be easily adopted in domestic scale (US\$ 2–5 m^{-3}), community systems of about 100MLD (US\$ 1–3.75 m^{-3}) and in large scale systems involving over 100 MLD (US\$0.05 m^{-3}) (Pandit & Kumar, 2019).

8.6 Filtration techniques

Filtration has a prominent place among the conventional methods to purify water. It is particularly a preferred choice among water purification techniques owing to its simplicity, ease of operation, low cost and perception. Based on the natural law of gravity, filtration has been employed by humans since ancient times. Most of these filtration techniques are conventional methods that are discussed below:

8.6.1 Traditional filtration methods

Slow sand filtration (SSF) is one of the oldest traditional methods of water filtration commonly used worldwide. Sand, which is a natural resource and being available in plenty has always been the most used filtration media across the globe. This method basically consists of passing the contaminated water through a bed of sand and gravel. As the filtration process is used, over time it results in the formation of a biological layer called "schmutzdecke." This thin biological layer along with the physical action of filtration together contribute to the water purification process. SSF is known to remove a spectrum of pollutants from untreated water such as turbidity, several organic and inorganic contaminants, bacteria, and even protozoan oocysts such as *Giardia* and *Cryptosporidium*. However, it cannot remove all pathogenic bacteria and viruses and often requires a chlorination step after filtration to ensure complete removal of contaminants. Sometimes, when the source water is turbid, a pretreatment step (sedimentation or coagulation) may be needed to prevent clogging of the filter. It is mostly used to treat surface water in small rural regions where land is not a constraining cause for SSF. Since sand and gravel are easily available and the process is gravity based, not requiring electricity, it can be easily adopted in developing countries. The typical construction expense of SSF is about US$15-US$60 and the cost per liter of treated water is reported as 0.068US cents [https://www.cdc.gov/safewater/sand-filtration.html (accessed on August 2021)].

Rapid sand filtration (RSF) is yet another conventional method of water filtration which is like SSF. This technique involves the use of coarse sand and various graded granular materials through which contaminated water is passed either by gravity or by applying pressure to enhance the flux of water/m^2 of filtration area. This method is most frequently adopted by municipal treatment facilities worldwide and requires less land as compared to SSF. The mechanism of action generally involves homogeneous, heterogeneous, and biological oxidation processes which results in the removal of bacteria, protozoa and some viruses. Pollutants such as manganese and iron are also eliminated successfully by this process. The main drawback is requirement of higher energy, more maintenance (due to clogging of filters) and higher cost (almost double the cost of SSF) (http://www.ircwash.org) as compared to SSF and hence may not be as suitable as SSF for low-income nations.

Activated carbon filtration is also a common and well-known technique of water purification in most parts of the world. Carbon is easily available from various sources such as coal and wood, and activated carbon is obtained by exposing carbon to an activation process by using chemical methods or steam processing. This results in the introduction of a slightly positive charge which helps in removing impurities during filtration (Pandit & Kumar, 2019). Activated carbon filtration results in the removal of most chemical, physical and biological pollutants mainly by the process of physical absorption. Moreover, the high surface area offered by the activated carbon is highly favorable for filtration and is flexible to alterations too. However, due to global deforestation issues, wood as a carbon source is not a good choice and on long-term use the activated carbon filters are prone to clogging. Granulated carbon is better than activated carbon as it can be reused and does not require constant replacement. These are one type of filters that can be easily adopted for water purification right from household to large scale systems in developing countries. The use of waste-agricultural biomass residue, replacing wood to make activated carbon could be a sustainable practice. The approximate cost of granulate activated carbon is around 10 cents to US$1.00 per 1000 gallons of water (http://www.wateronline.com).

Ceramic filtration is a popular technique routinely used in the developing countries for water purification. They are generally referred to as ceramic water filters (CWFs) and can be employed on a household or pilot scale level to purify water. These filters usually have a pore size of 0.2 microns and impurities larger than this size are easily deposited on the filter surface thereby resulting in the removal of water pollutants. The filter is typically in the shape of a flowerpot (8–10 L capacity) and is placed within a plastic receptacle. Water to be filtered is poured into the ceramic pot which gets filtered and treated water is collected in a storage container, fitted with an outlet tap (Fig. 8.2). Ceramic filters are usually coated

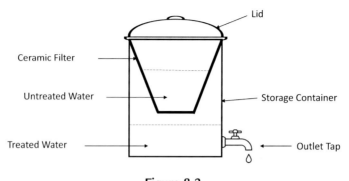

Figure 8.2
Ceramic filtration.

with colloidal silver to enhance the microbial disinfection efficacy and to prevent regrowth of bacteria within the filter [https://www.cdc.gov/safewater/ceramic-filtration.html (accessed August 2021)]. These filters have been effectively used to remove a spectrum of microbes from polluted water such as Fecal coliforms, *E. coli*, protozoa, and few viruses. The main attraction of these type of filters is that the community population can be involved in the production of these filters by utilizing locally available materials. This makes ceramic filters amazingly effective, simple, and economical. The cost per liter of treated water (inclusive of filter cost) is around 0.034–0.14 US cents [https://www.cdc.gov/safewater/ceramic-filtration.html (accessed May 2021)]. However, community compliance and filter maintenance are few challenges that needs to be addressed.

Over the years plenty of research investigations in the field of filtration has resulted in newer developments. Some studies have focused on ways to enhance the efficiency of filtration, while few investigations were made on modifications of the existing ways to make filtration economically feasible. Better and novel filtration materials with suitability to remove specific class of pollutants such as chemicals, physical and microbiological are yet another area of research that has received attention. To improve the efficacy and economic feasibility in some cases, these traditional methods have been improvised by several experts worldwide. Some of these recent research studies are discussed in the following section.

8.6.2 Recent advances/modifications in traditional filtration methods

In an interesting study, an economic and nonconventional SSF system with a cross sectional bed area of 0.0576 m^2 was operated under the hydraulic loading of 4.6296 L s m^{-2} for a period of one hour. The filtered water was tested for various parameters like salinity, conductivity, pH, and turbidity to name some and the results were compared with Indian standard parameters. This was found to be highly effective for improving household water quality as the designed SSF could remove several impurities (suspended solids, natural and organic pollutants) as compared to other filters. Moreover, the authors state that it was socially acceptable too (Agrawal et al., 2021).

In the field of RSF, it is interesting to note that recently Kramer et al. (2021) have highlighted some hydraulic insights on the RSF bed backwashing by using Carman-Kozeny model. This model is a standard model that is used for assessing the resistance (pressure head loss) through filter beds. The authors state that up-flow filtration in rapid sand filters under backwash conditions triggers the particle bed to collapse almost unnoticeably. Further, they state that the Carman-Kozeny model constant was not constant for increasing flow rates and the authors also put forward a new pseudo 3D image evaluation for particles with irregular shape. Thus the authors report a hydraulic reason for the extended terminal subfluidization (expanded bed) wash (ETSW) filter backwashing procedure in which turbidity and peaks in the number of particles are lowered thereby improving the water quality (Kramer et al., 2021).

In an innovative attempt to modify filtration media for point of use filtration (PoU), Jakubczak et al. (2021) have reported the use of a 2D nanocomposite-modified filtration material (polypropylene base). This modified material showed improved filtration velocity, effective microbial elimination (10^5 bacterial cells/cm^2 of filter material) and self-disinfecting potential (could eliminate 99.6 wt.% of bacteria collected on the filter) as compared to pristine polypropylene (unmodified reference material used in the study). Moreover, the material was found to be stable and environmentally safe as no secondary release of nanocomposites into the filtrate was observed making it potentially appropriate for PoU water treatments (Jakubczak et al., 2021).

In yet another outstanding investigation, Jiao et al. (2020) fabricated a 3D-activated wood filter that had an adsorption capacity of 198.64 mg g^{-1} and adsorption rate of almost 99.52% within 5 mins, for methylene blue (conducted in a proof-of-concept study). The gigantic vertically aligned channels found in the carbonized wood facilitated a high flow rate of water and the pollutants were successfully absorbed on the nanoporous channel walls of the activated wood. These filters could be regenerated for reuse by a simple carbonization process and were found to be a highly suitable replacement for portable commercial filters particularly in developing countries (Jiao et al., 2020).

In a unique and innovative study, Raimann et al. (2020) demonstrated that repurposed hemodialyzers that were once used as renal replacement therapy to save lives could be used again as an effective and affordable water purifying means to protect the residents in villages in rural Ghana. Surprisingly, 72% reduction in cases of diarrhea were noticed in the population using the repurposed hemodialyzers as a water treatment method as compared to the control group, that did not use this device. This hemodialysis membrane filter had hollow fibers with pore size of 0.003 μm that could effectively remove coliform bacteria from the source water (558 CFU/100 mL and zero CFU in the filtrate). The authors have found this unique filtration device potentially suitable as a good means of water purification for the world populace particularly in developing nations (Raimann et al., 2020).

The preceding discussion on filtration and its recent modifications highlights the potential suitability of these techniques for water purification in developing nations. Moreover, filtration method is also amenable for combination with other water purifying technologies to give a synergistic effect in the treatment of water. Such methods, generally termed as "Hybrid Filtration" are briefly discussed in the following section.

8.6.3 Hybrid filtration methods

Water filters attached to hand pumps have been used to remove arsenic from drinking water in West Bengal and Chhattisgarh, India. In this unique hybrid method, water to be treated passes through a muslin cloth which removes debris and subsequently through a bed of adsorbents consisting of iron oxide and charcoal that adsorb the arsenic present in water. The inlet of the filter (fitted in a simple PVC pipe) is designed to fit into outlet of a handpump. Almost 70%–80% arsenic removal coupled with no energy requirement due to the use of hand pumps makes this very economical (INR 2000 with filtration capacity of 20,000 L/day) for developing countries (Pandit & Kumar, 2019).

In another interesting hybrid filtration technique, Yamaguchi et al. (2017) studied the use of polymeric microfiltration membrane in a gravitational filtration mode (without electric driven pumps) along with GAC impregnated with copper (1 wt.%) to improve the quality of drinking water in developing countries. The results showed that the hybrid system provided superior chlorine removals (>90%), better permeate flux and higher *E. coli* removal (99.99%) as compared to the filtration system without GAC. Thus the authors suggest that this hybrid system has a promising potential as a point of use water treatment method in developing nations (Yamaguchi et al., 2017).

In a similar noteworthy research of hybrid water treatment technique involving decentralized ultrafiltration (UF) and coagulation with polyaluminum chloride (PACl), it was observed that the use of hybrid method resulted in 21% augmentation in the flux rate permitting for a 27% reduction in the membrane area thereby providing cost benefits as compared to the UF system used alone without coagulation. Moreover, the PACl coagulation decreased membrane fouling and improved the permeate water quality by augmenting removal of color, turbidity, and dissolved organic carbon (DOC). The researchers state that the hybrid system was inexpensive in terms of both the capital and operating costs and recommend PACl coagulation as a promising and economical antifouling technique for decentralized membrane-based water schemes (Arhin et al., 2019).

Yet another study conducted by Akosile et al. (2020) in the Ekiti state, Nigeria highlights the use of a mixture of different clay materials and sawdust in various combinations to produce low cost ceramic filtration units suitable as household water purification devices. They report that among several mixtures studied, the 50%−50% ratio of Igbara odo clay to sawdust was the best combination and this hybrid method gave a flow rate of 1.9 L h^{-1} and resulted in 80% and 100% removal of coliform and *E. coli*, respectively. The authors state that this combination can be used to produce ceramic filters on a large scale and that the addition of sawdust (in the range of 30%−70%) provides an opportunity to create insulating products by increasing the porosity of the filter (Akosile et al.,2020). These types of ceramic filters that employ hybrid filter materials appear to be promising in improving the physical and bacteriological water quality in low-income countries.

The hybrid methods described here are presented in Table 8.4. Thus it can be concluded that hybrid water treatment methods are often beneficial on account of the synergistic mechanisms involved in the process of water purification which yield higher efficiency in terms of pollutant removal. Additionally, most of the methods are also economical thereby being potentially suitable for water treatment in developing nations.

8.7 Natural treatment methods

Herbs are plants that are naturally occurring in all parts of the world and most of its parts such as seeds, leaves, stems, and roots have been routinely used since ancient times for medicinal purposes. For instance, Tulsi (*Ocimum sanctum*) has been traditional used in South Asia especially in India for its antimicrobial action and is often added to water to disinfect it. Similarly, drumstick seeds (*Moringa oleifera*) has been commonly used to clarify water in North Africa for decades. A spectrum of herbs and their active ingredients available from a plethora of plants are available globally for water purification. Some of these have coagulating properties while other exhibit antimicrobial and antiviral action on

Table 8.4: Few hybrid filtration methods potentially applicable for developing countries.

Sr. no	Hybrid filtration method	Process involved	Contaminants removed	Developing country
1.	Water filter made of iron oxide and charcoal + hand pump	Filtration and adsorption	70%−80% arsenic removal	West Bengal and Chhattisgarh, India
2.	Polymeric microfiltration membrane + GAC	Microfiltration and adsorption	>90% Chlorine removal 99.99% *Escherichia coli* removal	Highly recommended for developing countries
3.	Decentralized ultrafiltration (UF) and coagulation with polyaluminum chloride (PACl),	UF and coagulation	Enhanced removal of turbidity, color DOC	Kampala, Uganda
4.	Ceramic filters made of different combinations of clay and sawdust	Filtration and absorption	80% coliform removal and 100% *E. coli* removal	Ekiti State, Nigeria

water. As herbs are locally available and inexpensive, they can be easily used for water treatment in developing countries. There are a lot of studies that have been carried out on use of herbs for water treatment and among all the herbs, *M. oleifera* appears to be extensively researched. This may be attributed to its abundant availability, low cost, reduced by-product production, easy biodegradability, nontoxicity, and multifunctional behavior in removing several pollutants from water (Yamaguchi et al., 2021). Herbs such as Tamarind (*Tamarindus indica*), Okra (*Hibiscus esculentus*) and Neem (*Azadirachta indica*) have also received some attention in the field of water treatment.

It is fascinating to note that over last few years, herbal method of water purification are also being considered as hybrid technologies like filtration (discussed in preceding section) either in combination with other chemical coagulants, as mixture of two herbs or as a part of treatment train along with other processes like adsorption, filtrations, and solar disinfection. In one such especially useful and interesting study on hybrid water treatment with herbs, Varkey (2020) treated river water samples with the combination of *M. oleifera* and copper. Essentially, *M. oleifera* seed powder (as a coagulant and flocculant) was mixed with river water sample and a copper wire mesh (as a disinfectant) was placed in it. After 4 hours, the supernatant was decanted using a muslin cloth and analyzed for turbidity and *E. coli* counts. Amazingly, no *E. coli* was detected in the treated water and the turbidity levels was in the range of 3 NTU to 5 NTU where NTU refers to the nephelometric turbidity unit. The author recommends this combination as a point of use (PoU) method for water treatment in developing countries especially after natural disasters as this technique is easy to use without the need for electricity and is also cost effective (approximately 0.5 c per liter of potable water) (Varkey, 2020). Many such experiments have been reported in the literature and few instances of herbal water treatment methods alone or in combination with other techniques investigated in recent times is presented in Table 8.5.

Although there are several research investigations in hybrid treatment of water with herbs, there appears to be a lack of studies on a pilot scale to enable its implementation on a larger scale especially in developing countries. Moreover, the toxicity aspect of herbs (except in case of *M. oleifera* where some research has been done) in drinking water must be ascertained and more research is needed in this area. However, herbal methods appear to be a promising and environmentally friendly water treatment technology both as a stand-alone method or when used as hybrid processes for developing nations especially because they are economical and amply available in nature.

8.8 Cavitation-based water hand pumps

Cavitation is a phenomenon that results in the nucleation, growth and collapse of gas or vapor filled cavities in a liquid medium. This can be achieved by modification in the flow and pressure of the liquid as it goes via a narrow constriction (hydrodynamic cavitation) or

Table 8.5: Water treatment by herbs and its combination with other treatment methods.

S. no.	Herb/hybrid process	Mechanism involved	Pollutants removed	Developing country	Reference
1.	*Moringa oleifera* + filtration + SODIS	Coagulation, filtration and disinfection	Turbidity (99%) suspended solids (66%), total and fecal coliforms (97%–99%)	Rural Tanzania	Nancy Jotham Marobhe (2021)
2.	*M. oleifera* + Copper	Coagulation and disinfection	Turbidity (3NTU–5NTU), *Escherichia coli* (100%)	Recommended for rural communities	Varkey (2020)
3.	*M. oleifera* + *Azadirachta indica* (Neem)	Coagulation	Turbidity (15.70–5.10), Fluoride (2.80–1.00), *E. coli* (325–30)	Uttar Pradesh, India	Pandey et al. (2020)
4.	Filter + Herbs (Neem, Tulsi, Lemon) + Adsorption	Filtration, coagulation and disinfection	Turbidity (55%), BOD (85%), COD (84%) and chromium (60%)	Uttar Pradesh, India	Sharma et al. (2020)
5.	*M. oleifera, Tamarindus indica*	Coagulation	92.1% to 93% removal of fluoride from water	Recommended for developing countries	Dhama and Kuriakose (2021)

by the passage of ultrasound (acoustic cavitation). The main emphasis is on hydrodynamic cavitation where the collapse of cavities results in high temperatures (10,000K) and pressures (5000 atm) which brings about microbial inactivation. Therefore hydrodynamic cavitation has been investigated for microbial cell disruption, waste-water treatment and drinking water disinfection. Several studies on the use of hydrodynamic cavitation for water disinfection have been conducted by experts globally including the authors and it has been found to be an economical and cost effective water treatment process especially for the developing countries.

Hand pumps are widely used in rural areas of developing nations to draw ground water. In a pioneering work conducted by the author, the India Mark II Hand Pump (commonly used hand pump in Indian villages) was modified to create cavitating conditions which could successfully disinfect borewell water in rural areas of Maharashtra, India. In the field trials conducted by the author and team, different modified check valves were fitted in the hand pump assembly. They report that 40%–60% flow area opening as a function of lift is the optimum dimension of the check valve that could inactivated almost 85%–90% of *E. coli*. Moreover, since the cell wall strength of most pathogenic microorganisms are lower than *E. coli*, it is expected that the treated water will have fewer pathogenic bacteria. Thus use of modified handpumps result in cavitation that can efficiently kill microorganisms and disinfect water in rural regions where there is no access to electricity and safe drinking water (Pandit & Kumar, 2019).

Hydrodynamic cavitation has also been used to clean river water and in one such innovative initiative by the author and team, Rankala Lake located in Kolhapur, Maharashtra, India was treated with hydrodynamic cavitation in about a period of 3 weeks (8 hours/day). The treatment essentially consisted of passing the Rankala Lake water thorough a cavitating reactor with the help of a centrifugal pump (Fig. 8.3). The set-up was operated at an operating pressure of 4 bar and a volumetric flow rate of 100 m^3 hour^{-1}. The hydrodynamic cavitation generated in the set-up resulted in intense collapse of cavities which brought about the inactivation and/or removal of various chemical, physical and microbiological contaminants. It was observed that the lake water improved substantially in color and odor within a week after treatment. Moreover, the COD and BOD also reduced from 110 ppm to 30 ppm and 40 ppm to zero value, respectively. Amazingly, the microbial content of the lake water also dropped from 10^5 CFU/mL to 10^2 CFU/mL. The author reports the power requirement for this process to be around 8.34 kW h and the cost of treatment to be approximately 17.52 INR per m^3 of water. The total cost of treatment is reported to be INR 262710 (Pandit & Kumar, 2019). Thus hydrodynamic cavitation appears to a promising, feasible and cost- effective water treatment method for low-income countries with an access to electricity.

Hydrodynamic cavitation has also been explored as a hybrid technology with other treatment methods for water disinfection. The authors have investigated the use of hydrodynamic cavitation with chemical disinfectants such as hydrogen peroxide and ozone for treating bore well water. They recommend the use of hybrid method as an excellent alternative to single methodologies as the synergistic effects of cavitation and chemicals were able to inactivate heterotrophic bacteria, *E. coli*, fecal coliforms and fecal streptococci (Jyoti & Pandit, 2003).

Recently, similar attempts have been made to combine hydrodynamic cavitation with plasma discharge for treatment of polluted water. In the prototype investigated, the hybrid technique brought about intense generation of radicals, shock waves, UV light, and charged

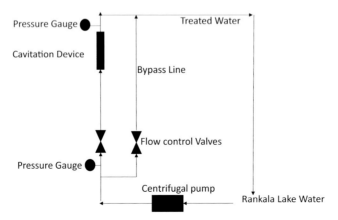

Figure 8.3
Hydrodynamic cavitation for Rankala Lake water treatment.

particles which could suppress *E. coli* and degrade organic pollutants. The energy requirement is reported as 1–4 kW, the estimated internal costs of discharge chambers, power blocks and electrodes were 100–5500 euro, and the maintenance cost was stated as 50 to 100 euro per year (The authors referred to costs in the Russian Federation at 1 kW per hour costs about 0.05 euro) (Abramov et al., 2021).

Yet another recent and pioneering attempt in the use of hybrid water treatment method using hydrodynamic cavitation and natural oil was carried out by Mane et al. (2020). They used two devices for hydrodynamic cavitation, a vortex diode (vortex flow) and an orifice-based device (linear flow) in combination with peppermint oil (0.1%) to inactivate opportunistic bacteria (*Pseudomonas aeruginosa* and methicillin resistant *Staphylococcus aureus*) and antimicrobial resistant bacteria (AMR). Interestingly, greater than 99% disinfection was achieved using the combination of vortex diode (operating at 1 bar) and 0.1% peppermint oil in less than 10 min of treatment. The disinfection time needed in the case of the orifice device (operation at 2 bars) was relatively more (about 20 min). The cost of operation of hydrodynamic cavitation using vortex diode was reported as 0.22\$ m^{-3}, which reduced substantially to about 0.0036\$ m^{-3} when used in combination with 0.1% peppermint oil for total disinfection of water. Thus due to high disinfection efficacy and cost effectiveness, the authors suggest the commercialization of this hybrid cavitation method by integrating it along with other treatment steps involved in a typical water treatment scheme to provide safe drinking water (Mane et al., 2020).

Therefore hydrodynamic cavitation can be considered as a potential scalable method to treat water in developing countries either alone or as a hybrid technique with other disinfectants/methods as discussed in the preceding section. This can be attributed to its negligible power consumption, ease of operation, environmentally friendly and easily affordable methodology for rural areas in low-income countries.

8.9 Sustainable disinfection of harvested rainwater

Rainwater harvesting is an ancient and popular technique used globally to conserve water. Rain, being a natural resource is nature's gift to mankind and this precious source of water, if collected can be employed as a good supply of water. This is especially true for people living in geographical regions that receive ample rainfall and most developing countries appear to fall in this category. Rainwater harvesting is defined as the collection of water from surfaces on which rain falls, and the subsequent storage of this water for later use (http://www.sustainable.com.au). A typical rainwater treatment system essentially involves collecting rain on surfaces called "catchment area" which could be a rooftop. A conveyance system (guttering) transports this collected water to storage tanks where the harvested rainwater is stored until further use. A vital part of the conveyance system is the "First Flush Device" which diverts the initial part of rainwater (containing dirt and impurities) away from the storage tank. This significantly enhances the water quality and reduces the burden on subsequent water treatment methods. Finally, the harvested rainwater quality is improved by treatments like filtration and disinfection (Fig. 8.4).

Most common uses of the collected rainwater are for domestic and agricultural purposes. Among, the rural communities in low-income regions, the harvested rainwater is commonly used for drinking apart from other uses such as bathing, cooking, etc. The level of treatment of harvested rainwater depends on its end use and needless to state that it should be disinfected if used for potable purposes. Among the spectrum of methods used for treatment of harvested rainwater, those that are typically suited to developing countries are briefly discussed here.

Chlorination is a common method employed to disinfect harvested rainwater as it is an easy, affordable, and effective technique to inactivate most microorganisms. Several aspects of chlorination have been already discussed in the preceding sections and only chlorination with respect to harvested rainwater treatment is described here. In one of the most recent initiative, Richards et al. (2021) investigated a roof top rainwater harvesting system (RWHS) to alleviate the water demands in a remote school in rural India. It was observed that the stored rainwater could be used devoid of treatment at the beginning of collection. However, it could not be used for a long time without treatment owing to microbial growth. Therefore the authors used chlorination which was effective in reducing the microbial growth in stored water for more than a month. The work demonstrated that the installed RWHS could reduce the burden on the prevailing water source at the school by up to 25% of the water that was utilized for washing and flushing with no treatment. Moreover, the authors state that with chlorination, numerous uses of water could be achieved (Richards et al., 2021).

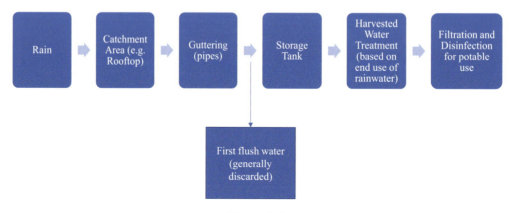

Figure 8.4
A typical rainwater harvesting process.

Filtration is yet another popular and accepted method of purifying harvested rainwater that is generally applied in almost all the RWHS worldwide. Most conventional filtration techniques described in the preceding section for purifying water are also applicable here. One of the recent innovations in user-friendly and low cost (150 US$) filtration techniques for harvested rainwater was demonstrated by da Costa et al. (2021). The study involved the use of Acrylon (an acrylic blanket) as an inexpensive and easy to use filtering medium to treat harvested rainwater in Itaborai, Rio de Janeiro, Brazil. Interestingly, almost 81.70% color removal efficiency was obtained in the case of the first flush (sample collected within 10 min of rain) and 49.54% for continuous samples (sample collected within 15–30 mins and after 30 mins of rainfall). The turbidity removal was 74.28% and 52.19%, respectively. The treatment efficacy is similar to conventional filtration systems, and the authors observed that the results met the requirements of the Brazilian standard for usage of rainwater based on turbidity, residual chlorine, and absence of coliforms. However, it did not meet the requirements for hydrogen potential and color removal (da Costa et al., 2021).

Solar disinfection is also frequently used to treat harvested rainwater and in one such study, it was found that solar pasteurization (SOPAS at 70°C–79°C; 80°C–89°C; >90°C) and solar disinfection (SODIS for 6–8 hours.) both could reduce the indicator microorganisms in harvested rainwater to almost below detection levels (>99%). Pathogenic organisms such as viable *Legionella* (2 log reduction by SOPAS and SODIS after 8 hours.) and *Pseudomonas* counts (3 log reduction by SOPAS and 1 log reduction by SODIS after 8 hours.) could also be considerably reduced (Strauss et al., 2016).

Likewise, *hybrid techniques* have been exploited successfully for treating harvested rainwater. In one such interesting study conducted at Ton Duc Thang University, Vietnam, rainwater was collected from rooftop made of galvanized iron which was followed by a 500 L First Flush Device and collected in four stainless steel storage tanks (1 m^3). The treatment encompassed sediment filtration (5 μm), nanofiltration and UV light (29 W Aqua Pro). Turbidity decreased from 0.15–1.63 NTU to 0.07–0.9 NTU after the hybrid treatment and amazingly the total coliforms and *E. coli* were completely inactivated indicating that the rainwater could be used for potable purposes (Bui et al., 2020).

Thus sustainable rainwater harvesting is indeed particularly useful globally, especially in the developing nations as it reduces the dependency on other water sources such as surface and ground water and contributes to water conservation. However, challenges such as dependency on climatical conditions, storage facilities needed and contamination of collected rainwater either during harvesting or storage are some impediments that need attention. Nevertheless, RWH appears to be very much in place across the world owing to its ease of use and cost effectiveness. For instance, a passive RWH system (that use simple plastic barrels) of 50 gallons (230 L) capacity costs approximately $70 (http://www.epa.gov) and active systems (using cisterns or tanks) cost between $1.50 to $ 3.00 per gallon (4.5 L) of storage with an additional $2-$5 per gallon (4.5 L) of harvesting system capacity for larger systems (http://www.epa.gov).

8.10 Recent research on emerging methods

The augmenting demand for safe and clean water has spurred a lot of research on novel water treatment technologies globally. Experts and researchers across the world are coming up with new yet simple and inexpensive strategies to provide clean water to the people in the developing nations. Most of these emerging methods appear to be based either on filtration, solar disinfection or hybrid methods involving both. Few instances under each of these categories are briefly discussed:

8.10.1 Filtration

"Lifestraw" is a popular portable water filter designed for individual and family use that employs hollow filters of less than 0.2 micron pore size in a simple plastic tube to treat unclean water. It is used like a straw by dipping it into the water (untreated) and sucking filtered water into the mouth by the user. Almost 99.99% bacterial removal and protozoan cysts have been reported and it costs around US$15 per 1000 L of water [http://www.lifestraw.com (accessed August 2021)]. Similarly, another unique household level water treatment innovation called the "Drinkable book" made of special filter paper (silver nanoparticles coated on cellulose fibers of absorbent blotting paper) has been used for removal of *E. coli* and *Enterococcus faecalis* (99.9%) (Dankovich, 2014). This easy to use, simple, and cost effective technology (10 cents per paper and can treat about 100 L of water) which has been field tested in South Africa, Ghana, and Kenya appears to be promising for the developing countries. "Cycleclean," a water purifying bicycle is a fascinating invention that produces clean water by merely pedaling it. It essentially uses a water purification system consisting of four filters attached to the bicycle which eliminates most physical, chemical and biological impurities from water and is designed to provide about 5 L of water per minute. It is well suited for remote areas without access to electric supply or in emergencies like natural disasters. Although it costs around US$6600, with some efforts on its price reduction, it can be an extremely useful way for proving clean water in developing nations (http://www.nipponbasic.ecnet.jp).

8.10.2 Solar disinfection

"Solar ball," an innovate, simple, easy to use spherical device designed by a Monash University graduate is a unique technique to treat (3 L of clean water per day) water based on solar evaporation. The water to be treated is placed in the spherical device and exposed to sunlight. This causes evaporation of the water, separation of impurities and results in generation of clean condensation on the roof of the device. This is then collected and stored as safe drinking water. This is at the prototype stage and costs around US$35 to US$50 per unit (http://www.livingnow.com.au). Recently, a similar device based on solar evaporation was developed by Haolan Xu, Associate Professor, University of South Australia. This set-up uses simple materials with a specialized film (photothermal sheet assembled in a compact frame) that converts solar energy to heat. The professor and his team state that a surface area of just 1 square meter of polluted water could generate between 10 to 20 L of potable water everyday which can suffice for a family of four. This technology which is still in the early developmental stages has immense potential to be exploited in the developing countries (http://www.abc.net.au).

8.10.3 Hybrid techniques

Life sack, a simple recycled plastic bag made of polyvinyl chloride (PVC) has been used to treat water in rural areas. The combined effect of solar rays that fall on the bag (containing untreated water) along with filtration through a 15 nm filter embedded in the bag results in the inactivation of most microorganisms. Although, the exact price of these bags is not known, it appears to be inexpensive and suitable for providing water for a family (approximately 20 L) (http://www.causetech.net). A similar hybrid method using filters and UV in a dual chamber compact bottle called "Pure Water Bottle" is yet another simple, portable and easy to use water treatment technique. Although it is still in the prototype stage, it has been proved useful for removing most pollutants (99.9%) from any water source and is well suited for developing countries (http://www.inhabitat.com).

Lack of electric supply and access to safe and clean drinking water are two main challenges faced by most developing countries. Both these limitations were suitably addressed in a pioneering field trial conducted in rural Mexico by Pichel et al. (2020). They used an innovative hybrid technology comprising of photovoltaic and solar disinfection system (SolWat) that essentially includes a photovoltaic component and a water disinfection device totally combined into a single unit. The hybrid unit was able to simultaneously produce clean water and electricity. The authors report total inactivation of *E. coli* and total coliforms present in water after 3 hours of sun exposure (Pichel et al., 2020). This is indeed a promising and valuable investigation in the arena of water treatment for low-income nations.

Thus most of the emerging techniques described here appear to be convenient, sustainable and cost effective for application in any rural setting. However, many of them are still in the prototype stage except for Lifestraw which are available and marketed in large numbers. With the involvement of various stakeholders, NGOs and with end user compliance these and such similar innovations will definitely see the light of the day.

8.11 Impact of Covid 19 pandemic on the water sector in developing countries

The Covid 19 pandemic occurred around late 2019 and has severely impacted the whole world. This acute respiratory disease caused by Covid 19 virus (SARS-CoV-2), a member of the coronavirus family has not only affected many lives but also impacted several industrial segments globally out of which the water sector is one major part. In response to the outbreak of the pandemic and to prevent its spread, the WHO issued guidelines which involved social distancing, wearing masks and frequent handwashing (http://www.who.int/emergencies/diseases/novel-coronavirus-2019/advice-for-public). Needless to state, plenty of water is required to ensure that every individual can wash his/her hands frequently. This appears to be a huge challenge in the developing countries which lack adequate water supply even for their routine and regular needs of drinking and other domestic uses.

Although the transmission of the Covid 19 virus is mainly through respiratory droplets and direct contact, the possibility of infection through fecal matter is also possible. Bhowmick et al. (2020) state that conventional sewage treatment involving chlorination and UV irradiation are likely to eradicate the Covid 19 virus. However, in densely populous countries like India, where there are lower sewage treatment facilities, the virus can survive up to several days in untreated sewage and pose a health risk to mankind. Fecal contamination of drinking water is a common challenge faced globally especially in the developing nations and the possibility of the Covid 19 virus spreading through fecal matter will only worsen the situation. The authors recommend preventing water pollution at consumption point, implementation of WHO guidelines and ensuring sufficient free residual chlorine in the drinking water as some ways to control the pandemic (Bhowmick et al., 2020). Furthermore, Bandala et al. (2021) have recently reviewed the impact of Covid-19 pandemic on the wastewater pathway into surface water and state that new Covid 19 virus quantification methods are required for wastewater. They recommend that innovative, low cost easy to use sensors for identification of the virus should be investigated by experts (Bandala et al., 2021). Similar views are expressed by Panchal et al. (2021) on the need for supervising the Covid 19 virus through sewage-based epidemiology. They state that this could provide useful evidence on the level of contamination and tracking the sewage systems could play a pivotal role as an early warning method for alerting the public health agencies. This in turn could help them take necessary actions to prevent the spread of the Covid 19 infection (Panchal et al., 2021).

The Pandemic has impacted the water sector in several ways. (see Fig. 8.5) Lack of water for drinking and hygiene practices (especially for handwashing in the pandemic) appears to have further exacerbated in the rural and urban areas of the developing countries. Apart from this, the increasing unemployment due to the impact of lockdown has put financial pressure on several households as they are unable to pay their water bills. However, many countries are waiving off the water bills to help people in the pandemic scenario. For instance, in the US, about 57 million citizens in numerous states have been permitted to get water from their utilities even if they cannot presently reimburse for it (Tortajada & Biswas, 2020). A recent article, by the International Finance Corporation (IFC), World Bank Group, discusses the impact of Covid 19 on water and sanitation sector and states that the pandemic could slow down the advancement in meeting the GDC 6 development goal (Ensure availability and sustainable management of water and sanitation for all by 2030) as water services face huge revenue losses which further accentuate their capability to make critical capital investments. In response to the pandemic, the IFC has been proactive in proving liquidity financing provision to their present clients and long-term financing to assist water utilities embarking on projects that guarantee steadiness of service in the short to medium duration. Moreover, several webinars on crisis recuperation are held by the IFC for water utilities to improve their capacity building.

It is indeed heartening to note the efforts made by several companies to ease the effects of the pandemic on the public at large. For instance, the Grifaid Community Filter (made in United Kingdom by a not-for-profit company, "The Safe Water Trust") is helping to

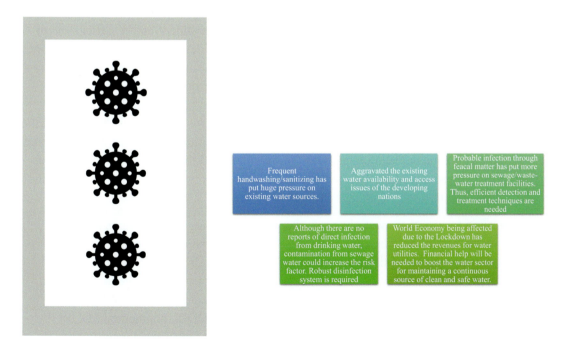

Figure 8.5
Few impacts of the Covid 19 pandemic on the water sector.

provide safe water for cleaning and drinking in hospitals, clinics and vaccination centers in the developing countries. This compact and portable filter works on advanced membrane technology that does not work on electricity making it suitable for healthcare centers in developing countries. It is reported to remove most bacteria and viruses from water and can provide around 300 L of treated water per hour. Moreover, the filter has a unique built-in cross flow cleaning system that prevents clogging of the membrane making it an overall cost effective and sustainable method for the developing nations (http://www.filtsep.com).

Thus it can be observed that the Pandemic has affected human lives, health and world economy in many ways and several such areas of impact are slowly surfacing as the Pandemic continues to shake the world. However, the Covid 19 pandemic has also been an eye opener for individuals, communities, and nations to be more alert and conscious about hygiene and this will also motivate the scientific community to bring out many innovations in the water and wastewater treatment methods to combat the spread of the virus in times to come.

8.12 Comparison of various purification techniques

Some of the important aspects on which water treatment methods can be compared is its ability to remove impurities from source water, the scale at which it can be operated and most importantly its cost effectiveness. In fact, the cost of treatment and its efficacy is of paramount importance when water treatment is applied in a developing country. Moreover, it is crucial to note that the selection of a suitable water treatment technology depends on a plethora of factors such as the source water quality, types of pollutants, efficiency, and affordability of the treatment process. It is also imperative to understand that there is no single universal water treatment method and a proper judicious choice must be made on the aspects discussed above. Table 8.6 compares most of the methods discussed in this chapter based on the contaminants eliminated, scale of treatment and the cost. A brief, discussion on the same is presented in the following section. These recommendations only denote the authors views and may not be always suitable in all cases.

It is interesting to note that most physical, chemical and biological pollutants are effectively removed by the traditional filtration methods (SSF, RSF, ceramic and activated carbon filters), hydrodynamic cavitation and herbal treatment methods. Among emerging techniques, the pure water bottle, water purifying bicycle, solar ball and device based on solar evaporation are also efficient approaches to remove most impurities. When only biological contaminants must be eliminated, conventional SODIS, augmented SODIS (can remove viruses also), the drinkable book, Life sack and Lifestraw appear to be good choices. However, they are not ideal methods to remove physical and chemical impurities. When opportunistic and antibiotic resistant bacteria are to be inactivated, hybrid methods like hydrodynamic cavitation and peppermint oil are suitable techniques.

Table 8.6: Comparison of water treatment methods suited for developing countries discussed in this chapter.

S. no	Method	Contaminants removed	Treatment scale	Approximate cost
1.	Conventional SODIS	Bacteria, fungi and protozoa	Household level	US$ 0.5 m^{-3} (100 uses of PET bottles) (Plappally et al., 2013)
2.	Augmented SODIS	Bacteria, fungi, protozoa and some viruses	Can be scaled to community level	Costlier than conventional SODIS due to added cost of additives/modification
3.	Boiling	Bacteria, protozoa, viruses	Point of use, household level	Monthly fuel cost US$0.88 (LPG) US$0.69 (wood) (Clasen et al., 2008)
4.	Chlorination	Bacteria and most viruses	Domestic, community (100 MLD) and large scale use (>100 MLD)	Domestic scale (US$ 2–5 m^{-3}), community (US$ 1–3.75 m^{-3}) large scale (US $0.05 m^{-3}) (Pandit & Kumar, 2019)
5.	Traditional filtration methods [slow sand filtration (SSF), RSF, ceramic and activated carbon filters]	Most physical, chemical and biological pollutants	Household to large scale except RSF suited for large scale only	SSF- 0.068US cents L^{-1} [https://www.cdc.gov/safewater/sand-filtration.html (accessed on August 2021)]. Activated Carbon Filters-10 cents to US $1.00/1000 gallons (http://www.wateronline.com) Ceramic filters- 0.034–0.14 US cents L^{-1} [https://www.cdc.gov/safewater/ceramic-filtration.html (accessed August 2021)].
6.	Hybrid filtration (filters fitted to handpumps)	Arsenic	Community level	INR 2000 with filtration capacity of 20,000 L per day (Pandit & Kumar, 2019)
7.	Herbs	Most physical, chemical and biological pollutants	Household to large scale	Herbs are naturally available hence cost is insignificant
8.	*Moringa oleifera* and copper	Turbidity and *Escherichia coli*	PoU	0.5 c per liter of potable water (Varkey, 2020)
9.	Hydrodynamic cavitation (Rankala Lake treatment)	Most physical, chemical and biological pollutants	Community level	17.52 INR per m^3 of water. The total cost = INR 262,710 (Pandit & Kumar, 2019)
10.	Hydrodynamic cavitation with plasma discharge	*E. coli* and organic pollutants	Prototype stage	Internal costs - 100 to 5500 euro and maintenance cost - 50 to 100 euro per year (Abramov et al., 2021)

(*Continued*)

Table 8.6: (Continued)

S. no	Method	Contaminants removed	Treatment scale	Approximate cost
11.	Hydrodynamic cavitation and 0.1% peppermint oil	Opportunistic and antimicrobial resistant bacteria bacteria	The set-up storage tank had 50 L capacity	0.0036$/m^3 (Mane et al., 2020)
12.	Treatment of harvested rainwater filtration using "Acrylon" (an acrylic blanket)	Turbidity and coliforms	Household and industrial use	The system cost was 150 USD (da Costa et al., 2021)
13.	Lifestraw	Bacterial and protozoan cysts	Individual and household level	US$15 per 1000 L of water [http://www.lifestraw.com (accessed August 2021)]
14.	Drinkable book	*E. coli* and *Enterococcus faecalis*	Household Level	10 cents per paper (treats about 100 L of water) (Dankovich, 2014)
15.	"Cycleclean," water purifying bicycle	Most physical, chemical and biological impurities	Household and community level	US$6600 (http://www.nipponbasic.ecnet.jp)
16.	Solar ball	Most physical, chemical and biological impurities	Prototype phase	US$35 to US$50 per unit (http://www.livingnow.com.au)
17.	Device based on solar evaporation	Most physical, chemical and biological impurities	In developmental stage (household level)	Exact cost not available (http://www.abc.net.au)
18.	Life sack	Most microorganisms	Household level	Exact cost not available (http://www.causetech.net)
19.	Pure water bottle	Most physical, chemical and biological impurities	Prototype phase	Exact cost not available (http://www.inhabitat.com)

Similarly, when the scale of treatment is considered, SODIS (conventional and augmented), The Drinkable book, Life sack, Life straw, water purifying bicycle and boiling are appropriate methods for household level. Interestingly some methods like the augmented SODIS and water purifying bicycle can be scaled up to the community level to cater to the needs of a larger population. Hydrodynamic cavitation and hybrid methods using hand pump also fall under the category of community level. Traditional filtration methods except the RSF (which is mostly suitable to treat water at municipal scale) are all capable of being used right from household to large scale levels. Similarly, chlorination and herbal water treatment techniques can be easily tweaked to fulfill the needs from household to community and large scale levels. Beguilingly, several innovative methods such as the pure water bottle, hydrodynamic cavitation with plasma discharge, and devices based on solar evaporation are in the prototype stage which are also potential methodologies for the developing nations.

When it comes to cost effectiveness which is one of the foremost aspects for selecting a water treatment method, SODIS, boiling, chlorination, traditional filtration, hydrodynamic cavitation, drinkable book and herbal disinfection methods are relatively inexpensive (<1US \$ m^{-3}) and most suitable for developing countries. Some of the hybrid methods like hydrodynamic cavitation with peppermint oil are also cost effective (US\$0.0036 m^{-3}). Techniques such as Lifestraw (US\$15 m^{-3}) and solar ball (US\$35-US\$50 per unit) are in the medium cost range and can be used if affordable. However, few techniques such as the water purifying bicycle are expensive at US\$6600 and can be employed only if sufficient funding from concerned stakeholders and/or the government is available. Many methods discussed here are in the developmental or prototype stage which can be viewed as promising methodologies provided, they are cheap and available easily in future.

In the case of treating harvested rainwater, the first choice would be the use of the "First Flush Device" as a pretreatment to remove coarse particulate matter. This will not only ensure a better quality of harvested water to be collected in the storage tank but will also reduce the burden on the subsequent treatment techniques. After this step, for further treating the harvested water, hybrid techniques (e.g., filtration + adsorption + UV treatment) appear to be the best choice when all types of pollutants (physical, chemical and biological) must be eradicated. However, they may not be cost effective due to the combination of processes involved. For developing countries where the cost factor is of paramount importance, again SODIS and chlorination emerge to be most appropriate. When scalability is considered most methods such as SODIS, filtration and UV treatment are suitable for household level and only chlorination appears to be amenable to be scaled from household to community and large scale level of treatment.

8.13 Conclusions

Safe and clean drinking water is vital for human life and the quality of water is of supreme importance to avert health risks. This is even more crucial in the developing countries as there is always a lack of access to basic water facilities. Moreover, contamination of the existing water sources with sewage and a spectrum of pollutants further impairs the scenario. Therefore suitable water treatment techniques as discussed in this chapter are required to mitigate this problem. The choice of an appropriate water treatment method should be made prudently based on its effectiveness in removing impurities, scale of operation, and cost effectiveness. Budding novel methods to treat water are necessary and provide hope to ensuring clean and safe water for all. Conservation of water is yet another major aspect that needs to be considered globally especially in the developing nations. With the ongoing pandemic, there is an imperative need of frequent handwashing and hygiene to prevent the spread of infection. Hence, conservation of water for such purposes over and above the normal needs (drinking and other domestic uses) of an individual/family has become particularly important and use of harvested rainwater appears to be a good way to do that. Thus the Covid 19 pandemic has emphasized the need for saving water and as rightly quoted by Benjamin Franklin, "When the well's dry, we know the worth of water."

8.14 Recommendations for future work

1. Existing water treatment methods need to be evaluated to remove additional and varied types of pollutants present in source water.
2. Further research should be focused on hybrid water treatment technologies that are also cost effective and scalable so that it can be used in developing nations.
3. Novel and effective ways to conserve water (like rainwater harvesting) need to be identified to cater to the needs of the people in developing countries, especially in the present pandemic scenario.
4. Stringent and effective ways to identify and treat the Covid 19 virus in sewage systems (due to the suspected possibility of its spread through the fecal route) are required to prevent contamination of the surface and groundwater sources. Wastewater reuse also needs rigorous monitoring and treatment on account of the same reason.
5. Even though it is believed that the Covid 19 virus does not spread through drinking water, it will be meaningful to focus more on new/existing water treatment methods that can also inactivate different types of viruses.

References

Abramov, V. O., Abramova, A. V., Cravotto, G., Nikonov, R. V., Fedulov, I. S., & Ivanov, V. K. (2021). Flow-mode water treatment under simultaneous hydrodynamic cavitatsion and plasma. *Ultrasonics Sonochemistry*, 70, 105323, ISSN 1350-4177.

Agrawal, A., Sharma, N., & Sharma, P. (2021). Designing an economical slow sand filter for households to improve water quality parameters. *Materials Today: Proceedings*, 43(Part 2), 1582–1586.

Akosile, S. I., Ajibade, F. O., Lasisi, K. H., Ajibade, T. F., Adewumi, J. R., Babatola, J. O., & Oguntuase, A. M. (2020). Performance evaluation of locally produced ceramic filters for household water treatment in

Nigeria. *Scientific African, 7*, e00218. Available from https://doi.org/10.1016/j.sciaf.2019.e00218, ISSN 2468-2276.

Alkhalidi, A., Arabasi, S., Othman, A. A., Sabanikh, T., Mahmood, L., & Abdelal, Q. (2021). Using Wood's glass to enhance the efficiency of a water solar disinfection (SODIS) apparatus with a Fresnel lens. *International Journal of Low-Carbon Technologies, 16*(2), 628–634. Available from https://doi.org/10.1093/ijlct/ctaa096.

Arhin, S. G., Banadda, N., Komakech, A. J., Pronk, W., & Marks, S. J. (2019). Application of hybrid coagulation–ultrafiltration for decentralized drinking water treatment: Impact on flux, water quality and costs. *Water Supply 1, 19*(7), 2163–2171. Available from https://doi.org/10.2166/ws.2019.097.

Bandala, E. R., Kruger, B. R., Cesarino, I., Leao, A. L., Wijesiri, B., & Goonetilleke, A. (2021). Impacts of COVID-19 pandemic on the wastewater pathway into surface water: A review. *Science of The Total Environment, 774*.

Bhowmick, G. D., Dhar, D., Nath, D., et al. (2020). Coronavirus disease 2019 (COVID-19) outbreak: Some serious consequences with urban and rural water cycle. *NPJ Clean Water, 3*, 32. Available from https://doi.org/10.1038/s41545-020-0079-1.

Bradley, E. C., Victoria, P., & Neil, R. (2020). Solar Disinfection (SODIS) Provides a Much Underexploited Opportunity for Researchers in Photocatalytic Water Treatment (PWT). *ACS Catal, 10*, 11779–11782.

Braun, L., Sylivester, Y. D., Zerefa, M. D., Maru, M., Allan, F., Zewge, F., et al. (2020). Chlorination of *Schistosoma mansoni* cercariae. *PLoS Neglected Tropical Diseases, 14*(8), e0008665. Available from https://doi.org/10.1371/journal.pntd.0008665.

Bui, T., Nguyen, D., Han, M., Kim, M., & Park, H. (2020). Rainwater as a source of drinking water: A resource recovery case study from Vietnam. *Journal of Water Process Engineering, 39*. Available from https://doi.org/10.1016/j.jwpe.2020.101740.

Byrne, J. A., Fernandez-Ibañez, P. A., Dunlop, P. S. M., Alrousan, D. M. A., & Hamilton, J. W. J. (2011). Photocatalytic enhancement for solar disinfection of water: A review. *International Journal of Photoenergy*, 1–12.

Clasen, T., McLaughlin, C., Nayaar, N., Boisson, S., Gupta, R., Desai, D., & Shah, N. (2008). Microbiological effectiveness and cost of disinfecting water by boiling in semi-urban India. *The American Journal of Tropical Medicine and Hygiene, 79*(3), 407–413, PMID: 18784234.

da Costa, P. C. L., de Azevedo, A. R. G., da Silva, F. C., Cecchin, D., & do Carmo, D. d. F. (2021). Rainwater treatment using an acrylic blanket as a filtering media. *Journal of Cleaner Production, 303*, 126964. Available from https://doi.org/10.1016/j.jclepro.2021.126964, ISSN 0959-6526.

Dankovich, T. A. (2014). Microwave-assisted incorporation of silver nanoparticles in paper for point-of-use water purification. *Environmental Science: Nano, 1*, 367–378.

Dhama, P., & Kuriakose, T. (2021). Comparison of fluoride uptake between *Moringa oleifera* and *Tamarindus indica*. *International Journal of Innovations in Engineering Research and Technology, 8*(2), 33–40, Retrieved from. Available from https://repo.ijiert.org/index.php/ijiert/article/view/1332.

Gadgil, A. (1998). Drinking water in developing countries. *Annual Review of Energy and the Environment, 23*, 253–286.

Heitzinger, K., Hawes, S. E., Rocha, C. A., Alvarez, C., & Evans, C. A. (2020). Assessment of the feasibility and acceptability of using water pasteurization indicators to increase access to safe drinking water in the Peruvian Amazon. *The American Journal of Tropical Medicine and Hygiene, 103*(1), 455–464. Available from https://doi.org/10.4269/ajtmh.18-0963.

<https://www.who.int/news-room/fact-sheets/detail/drinking-water> (accessed August 2021).

<https://www.who.int/water_sanitation_health/dwq/Boiling_water_01_15.pdf> (accessed August 2021).

<https://www.cdc.gov/healthywater/emergency/making-water-safe.html> (accessed August 2021).

<https://www.cdc.gov/safewater/sand-filtration.html> (accessed August 2021).

<https://www.cdc.gov/safewater/ceramic-filtration.html> (accessed August 2021).

<http://www.lifestraw.com> (accessed August 2021).

< https://www.who.int/emergencies/diseases/novel-coronavirus-2019/advice-for-public > (accessed August 2021).
< https://causetech.net/innovation-zone/tech-inspirations/life-sack > (accessed August 2021).
< https://livingnow.com.au/solarball/ > (accessed August 2021).
< https://nipponbasic.ecnet.jp/en/ > (accessed August 2021).
< https://solarcooking.fandom.com/wiki/Water_Pasteurization_Indicator > (accessed April 2021).
< https://www.sustainable.com.au/rainwater-harvesting > (accessed June 2021).
< https://inhabitat.com/pure-water-bottle-filters-99-9-of-bacteria-with-uv-light/ > (accessed August 2021).
< https://www.abc.net.au/news/2021-04-28/water-purification-research-drought-remote-rural-australia-unisa/100099816 > (accessed june 2021).
< https://www.epa.gov/sites/production/files/2015-11/documents/rainharvesting.pdf > (accessed June 2021).
< https://www.filtsep.com/water-and-wastewater/news/water-filters-fight-covid19-in-developing-world/ > (accessed August 2021).
< https://www.ircwash.org/sites/default/files/255.1-87SL-3062.pdf > (accessed May 2021).
< https://www.wateronline.com/doc/the-real-cost-of-treating-drinking-water-with-0001 > (accessed May 2021).
Jakubczak, M., Karwowska, E., Rozmysłowska-Wojciechowska, A., Petrus, M., Woźniak, J., Mitrzak, J., & Jastrzębska, A. M. (2021). Filtration materials modified with 2Dnanocomposites—A new perspective for point-of-use water treatment. *Materials*, *14*, 182. Available from https://doi.org/10.3390/ma14010182.
Jiao, M., Yao, Y., Chen, C., Jiang, B., Pastel, G., Lin, Z., Wu, Q., Cui, M., He, S., & Hu, L. (2020). Highly ecient water treatment via a wood-based and reusable filter. *ACS Materials Letters*, *2*(4), 430–437. Available from https://doi.org/10.1021/acsmaterialslett.9b00488.
Jyoti, K. K., & Pandit, A. B. (2003). Hybrid cavitation methods for water disinfection: Simultaneous use of chemicals with cavitation. *Ultrasonics Sonochemistry*, *10*(4–5), 255–264.
Karim, M. R., Khan, M. H. R. B., Akash, M. A.-S.-A., & Shams, S. (2021). Effectiveness of solar disinfection for household water treatment: An experimental and modeling study. *Journal of Water, Sanitation and Hygiene for Development*.
Kausley, S., Dastane, G., Kumar, J., Desai, K., Doltade, S., & Pandit, A. (2018). Clean water for developing countries: Feasibility of different treatment solutions. *Encyclopedia of Environmental Health*, 2nd Edition. Available from https://doi.org/10.1016/B978-0-12-409548-9.11079-610.1016/B978-0-12-409548-9.11079-6.
Kramer, O. J. I., de Moel, P. J., Padding, J. T., Baars, E. T., Rutten, S. B., Elarbab, Awad. H. E., Hooft, J. F. M., Boek, E. S., & van der Hoek, J. P. (2021). New hydraulic insights into rapid sand filter bed backwashing using the Carman−Kozeny model. *Water Research*, *197*, 117085. Available from https://doi.org/10.1016/j.watres.2021.117085.
Makuba, T. S., Taba, K., Rosillon, F. & Wathelet, B. (2016). Accounts rendered Chemistry (Weinheim an der Bergstrasse, Germany) 19, 827–831. Available from https://doi.org/10.1016/j.crci.2016.01.012.
Mane, M. B., Bhandari, V. M., Balapure, K., & Ranade, V. V. (2020). Destroying antimicrobial resistant bacteria (AMR) and difficult, opportunistic pathogen using cavitation and natural oils/plant extract. *Ultrasonics Sonochemistry*, *69*.
Mazhar, M. A., Khan, N. A., Ahmed, S., Khan, A. H., Hussain, A., Rahisuddin., Changani, F., Yousefi, M., Ahmadi, S., & Vambol, V. (2020). Chlorination disinfection by-products in municipal drinking water − A review. *Journal of Cleaner Production*, *273*, 123159. Available from https://doi.org/10.1016/j.jclepro.2020.123159, ISSN 0959−6526.
Nancy Jotham Marobhe, S. M. S. (2021). Treatment of drinking water for rural households using Moringa seed and solar disinfection. *Journal of Water, Sanitation and Hygiene for Development*, *11*(4), 579–590. Available from https://doi.org/10.2166/washdev.2021.253.
Panchal, D., Prakash, O., Bobde, P., & Pal, S. (2021). SARS-CoV-2: Sewage surveillance as an early warning system and challenges in developing countries. *Environmental Science and Pollution Research International*, *28*(18), 22221–22240. Available from https://doi.org/10.1007/s11356-021-13170-8, Epub 2021 Mar 17. PMID: 33733417; PMCID: PMC7968922. ISSN 0048-9697. Available from https://doi.org/10.1016/j.scitotenv.2021.145586.

Pandey, P., Khan, F., Ahmad, V., et al. (2020). Combined efficacy of *Azadirachta indica* and *Moringa oleifera* leaves extract as a potential coagulant in ground water treatment. *SN Applied Sciences, 2*, 1300. Available from https://doi.org/10.1007/s42452-020-3124-2.

Pandit, A. B., & Kumar, J. K. (2019). *Drinking water treatment for developing countries: Physical Chemical and Biological Pollutants*. UK: The Royal Society of Chemistry.

Pichel, N., Vivar, M., Fuentes, K. M., & Eugenio-Cruz, K. (2020). Study of a hybrid photovoltaic-photochemical technology for meeting the needs of safe drinking water and electricity in developing countries: First field trial in rural Mexico. *Journal of Water Process Engineering, 33*, 101056. Available from https://doi.org/10.1016/j.jwpe.2019.101056, ISSN 2214−7144.

Pichel, N., Vivar, M., & Fuentes, M. (2019). The problem of drinking water access: A review of disinfection technologies with an emphasis on solar treatment methods. *Chemosphere, 218*, 1014−1030.

Plappally, A. K. & John H. Lienhard V. (2013). Desalination and water treatment, 51:1−3, 200−232.

Raimann, J. G., Boaheng, J. M., Narh, P., et al. (2020). Public health benefits of water purification using recycled hemodialyzers in developing countries. *Scientific Reports, 10*, 11101. Available from https://doi.org/10.1038/s41598-020-68408-1.

Richards, S., Rao, L., Connelly, S., Raj, A., Raveendran, L., Shirin, S., Jamwal, P., & Helliwell, R. (2021). Sustainable water resources through harvesting rainwater and the effectiveness of a low-cost water treatment. *Journal of Environmental Management, 15*(286), 112223. Available from https://doi.org/10.1016/j.jenvman.2021.112223, May Epub 2021 Mar 5. PMID: 33684801.

Sharma, P. K., Sharma, R. L., & Pandit, S. (2020). Sustainable house hold water purification integrated with natural coagulants and solar technology. *Pollution Research, 39*(February Suppl. Issue), S119−S122.

Strauss, A., Dobrowsky, P. H., Ndlovu, T., et al. (2016). Comparative analysis of solar pasteurization vs solar disinfection for the treatment of harvested rainwater. *BMC Microbiology, 16*, 289. Available from https://doi.org/10.1186/s12866-016-0909-y.

Tortajada, C., & Biswas, A. K. (2020). COVID-19 heightens water problems around the world. *Water International, 45*(5), 441−442. Available from https://doi.org/10.1080/02508060.2020.1790133.

Varkey, A. J. (2020). Purification of river water using *Moringa oleifera* seed and copper for point-of-use household application. *Scientific African, 8*, e00364. Available from https://doi.org/10.1016/j.sciaf.2020.e00364, ISSN 2468-2276.

Yamaguchi, N. U., Cusioli, L. F., Quesada, H. B., Ferreira, M. E. C., Fagundes-Klen, M. R., Vieira, A. M. S., Gomes, R. G., Vieira, M. F., & Bergamasco, R. (2021). A review of *Moringa oleifera* seeds in water treatment: Trends and future challenges. *Process Safety and Environmental Protection, 147*, 405−420. Available from https://doi.org/10.1016/j.psep.2020.09.044, ISSN 0957−5820.

Yamaguchi, U. N., Abe, S., Medeiros, F. V. d. S., Vieira, A., & Bergamasco, R. (2017). Hybrid gravitational microfiltration system for drinking water purification. *Ambiente e Agua - An interdisciplinary Journal of Applied Science, 12*, 168. Available from https://doi.org/10.4136/ambi-agua.2047.

CHAPTER 9

Plastic pollution in waterways and in the oceans

Lei Mai, Hui He and Eddy Y. Zeng

Guangdong Key Laboratory of Environmental Pollution and Health, Center for Environmental Microplastics Studies, School of Environment, Jinan University, Guangzhou, P.R. China

Chapter Outline
9.1 Introduction 179
9.2 Global marine plastic pollution 180
    9.2.1 Sources of marine plastic debris 180
    9.2.2 Distribution of microplastics in the marine environment 182
9.3 Plastic pollution in rivers 183
    9.3.1 Distribution of plastics in global rivers 183
    9.3.2 Riverine transport of microplastics to coastal oceans/seas 187
9.4 Riverine plastic outflows 188
    9.4.1 Field measured riverine microplastic outflows 188
    9.4.2 Comparison of riverine plastic outflows between measurements and model estimates 190
References 192

9.1 Introduction

Since the invention of plastic in the early 1900s, plastics have been widely used to produce a variety of plastic products. In recent years, the annual plastic production has steady increased and reached 368 million tons in 2019 (Plastics Europe, 2020). The vast use of plastic products and the mismanagement of plastic waste have led to widespread occurrence of plastic wastes in the environment. Because of their degradation-resistant nature, plastic wastes can exist in the environment for a long time. Plastic wastes can be gradually fragmented into fine particles under physical, chemical, or biological degradation.

In 1972 marine plastic pollution was first reported (Carpenter & Smith, 1972). Since then, plastic pollution in aquatic environments has attracted increasing concerns from the scientific community and the public. A class of fine plastic particles was defined as "Microplastics (MPs)" in 2004 (Thompson et al., 2004), which has become a hot topic in environmental research. Although there is no unified agreement on the definition of MPs, plastic particles with aerodynamic diameters smaller than 5 mm are generally designated as MPs. Depending on their sources, MPs can be classified into primary and secondary ones. Primary MPs are mainly detected in personal care products which contain plastic beads, such as MPs in facial cleansers (Fendall & Sewell, 2009). Secondary MPs are generally derived from the fragmentation of large plastic debris. Because MPs can float and are degradation-resistant, they can transport and transfer among various environmental compartments. MPs can easily be captured and ingested by aquatic organisms, as they are small in size. Toxic effects induced by MPs are generally problematic for aquatic ecosystems and human health.

MPs widely distribute in various environmental compartments, including water (Mai et al., 2019), sediment (Dou et al., 2021), and organisms (Tien et al., 2020), from marine to terrestrial environments. Jambeck et al. (2015) emphasized that land-based inputs are the main source of plastics in the marine environment. To track the potential sources of MPs, researchers have gradually turned to the terrestrial environment (Rochman, 2018). Large amount of plastic wastes from land is continuously transported to coastal seas via rivers, and even to deep seas through oceanic circulation. Rivers constantly receive anthropogenic plastic wastes from humans residing within river basins. Once entering riverine environments, low density MPs generally float on the water surface, while high density MPs gradually sink to the river bottom and deposit on the river bed. If water discharge suddenly increases due to certain extreme events, such as flood (Hurley et al., 2018), portions of MPs trapped in river banks and deposited in river beds may re-float in water column and enter the coastal seas via riverine runoff. River networks have become an important pathway for land-based MPs to enter the marine environment.

Riverine MP outflows can be estimated by measuring the concentrations of MPs in the related river outlet combined with water discharge data. Because field measurements of riverine MP outflows are limited, models are generally built and verified with field measurements. Verified models can be used to predict riverine plastic outflows from global rivers. This chapter aims to depict the distribution patterns of MPs in global oceans/seas and rivers and estimate the annual global riverine plastic outflows with modeling efforts.

9.2 Global marine plastic pollution

9.2.1 Sources of marine plastic debris

Plastic debris in global oceans are mainly derived from marine and terrestrial sources. Marine sources mainly include disposed materials from marine shipping, fishing, and aquaculture, as well as fragments from aging facilities. Land-based plastic debris are generated from obsolete plastic products, are washed off to rivers by surface runoff, and enter the coastal oceans/seas via riverine discharge. In addition, plastic wastes produced by local residents along the coastlines and tourists at beach resorts can also enter the oceans by tidal flushing.

Sea-derived plastic wastes account for approximately one-fifth of the total marine plastic debris. Discarded plastic fishing gears and foam containers used in marine aquaculture are the two common types of sea-derived plastic wastes (Fig. 9.1) (Mai et al., 2018). Discarded fishing gears mainly include fishing nets, fishing lines, and nylon ropes, polymer types of which are polyethylene, polypropylene, and nylon. Aged fishing gears can result in numerous plastic fragments, which can easily enter the marine system. Broken fishing gears are often discarded into oceanic water and have become one of the main sources of marine plastic debris. Plastic wastes derived from fishing gears account for approximate 10% of the total marine debris (Good et al., 2010).

As the quantity of wild fish has become progressively depleted partly due to overfishing, the growing demand for proteins has led to rapid development of mariculture. Plastic products are widely used in mariculture, including plastic foam containers and buoys, the main component of which is polystyrene (Xue et al., 2020). These plastic foam containers and buoys can be fragmented to small pieces with aging. They can float on the seawater surface and become an important source of marine plastic wastes. In addition, maritime activities and offshore operations are important sources of marine plastic wastes.

The main pathways for land-based plastic wastes to enter the marine environment include riverine transport, illegal dumping, and disposal by tourists on coastal beaches or estuaries. For example, a large amount of plastic wastes was found in a tourism-oriented estuary in the Gulf of Mexico (Shruti et al., 2019). Similarly, tiny plastic particles prevalently occurred at a tourism beach near the South China Sea (Zhao et al., 2015). Aside from

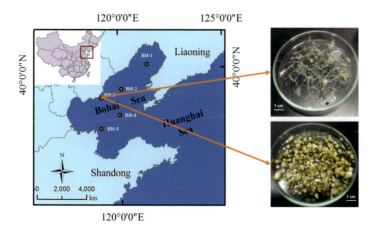

Figure 9.1
Plastic fishing lines and foams collected from Bohai Sea in June 2017.

coastal inputs, rivers are considered as the most important pathway for land-based plastic wastes to enter the marine environment, accounting for approximately 80% of the total land-based input.

The sources of riverine plastic wastes are considerably complex, including flushes of scattering plastic wastes by surface runoff, discharge of wastewater containing plastic particles from sewage treatment plants (Li et al., 2018), and leaches mixed with plastic wastes from landfills. For example, plastic beads in some cosmetics and personal care products can enter sewage treatment plants by face washing (Fendall & Sewell, 2009). With low removal efficiencies, these plastic beads can be discharged to rivers and then to the coastal seas. The detachment of microfibers from textile garments by washing is also an important pathway for MPs to migrate to rivers (Yang et al., 2019). Automobile tire wear on road surfaces can also produce a large amount of rubber debris, some of which is suspended into the air or flushed into freshwaters or oceans (Leads & Weinstein, 2019). Apparently, land-based plastic waste input is the most important cause for marine plastic pollution.

9.2.2 Distribution of microplastics in the marine environment

Because of their small particle sizes, MPs are more harmful to the environment than large plastic debris. For this reason, marine MP pollution has attracted increasing concerns from the scientific community and the public. The occurrence and distribution of marine MPs have been abundantly reported. In previous studies, samples were commonly collected by net trawling, identified by spectroscopic technologies, and quantified by manual counting with the aid of a microscope. The data we compiled indicate that the mean MPs concentration in the Pacific Ocean was higher than those in the Indian Ocean and Atlantic Ocean (Fig. 9.2).

Compared to the open oceans, the inland seas had higher MP concentrations (Fig. 9.2). The Black Sea, a semiinland sea surround by several industrialized countries, receives discharge from several well-known rivers such as Danube, Kuban, and Buger, which contain a variety of pollutants. Lechner et al. (2014) estimated that Danube River alone delivers 4.2 tons of plastic wastes to the Black Sea every day, with an annual input of up to 1530 tons. The Mediterranean Sea, which also suffers from extensive industrial and population pressure, has serious plastic pollution problem. The plastic concentration in the Mediterranean Sea can reach 240,000 particles km^{-2}, of which MPs account for 82% (Alomar et al., 2016). Therefore, the Mediterranean Sea is known as the sixth largest accumulation area of marine waste (Constant et al., 2020; Cózar et al., 2014; Lucia et al., 2014). Attention should be paid to plastic pollution in these inland seas as well as open oceans.

Plastic concentrations also varied greatly in different oceans, and the pattern was closely related to the spatial distribution of floating plastics on the water surface. Field studies have shown that plastics on river water surface tend to float as "ribbons." The transport of

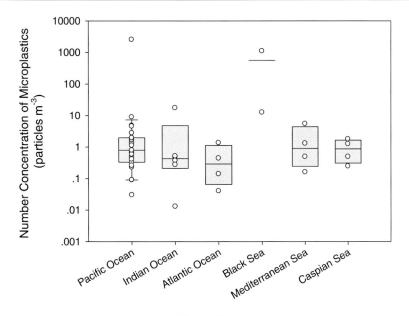

Figure 9.2
Microplastic concentrations in the open oceans and inland seas. Samples were collected using trawling nets.

floating marine plastics is highly influenced by oceanic currents and winds, which may drive the exchange between ocean layers. As widely reported, plastic debris tends to accumulate in the subtropical ocean gyres (Law et al., 2010), but field measured data on the temporal and spatial distribution of plastics in oceans are limited. Visual observation and satellite tracking have shown that plastics in the ocean surface are not evenly distributed and plastic accumulation zones can be clearly observed (Moore, 2003). Law et al. (2014) also found floating plastic debris accumulated in the surface of the North Pacific Ocean. Is this accumulation pattern universally applied? Can plastic samples collected from a certain area represent the level of plastic pollution in the entire area? How to avoid sampling errors caused by the spatial heterogeneity of plastics and select the proper parameters to evaluate plastic pollution in oceans or seas remains to be further explored.

9.3 Plastic pollution in rivers

9.3.1 Distribution of plastics in global rivers

There are numerous reports on plastic or MP pollution in rivers, and Asian rivers have been hot spots for MPs studies. To alleviate the difference in sampling methods for collecting MPs, we summarized the concentrations of riverine MPs collected by Manta net trawling (Table 9.1) and bulk water sampling (Table 9.2) separately. Riverine MP pollution in

Table 9.1: Concentrations of riverine microplastics collected by Manta trawling nets.

Country/region	River	Net pore size (µm)	Sampling time	Concentration (particle m^{-3})	References
China	Haihe	333	2018	14.2 ± 14.6	Liu et al. (2020)
China	Pearl River main streams	160	2016–17	Spring: 370 ± 50 Summer: 350 ± 80 Winter: 1960 ± 900	Fan et al. (2019)
China	Beijing	160	2016–17	Spring: 140 ± 10 Summer: 140 ± 10 Winter: 360 ± 10	Fan et al. (2019)
China	Dongjiang	160	2016–17	Spring: 250 ± 50 Summer: 240 ± 80 Winter: 640 ± 330	Fan et al. (2019)
China	Xijiang	160	2016–17	Spring: 220 ± 20 Summer: 270 ± 110 Winter: 430 ± 120	Fan et al. (2019)
China	Human Outlet Jiaomen Outlet Hongqimen Outlet Hengmen Outlet Modaomen Outlet Jitimen Outlet Hutiaomen Outlet Yamen Outlet	333	2018	0.043–0.711 0.063–0.148 0.019–0.122 0.131–0.361 0.041–0.556 0.027–0.083 0.005–0.133 0.040–0.129	Mai et al. (2019)
China	Yangtze River	330	2017	211	Xiong et al. (2019)
Thailand	Chao Phraya River	300	–	80 ± 65	Ta and Babel (2020)
Malaysia	Cherating River	100	2017–18	0.0042 ± 0.0033	Pariatamby et al. (2020)
Indonesia	Citarum River	300	–	0.0574 ± 0.025	Sembiring et al. (2020)
Indonesia	Surabaya River	333	2019	1.47–43.1	Lestari et al. (2020)
India	Lower Ganga River	300	2019	0.466	Singh et al. (2021)
England	Thames River	250	2017	14.2–24.8	Rowley et al. (2020)
Germany	Elbe River	150	2015	5.57 ± 4.33	Scherer et al. (2020)
France	Rhône River	330	2015–16	12 ± 18	Constant et al. (2020)
France	Rhône River	330	2016–17	19 ± 28	Constant et al. (2020)
France	Têt River	330	2015–16	42 ± 18	Constant et al. (2020)
France	Gave de Pau River	330	2018	3.34 ± 0.2	Bruge et al. (2020)
Spain	Ebro River	5	2017	3.5 ± 1.4	Simon-Sanchez et al. (2019)
United States	Menomonee River	333	2016	1.58–2.71	Lenaker et al. (2019)
United States	Kinnickinnic River	333	2016	1–5.67	Lenaker et al. (2019)
United States	29 Great Lakes Tributaries	333	2014–15	0.3–12.2	Baldwin et al. (2016)
Canada	North Saskatchewan River	53	2017	26.2 ± 18.4	Bujaczek et al. (2021)

Table 9.2: Concentrations of riverine microplastics collected by bulk water sampling.

Country/region	River	Water volume	Membrane pore size (μm)	Sampling time	Concentration (particle m^{-3})	References
China	Haihe	20 L	1.2	2018	9300 ± 4700	Liu et al. (2020)
China	Qiangtangjiang	20 L	45	2018	1180 ± 269	Zhao et al. (2020)
China	Minjiang	50 L	300	2018	700 ± 700	Huang et al. (2020)
China	Yellow River	5 L	0.45	2018–19	497,000–930,000	Han et al. (2020)
China	Yongjiang	10 L	50	2018	2350 ± 1860	Zhang et al. (2020)
China	Huangpujiang	5 L	10	2018	38,900 ± 14,100	Zhang et al. (2019)
China	Yangtze River	100 L	60	2017	157 ± 75.8	Zhao et al. (2019)
China	Yangtze River Hanjiang	20 L	50	2016	2520 ± 912 2930 ± 306	Wang et al. (2017)
China	Pearl River Outlets	5 L	50	2017	7850–10,950	Yan et al. (2019)
Korea	Nakdong River upstream Nakdong River downstream	100 L	20	2017	293 ± 83 4760 ± 5240	Eo et al. (2019)
India	River Alaknanda	50 L	100	2019	2220	Chauhan et al. (2021)
Pakistan	Ravi River	100 L	150	–	2070 ± 3650	Irfan et al. (2020)
Netherlands	Dommel River Meuse River	1.3–8 m^3	20	2017–18	654 867	Mintenig et al. (2020)
United States	Hudson River	3 L	0.45	2016	980	Miller et al. (2017)
United States	Gallatin River	1.3 L	0.45	2015–17	1200	Barrows et al. (2018)
Mexico	Tecolutla Estuary	5 L	1.2	2016–17	151,000	Sánchez-Hernández et al. (2021)
South Africa	Orange—Vaal River	30–60 L	25	2018	210 ± 270	Weideman et al. (2019)

Chinese rivers has been abundantly reported and most of the studies focus mainly on rivers located in the east and south of China, including the Yangtze River (Zhao et al., 2019), Yellow River (Han et al., 2020), and Pearl River (Mai et al., 2019) among others.

The Pearl River network is dominated by three main streams: the Dongjiang, Xijiang, and Beijing, which converge in the upper reach of the Pearl River Delta (PRD). Together with other relatively small rivers in the PRD, they enter the South China Sea via eight main riverine outlets. The concentrations of MPs in the outlets (Mai et al., 2019) were much lower than those in the three main streams (Fan et al., 2019) which flow through highly urbanized cities around the PRD. After entering the rivers, MPs can gradually sink to the river bottom, are ingested by organisms, and are fragmented to extremely tiny particles that are not easily collected (Cózar et al., 2014). These factors may lead to the lower concentrations of MPs near the outlets than within the main streams.

The data we compiled indicate that the concentration of MPs in the Yellow River obtained by bulk water sampling, with nearly 1 million MP particles per cubic meter of water (Han et al., 2020), was much higher than those in other rivers. Two plausible reasons may be given for the difference. First, the water discharge of the Yellow River is relatively small compared to the Yangtze River and Pearl River and contains high sediment contents, collectively slowing the riverine discharge of MPs. Second, that study only collected 5 L of water from the Yellow River and filtered it through a membrane with pore size of 0.45 μm. All plastic particles with sizes greater than 0.45 μm were counted. Concentrations of MPs tend to increase with decreasing particle size (Cózar et al., 2014), leading to greater concentration of MPs in the Yellow River than those in other Chinese rivers.

With similar sampling methods the concentrations of MPs in rivers illustrated a certain regional distribution pattern, reflecting in Asia > Europe > America (Tables 9.1 and 9.2). Most countries in Europe and America are moderately to highly developed, while those in Asia are mostly underdeveloped but experiencing rapid economic growth. As a result, the amounts of plastic wastes induced by industrial activities in Asian countries are much greater than those in European and American countries. In addition, most Asian countries are densely populated, and population density has proven to be positively correlated with concentrations of MPs in rivers. The influence of urban development and population density on riverine MPs has been widely confirmed (Di & Wang, 2018; Mai et al., 2019; Yonkos et al., 2014).

The use of various sampling methods makes it difficult to objectively compare the concentrations of riverine MPs in different regions. Although trawling nets are often used, the pore sizes of these nets may vary, with the most commonly adopted net pore size being 330 μm (Table 9.1). In bulk water sampling, sample volumes and pore sizes of membranes used for filtration also varied substantially with different studies (Table 9.2). Hence, a robust conversion method between different sampling methods needs to be developed in order to effectively compare concentrations of MPs collected by different sampling methods.

9.3.2 Riverine transport of microplastics to coastal oceans/seas

Large amounts of MPs discharged from land into rivers are further transported to adjacent marine environments. In this chapter, the contribution of riverine MPs to marine MP pollution is analyzed using data of MP concentration in rivers and adjacent oceans/seas compiled from the literature. Consistent with the distribution pattern of MPs in the marine environment, rivers flowing to the Pacific Ocean had the highest mean MP concentration, followed by those draining to the Indian Ocean, Mediterranean Sea, and Atlantic Ocean (Fig. 9.3). Obviously, marine MP pollution is closely related to the riverine MPs outflows, which are therefore deemed important sources of marine MP pollution.

The ocean is often considered as a sink for land-based plastics. However, the number concentrations of MPs floating in ocean surface were much lower than those in the rivers draining to the oceans (Fig. 9.3). The missing plastics may have sunk to sediment, been ingested by aquatic organisms, and/or been retained at river banks or shorelines (Cózar et al., 2014; Tien et al., 2020) detected 334–1060 particles m^{-3} of MPs on the water surface of Fengshan River in Taiwan, and the concentration of MPs in sediment was 508–3990 particles kg^{-1} dry sediment. Similarly, a large amount of MPs (2050 ± 750 particles kg^{-1} dry sediment) was found in the sediment of Ebro River (Simon-Sanchez et al., 2019). River or deep ocean/sea beds may not just be a habitat for aquatic organisms, but also a sink for plastic wastes. Unfortunately, no reasonable estimation of this missing amount of plastics has been conducted.

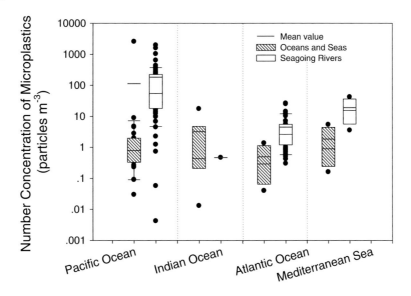

Figure 9.3
Concentration of MPs in oceans/seas and the discharging rivers. Samples were collected by trawling nets. *MPs*, Microplastics.

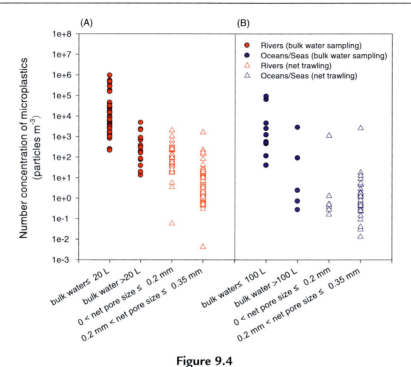

Figure 9.4
Concentration of MPs in (A) the discharging rivers and (B) oceans/seas using different sampling methods. *MPs*, Microplastics.

Different sampling methods may result in large differences in the concentrations of floating MPs detected in different regions. Whether in rivers or oceans/seas, concentrations of MPs collected by bulk water sampling were higher than those by net trawling (Fig. 9.4). The concentrations of MPs increased with decreasing water volume collected. The Manta net with pore size of 330 μm is the most commonly used sampling tool for MPs collection. Current available nets have various pore sizes, ranging from 80 μm to 1 mm. The measured concentration of MPs generally increase with decreasing net pore size. Hence, some of the missing MPs may be plastic particles with sizes smaller than the pore sizes of sampling tools.

9.4 Riverine plastic outflows

9.4.1 Field measured riverine microplastic outflows

Rivers are an important route for plastic waste entering the ocean. Moore et al. (2011) estimated that plastic waste flowing into the adjacent sea within three days from two rivers in California was approximately 30 tons. A large amount of land-based plastic wastes enter the oceans through rivers, causing serious plastic pollution in the marine environment. Accurate estimate of global riverine plastic outflows is meaningful for clarifying the contribution of river inputs to marine plastic pollution.

Currently, riverine MP outflows from only a few rivers have been measured, including Pearl River (Mai et al., 2019), Yangtze River (Zhao et al., 2019), and Nakdong River (Eo et al., 2019). Microplastics in the Pearl River system are discharged into the coastal ocean via eight main riverine outlets. The sum of MP outflows from these outlets was designated as the riverine MP outflow of the Pearl River system, and the measured value was approximately 39 billion particles or 66 tons (Mai et al., 2019). The MP outflows of the Pearl River system showed seasonal variation, with the highest in summer and the lowest in winter (Fig. 9.5A).

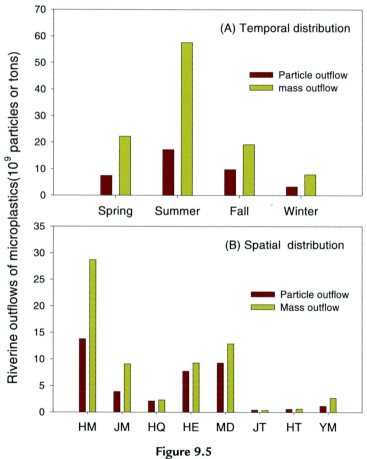

Figure 9.5
(A) Temporal distribution and (B) spatial distribution of riverine outflows of microplastics from the eight main outlets of the Pearl River (Mai et al., 2019). *HE*, Hengmen; *HM*, Humen; *HQ*, Hongqimen; *HT*, Hutiaomen; *JM*, Jiaomen; *JT*, Jitimen; *MD*, Modaomen; *YM*, Yamen.

This is mainly attributed to the different runoff amounts between summer and winter, which accounts for 37% and 14.2% of the annual runoff, respectively. The annual MP outflows varied greatly among the eight outlets, with the Humen outlet having the highest number and mass fluxes (Fig. 9.5B). The Humen outlet not only receives riverine runoff flowing through Guangzhou and Dongguan, which are highly urbanized cities, but also has a relatively large total runoff. In general, four eastern outlets contributed more to the amount of MPs than three western outlet, except for the Modaomen outlet with the largest runoff among all outlets. This pattern is consistent with the pattern for riverine outflows of persistent organic pollutants observed in a previous study (Guan et al., 2009). The measured riverine outflow of MPs from the Pearl River was far lower than the model estimates by Lebreton et al. (2017) and Schmidt et al. (2017). Due to the limited number of measured data, model estimates deviate substantially from measured values. Hence, it is necessary to conduct long-term field surveys for model calibration and validation.

9.4.2 Comparison of riverine plastic outflows between measurements and model estimates

Lebreton et al. (2017) used mismanaged plastic waste (MPW) as the main predictor to estimate global riverine plastic outflows. However, the assignment of MPW values is obviously unreasonable. Low-income countries in Asia and Africa are assigned an MPW value of 89%, while high-income countries in Europe and North America are assigned with a value of 2%. This should be the main reason for the imbalance between model estimates and the measurements. Our latest study used the Human Development Index (HDI) reported by the United Nations Development Program as the main predictor to estimate the global riverine plastic outflows. The HDI model estimates are close to the measured values (Mai et al., 2020).

The estimates by three models are compared and analyzed. Based on MPW, Schmidt et al. (2017) established two models, model 1 and model 2, which predicted the median annual plastic outflows at 0.47 and 2.75 million tons, respectively. The median estimate predicted by Lebreton model was approximately 1.42 million tons (Lebreton et al., 2017). These estimates are much higher than the median estimate by HDI model of 0.134 million tons (Mai et al., 2020). Our study found that HDI is better than MPW in predicting global riverine plastic outflows. The much lower estimates by HDI model can greatly ease public concerns on marine plastic pollution and alleviate the financial pressure caused by plastic pollution control.

Eight of the top 10 rivers with largest riverine plastic outflows are located in Asia. Although three models were consistent with each other on this conclusion, the specific top ten rivers varied among the models (Fig. 9.6). It should be noted that Asian rivers have relatively large runoff with high population density within the river basins, and low HDI values are assigned to Asian countries. The combined effect of these factors leads to higher plastic riverine outflows in Asian

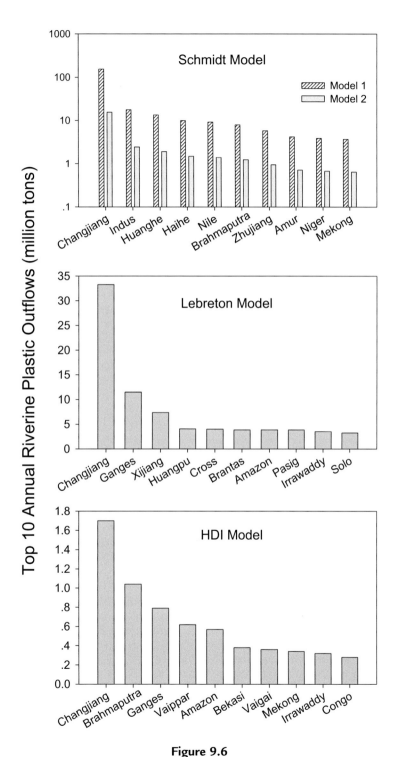

Figure 9.6
Comparison of top 10 rivers with the largest riverine plastic outflows estimated by the Schmidt model, Lebreton model, and HDI model. *HDI*, Human Development Index.

rivers than in other rivers. The results suggest that per capita contribution may be a better indicator for assessing the contributions of individual countries to global marine plastic pollution.

References

Alomar, C., Estarellas, F., & Deudero, S. (2016). Microplastics in the Mediterranean Sea: Deposition in coastal shallow sediments, spatial variation and preferential grain size. *Marine Environmental Research*, *115*, 1–10.

Baldwin, A. K., Corsi, S. R., & Mason, S. A. (2016). Plastic debris in 29 Great Lakes tributaries: Relations to watershed attributes and hydrology. *Environmental Science & Technology*, *50*(19), 10377–10385.

Barrows, A. P. W., Christiansen, K. S., Bode, E. T., & Hoellein, T. J. (2018). A watershed-scale, citizen science approach to quantifying microplastic concentration in a mixed land-use river. *Water Research*, *147*, 382–392.

Bruge, A., Dhamelincourt, M., Lanceleur, L., Monperrus, M., Gasperi, J., & Tassin, B. (2020). A first estimation of uncertainties related to microplastic sampling in rivers. *Science of the Total Environment*, *718*, 137319.

Bujaczek, T., Kolter, S., Locky, D., & Ross, M. S. (2021). Characterization of microplastics and anthropogenic fibers in surface waters of the North Saskatchewan River, Alberta, Canada. *FACETS*, *6*, 26–43.

Carpenter, E. J., & Smith, K. L. (1972). Plastics on the Sargasso Sea surface. *Science*, *175*(4027), 1240–1241.

Chauhan, J. S., Semwal, D., Nainwal, M., Badola, N., & Thapliyal, P. (2021). Investigation of microplastic pollution in river Alaknanda stretch of Uttarakhand. *Environment, Development and Sustainability*.

Constant, M., Ludwig, W., Kerhervé, P., Sola, J., Charrière, B., Sanchez-Vidal, A., Canals, M., & Heussner, S. (2020). Microplastic fluxes in a large and a small Mediterranean river catchments: The Têt and the Rhône, Northwestern Mediterranean Sea. *Science of the Total Environment*, *716*, 136984.

Cózar, A., Echevarría, F., González-Gordillo, J. I., Irigoien, X., Ubeda, B., Hernández-León, S., Palma, A. T., Navarro, S., García-De-Lomas, J., & Ruiz, A. (2014). Plastic debris in the open ocean. *Proceedings of the National Academy of Sciences of the United States of America*, *111*(28), 10239–10244.

Di, M., & Wang, J. (2018). Microplastics in surface waters and sediments of the Three Gorges Reservoir, China. *Science of the Total Environment*, *616–617*, 1620–1627.

Dou, P.-C., Mai, L., Bao, L.-J., & Zeng, E. Y. (2021). Microplastics on beaches and mangrove sediments along the coast of South China. *Marine Pollution Bulletin*, *172*, 112806.

Eo, S., Hong, S. H., Song, Y. K., Han, G. M., & Shim, W. J. (2019). Spatiotemporal distribution and annual load of microplastics in the Nakdong River, South Korea. *Water Research*, *160*, 228–237.

Fan, Y., Zheng, K., Zhu, Z., Chen, G., & Peng, X. (2019). Distribution, sedimentary record, and persistence of microplastics in the Pearl River catchment, China. *Environmental Pollution*, *251*, 862–870.

Fendall, L. S., & Sewell, M. A. (2009). Contributing to marine pollution by washing your face: Microplastics in facial cleansers. *Marine Pollution Bulletin*, *58*(8), 1225–1228.

Good, T. P., June, J. A., Etnier, M. A., & Broadhurst, G. (2010). Derelict fishing nets in Puget Sound and the Northwest Straits: Patterns and threats to marine fauna. *Marine Pollution Bulletin*, *60*(1), 39–50.

Guan, Y.-F., Wang, J.-Z., Ni, H.-G., & Zeng, E. Y. (2009). Organochlorine pesticides and polychlorinated biphenyls in riverine runoff of the Pearl River Delta, China: Assessment of mass loading, input source and environmental fate. *Environmental Pollution*, *157*(2), 618–624.

Han, M., Niu, X., Tang, M., Zhang, B.-T., Wang, G., Yue, W., Kong, X., & Zhu, J. (2020). Distribution of microplastics in surface water of the lower Yellow River near estuary. *Science of the Total Environment*, *707*, 135601.

Huang, Y., Tian, M., Jin, F., Chen, M., Liu, Z., He, S., Li, F., Yang, L., Fang, C., & Mu, J. (2020). Coupled effects of urbanization level and dam on microplastics in surface waters in a coastal watershed of Southeast China. *Marine Pollution Bulletin*, *154*, 111089.

Hurley, R., Woodward, J., & Rothwell, J. J. (2018). Microplastic contamination of river beds significantly reduced by catchment-wide flooding. *Nature Geoscience*, *11*(4), 251–257.

Irfan, M., Qadir, A., Mumtaz, M., & Ahmad, S. R. (2020). An unintended challenge of microplastic pollution in the urban surface water system of Lahore, Pakistan. *Environmental Science and Pollution Research, 27* (14), 16718−16730.

Jambeck, J. R., Geyer, R., Wilcox, C., Siegler, T. R., Perryman, M., Andrady, A., Narayan, R., & Law, K. L. (2015). Plastic waste inputs from land into the ocean. *Science, 347*(6223), 768−771.

Law, K. L., Morét-Ferguson, S. E., Goodwin, D. S., Zettler, E. R., Deforce, E., Kukulka, T., & Proskurowski, G. (2014). Distribution of surface plastic debris in the eastern Pacific Ocean from an 11-year data set. *Environmental Science & Technology, 48*(9), 4732−4738.

Law, K. L., Morét-Ferguson, S., Maximenko, N. A., Proskurowski, G., Peacock, E. E., Hafner, J., & Reddy, C. M. (2010). Plastic accumulation in the North Atlantic Subtropical Gyre. *Science, 329*(5996), 1185−1188.

Leads, R. R., & Weinstein, J. E. (2019). Occurrence of tire wear particles and other microplastics within the tributaries of the Charleston Harbor Estuary, South Carolina, USA. *Marine Pollution Bulletin, 145*, 569−582.

Lebreton, L. C. M., Van Der Zwet, J., Damsteeg, J.-W., Slat, B., Andrady, A., & Reisser, J. (2017). River plastic emissions to the world's oceans. *Nature Communications, 8*, 15611.

Lechner, A., Keckeis, H., Lumesberger-Loisl, F., Zens, B., Krusch, R., Tritthart, M., Glas, M., & Schludermann, E. (2014). The Danube so colourful: A potpourri of plastic litter outnumbers fish larvae in Europe's second largest river. *Environmental Pollution, 188*, 177−181.

Lenaker, P. L., Baldwin, A. K., Corsi, S. R., Mason, S. A., Reneau, P. C., & Scott, J. W. (2019). Vertical distribution of microplastics in the water column and surficial sediment from the Milwaukee River basin to Lake Michigan. *Environmental Science & Technology, 53*(21), 12227−12237.

Lestari, P., Trihadiningrum, Y., Wijaya, B. A., Yunus, K. A., & Firdaus, M. (2020). Distribution of microplastics in Surabaya River, Indonesia. *Science of the Total Environment, 726*, 138560.

Li, X., Chen, L., Mei, Q., Dong, B., Dai, X., Ding, G., & Zeng, E. Y. (2018). Microplastics in sewage sludge from the wastewater treatment plants in China. *Water Research, 142*, 75−85.

Liu, Y., Zhang, J., Cai, C., He, Y., Chen, L., Xiong, X., Huang, H., Tao, S., & Liu, W. (2020). Occurrence and characteristics of microplastics in the Haihe River: An investigation of a seagoing river flowing through a megacity in northern China. *Environmental Pollution, 262*, 114261.

Lucia, G. A. D., Caliani, I., Marra, S., Camedda, A., Coppa, S., Alcaro, L., Campani, T., Giannetti, M., Coppola, D., & Cicero, A. M. (2014). Amount and distribution of neustonic micro-plastic off the western Sardinian coast (Central-Western Mediterranean Sea). *Marine Environmental Research, 100*(3), 10−16.

Mai, L., Bao, L.-J., Shi, L., Liu, L.-Y., & Zeng, E. Y. (2018). Polycyclic aromatic hydrocarbons affiliated with microplastics in surface waters of Bohai and Huanghai Seas, China. *Environmental Pollution, 241*, 834−840.

Mai, L., Sun, X.-F., Xia, L.-L., Bao, L.-J., Liu, L.-Y., & Zeng, E. Y. (2020). Global riverine plastic outflows. *Environmental Science & Technology, 54*(16), 10049−10056.

Mai, L., You, S.-N., He, H., Bao, L.-J., Liu, L., & Zeng, E. Y. (2019). Riverine microplastic pollution in the Pearl River Delta, China: Are modeled estimates accurate? *Environmental Science & Technology, 53*, 11810−11817.

Miller, R. Z., Watts, A. J. R., Winslow, B. O., Galloway, T. S., & Barrows, A. P. W. (2017). Mountains to the sea: River study of plastic and non-plastic microfiber pollution in the northeast USA. *Marine Pollution Bulletin, 124*(1), 245−251.

Mintenig, S. M., Kooi, M., Erich, M. W., Primpke, S., Redondo- Hasselerharm, P. E., Dekker, S. C., Koelmans, A. A., & Van Wezel, A. P. (2020). A systems approach to understand microplastic occurrence and variability in Dutch riverine surface waters. *Water Research, 176*, 115723.

Moore, C., Lattin, G. L., & Zellers, A. F. (2011). Quantity and type of plastic debris flowing from two urban rivers to coastal waters and beaches of Southern California. *Journal of Integrated Coastal Zone Management, 11*(1), 65−73.

Moore, C. (2003). Trashed: Across the Pacific Ocean, plastics, plastics, everywhere. *Natural History, 112*, 46−51.

Pariatamby, A., Hamid, F. S., Bhatti, M. S., Anuar, N., & Anuar, N. (2020). Status of microplastic pollution in aquatic ecosystem with a case study on Cherating River, Malaysia. *Journal of Engineering and Technological Sciences, 52*(2), 222−241.

Plastics Europe. (2020). Plastics — The facts 2020. In: *Analysis of European plastics production, demand and waste data*.

Rochman, C. M. (2018). Microplastics research—From sink to source. *Science, 360*(6384), 28–29.

Rowley, K. H., Cucknell, A.-C., Smith, B. D., Clark, P. F., & Morritt, D. (2020). London's river of plastic: High levels of microplastics in the Thames water column. *Science of the Total Environment, 740*, 140018.

Sánchez-Hernández, L. J., Ramírez-Romero, P., Rodríguez-González, F., Ramos-Sánchez, V. H., Márquez Montes, R. A., Romero-Paredes Rubio, H., Sujitha, S. B., & Jonathan, M. P. (2021). Seasonal evidences of microplastics in environmental matrices of a tourist dominated urban estuary in Gulf of Mexico, Mexico. *Chemosphere, 277*, 130261.

Scherer, C., Weber, A., Stock, F., Vurusic, S., Egerci, H., Kochleus, C., Arendt, N., Foeldi, C., Dierkes, G., Wagner, M., Brennholt, N., & Reifferscheid, G. (2020). Comparative assessment of microplastics in water and sediment of a large European river. *Science of the Total Environment, 738*, 139866.

Schmidt, C., Krauth, T., & Wagner, S. (2017). Export of plastic debris by rivers into the sea. *Environmental Science and Technology, 51*(21), 12246–12253.

Sembiring, E., Fareza, A. A., Suendo, V., & Reza, M. (2020). The presence of microplastics in water, sediment, and milkfish (*Chanos chanos*) at the downstream area of Citarum River, Indonesia. *Water, Air, & Soil Pollution, 231*(7), 355.

Shruti, V. C., Jonathan, M. P., Rodriguez-Espinosa, P. F., & Rodriguez-Gonzalez, F. (2019). Microplastics in freshwater sediments of Atoyac River basin, Puebla City, Mexico. *Science of the Total Environment, 654*, 154–163.

Simon-Sanchez, L., Grelaud, M., Garcia-Orellana, J., & Ziveri, P. (2019). River Deltas as hotspots of microplastic accumulation: The case study of the Ebro River (NW Mediterranean). *Science of the Total Environment, 687*, 1186–1196.

Singh, N., Mondal, A., Bagri, A., Tiwari, E., Khandelwal, N., Monikh, F. A., & Darbha, G. K. (2021). Characteristics and spatial distribution of microplastics in the lower Ganga River water and sediment. *Marine Pollution Bulletin, 163*, 111960.

Ta, A. T., & Babel, S. (2020). Microplastic contamination on the lower Chao Phraya: Abundance, characteristic and interaction with heavy metals. *Chemosphere, 257*, 127234.

Thompson, R. C., Olsen, Y., Mitchell, R. P., Davis, A., Rowland, S. J., Anthony, W. G. J., Mcgonigle, D., & Russell, A. E. (2004). Lost at sea: Where is all the plastic? *Science, 304*(5672), 838.

Tien, C. J., Wang, Z. X., & Chen, C. S. (2020). Microplastics in water, sediment and fish from the Fengshan River system: Relationship to aquatic factors and accumulation of polycyclic aromatic hydrocarbons by fish. *Environmental Pollution, 265*, 11.

Wang, W., Ndungu, A. W., Li, Z., & Wang, J. (2017). Microplastics pollution in inland freshwaters of China: A case study in urban surface waters of Wuhan, China. *Science of the Total Environment, 575*, 1369–1374.

Weideman, E. A., Perold, V., & Ryan, P. G. (2019). Little evidence that dams in the Orange–Vaal River system trap floating microplastics or microfibres. *Marine Pollution Bulletin, 149*, 110664.

Xiong, X., Wu, C., Elser, J. J., Mei, Z., & Hao, Y. (2019). Occurrence and fate of microplastic debris in middle and lower reaches of the Yangtze River – From inland to the sea. *Science of the Total Environment, 659*, 66–73.

Xue, B. M., Zhang, L. L., Li, R. L., Wang, Y. H., Guo, J., Yu, K. F., & Wang, S. P. (2020). Underestimated microplastic pollution derived from fishery activities and "hidden" in deep sediment. *Environmental Science & Technology, 54*(4), 2210–2217.

Yan, M., Nie, H., Xu, K., He, Y., Hu, Y., Huang, Y., & Wang, J. (2019). Microplastic abundance, distribution and composition in the Pearl River along Guangzhou city and Pearl River estuary, China. *Chemosphere, 217*, 879–886.

Yang, L., Qiao, F., Lei, K., Li, H., Kang, Y., Cui, S., & An, L. (2019). Microfiber release from different fabrics during washing. *Environmental Pollution, 249*, 136–143.

Yonkos, L. T., Friedel, E. A., Perez-Reyes, A. C., Ghosal, S., & Arthur, C. D. (2014). Microplastics in four estuarine rivers in the Chesapeake Bay, U.S.A. *Environmental Science & Technology, 48*(24), 14195–14202.

Zhang, J., Zhang, C., Deng, Y., Wang, R., Ma, E., Wang, J., Bai, J., Wu, J., & Zhou, Y. (2019). Microplastics in the surface water of small-scale estuaries in Shanghai. *Marine Pollution Bulletin*, *149*, 110569.

Zhang, X., Leng, Y., Liu, X., Huang, K., & Wang, J. (2020). Microplastics' pollution and risk assessment in an urban river: A case study in the Yongjiang River, Nanning City, South China. *Exposure and Health*, *12*(2), 141–151.

Zhao, S., Wang, T., Zhu, L., Xu, P., Wang, X., Gao, L., & Li, D. (2019). Analysis of suspended microplastics in the Changjiang Estuary: Implications for riverine plastic load to the ocean. *Water Research*, *161*, 560–569.

Zhao, S., Zhu, L., & Li, D. (2015). Characterization of small plastic debris on tourism beaches around the South China Sea. *Regional Studies in Marine Science*, *62*, 55–62.

Zhao, W., Huang, W., Yin, M., Huang, P., Ding, Y., Ni, X., Xia, H., Liu, H., Wang, G., Zheng, H., & Cai, M. (2020). Tributary inflows enhance the microplastic load in the estuary: A case from the Qiantang River. *Marine Pollution Bulletin*, *156*, 111152.

CHAPTER 10

Desalination and sustainability

Anju Vijayan Nair and Veera Gnaneswar Gude

Richard A Rula School of Civil and Environmental Engineering, Mississippi State University, Starkville, MS, United States

Chapter Outline
10.1 Introduction 197
10.2 Description of desalination processes 198
    10.2.1 Solar stills 198
    10.2.2 Multi-stage flash desalination 199
    10.2.3 Multi-effect distillation 199
    10.2.4 Vapor compression 199
    10.2.5 Energy efficiency of thermal desalination 200
    10.2.6 Reverse osmosis 200
    10.2.7 Membrane distillation 201
10.3 Desalination and sustainability 203
    10.3.1 Environmental footprint 203
    10.3.2 Economics of desalination 205
    10.3.3 Social aspects of desalination 206
10.4 Renewable energy integration 207
    10.4.1 Selection process of desalination process 209
10.5 Future of sustainable desalination 210
    10.5.1 Desalination at household level 210
    10.5.2 Desalination at community or municipal scale 211
References 212

10.1 Introduction

As a freshwater supply option, desalination has become an important element of the water supply portfolio of many global communities in recent years (Gude, 2016). The need for desalination, along with other conventional and non-conventional water management options, stems from ever-growing challenges of population growth, climate change and freshwater scarcity (Gude, 2016). One in three people living on this planet are already living under water scarce conditions forcing many communities and municipalities to consider non-conventional water sources such as seawater or brackish groundwater or deep groundwater sources for water supplies. These sources are often high in dissolved solids concentrations requiring special treatment methods such as desalination technologies to produce freshwater.

Desalination of saline waters or freshwater production can be accomplished by three fundamental mechanisms: evaporation/condensation, separation/filtration, and ionic transfer under electric field (Gude et al., 2010). The phase change process includes evaporation of freshwater from saline water and condensation of water vapor over a cold surface to collect freshwater. Thermal energy is required to achieve evaporation of freshwater from saline water. Mechanical energy is also required to maintain the pressure (usually low pressure) required for evaporation which is supplied through conversion of electrical energy. Thus phase change processes (involving evaporation and condensation steps) may require both heat and electricity to produce freshwater. Non-phase change processes involve separation of water molecules from saline water by employing a semi-permeable membrane as a physical barrier. This barrier selectively allows water molecules to pass through while retaining the pollutants (including salts) behind, typically known as concentrate or brine or retentate (Gude, 2011). The non-phase change processes may also employ electric field separated by membranes to achieve selective removal of dissolved solids and ions resulting in desalination of saline waters. Hybrid desalination processes may simply involve integration of both phase change and non-phase change processes to offset the limitations setout by each of the technologies and to complement the unique capabilities of each of the processes.

There are many phase change processes, the most commonly known processes are: solar stills; multi-effect solar stills (MESS); multi-effect evaporation/distillation (MED); multi-stage flash (MSF) distillation; thermal vapor compression (TVC); and mechanical vapor compression (MVC). Separation processes may be based on ion transfer or solvent transfer phenomena as in electrodialysis (ED) and reverse osmosis (RO) processes, respectively (El-Dessouky et al., 1999; Ghalavand et al., 2015). Other emerging hybrid (involving both evaporation and separation principles) desalination processes are membrane distillation (MD) and RO combined with MSF or MED processes (Gude et al., 2010). Phase change desalination processes require both thermal and electrical energy. Non-phase change desalination processes require electrical energy only. Hybrid desalination processes may require both thermal and electrical energy sources.

10.2 Description of desalination processes

10.2.1 Solar stills

Solar still is considered the first desalination method dating back to 15th century (Gude, 2015). The still includes an evaporating surface where the saline water is exposed to harvest thermal energy from solar irradiation to cause evaporation of freshwater in the form of vapor. A condensing surface maintained at ambient temperature allows for condensation of freshwater vapor to result in freshwater production. The absorption of solar irradiation by saline water increases the water temperature and causes vapor formation. To increase the heat absorption, the evaporating surface is often coated with a black coating. To increase energy recovery and storage, solar stills may be combined with storage materials. This provides continued evaporation during non-sunlight hours. Multi-effect solar still desalination process allows for recovering condensing energy to put into evaporation effect in the next stage. Two or three effect solar stills are common.

10.2.2 Multi-stage flash desalination

Multi-stage flash (MSF) desalination process was widely installed in the 1950s and once dominated the desalination industry (Gude, 2016). This process consists of a number of flashing stages, a brine heater, pumps, venting system, and a cooling water control loop. Incoming seawater temperature is increased by passing it through the final heat exchanger (El-Dessouky et al., 1999). Then it is passed through the brine heater where steam from an external source heats the incoming seawater in the heat exchanger to the required maximum process temperature (80°C–90°C). Finally, it is released into the first vacuum chamber where the water vapor condenses into freshwater product on the cooling water control loop and this operation is repeated in a number of stages to achieve energy recovery and recycling and increase freshwater output.

10.2.3 Multi-effect distillation

MED process operates under falling pressure in successive distillation stages so that heat recovery can be achieved in a number of stages (Gude, 2015). The heat rejected in a previous stage serves as the heat source in the next stage. Feed water entering the first effect is heated to the boiling point. Both feed water and heating vapor to the evaporators flow in the same direction. The remaining water is pumped to the second effect, where it is once again applied to a tube bundle. This process continues for several stages called effects, about 4–21 effects in a large desalination plant. In recently designed plants, the feed water is divided into several parts before entering the flash drums. The rest of the process remains similar to the other design (Ghalavand et al., 2015).

10.2.4 Vapor compression

In this process, the heat for evaporating the feed water is generated by compressing the vapor. Two methods are used to condense the water vapor and to produce the amount of sufficient heat to evaporate the incoming feed seawater: a mechanical compressor and a steam jet (thermal compressor). In this method, seawater is evaporated and the vapor is passed through a compressor (Ghalavand et al., 2015). The vapor is compressed, which leads to an increase in vapor dew point (in this condition the compressed vapor dew point is higher than seawater boiling point), so vapor can be condensed by indirect contact with seawater and cause freshwater evaporation. The compression ratio should be maintained near unity to increase energy efficiency. In this process, the seawater temperature is held at 100°C. VC units are built in a variety of configurations in order to promote seawater evaporation. The compressor creates a vacuum in the evaporator and then compresses the vapor taken from the evaporator and condenses it in a tube bundle.

10.2.5 Energy efficiency of thermal desalination

Thermal energy required for desalination can be provided from different heating sources such as natural gas, and steam generated from power plants. The energy efficiency is either reported as gain output ratio (GOR) or performance ratio (PR). GOR is defined as the ratio of the mass of water produced through a desalination process over a fixed quantity of energy consumed. PR is defined as the mass, in pounds, of water produced by desalination per 1000 BTU of heat provided to the process (El-Dessouky et al., 1999; Ghalavand et al., 2015; Gude, 2015). This is equivalent to the number of kg of freshwater produced per 2326 kJ of heat. As shown in Fig. 10.1, the MED process has high GOR or PR compared to MSF and VC processes. MED is also known to be thermodynamically more efficient.

10.2.6 Reverse osmosis

Reverse osmosis (RO) is the most commonly used technology in desalination. In this process, the osmotic pressure is overcome by applying an external pressure higher than the osmotic pressure on the feed water (brackish/seawater); thus, water flows in the reverse direction to the natural flow across the membrane, leaving the dissolved salts behind the membrane with an increase in salt concentration called brine (Gude, 2011). Most of the energy consumption for membrane desalination is used for pressurizing the feed water. A typical large seawater RO plant consists of four major components: feed water pretreatment, high pressure pumping, membrane separation, and permeate post-treatment (Gude, 2011). Major design considerations for seawater RO plants are the flux, conversion or recovery ratio, permeate salinity, membrane life, energy consumption, and feed water temperature. Energy demand by membrane processes is reported as specific energy consumption in kWh/m^3 of freshwater produced. The specific energy consumption has been reduced by almost ten times over the past two decades as a result of developing energy recovery devices and energy-efficient and highly permeable membranes.

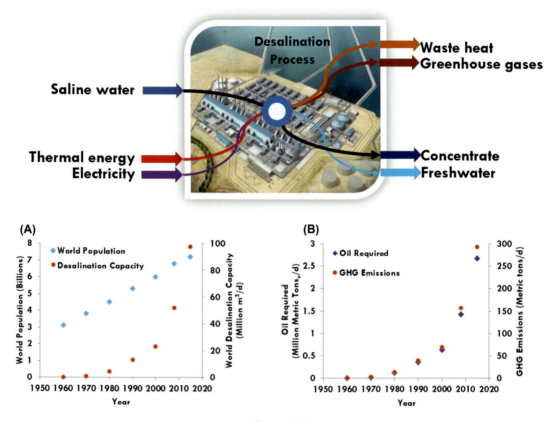

Figure 10.1
Desalination capacity trends (A); oil requirements and associated greenhouse gas emissions (B).

10.2.7 Membrane distillation

MD process consists of both phase change and non-phase change process characteristics. Hot feed water and cold permeate flow channels are separated by a semi-permeable membrane. Feed water (saline water) is heated to produce water vapor which will diffuse through the membrane to condense on the cold permeate side. Feed water can be heated using process waste heat sources. The permeate side can be maintained at ambient temperature. In the MD process, evaporation of freshwater takes place at temperatures as low as 40°C. The MD process can be operated under different configurations: direct contact MD; air-gap MD; sweeping gas MD; and vacuum MD. The application of each of these configurations depends on the feed water characteristics and energy sources.

Table 10.1 presents a list of desalination processes with process operating conditions and renewable energy applicability. Hybrid technologies like MD and low-temperature

Table 10.1: A comparison of various desalination (membrane and thermal) processes and potential applications (Ghalavand et al., 2015; Gude, 2015).

Process	Description	Renewable energy applications	Remarks
Solar still	Operating temperature and pressure: 40°C–80°C, atmospheric Source water: seawater/brackish water Recovery rate: 35%–45% Specific energy consumption (SEC): 5040 kJ/kg	Direct solar energy, solar collectors, PVT collectors, solar ponds, and process waste heat	Small and rural applications possible, low capital and little maintenance costs, no energy costs Not suitable for large scale applications due to lower efficiency
Multi-effect solar still	Operating temperature and pressure: 40°C–80°C, 1 atm Source water: seawater/brackish water SEC: 5040 kJ/kg	Direct solar energy, solar collectors, PVT collectors, solar ponds, waste heat source	
MSF	Operating temperature and pressure: 80°C–120°C, 1–2 atm Source water: seawater Recovery rate: 35%–45% SEC: 200–350 kJ/kg	Solar collectors and geothermal source	Large scale applications, reliable process and experience in operations Can be combined with power generation (cogeneration) Cost and energy intensive, not suitable for small scale applications
MED	50°C–90°C, 0.1–0.5 at, Source water: seawater Recovery rate: 35%–45% Specific energy consumption: 150–250 kJ/kg	Solar collectors, solar pond, geothermal source, and process waste heat	
Low-temperature MED	Operating temperature and pressure: 40°C–70°C, 0.1–0.4 atm, Source water: seawater Recovery rate: 35%–45%	Solar collectors, solar pond, PVT collectors, geothermal source, and process waste heat	High thermodynamic efficiency, low cost or free waste heat Less experience; large heat transfer areas
M/TVC	Source water: seawater Recovery rate: 25%–40% SEC: 150–240 kJ/kg	Solar collectors, solar pond, thermal collectors, geothermal source, and process waste heat	Low specific energy consumption Electrical and mechanical energy input
MD	Operating temperature and pressure: 40°C–80°C, atmospheric Source water: seawater Recovery rate: 35%–45%	Solar collectors, solar pond, PVT collectors, PV modules, geothermal source, and process waste heat	High product recovery, low temperature operation Limited commercial applications; require membranes
RO	Operating temperature and pressure: <45°C, 20–60 atm Source water: seawater/brackish water, Recovery rate: 35%–50% (SW), 50%–90% (BW) SEC: 120 kJ/kg	PVT collectors, PV modules, wind energy, wave energy	Reliable and most widely used technology High electricity requirements, capital and O&M costs
ED	Operating temperature and pressure: < 45°C Source water: BW Recovery rate: 50%–90% (BW) SEC:144 kJ/kg	PVT collectors, PV modules, wind energy, wave energy	Efficient for high quality product from BWRO Highly sensitive to raw water quality, pretreatment

Note: ED, electrodialysis; MD, membrane distillation; MED, multi-effect evaporation/distillation; MSF, multi-stage flash; RO, reverse osmosis; TVC, thermal vapor compression.

desalination technologies need more pilot-scale demonstrations before they can be considered for large scale operations.

10.3 Desalination and sustainability

Desalination processes are known as energy-demanding and cost-intensive. Significant quantities of energy are required by both phase change and non-phase change processes as shown in Table 10.1. The use of conventional and no-renewable energy sources results in serious environmental impacts. The energy costs are significant when considering the overall cost of freshwater from the desalination processes. For example, water costs for conventional water supplies (i.e., water supplies from surface water sources such as lakes, rivers and groundwater aquifers) ranges from 0.25 to 5 $/m^3 when compared to seawater RO with a range of 0.5 to 4 $/m^3 (Voutchkov, 2011; Voutchkov, 2014). The high range of conventional water supplies depends on factors such as transportation distance and the required head loss during water transfer. The energy requirements also vary in a similar trend. Specific energy consumption for conventional water supplies is between 0.1 and 4.5 kWh/m^3 while the range for seawater RO is 2.5 to 4.5 kWh/m^3 (Plappally, 2012).

10.3.1 Environmental footprint

Fig. 10.1 shows the inputs and outputs of a desalination process including the environmental emissions. The production of 1000 tons (m^3) per day of freshwater by desalination technologies requires 10,000 tons (toe) of oil per year and results in environmental degradation through greenhouse gas emissions and brine discharges (Gude, 2015). This level of consumption involves release of significant quantities of greenhouse gas (GHG) and other environmental emissions. As shown in Fig. 10.1A, worldwide desalination capacity is increasing with population growth and subsequent energy consumption results in environmental pollution. Table 10.2 shows a comparison of environmental footprints of two major desalination processes in industry.

Desalination plants can have significant impact (direct or indirect) on the environment. Increased use of fossil fuels for desalination may accelerate air pollution due to greenhouse gas emissions which include carbon monoxide (CO), nitric oxide (NO), nitrogen dioxide (NO_2), and sulfur dioxide (SO_2) known to produce detrimental factors on local public health and environment (Younos, 2005). As shown in Table 10.2, desalination plants utilize significant amounts of chemicals for pretreatment of saline water and post-treatment of desalinated water. The concentrates generated from brackish water via an RO process (with 60%–85% recovery) would have a concentration factor that is 2.5–7 times higher and the same for the seawater RO (with 30%–50% recovery) would be between 1.25 and 2.0 (National Water Commission, 2008). Thermal desalination technologies have a wide range

Table 10.2: Potential ecological/environmental impacts of the RO and MSF desalination process concentrates (Gude, 2016).

	RO plants	MSF plants	Environmental/ecological impacts
Salinity and temperature	Up to 65,000–85,000 mg/L ambient seawater temperature	About 50,000 mg/L +5°C to 15°C above ambient	Can be harmful; reduces vitality and biodiversity at higher values; harmless after good dilution
Chlorine	Chlorine or other oxidants are neutralized before the water enters the membranes to prevent membrane damage	Approx. 10%–25% of source water feed dosage, if not neutralized	• Very toxic to many organisms in the mixing zone, but rapidly degraded, trihalomethanes (THM) • RO • MSF
Halogenated organics	Typically low content below harmful levels	Varying composition and concentrations, typically THM	• Carcinogenic effects; possible chronic effects, more persistent, dispersal with current, main route of loss is thorough evaporation
Coagulants (e.g., iron-III-chloride)	May cause effluent coloration if not equalized prior to discharge	Not present (treatment not required)	• Non-toxic; increased local turbidity → may disturb
Coagulant aids (e.g., polyacrylamide)	May be present if source water is conditioned	Not present (treatment not required)	• Photosynthesis; possible accumulation in sediments
Antiscalants acid (H_2SO_4)	Not present (reacts with seawater to cause harmless compounds). Typically low content below toxic levels	Typically, low content below toxic levels (reacts with seawater to cause harmless compounds)	• Poor or moderate degradability + high total loads → accumulation, chronic effects, unknown side-effects
Antifoaming agents (e.g., polyglycol) heavy metals	Not present (treatment not required) May contain elevated levels of iron, chromium, nickel, molybdenum if low-quality stainless steel is used	Typically, low content below harmful levels May contain elevated copper and nickel concentrations if inappropriate materials are used for the heat exchangers	• Non-toxic in concentration levels; good degradability Copper • MSF (15–100 µg/L) • Low acute toxicity for most species; high danger of accumulation and long term effects; bioaccumulation
Cleaning chemicals	Alkaline (pH 11–12) or acidic (pH 2–3) solutions with additives such as: detergents (e.g., dodecylsulfate), complexing agents (e.g., EDTA), oxidants (e.g., sodium perborate), biocides (e.g., formaldehyde)	Acidic (pH 2) solution containing corrosion inhibitors such as benzotriazole derivates	• Highly acidic or alkaline cleaning solutions that may cause toxicity without neutralization, MSF cleaning solutions • Low pH, corrosion inhibitor • Highly acidic cleaning solutions that cause toxicity without neutralization low toxicity; poor degradability

Note: MSF, multi-stage flash; RO, reverse osmosis.

of recovery ratios between 15% and 50% with a concentration factor of 1.15 (due to cooling water mixing). Specific seawater concentrate properties such as salinity, temperature, and various chemicals used for coagulation, biocides for controlling biological growth, antifoaming, anti-corrosion and cleaning chemicals cause serious environmental concerns for their proper disposal.

About two units of concentrate are generated for every unit of desalinated water produced. Current practice of handling these concentrates is to discharge them into the coastal waters which could have detrimental effects on the aquatic life and coastal environment (Gude, 2016). Brine/concentrate disposal technologies and management options are not as robust and practical as desalination technologies. In many cases, brine discharge poses major concerns for disapproval of new inland desalination plants and some coastal desalination plants. Continuous recovery and discharge of the seawater at a specific location may also affect the quality of incoming supply water for desalination in the long run. To mitigate major environmental concerns related to brine/concentrate discharges, concentrates should be re-diluted with seawater to minimize the effects related to high salt concentrations.

Brine disposal alternative selection requires consideration of various factors which include location of the plant, land availability, and other treatment and beneficial applications (Younos & Lee, 2019). Surface discharge is the most feasible option for coastal desalination plants. Many options are available for inland communities which include evaporation ponds, sewer disposal, spray irrigation, deep well injection and zero liquid discharge (ZLD) systems for valuable resource recovery. Recovery of valuable chemicals such as gypsum from the concentrates is an attractive option. Reuse of treated brackish water in irrigation applications is another alternative to reduce the demands for freshwater sources. Finally, improving the desalination process schemes would enhance the energetic and environmental performance of the desalination processes. For instance, a SWRO plant with a capacity of 100 million m^3 (per year) would require an electrical load of 50 MW (Semiat, 2008). A desalination and power plant cogeneration scheme would operate at a higher efficiency than separate desalination and power generation plants. The cogeneration scheme operation is based on the water supply demands, which are subject to day-night and summer-winter fluctuations resulting in an energy-efficient process scheme. Gas turbines can provide higher efficiency with the use of high thermal energy released in off-gases reducing the actual energy utilization compared to other common uses.

10.3.2 Economics of desalination

Freshwater costs for desalination processes vary significantly depending on the process scheme, technology, feed water characteristics, energy sources, financial package, scale of operation and finally brine disposal and management. In general, large scale operation will result in lower freshwater costs but it may not be applicable in all scenarios. Fig. 10.2

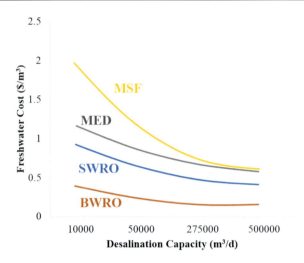

Figure 10.2
Relationship between the desalination capacity and freshwater costs.

shows the freshwater costs with respect to the desalination capacity (Gude et al., 2010; Gude, 2011; Plappally, 2012). MSF process is the most expensive due to high energy requirements and capital and maintenance costs. RO process ca deliver freshwater at significantly lower costs and the brackish water desalination by membrane processes result in lowest freshwater costs s blending can be performed to meet the water quality standards.

10.3.3 Social aspects of desalination

Small and large scale desalination plants face social issues and these vary quite significantly. For small scale operations, usually in remote settings, where the scarcity of freshwater sources justifies consideration of desalination processes. Financial affordability combined with lack of human resources present complex challenges in this scenario. Lack of technology know-how and operational skills often lead to failure of desalination plants. In some cases, the financial burden is not justified. In large scale applications, gaining public trust seems to be a major issue. Desalination processes are usually considered as the last option in this scenario followed by water transfers, rainwater harvesting, and water conservation practices. Yes, the environmental issues, economic burdens along with large energy demands for desalination processes pose long-standing coordination and compliance issues for these plants. Public trust and willingness to pay higher freshwater costs are also considered as major barriers. In some instances, location and the intended purpose of coastal regions are critical aspects. Other factors such as dust, noise, and visual amenity impacts, beach and ocean-based recreational activity impacts, disruption to the properties and activities along the pipeline, impact on the future residential development, impact on the industrial development in the local area, impact on local economic activity, risk to public safety, disruption of local businesses, and impact of construction on road users (Gude 2016a) also need consideration. Many communities do not like to have a physical plant installed in nearby locations or cities for various reasons often termed as "Not In My Backyard (NIMBY)." Construction of large desalination plants involves several inconveniences to the local communities including closure of recreational facilities, local transportation, noise and air pollution and aesthetical depreciation, inflow and outflow traffic transporting materials. For example, a survey conducted in Australia to assess the public choices regarding desalination plants has yielded interesting results as shown in Table 10.3.

10.4 Renewable energy integration

Currently, the world established desalination capacity is around 10% of total world's minimum freshwater demand (basis: A minimum of 100 L/cap/day for 7.8 billion current world population). It requires 1.63 million tons of oil/day (releases 179 metric tons of CO_2/day). It is clear that freshwater production through fossil fuel powered desalination technologies is not an environmentally sustainable alternative. It is necessary to develop alternatives to replace conventional energy sources used in the desalination process with renewable ones and reduce the energy requirements for desalination by developing innovative, low-cost, low-energy technologies and process hybridizations (Gude, 2015).

Renewable energy sources such as solar, wind, geothermal and wave energy sources can be considered for desalination processes. Selection of appropriate renewable energy

Table 10.3: Results from a social impact management plan survey (n = 333) (Southern Seawater Desalination Project—Social Impact Management Plan, 2009).

Type of impact	Phase	Mean	Negative %	Neither %	Positive %	No opinion %
Impact on local marine and coastal environments	Construction	2.35	76.5	17.7	3.0	2.7
	Operation	2.50	71.8	21.6	3.9	2.7
Impact on recreational beach use	Construction	2.50	67.3	28.2	3.3	1.2
	Operation	3.06	48.6	44.1	4.5	2.7
Impact on local residents' satisfaction with community	Construction	2.89	57.7	34.8	4.2	3.3
	Operation	3.13	51.1	37.5	7.2	4.2
Impact on the visual amenity of local community	Construction	3.04	45.9	50.5	2.1	1.5
	Operation	3.13	44.1	51.4	2.7	1.8
Impact on noise levels at your residence	Construction	3.08	49.8	44.4	2.1	3.6
	Operation	3.19	47.1	46.5	2.4	3.9
Impact on future residential development	Construction	3.12	54.1	28.8	14.1	3.0
	Operation	3.21	50.5	31.8	14.4	3.3
Impact on the local economy	Construction	4.56	16.2	33.9	46.5	3.3
	Operation	4.14	18.3	52.3	26.7	2.7

technologies depends on various factors. Some technologies such as PV and wind turbines have reached maturity making their application in desalination economical. It should be noted that the desalination processes can function regardless of the source of energy. Heat energy and electricity can be supplied from any renewable energy source. The energy production costs can be improved through process integration and cogeneration schemes (Gude, 2016). Renewable energy applications are more favored in regions where they are competitive with other energy production methods. Desalination technologies should be integrated with locally available renewable energy sources considering factors such as plant capacity, location and availability of renewable energy sources. Technologies that require thermal energy can be conveniently combined with locally available renewable energy sources such as geothermal sources (Gude, 2016). However, a proper site and source evaluation should be conducted to prevent any environmental issues. If the region has a strong solar insolation, concentrated solar panels (either parabolic troughs or tower type) can be used to support even large scale desalination operations. For membrane technologies which primarily require electrical energy, solar PV, wind or wave energy sources can be considered depending on their availability.

Integration of renewable energy sources with desalination technologies comes with some challenges. Integration of solar, wind, and wave energy sources may require storage facilities (Gude, 2015). For example, solar energy is a self-renewable resource, which can be easily captured and harvested as thermal energy for many beneficial uses. However, the concern with the solar energy is that it is intermittent in nature and its intensity depends on the hour of the day and local weather conditions. One of the solutions to utilize the fluctuating solar energy on a continuous basis is to incorporate thermal energy storage system. The energy input can be provided by solar collectors, parabolic trough collectors and process waste heat releases. Similar to solar energy, wind and wave energy sources also have high variability requiring battery energy storage. Photovoltaic panels also require battery energy storage to ensure round the clock operations (Gude, 2015).

Renewable energy integration can alleviate the economic and environmental issues. For example, in a solar energy driven desalination plant, the prime energy consumption to produce a solar collector of 1 m^2 area and the supporting frame is about 7 GJ (Ardente et al., 2005). If the annual useful thermal energy harvested by the total solar collector area is 5.5 GJ/m^2 year, which is used for desalination, then considering the pumping energy required in an active solar collector system (between the solar collector system-desalination system), the prime energy payback period of the solar collector system is under 2 years (Gude et al., 2012). The global warming potential (GWP) of a solar collector of 1 m^2 area is estimated to be around 721 kg eq. CO_2. Annual CO_2 eq. emission savings due to thermal energy generated by solar collectors is estimated as 407 kg eq. CO_2 (basis: specific global warming factor of 0.066 kg eq. CO_2 per MJ of useful thermal energy). Therefore the global warming potential of the solar collectors can be recovered in less than 2 years (Gude et al., 2012). This data supports the clean idea of utilizing renewable energy harvested by solar collectors for low temperature desalination.

10.4.1 Selection process of desalination process

The selection of appropriate renewable energy-powered desalination process requires complete analysis of the plant from start to end. Selection of a desalination process depends on various factors such as the water demand (desalination capacity), the intended use of water (product water quality), availability of energy sources, energy source compatible desalination technology and skills to operate and maintain the desalination process, feed water source and water quality characteristics, economic feasibility and environmental issues or considerations. The selection process is shown in Fig. 10.3. If the water source has a salinity of less than 2000 mg/L, nanofiltration can be considered. For source water with less than 5000 mg/L salinity, electrodialysis or nanofiltration process can be considered. If the water source salinity is higher than 5000 mg/L, then pretreatment and post-treatment schemes may be required

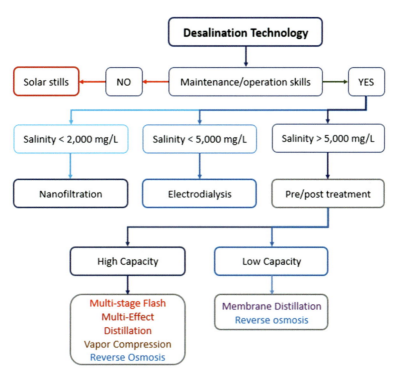

Figure 10.3
Criteria for selecting a desalination technology.

10.5 Future of sustainable desalination

Renewable energy integrated desalination systems are expected to be applied in wide ranges considering the water scarcity issues in many regions of the world (Hamed et al., 2016; Khan et al., 2018). Water scarcity impacts both food security and economic prosperity (Schewe et al., 2014). In rural and remote settings, it is feasible to implement desalination systems at household level while in some locations a community owned system is preferred depending on the freshwater demand. In both cases, it is important to understand the energy demands with respect to the total energy demands at the household level or community level.

10.5.1 Desalination at household level

The energy demands for domestic desalination plants are not quite alarming. A comparison between the desalination system energy utilization with that for other domestic uses may provide a useful insight on the overall energy budget of a household as shown in Fig. 10.4. A typical family consisting of four members would have a water consumption rate of nearly 18 m^3/month (150 liters per person for a family of four) and uses about 1200 kWh$_e$ of electricity or other equivalent energy sources per month (Younos & Lee, 2019). A seawater desalination process (RO process) can deliver the freshwater needs at an energy utilization rate of 140 (kWh$_{th}$)/month (fuel value) excluding water transportation energy needs. The transportation needs for an average distance of 1500 km/month can be met by a family car with 160 L of oil consumption per month which is equivalent to 1500 (kWh$_{th}$)/month while the electricity needs are around 1200 kWh$_e$/month which are equivalent to 2667 (kWh$_{th}$)/month (with a conversion efficiency of 45%; 1200 kWh$_e$/0.45). From this analysis,

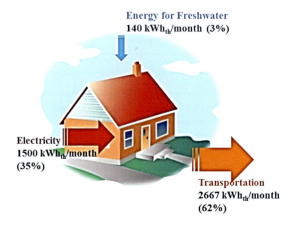

Figure 10.4
Energy budget for desalination at domestic scale.

desalination energy needs represent only 9.3% or 5.2% or 3.2% of the transportation energy, electricity, and total domestic energy needs respectively. This analysis clearly supports the use of desalination process for affordable freshwater supply at household levels.

10.5.2 Desalination at community or municipal scale

The inseparable connection between water and energy sources and food demands by growing populations force many arid communities to consider desalination centered food-energy-water systems as shown in Fig. 10.5. Both power generation and water supply systems can be combined and co-located in renewable energy driven desalination plants which eliminates the need for obtaining environmental permits separately (Gude, 2016). The combined footprint of the cogeneration plant would be much smaller than the individual power and water plants. While surface water and treated effluents (water recycling from wastewater treatment plants) can be used for agricultural applications, these sources are limited in some regions. The need for extracting deep groundwater sources becomes inevitable. These sources often contain high total dissolved solids concentrations. Desalination technologies such as nanofiltration and RO processes can be considered for

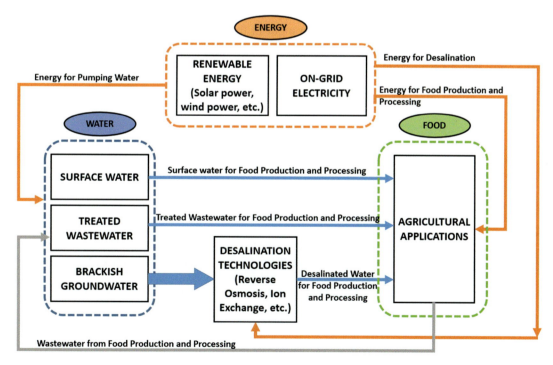

Figure 10.5
Desalination centered food-energy-water nexus for arid regions.

treating these brackish water sources. The same technologies can be considered for water reuse schemes as well. Agricultural production also requires mechanical and electrical energy sources for various operations. Renewable energy sources combined with desalination processes can meet both water and energy requirements for food production. Thus renewable energy integrated desalination systems may address the food-energy-water nexus more effectively with minimal environmental footprints and at low costs.

The desalination industry is experiencing an unprecedented growth in response to increasing freshwater demands exacerbated by water scarcity and climate change issues at global levels. While initially cost-prohibitive, with continuous improvements, desalination processes have become much more affordable than they were a few decades ago. Impressive cost and energy reductions have made the desalination technologies face the most challenging times and have become an acceptable alternative for addressing the water and energy issues (in some cases). Membrane systems are expected to expand over a broad range of applications (small and large scale, industrial and municipal applications), due to their elegant features, robustness, and simplicity. There is a limited scope in further minimizing the energy demands without sacrificing the permeation rates. There is a stronger need for optimizing both energy consumption and permeate production in these systems. Further, technological advances in novel membrane development, process configurations, and process hybridization of membrane and thermal processes are crucial for sustaining desalination industry development in the near future. A greater understanding of water scarcity scenarios and discoveries in desalination processes targeting local supply issues will benefit from strategic planning and management.

References

Ardente, F., Beccali, G., Cellura, M., & Brano, V. L. (2005). Life cycle assessment of a solar thermal collector. *Renewable energy, 30*(7), 1031–1054.

El-Dessouky, H. T., Ettouney, H. M., & Al-Roumi, Y. (1999). Multi-stage flash desalination: Present and future outlook. *Chemical Engineering Journal., 73*(2), 173–190.

Ghalavand, Y., Hatamipour, M. S., & Rahimi, A. (2015). A review on energy consumption of desalination processes. *Desalination and Water Treatment, 54*(6), 1526–1541.

Gude, V. G., Nirmalakhandan, N., Deng, S., & Maganti, A. (2012). Low temperature desalination using solar collectors augmented by thermal energy storage. *Applied Energy., 91*(1), 466–474.

Gude, V. G., Nirmalakhandan, N., & Deng, S. (2010). Renewable and sustainable approaches for desalination. *Renewable and sustainable energy reviews, 14*(9), 2641–2654.

Gude, V. G. (2016). Desalination and sustainability—An appraisal and current perspective. *Water Research, 89*, 87–106.

Gude, V. G. (2011). Energy consumption and recovery in reverse osmosis. *Desalination and water treatment, 36* (1–3), 239–260.

Gude, V. G. (2015). Energy storage for desalination processes powered by renewable energy and waste heat sources. *Applied Energy, 137*, 877–898.

Gude, V. G. (2016). Geothermal source potential for water desalination—Current status and future perspective. *Renewable and Sustainable Energy Reviews., 57*, 1038–1065.

Hamed, O. A., Kosaka, H., Bamardouf, K. H., Al-Shail, K., & Al-Ghamdi, A. S. (2016). Concentrating solar power for seawater thermal desalination. *Desalination.*, *396*, 70–78.

Khan, M. A., Rehman, S., & Al-Sulaiman, F. A. (2018). A hybrid renewable energy system as a potential energy source for water desalination using reverse osmosis: A review. *Renewable and Sustainable Energy Reviews*, *97*, 456–477.

National Water Commission (2008). Emerging trends in desalination: A review. Waterlines report, National Water Commission, Canberra.

Plappally, A. K. (2012). Energy requirements for water production, treatment, end use, reclamation, and disposal. *Renewable and Sustainable Energy Reviews*, *16*(7), 4818–4848.

Schewe, J., Heinke, J., Gerten, D., Haddeland, I., Arnell, N. W., Clark, D. B., Dankers, R., Eisner, S., Fekete, B. M., Colón-González, F. J., & Gosling, S. N. (2014). Multimodel assessment of water scarcity under climate change. *Proceedings of the National Academy of Sciences.*, *111*(9), 3245–3250.

Semiat, R. (2008). Energy issues in desalination processes. *Environmental Science & Technology*, *42*(22), 8193–8201.

Southern Seawater Desalination Project—Social Impact Management Plan (2009). Watercorporaton.com.au

Voutchkov, N. (2011). Current & future desalination trends wateReuse research foundation. *Strategic Planning Session*.

Voutchkov N. (2014). Overview of desalination trends and Jaffna project. Asian Development Bank.

Younos, T., & Lee, J. (2019). Desalination: opportunities and challenges. *Water Environ. Technol.*, 22–27.

Younos, T. (2005). Environmental issues of desalination. *Journal of contemporary water research and education*, *132*(1), 3.

CHAPTER 11

Groundwater sustainability in a digital world

Ahmed S. Elshall[1], Ming Ye[1,2] and Yongshan Wan[3]

[1]Department of Earth, Ocean, and Atmospheric Science, Florida State University, Tallahassee, FL, United States, [2]Department of Scientific Computing, Florida State University, Tallahassee, FL, United States, [3]Center for Environmental Measurement and Modeling, United States Environmental Protection Agency, Gulf Breeze, FL, United States

Chapter Outline

11.1 Introduction 215
11.2 Groundwater sustainability 217
11.3 Digital groundwater 220
11.4 Internet of Things–based data collection 221
11.5 Web-based data sharing 222
11.6 Workflow for data processing 225
11.7 Scenarios for data usage 228
11.8 Perspectives of web-based groundwater platforms 230
11.9 Disclaimer 233
Acknowledgments 233
References 234

11.1 Introduction

Beneath the Earth's surface, groundwater (GW) flows in the pores, fractures, and conduits of the aquifers, which are water-saturated soil and rock formations that contain and transmit significant quantities of water under normal field conditions (Hornberger et al., 2014). Making up more than 97% of the liquid freshwater on Earth, GW supplies more than half of the drinking water, approximately 40% of the irrigation water, and about one third of industrial freshwater (UN-Water, 2018). GW is critical for supporting many terrestrial, aquatic, and marine GW-dependent ecosystems. In addition, GW is a manageable buffer to floods, seasonal variations of surface water (SW), and droughts. However, water users' self-interests are leading to the depletion of more than half of the largest aquifers on the Earth, illustrating the tragedy of the commons (Elshall et al., 2021). GW systems are stressed through GW over-pumping and by contamination from point and nonpoint sources. Impacts of excessive pumping include the degradation of GW-dependent ecosystems, saltwater intrusion, mobilization of heavy metals, and land subsidence. Due to inaction, it is now globally recognized that the on-going GW over-pumping and water quality degradation could transform areas of economic expansion into regions of poverty (Elshall et al., 2021). These adverse impacts call for action to ensure GW sustainability (Gleeson et al., 2019). As a result, major policy reforms have been approved in many countries around the globe (Elshall et al., 2020, 2021) such as the State Water Code of Hawaii (1987), the National Water Act of South Africa (1998), European Union Water Framework Directive (EU-WFD, 2000), National Water Initiative in Australia (2004), Water Sustainability Act in British Columbia (2014), and Sustainable Groundwater Management Act in California (SGMA, 2014). Such policy reforms are crucial for achieving the Sustainable Development Goals (SDGs) of the United Nations' 2030 Agenda, particularly SDG 6 "Clean water and sanitation." GW sustainability is also important for SDG 2 "Zero hunger" through supporting sustainable agriculture, SGD 7 "Affordable and Clean Energy" through ground source heat pumps, SGD 13 "Climate action" as GW is important for mitigating impacts of climate change such as droughts and floods, SDG 14 "Life below water" through supporting marine GW-dependent ecosystems, and SDG 15 "Life on land" through supporting terrestrial GW-dependent ecosystems. GW sustainability in a digital world is also important for SDG 11 "Sustainable cities and communities" as digital GW can be a main component in smart city.

This chapter discusses how current digital transformation can shape the future of sustainable GW management and contribute to the success of GW policy reforms. We are in a period of human history where technological changes are happening at an exponential rate (Roser & Ritchie, 2013). These rapid technological changes are driving the current digital transformation of the fourth industrial revolution, which is characterized by artificial intelligence, big data, cloud computing, cyber-physical systems (CPS), Internet of Things (IoT), and other smart technologies. Each industrial revolution is characterized by a disruptive technology that creates a shift in how we run the world, leading to changes in the social framework and scientific research as shown in Fig. 11.1. The term *fourth industrial revolution* (a.k.a., 4IR and Industry 4.0) was first introduced by a team of scientists developing a high-tech infrastructure strategy for the government of German in 2011, and was the theme of the World Economic Forum Annual Meeting in 2016 (Schwab, 2016). With respect to scientific research, the reader is referred to Hey et al. (2009) for details about the paradigm shift to the nature of science, and the fourth paradigm of "data-intensive scientific discovery" (Hey et al., 2009). How the digital transformation of the fourth industrial revolution is advancing the operationalization of GW sustainability policies is the subject of this chapter.

	Paradigm 1.0 Mechanization 1765-	Paradigm 2.0 Industrialization 1870-	Paradigm 3.0 Information Technology 1969-	Paradigm 4.0 Digital Transformation 2011-
Paradigm-shifter	Steam	Electricity	Computer and internet	Artificial intelligence and internet-of-things
Industrial production	Mechanical production	Assembly line mass production	Automated production	Cyber-physical systems production
Mobility	Steamboat	Railroad	Vehicles, vessels, and aviation	Autonomous vehicles, ships, and drones
Communication	Telegraph	Telephone	Internet	Internet-of-things
Energy	Wood	Coal	Fossil fuel	Renewable
Water supply	Water system	Sewage system	Storm drainage system	Resilient, equitable, and digital infrastructure
Agriculture	Agricultural revolution	Agricultural mechanization	Agriculture technology	Smart agriculture
Society	Agrarian society	Industrial society	Information society	Digital society
Social change	Revolutions	Human rights	Globalization	Connectivity
Business model	Industrial cities	Industrial regions	Global production network	Commons-based peer production
Science	Empirical description of nature	Theoretical with models and generalization	Computational with complex system simulation	Data intensive scientific discovery and eScience

Figure 11.1
An inexact outline of four industrial revolutions and corresponding paradigms with respect to the society (Schwab, 2016) and scientific research (Hey et al., 2009).

11.2 Groundwater sustainability

GW sustainability can be defined as "maintaining long-term, dynamically stable storage [and flow] of high quality groundwater using inclusive, equitable, and long-term governance and management" (Gleeson et al., 2020). GW sustainability considers a combination of multiple aquifer performance and governance factors, and as illustrated in Fig. 11.2, requires a participation-based multiprocess approach with broad uncertainty analysis (Elshall et al., 2020, 2021; Pierce et al., 2013).

A basic component of GW sustainability evaluation is multiprocess modeling (Fig. 11.2). A multiprocess approach accounts for the coupled water—ecology—human systems. In the water system, modeling the SW—GW system generally involves the use of mechanistic numerical models (Henriksen et al., 2008), phenomenological models such as analytical functions (Miro & Famiglietti, 2018), and data-driven machine learning models (Salem et al., 2017). With respect to the ecology system, considering ecosystem services of GW-dependent ecosystems is generally through defining ecological and ecosystem service targets with indicators and thresholds established for each target. An adaptive management process, which is learning-by-doing, is generally used to update the targets and indicators (Rohde et al., 2020). Although developing predictive GW and ecological models to prioritize the most effective management strategies can

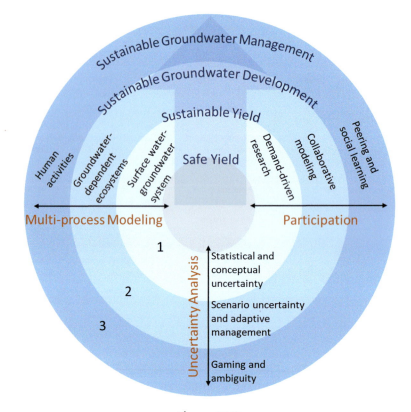

Figure 11.2
Multiprocess modeling, uncertainty analysis, and participation are three basic components of groundwater sustainability evaluation with sphere number reflecting the increasing degree of integration. *Source: Modified from Elshall, A. S., Arik, A. D., El-Kadi, A. I., Pierce, S., Ye, M., Burnett, K. K., Wada, C., Bremer, L. L., & Chun, G. (2020). Groundwater sustainability: A review of the interactions between science and policy.* Environmental Research Letters. <https://doi.org/10.1088/1748-9326/ab8e8c>.

be challenging, these models are recommended for high-capacity pumping (Saito et al., 2021). The human system focuses on GW supply needed for agricultural, municipal, and industrial purposes. Accounting for human activities can be achieved through simple approaches such as pumping projection based on population projection (Urrutia et al., 2018), and recharge projection based on "best-guess" land use and landcover changes and similar forms of scenario analysis (Bremer et al., 2021). More elaborate approaches include integrated water resources management (Feng et al., 2018), hydroeconomic modeling (Mulligan et al., 2014), and socio-hydrology (Castilla-Rho et al., 2019). While integrated water resources management promotes the joint management of water, land, and other natural resources, hydroeconomic modeling aims at optimizing the economic objectives of a water–ecology–human system, and socio-hydrology is a more descriptive approach that integrates human activities as an endogenous component of the water-ecology system (Elshall et al., 2021).

Modeling tools are advancing to account for coupled water−ecology−human systems. For example, MODFLOW One-Water Hydrologic Flow Model version 2 (MF-OWHM2, Boyce et al., 2020) is an integrated hydrologic model with tight coupling of GW, SW, and vertical unsaturated flows. MF-OWHM2 simulates multiple processes such as landscape processes (e.g., land use and crop simulation, root uptake of GW, and irrigation demand estimation), reservoir operations, aquifer compaction and subsidence, and saltwater intrusion (using sharp-interface). Also, MF-OWHM2 produces a binary flow file for use with MT3DMS and MT3DUSGS for transport simulation in the saturated and unsaturated zones, respectively. GSFLOW (Markstrom et al., 2008; Regan & Niswonger, 2021) couples the Precipitation-Runoff Modeling System (PRMS-V) with MODFLOW-2005 and MODFLOW-NWT, and can be used with the transport models of the MODFLOW family. Wei et al. (2019) present Soil and Water Assessment Tool (SWAT)-MODFLOW-RT3D that tightly couples the semi-distributed watershed model SWAT with the GW flow model MODFLOW and the GW solute reactive transport model RT3D in a watershed system. For aquifers with density-dependent flow (e.g., coastal aquifers), several coupled commercial models (e.g., FEFLOW, HydroGeoSphere, and MIKE-SHE) are available. Coupled public domain models with density-dependent flow are relatively immature, which is particularly true for karst aquifers with conduit flow.

Participation is the second basic component of effective GW sustainability evaluation (Fig. 11.2). This is particularly important because human behavior and societal preferences can be considered a root cause of unsustainability as well as part of the solution (Castilla-Rho et al., 2019; Elshall et al., 2021). For example, Castilla-Rho et al. (2017) examined human behavior, cooperation, and collective action, and illustrated tipping points where social norms toward GW conservation shift abruptly with changes in cultural values. On the other hand, Shalsi et al. (2019) present an Australian case study in which comanagement through local collective action was successful in recovering an aquifer that was at risk of depletion and subject to GW quality degradation including salinization. In addition, the authors suggest that comanagement through local collective action can improve the social acceptability of new GW initiatives such as managed aquifer recharge and conjunctive use of SW−GW.

Uncertainty analysis is the third component of effective GW sustainability evaluation (Fig. 11.2). Uncertainty analysis is an integral part of GW sustainability policy through provisions such as the precautionary principle in the EU-WFD, and adaptive management in SGMA. In practice, decisions on GW sustainability in a changing environment are difficult because our scientific knowledge about complex GW systems is inherently uncertain, and because societal preferences are difficult to elicit and may be conflicting. This requires a broad uncertainty analysis accounting for natural and societal aspects (Elshall et al., 2021; Refsgaard et al., 2007).

There are eight essential factors to consider when evaluating GW sustainability at local, regional, and transboundary scales (Elshall et al., 2020, 2021; Pierce et al., 2013). Among the eight factors, five factors related to aquifer performance are (1) recharge rates and storage conditions, (2) water quality, (3) GW capture, discharge rates, and environmental flows, (4) natural hazards and threats, and (5) facilities and technologies of water resources management. Examples of natural hazards and threats are land subsidence due to over-pumping, sinkholes, and severe prolonged droughts. Examples of facilities and technologies are pumping and water distribution systems as well as managed aquifer recharge. Factors (1−4) are related to the physical aquifer system, and Factor (5) is related the infrastructure system. Three factors related to aquifer governance are (6) legal and institutional constraints, (7) societal values and preferences, and (8) economic feasibility. Societal values and preferences include instrumental, intrinsic, relational, and esthetic values, intragenerational and intergenerational equity, public health, resilience, indigenous rights, and consensus as described by Elshall et al. (2020). Factor (6) is related to institutional system, and Factors (7−8) are related to the socioeconomic system.

While this section only provides a brief overview on the concept of GW sustainability, for detailed information about GW depletion, challenges, and sustainability the reader is referred to recent studies (Bierkens & Wada, 2019; Elshall et al., 2020, 2021; Gleeson et al., 2020; Lall et al., 2020; Rinaudo et al., 2020). Given this brief overview, effective application of such a participation-based and multiprocess approach with broad uncertainty analysis remains problematic for both the academic community and water managers (Elshall et al., 2020; Thompson et al., 2021). In the rest of this chapter we show how the current digital transformation can influence our GW sustainability practices.

11.3 Digital groundwater

Initiatives such as the INSPIRE Directive (https://inspire.ec.europa.eu) of the EU Commission, and the EarthCube initiative for geosciences (https://www.earthcube.org) of the National Science Foundation (NSF) in the US promote the increased use of information and communications technology (ICT) to build cyberinfrastructure for geosciences. This is mainly to assist in the creation, dissemination, and application of geoscientific data and research to the benefit of society. For example, to support sustainable development, the Infrastructure for Spatial Information in the European Community (INSPIRE) is a Directive to establish an EU spatial data infrastructure (SDI) to support environmental policies and environmental applications. The EarthCube aims at transforming the conduct of geoscience research, education, and accordingly their services to the society, by encouraging the geoscience community to systematically build geoscience cyberinfrastructure through community dialogue, governance, and a common vision (NSF, 2015). Research agendas for intelligent systems and digital initiatives are emerging to serve this digital transformation (Chen et al., 2020; Gil et al., 2018; Hubbard et al., 2020). For example, the Digital Water Program (https://iwa-network.org/programs/digital-water) of the International Water Association serves as a starting point to initiate the dialogue on digital water, and to share solutions and experiences in applying digital solutions for water utilities. In addition, strategy and recommendations for US Executive Presidential Order 13956 (10/16/2020)—Modernizing America's Water Resource Management and Water Infrastructure—emphasize the importance of the next generation water observation networks and water resources modeling capability, leveraging on ICT (Petty et al., 2021).

An example of this digital transformation is smart GW management, which can improve GW sustainability practices. Many aspects of geoscience domains such as GW hydrology pose novel and challenging problems for intelligent systems research, which would significantly transform intelligent systems and greatly benefit the geosciences in turn (Gil et al., 2018). Intelligent systems are becoming more common, and they can be "intelligent/smart anything" such as smart agriculture, smart city, intelligent transportation system, smart home, and smart health care. For example, smart agriculture includes precise agriculture and smart irrigation (i.e., to estimate and supply the fertilizer and water needs at the resolution of individual plants), and smart greenhouse by controlling the physical environment to increase crop yield. Smart city includes intelligent transportation systems, smart grids, smart mobility, smart buildings, smart street lighting, online banking, telehealth, a digital twin of the city, and smart devices to allow citizens to connect to the smart city services. A digital twin of a city is a virtual representation of the systems of the city to enable stakeholders to monitor and manage water, air quality, energy, mobility, and other services in the city. While certain areas such as smart city and smart agriculture have reached a certain level of maturity, smart GW management is still evolving. Yet certain cross-disciplinary technologies and elements are shared among all these smart applications such as artificial intelligence, big data, blockchain technology, cloud computing, CPS, digital twin, IoT, workflows, and web-based platforms. For example, IoT is a deeply interconnected ecosystem of sensors, cameras, computers, smart systems, connected devices, smart devices, and other technologies to share data, work together to make decisions, and operate autonomously in the background. A web-based platform (a.k.a., science gateway, citizen science website, hub, e-Science, e-Research, virtual community platform, virtual research environment, virtual laboratory, etc.) combines a variety of cyberinfrastructure components (e.g., sensors, cloud computing, high performance computing, workflows, data repository, visualization tools, analysis tools, simulation tools, and gaming tools.) to support data collection and applications such that users can access diverse resources and communicate. A workflow (a.k.a., workflow software, scientific workflow system, workflow management system, workflow engine) is a software that manages processes and automates a process or more. Putting these pieces together to develop smart GW management applications is an emerging field. In this chapter, we provide a brief overview on emerging technologies and discuss how these technologies can change our sustainable GW management practices.

11.4 Internet of Things—based data collection

Advancement of sensor devices, and the inexpensive easy-to-use IoT-based technologies, leads to low-cost, low-power, open source, and do-it-yourself (DIY) sensors and data logging solutions. Such high-quality scientific measurements for environmental monitoring applications can be made using inexpensive and off-the-shelf components (Horsburgh et al., 2019). Examples of these sensor technologies and data logging solutions include microcontroller units such as the Arduino suite of products, single-board computers like the Raspberry Pi, and the diverse array of IoT devices (Horsburgh et al., 2019). These IoT-based low-cost and DIY sensing systems are gradually emerging. For example, the Openly Published Environmental Sensing project (https://open-sensing.org) at Oregon State University, which focuses on developing environmental sensing projects and research, offers tutorials for students and practitioners on DIY sensor networks. Currently in academic labs, an IoT-based borehole sensor, which measures water pressure and quality, can be an order of magnitude cheaper than a typical commercial senor. Examples of low-cost, community-based, and real-time GW monitoring networks with sensor-to-web data streaming are presented by Drage and Kennedy (2020) in Nova Scotia, Canada and by Calderwood et al. (2020) in California. Sensor data can be streamed to a web-based platform through Wi-Fi (Drage & Kennedy, 2020), cellular connection (Drage & Kennedy, 2020), and satellite (Thomas et al., 2019). Note that a *web-based platform* is generally accessed over a network connection, using a web browser or as a client-based desktop and mobile application, with most of the processing occurring on external servers.

The applications of the IoT-based sensors are diverse. These IoT-based, low-cost, and DIY sensing networks that stream data to a web-based platform can be particularly useful for developing countries (Maroli et al., 2021; Narendran et al., 2017). For example, Thomas et al. (2019) demonstrate that sensor network implementation across large spatial scales in arid regions in Africa can provide both practical benefits such as real-time monitoring of pump malfunction, and more importantly GW pumping data that are otherwise difficult to collect. For details about IoT-based monitoring networks and their related applications in smart water management, the reader is referred to recent review articles (Jan et al., 2021; Salam, 2020; Varadharajan et al., 2019). As these IoT-based low-cost sensing systems are allowing easy collection of high-frequency data, this is accordingly changing our practices for data logging, transmission, storage, sharing, processing, and usage as discussed below.

11.5 Web-based data sharing

To improve GW SDI and to sustainably manage water resources, several data networks are already emerging in hydrology. SDI, which consists of spatial data, metadata, models, tools, and interactive user interfaces, is a digital infrastructure that enables the sharing and flexible use of data. The internet of water for sharing and integrating water data is emerging through diverse water platforms that enable open water data, integrate existing public water data, and connect regional data sharing communities (Patterson et al., 2017). These water platforms are increasing in number not only in response to technological advances, societal needs, and initiatives such as EarthCube, but also due to policy changes. For example, the Subcommittee on Water Availability and Quality of the US Office of Science and Technology Policy enacted the Open Water Data Initiative in 2014, which was chartered under the Department of the Interior's Advisory Committee on Water Information (http://acwi.gov/spatial/index.html) as an organized effort to provide water data, and community applications built on these data (Bales, 2016).

A water platform could be nonspatially bounded, forming around a unified purpose (Patterson et al., 2017). For example, HydroShare (https://www.hydroshare.org) of the Consortium of Universities for the Advancement of Hydrologic Science (CUAHSI) is a water platform with generic data model and content packaging scheme to allow the hydrologic science community to store, manage, share, publish, annotate, and collaborate around diverse types of data and models (Horsburgh et al., 2016; Morsy et al., 2017). Another example is WHYMAP (https://www.whymap.org)—the World-wide Hydrogeological Mapping and Assessment Program, which among its main purposes are to summarize GW information on the global scale, provide a GW resources map of the world, and provide map information for international discussion on water. The Water Information Network System of Intergovernmental Hydrological Program of the UNESCO (https://en.unesco.org/ihp-wins) is an open access platform for sharing water-related information and connecting water stakeholders from developed and developing countries. WaterShare (http://www.watershare.eu) is a platform for global to local collaboration, and knowledge sharing of water solutions and innovations to contribute to achieving the SDG6. In addition to the abovementioned discipline-specific repositories, there are general repositories for data sharing and collaboration. Stall et al. (2020) provide a comparison of general repositories such as the Open Science Framework (https://osf.io) for project management through the entire project life cycle based on open science best practices. The choice of the best repository for data sharing is case specific depending on the project size and features, funding agency requirements, community needs, and many other factors. Generally, choosing a well-established and trustworthy hydrology-specific repository would support the hydrology community.

A web-based water platform could be spatially bounded for data and model sharing and community collaboration (Patterson et al., 2017). For example, the online GW database of the Texas Water Development Board is a spatially bounded GW platform that allows access to information about Texas aquifers (Rosen et al., 2019), following the FAIR Data Principles such that data are Findable, Accessible, Interoperable, and Reusable (Wilkinson et al., 2016). This platform allows users to access Texas GW data, understand the implications of contamination events, and determine long-term GW availability trends (Rosen et al., 2019). An example of a more generic regional platform that is also based on FAIR Data Principles is the Gulf of Mexico Research Initiative Information and Data Cooperative (https://data.gulfresearchinitiative.org), which is a data management system for the full life cycle of data for researchers in the Gulf of Mexico. A spatially bounded GW platform is not necessarily regional but can range from a local to transboundary scale data networks that supply heterogeneous data to users. For example, through conformance to international standards, SDI architectures, and shared vocabularies, the Canadian Groundwater Information Network and the US National Groundwater Monitoring Network, are examples of large GW data networks that can be interoperable, allowing public access to harmonized GW data from shared international borders (Brodaric et al., 2016). A web-based platform for GW data could be spatially bounded not only to share data and models, but also to serve as a platform for community collaboration as discussed in Section 11.8.

Web-based data sharing requires SDI and standards that shape the data model. A *data model* determines the vocabulary and structure of how data is collected, stored, processed, queried, and visualized in a data network. *Interoperability* among data models permits a common language among data networks to make data quarriable and seamlessly usable across data networks regardless of its original heterogeneity. Accordingly, standardization across data models is needed. For example, GroundWaterML2 (GWML2, Brodaric et al., 2018) consists of data structures and encoding guidelines for hydrogeology web data exchange. GWML2 is a new global standard for international GW data representation that is developed by the Groundwater Standards Working Group of the Open Geospatial Consortium (OGC). GWML2 is compliant with the concepts, standards and technologies of SDI, and can be used in conjunction with a variety of web services as central structure for the query and transport of data. This enables online data interoperability at multiple levels amongst numerous and heterogeneous data sources. Similarly, the EU-wide INSPIRE addresses 34 spatial data themes (e.g., INSPIRE-Meteorology, INSPIRE-Geology, INSPIRE-Landcover, INSPIRE-Landuse, and INSPIRE-Hydrograph) that are relevant to the environment to ensure data accessibility and seamless data combination from different sources and across different scales and levels (Ilie & Gogu, 2019). The INSPIRE-Hydrogeology is part of the INSPIRE-Geology.

The hydrology community pursues achieving more compatibility among different data representations. This requires interoperability between their representations at the levels of syntactic, schematic, and semantic that include the differences in terminology and definitions (Hahmann et al., 2016). For example, by describing how terms (e.g., geologic unit or water body) are used in different data models, a conversion to and from GWML2 with INSPIRE-Hydrogeology is seamless. In addition, although a significant attention has been devoted to the sharing and reusing hydrological data than hydrological models, flexible and general metadata frameworks for model sharing are emerging across wide variety of models (Morsy et al., 2017). Finally, data configuration managers are needed to customize data requirements. For example, Wang et al. (2020) propose a universal data exchange model that provides data to users through data services rather than through downloading raw data files. This data configuration manager additionally provides interactive programming tools for data customization, and a component-based data viewer of different types of hydrological information (Wang et al., 2020).

11.6 Workflow for data processing

The scope of the workflow depends on the level of vertical integration and horizontal scope as shown in Fig. 11.3. For example, after manual or automatic data retrieval from sensors, data networks, collaborative data updates, and user data, the workflow for data processing can range from basic data analysis and visualization to model-based analysis and decision support. Dahlhaus et al. (2016) developed a GW web-based platform for the State of Victoria in Australia with tools for data querying and 3D visualizations. This includes simple data analysis such as the ability to query the predicted depth to water table, water

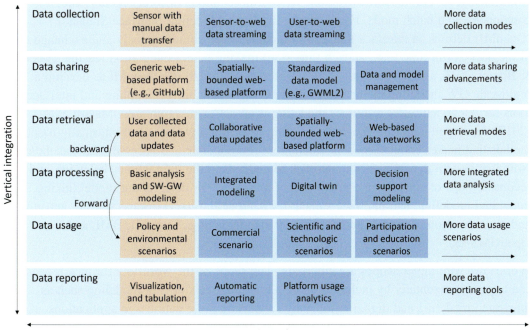

Figure 11.3
Different components of the workflow. The workflow can expand through forward and backward vertical integration of different components (y-axis), and through expanding the scope of each component by considering further subcomponents and advancements (x-axis). Blocks highlighted in orange represent an example of a basic workflow where the data is manually collected and stored on a generic web-based platform for data sharing, and then the workflow retrieves the user data, update and run a SW—GW model, analyze the model outputs with respect to a policy problem, and reports figures and tables. SW—GW, Surface water and groundwater.

quality and hydrostratigraphy at any selected point, and provides virtual borehole logs at any selected location in Victoria. A workflow with a wider data processing scope can include executing existing model instances or developing new model instances, given both phenomenological and process-based GW model programs. Note that a model program is a logical representation of ideas as source codes or compiled executables that can be used to execute many model instances with different input files and generate output files. An example of model instance is the MODFLOW input files representing the hydrological settings and conditions at a specific site and a given scenario. Automating the workflow for data processing requires an application programming interface (API) to communicate with the model in a batch mode that is without end user interaction. For example, the Python-based FloPy facilitates interacting with MODFLOW to configure input files, execute model, and analyze model outputs (Bakker et al., 2022). Similarly, PhreeqPy provides Python tools to work with PHREEQC for reactive transport modeling (Charlton & Parkhurst, 2011).

Additionally, several tools are available as interfaces among different models and components in a batch mode. For example, the Open Modeling Interface Standard (OpenMI) defines an interface that allows models to run simultaneously and exchange data in memory at run-time (e.g., at each time step) making model integration feasible (Becker & Burzel, 2016; Gregersen et al., 2007). Upton et al. (2020) use OpenMI to develop a multiscale GW modeling method to evaluate GW sustainability. Coupling and interfaces tools for integrated water resources modeling is an active research area. Malard et al. (2017) developed a tool for dynamic coupling of system dynamics and physically-based models that makes coupled models much more reproducible and accessible to stakeholders. Note that system dynamics is an approach to understand the interactions, behavior, and feedback loops of constituent components of a system. For detail about model linking, integrated modeling, and interface standard, Zhang et al. (2021) discuss sharing, reusing, and interoperation of models across different standards (e.g., OpenMI, BMI, and OpenGMS-IS).

Workflows with larger scopes are emerging. The scope of the workflow for data processing can expand horizontally by including more advanced levels of GW modeling such as integrated modeling and digital twin, or by supporting more data usage applications (Fig. 11.3). The scope of the workflow for data processing component can expand vertically with forward and backward integration with other components (e.g., data collection, data sharing, and data reporting). For example, De Filippis, Stevenazzi, et al. (2020) used commonly available standards and tools to develop a workflow to collect and process vadose-zone data from field sensors to simulate percolation to the water table with automatic generation of summary reports like plots and tables. Barnhart et al. (2010) develop a data assimilation method that integrates GW contaminant transport models with wireless sensor networks. Su et al. (2020) propose a similar a web-based workflow for GW simulation. In addition, a workflow can link the GW models to model development and decision-support tools for parameter estimation, data assimilation, surrogate modeling, sensitivity analysis, simulation optimization, agent-base modeling, and uncertainty quantification. These tools are generally based on optimization algorithms, sampling algorithms, and machine and deep learning algorithms.

Diverse workflows have been developed to process data for analyzing SW—GW resources. Workflows with IoT-based data acquisition and wide scope of data processing are nascent in GW hydrology relative to SW hydrology. Examples of simple to more involved workflows in SW hydrology are numerous. For example, to solve the lack of transparency on model creation, Chawanda et al. (2020) developed the workflow software (SWAT + AW), which is an automatic workflow (AW) that automates the creation of catchment hydrological model instances for the SWAT model program using user collected data. To increase the SWAT model usage for nontechnically trained stakeholders and decision makers, McDonald et al. (2019) developed a web-based SWATOnline workflow with modular web applications such as automatic climate data retrieval from the National Aeronautics and Space Administration (NASA) servers. Hou et al. (2019) review the input data preparation methods from manual data preparation to intelligent geoprocessing, which allows full automated data preparation for geospatial modeling and is adaptive to application contexts. An emerging trend is to move toward more vertical integration and horizontal advances and expansion (Fig. 11.3). For example, Taylor et al. (2021) developed a web-based platform with cloud computing that standardizes the user workflow and preintegrates models and data. This allows users such as water planners and educators to rapidly develop case studies for basin-scale water assessment and scenario investigation to assess changes arising from developments in agriculture, water storages, population growth, and climate changes.

In GW hydrology, more integrated workflows are also developing. For example, FREEWAT (Rossetto et al., 2018), which is funded by the EU Commission, integrates open source models and tools for SW—GW management supporting data collection, data sharing, and data analysis for supporting model-based planning and decision-making. FREEWAT uses the FloPy Python library to connect SW—GW models with a toolbox for model calibration and uncertain quantification in a QGIS-enabled and integrated environment for spatial data management, processing, and visualization. QGIS is an open-source cross-platform geographic information system software. The FREEWAT supports a number of models, including the integrated SW—GW MF-OWHM, a Crop Growth Module for crop yield modeling, MT3DUSGS and MT3DMS for solute transport in the unsaturated and saturated zones, and SEAWAT for density-dependent GW flow. De Filippis, Pouliaris, et al. (2020) displays a total of 13 case studies in European and non-European countries where the FREEWAT platform along with ICTs were applied for SW—GW management, transboundary aquifer management, protection of GW-dependent ecosystems, and rural water management. The case studies show that improved access to data and the portability of models and model results can help promote water sustainability from the local to basin scales.

In addition to data analysis with respect to SW−GW resources, workflows are emerging to address the other components of GW sustainability evaluation (Fig. 11.2), such as GW-dependent ecosystems, human activities, uncertainty analysis, and participation. Workflows for the protection of GW-dependent ecosystems are emerging (Tague & Frew, 2021; Turner et al., 2020; Wohner et al., 2020). Workflows generally accounts for human activities using simple techniques such as scenario analysis. For example, Bojovic et al. (2018) developed a web-based platform to facilitate stakeholder collaboration in the analysis of water management adaptation options in the Alps. As integrating social and hydrologic data in an elaborate socio-hydrology analysis can be generally challenging, Flint et al. (2017) provide social water science data classification and recommendations on data management considerations for cyberinfrastructure purposes. With respect to uncertainty analysis, White et al. (2020) provided a MODFLOW-based GW modeling framework that uses FloPy to develop a workflow for parameter estimation and uncertainty quantification. With respect to participation, several levels and forms of participation exist such as collaborative modeling with key stakeholders, general public participation for planning and decision-making, peering for scientific research, commons-based peer management including codesign and codecision-making, among many other forms. Examples of mature workflows and platforms for spatially bounded and community centered collaboration include the Bay Delta Live for the San Francisco Bay Delta estuary in California (https://www.baydeltalive.com) for water and ecosystem services management. Bay Delta Live, in which Southern California's Metropolitan Water District invested two million dollars over five years to help launch the platform, includes modules for collaboration with a desktop and phone applications providing real-time information for daily decision-making (Jooste, 2017).

11.7 Scenarios for data usage

Web-based platforms for GW sustainability can support a number of application scenarios. For example, Brodaric et al. (2018) identify five data usage scenarios that motivate GWML2, which are commercial scenario, policy scenario, environmental scenario, scientific scenario, and technologic scenario. The policy scenario includes administrative reporting on withdrawal limits, sharing information across different water authorities, and incentive mechanisms for GW policy implementation. For example, in response to SGMA a pilot project in Sacramento-San Joaquin River Delta, California is developing a low-cost satellite connected sensors with real-time GW data streaming to IBM Blockchain Platform (IBM Research, 2019). This respones to SGMA mandate to the creation of local groups to develop and implement solutions to make their local GW usage sustainable by 2040. The environmental scenario includes components such as monitoring, protection, and management GW-dependent ecosystems. For example, the Bay Delta Live web-based platform focuses on understanding the complex and dynamic ecosystem of the Sacramento-San Joaquin Bay Delta (Patterson et al., 2017).

An example of the commercial scenario is to estimate the cost and timeline for drilling a new well. A well driller can use a web platform to explore the local geology, which can inspect wells located near the target area in terms of the lithology, water level, yield, and total depth at each well (Brodaric et al., 2018).

Scientific scenario includes data sharing, data usage, and peering to perform multiple scientific activities such as data analysis, model development, data worth analysis, and model-based analysis. For example, the HydroShare of CUAHSI was primarily developed to facilitate data and model sharing in the academic community. A spatially bounded example is a web-based workflow developed by Shuler and Mariner (2020) for collaborative GW modeling with key stakeholders at the American Samoa. Bandaragoda et al. (2019) discuss how the existing cyberinfrastructure tools and resources can enable collaborative numerical modeling in Earth sciences. Additionally, this can serve as a platform for peering within the scientific community and with stakeholders, and broadens the use of scientific applications such as integrated-model simulation and gaming to stakeholders. Moreover, these web-based platforms with monitoring data can support machine learning-based data exploration (Rau et al., 2020), and GW representation in continental to global scale models (Gleeson et al., 2021).

The technologic scenario involves data delivery situations, which require compatibility with other hydrogeological data representations to enable data interoperability within a GW data network, and between different data networks (Brodaric et al., 2018). In addition, this includes communication with other devices to create a CPS, in which a process is monitored and controlled through the IoT. For example, Wang et al. (2013) present a case study in Xiamen City, China, in which an IoT-based online water quality management system maintains water level of the Scenic river by automatically supplementing reclaimed domestic wastewater and fresh SW from the Xinglin Bay, and cycling landscape water to stabilize the water quality. CPS is a widely used technology in smart water applications to manage water supply networks (Kulkarni & Farnham, 2016; Pan et al., 2015).

Other data usage scenarios are participation and education. Web-based GW platforms allow open community participation from different physical locations and domains of expertise to join the GW exploration, to easily exchange ideas with less thresholds as compared to centralized systems, and to perform comprehensive modeling and analysis tasks collaboratively (Chen et al., 2020). For example, to help communicate complex hydrogeological concepts to improve confidence in decision-making, Wolhuter et al. (2020) developed the 3D Water Atlas of the Surat Basin, Queensland, Australia. This is an interactive web-based platform based on a three-dimensional geological model for visualizing and analyzing hydraulic and hydrogeochemical data from boreholes in a way that is accessible to a wide audience. Dahlhaus et al. (2016) show that the web-based GW platform of the State of Victoria, which was developed outside of the government to meet end user needs and educate a broader community, has increased the end user interaction and participation, empowering society with the value of big data to guide future planning for sustainable and equitable GW.

Education is a valuable data usage scenario. Bridging sustainability science, Earth science, and data science requires interdisciplinary education with competencies in data processing and model development (Pennington et al., 2020). To prepare future hydrologists and engineers, Lane et al. (2021) present HydroLearn, an open web-based educational platform to provide a formal pedagogical structure for developing effective problem-based learning activities. Targeting upper-level undergraduate and early graduate students in hydrology and engineering, HydroLearn allows the students to explore how well models or equations work in particular settings or to answer specific problems using real data (Lane et al., 2021).

11.8 Perspectives of web-based groundwater platforms

The advances in technology are enabling the transition toward smart GW management, and accordingly the increase in the number and scope of web-based GW platforms. This will change the paradigm for managing GW as these web-based GW platforms enable open resource and data sharing, open integrated modeling and simulation to be performed, and open community to grow and expand organically (Chen et al., 2020). Given the technologies of the fourth industrial revolution, we discuss a spatially bounded, web-based GW platform with a digital twin for commons-peer management of GW resources. We also discuss how these platforms can improve GW sustainability practices.

The web-based GW platform connects the workflow with available computing resources and provides a web-based interface for users. As many of the workflows require extensive processing power, storage space, and communication speed, they are executed over a large-scale platform with user interfaces. Users include expert users, who have certain technical and scientific knowledge about parts of the topic, and general stakeholders, who might know less about the topic from a technical perspective though they could have ample indigenous and intuitive knowledge about the topic (Chen et al., 2020). A web-based platform has several advantages such that the platform users do not need to install software and manage updates, the platform connects to cloud-based high performing computing resources, and the platform automatically retrieves data from data networks. Currently, cyberinfrastructure is reaching a point where it is possible to build open and transparent environmental modeling systems (Choi et al., 2021). As such, computational environments are interlinked with sensors and data repositories, supported by APIs for programmatic control of the modeling activities and the workflow, serve as gateways to high performance computing resources, and provide user web-based interfaces (Chen et al., 2020; Choi et al., 2021; Hubbard et al., 2020).

A spatially bounded web-based GW platform can mature to a digital twin for commons-based peer management of GW resources. GW workflows are developing to allow for more advanced data processing. Goodall et al. (2011) discuss modeling water resource systems using a service-oriented computing paradigm, which is software engineering paradigm that deals with a complex software system as an interconnected collection of distributed computational components. An advanced web-based workflow such as the one shown in Fig. 11.4 is nascent in GW hydrology. The workflow shown in Fig. 11.4 allows for data search, collection, and harmonization from data networks and sensor-to-web data streaming, data and model representation following international standards (e.g., OGC standards), and model programs and instances linkage with decision-support tools. One of the existing workflows that includes several of the components shown in Fig. 11.4 is FREEWAT, which can be used and extended for smart GW management especially in regions with historical water scarcity exacerbated by climate change (Theuma et al., 2017). Other existing tools include Delft-FEWS (https://oss.deltares.nl/web/delft-fews) that is an open platform that integrates data and models. For example, van der Vat et al. (2019) use Delft-FEWS to develop the Ganga

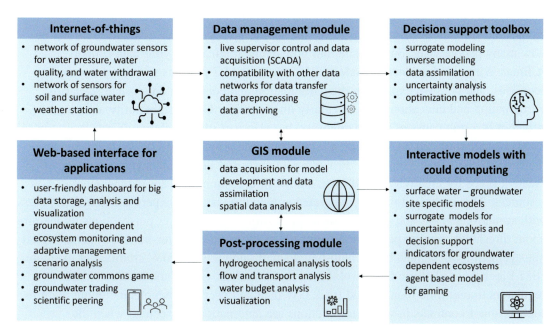

Figure 11.4
Components of a web-based groundwater platform with a digital twin for commons-based peer management of groundwater resources.

Water Information System (GangaWIS) that contains model inputs and relevant outputs of SW−GW models that support strategic planning in the Ganga Basin in India.

The workflow scope can be expanded to include a digital twin, which is a virtual presentation of the aquifer system to enable stakeholders to monitor and manage the SW−GW system, GW -dependent ecosystems, and GW-dependent human activities (e.g., irrigation and municipal water supply). A digital twin of the aquifer changes as the physical aquifer changes. For example, sensor-to-web data streaming and data assimilation techniques update the GW flow model, water budget calculation, solute transport model, GW-dependent ecosystem indicators, and GW sustainability indicators. The digital twin can expand in scope to serve as an integrated GW system in the digital world, which changes as its natural and social counterparts change. This can be particularly useful in an agricultural setting where this GW digital twin works together with a smart agriculture system.

The web-based GW platform has a web-based interface for applications. This web-based interface will typically include a website for static information such site-specific documents and tutorials, data viewer to present data from the database, and dashboard with different options related to data viewing, processing, and post-analysis. The web-based platform can be accessed through a web browser, and as a client-based desktop and mobile application as needed.

A spatially bounded, web-based GW platform can extend beyond merely sharing data, models, and computational resources, to provide multiple water resources and environmental management services. "In the future, your performance metric will not be how many people visit your website, but how many applications your data support" (ClimateWire, 2015 as cited in Bales, 2016). Section 11.7 presents several data usage scenarios, and here we expand on the participation and education data usage scenarios. Participation and education are particularly important for GW sustainability as human behavior is a root cause of unsustainability, but also part of the solution (Castilla-Rho et al., 2019; Elshall et al., 2021). Creating a digital representation of the GW system provides a platform to create a conversation. Providing analysis and digital management tools can create a space for creativity, build trust, and facilitate commons-peer management of GW resources. This includes tools for integrated simulation tasks for expert and nonexpert users with scenario analysis and gaming. Such applications can shift user perspectives by learning about thresholds, tipping points, and pathways of the coupled water-human system. Digital management tools include blockchain-based smart contracts and GW credits. For example, IBM Blockchain Platform has a web-based dashboard for GW users, financers, and regulators to real-time monitor and track GW data and user transactions including features such as smart contracts in which transactions are automatically executed when the conditions are matched (IBM Research, 2019). This platform supports policy and market mechanisms such as GW individual users share cap, GW credit, GW share purchase, and GW trading. Thus, these web-based platforms facilities not only the coproduction of data (e.g., hydraulic and geochemical data), information (e.g., water budget), and knowledge (e.g., sustainable pumping limits), but also codecision-making and joint action with respect to GW pumping and GW trading.

The advancements of these web-based GW platforms can eventually lead to commons-peer management of GW resources. The term *commons-based peer production* was coined by Benkler (2002) to refer to a social-economic phenomenon emerging in which a large number of people work cooperatively without the traditional firm-based or market-based ownership of the resulting product. Examples include working on large and small-scale projects, generally, online (e.g., Wikipedia, Python, Linux, and open source software), but sometimes offline (e.g., community gardening). To produce data, knowledge, and goods, commons-based peer production follows motivational drives and social signals, rather than market prices and managerial commands (Benkler & Nissenbaum, 2006; Benkler, 2002). Accordingly, commons-based peer production can be regarded as a virtuous behavior, and a society that provides opportunities for virtuous behavior is one that is more conducive to virtuous individuals (Benkler & Nissenbaum, 2006). This is particularly important because among the reasons of GW unsustainability are the influence of some social groups over less privileged social groups in water resources governance (Baldassarre et al., 2021; Méndez-Barrientos et al., 2020), and when individual profits prevail over the need to preserve a common resource (Leduc et al., 2017). Thus, commons-based peer management improves equity and inclusivity, and can be regarded as a form of managing the GW as a common-pool resource (Ostrom, 1990). A common-pool resource (e.g., GW and fishpond) is an economic term referring to a resource that is shared and available to everyone like public goods, but with variable and limited stock that is subject to rivalrous consumption like private goods such that each unit consumption subtracts from the total stock (Hayes, 2021). The common-pool resource is subject to the tragedy of the commons (Hardin, 1968) that is when individuals try to maximize their self-interest regardless of the social cost. Unlike public goods that can be utilized without reducing availability for others, a common-pool resource requires protection to prevent overuse and congestion, and to ensure continuous and nonexcludable supply. While this generally calls for government regulation, it additionally calls for commons-based peer management.

11.9 Disclaimer

The views expressed in this article are those of the authors and do not necessarily reflect the views or policies of the US Environmental Protection Agency.

Acknowledgments

The first two authors of the chapter were supported by NSF Award #1939994 and NSF Grant EAR-1552329.

References

Bakker, M., Post, V., Hughes, J. D., Langevin, C. D., White, J. T., Leaf, A. T., Paulinski, S. R., Bellino, J. C., Morway, E. D., Toews, M. W., Larsen, J. D., Fienen, M. N., Starn, J. J., & Brakenhoff, D. (2022). FloPy v3.3.6 — release candidate: U.S. Geological Survey Software Release, 08 March 2022, http://dx.doi.org/10.5066/F7BK19FH

Baldassarre, G. D., Cloke, H., Lindersson, S., Mazzoleni, M., Mondino, E., Mård, J., Odongo, V., Raffetti, E., Ridolfi, E., Rusca, M., Savelli, E., & Tootoonchi, F., (2021). Integrating multiple research methods to unravel the complexity of human-water systems. *AGU Advances, 2,* e2021AV000473. Available from https://doi.org/10.1029/2021AV000473.

Bales, J. (2016). Featured collection introduction: Open water data initiative. *JAWRA Journal of the American Water Resources Association, 52,* 811–815. Available from https://doi.org/10.1111/1752-1688.12439.

Bandaragoda, C., Castronova, A., Istanbulluoglu, E., Strauch, R., Nudurupati, S. S., Phuong, J., Adams, J. M., Gasparini, N. M., Barnhart, K., Hutton, E. W. H., Hobley, D. E. J., Lyons, N. J., Tucker, G. E., Tarboton, D. G., Idaszak, R., & Wang, S. (2019). Enabling collaborative numerical modeling in earth sciences using knowledge infrastructure. *Environmental Modelling & Software, 120,* 104424. Available from https://doi.org/10.1016/j.envsoft.2019.03.020.

Barnhart, K., Urteaga, I., Han, Q., Jayasumana, A., & Illangasekare, T. (2010). On integrating groundwater transport models with wireless sensor networks. *Groundwater, 48,* 771–780. Available from https://doi.org/10.1111/j.1745-6584.2010.00684.x.

Becker, B., & Burzel, A. (2016). Model coupling with OpenMI introduction of basic concepts. In R. Setola, V. Rosato, E. Kyriakides, & E. Rome (Eds.), *Managing the complexity of critical infrastructures: A modelling and simulation approach* (pp. 279–299). Cham: Springer International Publishing Ag. Available from https://doi.org/10.1007/978-3-319-51043-9_11.

Benkler, Y. (2002). Coase's penguin, or, Linux and the nature of the firm. *The Yale Law Journal, 112,* 369. Available from https://doi.org/10.2307/1562247.

Benkler, Y., & Nissenbaum, H. (2006). Commons-based peer production and virtue. *The Journal of Political Philosophy, 14,* 394–419. Available from https://doi.org/10.1111/j.1467-9760.2006.00235.x.

Bierkens, M. F. P., & Wada, Y. (2019). Non-renewable groundwater use and groundwater depletion: A review. *Environmental Research Letters, 14,* 063002. Available from https://doi.org/10.1088/1748-9326/ab1a5f.

Bojovic, D., Giupponi, C., Klug, H., Morper-Busch, L., Cojocaru, G., & Schoerghofer, R. (2018). An online platform supporting the analysis of water adaptation measures in the Alps. *Journal of Environmental Planning and Management, 61,* 214–229. Available from https://doi.org/10.1080/09640568.2017.1301251.

Boyce, S. E., Hanson, R. T., Ferguson, I., Schmid, W., Henson, W. R., Reimann, T., Mehl, S. W., & Earll, M. M. (2020). *One-water hydrologic flow model: A MODFLOW based conjunctive-use simulation software (USGS numbered series no. 6-A60), one-water hydrologic flow model: A MODFLOW based conjunctive-use simulation software, techniques and methods.* Reston, VA: US Geological Survey. Available from https://doi.org/10.3133/tm6A60.

Bremer, L. L., Elshall, A. S., Wada, C. A., Brewington, L., Delevaux, J. M. S., El-Kadi, A. I., Voss, C. I., & Burnett, K. M. (2021). Effects of land-cover and watershed protection futures on sustainable groundwater management in a heavily utilized aquifer in Hawai'i (USA). *Hydrogeology Journal, 29,* 1749–1765. Available from https://doi.org/10.1007/s10040-021-02310-6.

Brodaric, B., Boisvert, E., Chery, L., Dahlhaus, P., Grellet, S., Kmoch, A., Letourneau, F., Lucido, J., Simons, B., & Wagner, B. (2018). Enabling global exchange of groundwater data: GroundWaterML2 (GWML2). *Hydrogeology Journal, 26,* 733–741. Available from https://doi.org/10.1007/s10040-018-1747-9.

Brodaric, B., Booth, N., Boisvert, E., & Lucido, J. (2016). Groundwater data network interoperability. *Journal of Hydroinformatics, 18,* 210–225. Available from https://doi.org/10.2166/hydro.2015.242.

Calderwood, A. J., Pauloo, R. A., Yoder, A. M., & Fogg, G. E. (2020). Low-cost, open source wireless sensor network for real-time, scalable groundwater monitoring. *Water, 12,* 1066. Available from https://doi.org/10.3390/w12041066.

Castilla-Rho, J. C., Rojas, R., Andersen, M. S., Holley, C., & Mariethoz, G. (2019). Sustainable groundwater management: How long and what will it take? *Global Environmental Change*, *58*, 101972. Available from https://doi.org/10.1016/j.gloenvcha.2019.101972.

Castilla-Rho, J. C., Rojas, R., Andersen, M. S., Holley, C., & Mariethoz, G. (2017). Social tipping points in global groundwater management. *Nature Human Behaviour*, *1*, 640−649. Available from https://doi.org/10.1038/s41562-017-0181-7.

Charlton, S. R., & Parkhurst, D. L. (2011). Modules based on the geochemical model PHREEQC for use in scripting and programming languages. *Computers & Geosciences*, *37*, 1653−1663. Available from https://doi.org/10.1016/j.cageo.2011.02.005.

Chawanda, C. J., George, C., Thiery, W., Griensven, A., van Tech, J., Arnold, J., & Srinivasan, R. (2020). User-friendly workflows for catchment modelling: Towards reproducible SWAT + model studies. *Environmental Modelling & Software*, *134*, 104812. Available from https://doi.org/10.1016/j.envsoft.2020.104812.

Chen, M., Voinov, A., Ames, D. P., Kettner, A. J., Goodall, J. L., Jakeman, A. J., Barton, M. C., Harpham, Q., Cuddy, S. M., DeLuca, C., Yue, S., Wang, J., Zhang, F., Wen, Y., & Lü, G. (2020). Position paper: Open web-distributed integrated geographic modelling and simulation to enable broader participation and applications. *Earth-Science Reviews*, *207*, 103223. Available from https://doi.org/10.1016/j.earscirev.2020.103223.

Choi, Y.-D., Goodall, J. L., Sadler, J. M., Castronova, A. M., Bennett, A., Li, Z., Nijssen, B., Wang, S., Clark, M. P., Ames, D. P., Horsburgh, J. S., Yi, H., Bandaragoda, C., Seul, M., Hooper, R., & Tarboton, D. G. (2021). Toward open and reproducible environmental modeling by integrating online data repositories, computational environments, and model application programming interfaces. *Environmental Modelling & Software*, *135*, 104888. Available from https://doi.org/10.1016/j.envsoft.2020.104888.

Dahlhaus, P., Murphy, A., MacLeod, A., Thompson, H., McKenna, K., & Ollerenshaw, A. (2016). Making the invisible visible: The impact of federating groundwater data in Victoria, Australia. *Journal of Hydroinformatics*, *18*, 238−255. Available from https://doi.org/10.2166/hydro.2015.169.

De Filippis, G., Pouliaris, C., Kahuda, D., Vasile, T. A., Manea, V. A., Zaun, F., Panteleit, B., Dadaser-Celik, F., Positano, P., Nannucci, M. S., Grodzynskyi, M., Marandi, A., Sapiano, M., Kopač, I., Kallioras, A., Cannata, M., Filiali-Meknassi, Y., Foglia, L., Borsi, I., & Rossetto, R. (2020). Spatial data management and numerical modelling: Demonstrating the application of the QGIS-integrated FREEWAT platform at 13 case studies for tackling groundwater resource management. *Water*, *12*, 41. Available from https://doi.org/10.3390/w12010041.

De Filippis, G., Stevenazzi, S., Camera, C., Pedretti, D., & Masetti, M. (2020). An agile and parsimonious approach to data management in groundwater science using open-source resources. *Hydrogeology Journal*, *28*, 1993−2008. Available from https://doi.org/10.1007/s10040-020-02176-0.

Drage, J., & Kennedy, G. (2020). Building a low-cost, Internet-of-Things, real-time groundwater level monitoring network. *Groundwater Monitoring & Remediation*, *40*, 67−73. Available from https://doi.org/10.1111/gwmr.12408.

Elshall, A. S., Arik, A. D., El-Kadi, A. I., Pierce, S., Ye, M., Burnett, K. K., Wada, C., Bremer, L. L., & Chun, G. (2020). Groundwater sustainability: A review of the interactions between science and policy. *Environmental Research Letters*. Available from https://doi.org/10.1088/1748-9326/ab8e8c.

Elshall, A. S., Castilla-Rho, J., El-Kadi, A. I., Holley, C., Mutongwizo, T., Sinclair, D., & Ye, M. (2021). Sustainability of groundwater. *Reference module in earth systems and environmental sciences*. Elsevier. Available from https://doi.org/10.1016/B978-0-12-821139-7.00056-8.

Feng, D., Zheng, Y., Mao, Y., Zhang, A., Wu, B., Li, J., Tian, Y., & Wu, X. (2018). An integrated hydrological modeling approach for detection and attribution of climatic and human impacts on coastal water resources. *Journal of Hydrology*, *557*, 305−320. Available from https://doi.org/10.1016/j.jhydrol.2017.12.041.

Flint, C. G., Jones, A. S., & Horsburgh, J. S. (2017). Data management dimensions of social water science: The iUTAH experience. *JAWRA Journal of the American Water Resources Association*, *53*, 988−996. Available from https://doi.org/10.1111/1752-1688.12568.

Gil, Y., Pierce, S. A., Babaie, H., Banerjee, A., Borne, K., Bust, G., Cheatham, M., Ebert-Uphoff, I., Gomes, C., Hill, M., Horel, J., Hsu, L., Kinter, J., Knoblock, C., Krum, D., Kumar, V., Lermusiaux, P., Liu, Y., North, C., ... Zhang, J. (2018). Intelligent systems for geosciences: An essential research agenda. *Communications of the ACM*, 62, 76–84. Available from https://doi.org/10.1145/3192335.

Gleeson, T., Cuthbert, M., Ferguson, G., & Perrone, D. (2020). Global groundwater sustainability, resources, and systems in the anthropocene. *Annual Review of Earth and Planetary Sciences*, 48. Available from https://doi.org/10.1146/annurev-earth-071719-055251.

Gleeson, T., Villholth, K., Taylor, R., Perrone, D., & Hyndman, D. (2019). Groundwater: A call to action. *Nature Research*. Available from https://doi.org/10.1038/d41586-019-03711-0.

Gleeson, T., Wagener, T., Döll, P., Zipper, S. C., West, C., Wada, Y., Taylor, R., Scanlon, B., Rosolem, R., Rahman, S., Oshinlaja, N., Maxwell, R., Lo, M.-H., Kim, H., Hill, M., Hartmann, A., Fogg, G., Famiglietti, J. S., Ducharne, A., ... Bierkens, M. F. P. (2021). GMD perspective: The quest to improve the evaluation of groundwater representation in continental to global scale models. *Geoscientific Model Development Discussions*, 1–59. Available from https://doi.org/10.5194/gmd-2021-97.

Goodall, J. L., Robinson, B. F., & Castronova, A. M. (2011). Modeling water resource systems using a service-oriented computing paradigm. *Environmental Modelling & Software*, 26, 573–582. Available from https://doi.org/10.1016/j.envsoft.2010.11.013.

Gregersen, J. B., Gijsbers, P. J. A., & Westen, S. J. P. (2007). OpenMI: Open modelling interface. *Journal of Hydroinformatics*, 9, 175–191. Available from https://doi.org/10.2166/hydro.2007.023.

Hahmann, T., Stephen, S., & Brodaric, B. (2016). Semantically refining the groundwater markup language (GWML2) with the help of a reference ontology. *International Conference on GIScience Short Paper Proceedings*, 1. Available from https://doi.org/10.21433/B3118cz973mw.

Hardin, G. (1968). The tragedy of the commons. *Science (New York, NY)*, 162, 1243–1248. Available from https://doi.org/10.1126/science.162.3859.1243.

Hayes, A. (2021). *Common-pool resource definition* [WWW document]. Investopedia. <https://www.investopedia.com/terms/c/common-pool.asp> Accessed 09.13.21.

Henriksen, H. J., Troldborg, L., Hojberg, A. L., Refsgaard, J. C., Højberg, A. L., & Refsgaard, J. C. (2008). Assessment of exploitable groundwater resources of Denmark by use of ensemble resource indicators and a numerical groundwater-surface water model. *Journal of Hydrology*, 348, 224–240. Available from https://doi.org/10.1016/j.jhydrol.2007.09.056.

Hey, T., Tansley, S., & Tolle, K. (2009). *The fourth paradigm: Data-intensive scientific discovery*.

Hornberger, G. M., Wiberg, P. L., Raffensperger, J. P., & D'Odorico, P. (2014). *Elements of physical hydrology* (2nd ed.). Baltimore, MD: JHU Press.

Horsburgh, J. S., Caraballo, J., Ramírez, M., Aufdenkampe, A. K., Arscott, D. B., & Damiano, S. G. (2019). Low-cost, open-source, and low-power: But what to do with the data? *Frontiers in Earth Science*, 7, 67. Available from https://doi.org/10.3389/feart.2019.00067.

Horsburgh, J. S., Morsy, M. M., Castronova, A. M., Goodall, J. L., Gan, T., Yi, H., Stealey, M. J., & Tarboton, D. G. (2016). HydroShare: Sharing diverse environmental data types and models as social objects with application to the hydrology domain. *JAWRA Journal of the American Water Resources Association*, 52, 873–889. Available from https://doi.org/10.1111/1752-1688.12363.

Hou, Z.-W., Qin, C.-Z., Zhu, A.-X., Liang, P., Wang, Y.-J., & Zhu, Y.-Q. (2019). From manual to intelligent: A review of input data preparation methods for geographic modeling. *ISPRS International Journal of Geo-Information*, 8, 376. Available from https://doi.org/10.3390/ijgi8090376.

Hubbard, S. S., Varadharajan, C., Wu, Y., Wainwright, H., & Dwivedi, D. (2020). Emerging technologies and radical collaboration to advance predictive understanding of watershed hydrobiogeochemistry. *Hydrological Processes*, 34, 3175–3182. Available from https://doi.org/10.1002/hyp.13807.

IBM Research. (2019). *State of California tackles drought with IoT & blockchain* [WWW document]. IBM Newsroom. <https://newsroom.ibm.com/2019-02-08-State-of-California-Tackles-Drought-with-IoT-Blockchain> Accessed 09.07.21.

Ilie, C. M., & Gogu, R. C. (2019). Current trends in the management of groundwater specific geospatial information. *E3S Web of Conferences*, *85*, 07020. Available from https://doi.org/10.1051/e3sconf/20198507020.

Jan, F., Min-Allah, N., & Düştegör, D. (2021). IoT based smart water quality monitoring: Recent techniques, trends and challenges for domestic applications. *Water*, *13*, 1729. Available from https://doi.org/10.3390/w13131729.

Jooste, M. (2017). *A collaborative approach to open water data* [WWW document]. Redstone Strategy Group. <https://www.redstonestrategy.com/2017/07/25/bay-delta-live-water-data/> Accessed 08.30.21.

Kulkarni, P., & Farnham, T. (2016). Smart city wireless connectivity considerations and cost analysis: Lessons learnt from smart water case studies. *IEEE Access*, *4*, 660–672. Available from https://doi.org/10.1109/ACCESS.2016.2525041.

Lall, U., Josset, L., & Russo, T. (2020). A snapshot of the world's groundwater challenges. *Annual Review of Environment and Resources*, *45*, 171–194. Available from https://doi.org/10.1146/annurev-environ-102017-025800.

Lane, B., Garousi-Nejad, I., Gallagher, M. A., Tarboton, D. G., & Habib, E. (2021). An open web-based module developed to advance data-driven hydrologic process learning. *Hydrological Processes*, *35*, e14273. Available from https://doi.org/10.1002/hyp.14273.

Leduc, C., Pulido-Bosch, A., & Remini, B. (2017). Anthropization of groundwater resources in the Mediterranean region: Processes and challenges. *Hydrogeology Journal*, *25*, 1529–1547. Available from https://doi.org/10.1007/s10040-017-1572-6.

Malard, J. J., Inam, A., Hassanzadeh, E., Adamowski, J., Tuy, H. A., & Melgar-Quinonez, H. (2017). Development of a software tool for rapid, reproducible, and stakeholder-friendly dynamic coupling of system dynamics and physically-based models. *Environmental Modelling & Software*, *96*, 410–420. Available from https://doi.org/10.1016/j.envsoft.2017.06.053.

Markstrom, S. L., Niswonger, R. G., Regan, R. S., Prudic, D. E., & Barlow, P. M. (2008). *GSFLOW – Coupled ground-water and surface-water flow model based on the integration of the precipitation-runoff modeling system (PRMS) and the modular ground-water flow model (MODFLOW-2005) (USGS numbered series no. 6-D1)*. U.S. Geological Survey Techniques and Methods. Available from https://doi.org/10.3133/tm6D1.

Maroli, A. A., Narwane, V. S., Raut, R. D., & Narkhede, B. E. (2021). Framework for the implementation of an Internet of Things (IoT)-based water distribution and management system. *Clean Technologies and Environmental Policy*, *23*, 271–283. Available from https://doi.org/10.1007/s10098-020-01975-z.

McDonald, S., Mohammed, I. N., Bolten, J. D., Pulla, S., Meechaiya, C., Markert, A., Nelson, E. J., Srinivasan, R., & Lakshmi, V. (2019). Web-based decision support system tools: The Soil and Water Assessment Tool Online visualization and analyses (SWATOnline) and NASA earth observation data downloading and reformatting tool (NASAaccess). *Environmental Modelling & Software*, *120*, 104499. Available from https://doi.org/10.1016/j.envsoft.2019.104499.

Méndez-Barrientos, L. E., DeVincentis, A., Rudnick, J., Dahlquist-Willard, R., Lowry, B., & Gould, K. (2020). Farmer participation and institutional capture in common-pool resource governance reforms. The case of groundwater management in California. *Society & Natural Resources*, *33*, 1486–1507. Available from https://doi.org/10.1080/08941920.2020.1756548.

Miro, M. E., & Famiglietti, J. S. (2018). A framework for quantifying sustainable yield under California's Sustainable Groundwater Management Act (SGMA). *Sustainable Water Resources Management*. Available from https://doi.org/10.1007/s40899-018-0283-z.

Morsy, M. M., Goodall, J. L., Castronova, A. M., Dash, P., Merwade, V., Sadler, J. M., Rajib, M. A., Horsburgh, J. S., & Tarboton, D. G. (2017). Design of a metadata framework for environmental models with an example hydrologic application in HydroShare. *Environmental Modelling & Software*, *93*, 13–28. Available from https://doi.org/10.1016/j.envsoft.2017.02.028.

Mulligan, K. B., Brown, C., Yang, Y.-C. E., & Ahlfeld, D. P. (2014). Assessing groundwater policy with coupled economic-groundwater hydrologic modeling. *Water Resources Research*, *50*, 2257–2275. Available from https://doi.org/10.1002/2013WR013666.

Narendran, S., Pradeep, P., & Ramesh, M. V. (2017). An Internet of Things (IoT) based sustainable water management. In: *2017 IEEE Global Humanitarian Technology Conference (GHTC). Presented at the 2017 IEEE Global Humanitarian Technology Conference (GHTC)* (pp. 1–6). Available from https://doi.org/10.1109/GHTC.2017.8239320.

NSF. (2015). *EarthCube: (nsf21515) | NSF – National Science Foundation* [WWW document]. <https://www.nsf.gov/pubs/2021/nsf21515/nsf21515.htm> Accessed 09.04.21.

Ostrom, E. (1990). *Governing the commons: The evolution of institutions for collective action*. Cambridge: Cambridge University Press.

Pan, Q., Jonoski, A., Castro-Gama, M. E., & Popescu, I. (2015). Application of a web-based decision support system for water supply networks. *Environmental Engineering and Management Journal, 14*, 2087–2094.

Patterson, L., Doyle, M., King, K., & Monsma, D. (2017). *Internet of water: Sharing and integrating water data for sustainability*. The Aspen Institute.

Pennington, D., Ebert-Uphoff, I., Freed, N., Martin, J., & Pierce, S. A. (2020). Bridging sustainability science, earth science, and data science through interdisciplinary education. *Sustainability Science, 15*, 647–661. Available from https://doi.org/10.1007/s11625-019-00735-3.

Petty, T., Wildeman, A., Northey, B., James, R. D., Simmons, D. R., & Gallaudet, T. (2021). Strategy and recommendations for modernizing America's water resource management and water infrastructure: Initial reports and recommendations. *Water Subcabinet Pursuant to Executive Order, 13956*. Available from https://www.doi.gov/sites/doi.gov/files/modernizing-americas-water-resource.pdf, 35.

Pierce, S. A., Sharp, J. M., Guillaume, J. H. A., Mace, R. E., & Eaton, D. J. (2013). Aquifer-yield continuum as a guide and typology for science-based groundwater management. *Hydrogeology Journal, 21*, 331–340. Available from https://doi.org/10.1007/s10040-012-0910-y.

Rau, G. C., Cuthbert, M. O., Post, V. E. A., Schweizer, D., Acworth, R. I., Andersen, M. S., Blum, P., Carrara, E., Rasmussen, T. C., & Ge, S. (2020). Future-proofing hydrogeology by revising groundwater monitoring practice. *Hydrogeology Journal, 28*, 2963–2969. Available from https://doi.org/10.1007/s10040-020-02242-7.

Refsgaard, J. C., van der Sluijs, J. P., Højberg, A. L., & Vanrolleghem, P. A. (2007). Uncertainty in the environmental modelling process – A framework and guidance. *Environmental Modelling & Software, 22*, 1543–1556. Available from https://doi.org/10.1016/j.envsoft.2007.02.004.

Regan, R. S., & Niswonger, R. G. (2021). *GSFLOW version 2.2.0: Coupled groundwater and surface-water FLOW model: U.S. Geological Survey Software Release, 18 February 2021*.

Rinaudo, J.-D., Holley, C., Barnett, S., & Montginoul, M. (2020). *Sustainable groundwater management: A comparative analysis of French and Australian policies and implications to other countries, Global issues in water policy*. Springer International Publishing. Available from https://doi.org/10.1007/978-3-030-32766-8.

Rohde, M. M., Saito, L., & Smith, R. (2020). *Groundwater thresholds for ecosystems: A guide for practitioners*. Global Groundwater Group, the Nature Conservancy.

Rosen, R., Hermitte, S. M., Pierce, S., Richards, S., & Roberts, S. V. (2019). An internet for water: Connecting Texas water data. *Texas Water Journal, 10*, 24–31. Available from https://doi.org/10.21423/twj.v10i1.7086.

Roser, M., & Ritchie, H. (2013). *Technological progress*. Our World in Data.

Rossetto, R., De Filippis, G., Borsi, I., Foglia, L., Cannata, M., Criollo, R., & Vázquez-Suñé, E. (2018). Integrating free and open source tools and distributed modelling codes in GIS environment for data-based groundwater management. *Environmental Modelling & Software, 107*, 210–230. Available from https://doi.org/10.1016/j.envsoft.2018.06.007.

Saito, L., Christian, B., Diffley, J., Richter, H., Rohde, M. M., & Morrison, S. A. (2021). Managing groundwater to ensure ecosystem function. *Groundwater, 59*, 322–333. Available from https://doi.org/10.1111/gwat.13089.

Salam, A. (2020). Internet of Things for water sustainability. *Internet of Things for sustainable community development: Wireless communications, sensing, and systems, Internet of Things* (pp. 113–145). Cham: Springer International Publishing. Available from https://doi.org/10.1007/978-3-030-35291-2_4.

Salem, G. S. A., Kazama, S., Komori, D., Shahid, S., & Dey, N. C. (2017). Optimum abstraction of groundwater for sustaining groundwater level and reducing irrigation cost. *Water Resources Management*, *31*, 1947−1959. Available from https://doi.org/10.1007/s11269-017-1623-8.

Schwab, K. (2016). *The fourth industrial revolution*.

Shalsi, S., Ordens, C. M., Curtis, A., & Simmons, C. T. (2019). Can collective action address the "tragedy of the commons" in groundwater management? Insights from an Australian case study. *Hydrogeology Journal*, *27*, 2471−2483. Available from https://doi.org/10.1007/s10040-019-01986-1.

Shuler, C. K., & Mariner, K. E. (2020). Collaborative groundwater modeling: Open-source, cloud-based, applied science at a small-island water utility scale. *Environmental Modelling & Software*, *127*, 104693. Available from https://doi.org/10.1016/j.envsoft.2020.104693.

Stall, S., Martone, M. E., Chandramouliswaran, I., Crosas, M., Federer, L., Gautier, J., Hahnel, M., Larkin, J., Lowenberg, D., Pfeiffer, N., Sim, I., Smith, T., Van Gulick, A. E., Walker, E., Wood, J., Zaringhalam, M., & Zigoni, A. (2020). *Generalist repository comparison chart*. Available from https://doi.org/10.5281/zenodo.3946720.

Su, Y.-S., Ni, C.-F., Li, W.-C., Lee, I.-H., & Lin, C.-P. (2020). Applying deep learning algorithms to enhance simulations of large-scale groundwater flow in IoTs. *Applied Soft Computing*, *92*, 106298. Available from https://doi.org/10.1016/j.asoc.2020.106298.

Tague, C., & Frew, J. (2021). Visualization and ecohydrologic models: Opening the box. *Hydrological Processes*, *35*, e13991. Available from https://doi.org/10.1002/hyp.13991.

Taylor, P., Rahman, J., O'Sullivan, J., Podger, G., Rosello, C., Parashar, A., Sengupta, A., Perraud, J.-M., Pollino, C., & Coombe, M. (2021). Basin futures, a novel cloud-based system for preliminary river basin modelling and planning. *Environmental Modelling & Software*, *141*, 105049. Available from https://doi.org/10.1016/j.envsoft.2021.105049.

Theuma, N., Rossetto, R., & Calabro, G. (2017). Final report on the focus groups integrating the participatory approach to technical modelling activities: FREE and open source software tools for WATer resource management. In: *EU Horizon 2020 project, version 2*.

Thomas, E. A., Needoba, J., Kaberia, D., Butterworth, J., Adams, E. C., Oduor, P., Macharia, D., Mitheu, F., Mugo, R., & Nagel, C. (2019). Quantifying increased groundwater demand from prolonged drought in the East African Rift Valley. *Science of the Total Environment*, *666*, 1265−1272. Available from https://doi.org/10.1016/j.scitotenv.2019.02.206.

Thompson, Jr., B. H., Rohde, M. M., Howard, J. K., & Matsumoto, S. (2021). *Mind the gaps: The case for truly comprehensive sustainable groundwater management, water in the west*. Stanford Digital Repository. Available from: <https://purl.stanford.edu/hs475mt1364>.

Turner, B., Hill, D. J., & Caton, K. (2020). Cracking "Open" technology in ecohydrology. In D. F. Levia, D. E. Carlyle-Moses, S. Iida, B. Michalzik, K. Nanko, & A. Tischer (Eds.), *Forest-water interactions, ecological studies* (pp. 3−28). Cham: Springer International Publishing. Available from https://doi.org/10.1007/978-3-030-26086-6_1.

UN-Water. (2018). *Groundwater overview − Making the invisible visible (UN-Water category III publication)*. UN-Water, UN-Water Category III Publication.

Upton, K. A., Jackson, C. R., Butler, A. P., & Jones, M. A. (2020). An integrated modelling approach for assessing the effect of multiscale complexity on groundwater source yields. *Journal of Hydrology*, *588*, 125113. Available from https://doi.org/10.1016/j.jhydrol.2020.125113.

Urrutia, J., Jodar, J., Medina, A., Herrera, C., Chong, G., Urqueta, H., & Luque, J. A. (2018). Hydrogeology and sustainable future groundwater abstraction from the Agua Verde aquifer in the Atacama Desert, Northern Chile. *Hydrogeology Journal*, *26*, 1989−2007. Available from https://doi.org/10.1007/s10040-018-1740-3.

van der Vat, M., Boderie, P., Bons, K. C. A., Hegnauer, M., Hendriksen, G., van Oorschot, M., Ottow, B., Roelofsen, F., Sankhua, R. N., Sinha, S. K., Warren, A., & Young, W. (2019). Participatory modelling of surface and groundwater to support strategic planning in the Ganga Basin in India. *Water*, *11*, 2443. Available from https://doi.org/10.3390/w11122443.

Varadharajan, C., Agarwal, D. A., Brown, W., Burrus, M., Carroll, R. W. H., Christianson, D. S., Dafflon, B., Dwivedi, D., Enquist, B. J., Faybishenko, B., Henderson, A., Henderson, M., Hendrix, V. C., Hubbard, S. S., Kakalia, Z., Newman, A., Potter, B., Steltzer, H., Versteeg, R., ... Wu, Y. (2019). Challenges in building an end-to-end system for acquisition, management, and integration of diverse data from sensor networks in watersheds: Lessons from a mountainous community observatory in East River, Colorado. *IEEE Access*, 7, 182796–182813. Available from https://doi.org/10.1109/ACCESS.2019.2957793.

Wang, J., Chen, M., Lü, G., Yue, S., Wen, Y., Lan, Z., & Zhang, S. (2020). A data sharing method in the open web environment: Data sharing in hydrology. *Journal of Hydrology*, 587, 124973. Available from https://doi.org/10.1016/j.jhydrol.2020.124973.

Wang, S., Zhang, Z., Ye, Z., Wang, X., Lin, X., & Chen, S. (2013). Application of environmental internet of things on water quality management of urban scenic river. *International Journal of Sustainable Development and World Ecology*.

Wei, X., Bailey, R. T., Records, R. M., Wible, T. C., & Arabi, M. (2019). Comprehensive simulation of nitrate transport in coupled surface-subsurface hydrologic systems using the linked SWAT-MODFLOW-RT3D model. *Environmental Modelling & Software*, 122, 104242. Available from https://doi.org/10.1016/j.envsoft.2018.06.012.

White, J. T., Foster, L. K., Fienen, M. N., Knowling, M. J., Hemmings, B., & Winterle, J. R. (2020). Toward reproducible environmental modeling for decision support: A worked example. *Frontiers in Earth Science*, 8, 50. Available from https://doi.org/10.3389/feart.2020.00050.

Wilkinson, M. D., Dumontier, M., Aalbersberg, Ij. J., Appleton, G., Axton, M., Baak, A., Blomberg, N., Boiten, J.-W., da Silva Santos, L. B., Bourne, P. E., Bouwman, J., Brookes, A. J., Clark, T., Crosas, M., Dillo, I., Dumon, O., Edmunds, S., Evelo, C. T., Finkers, R., ... Mons, B. (2016). The FAIR guiding principles for scientific data management and stewardship. *Scientific Data*, 3, 160018. Available from https://doi.org/10.1038/sdata.2016.18.

Wohner, C., Peterseil, J., Genazzio, M. A., Guru, S., Hugo, W., & Klug, H. (2020). Towards interoperable research site documentation – Recommendations for information models and data provision. *Ecological Informatics*, 60, 101158. Available from https://doi.org/10.1016/j.ecoinf.2020.101158.

Wolhuter, A., Vink, S., Gebers, A., Pambudi, F., Hunter, J., & Underschultz, J. (2020). The 3D Water Atlas: A tool to facilitate and communicate new understanding of groundwater systems. *Hydrogeology Journal*, 28, 361–373. Available from https://doi.org/10.1007/s10040-019-02032-w.

Zhang, F., Chen, M., Kettner, A. J., Ames, D. P., Harpham, Q., Yue, S., Wen, Y., & Lu, G. (2021). Interoperability engine design for model sharing and reuse among OpenMI, BMI and OpenGMS-IS model standards. *Environmental Modelling & Software*, 144, 105164. Available from https://doi.org/10.1016/j.envsoft.2021.105164.

CHAPTER 12

Economics of water productivity and scarcity in irrigated agriculture

James F. Booker

Siena College, Loudonville, NY, United States

Chapter Outline
12.1 Introduction   241
12.2 Productivity concepts as indicators   243
12.3 Production and scarcity in economics   248
12.4 Constructing water productivity indicators: physical and economic considerations   249
    12.4.1 Physical considerations   250
    12.4.2 Economic considerations   251
    12.4.3 Net water productivity   252
12.5 Trends in water productivity   253
    12.5.1 Physical and economic water productivity trends   253
    12.5.2 Economic net water productivity trends: value added, optimization, and hydroeconomic models   254
    12.5.3 Other economic approaches to water productivity and efficiency   256
12.6 Conclusions   257
References   258

12.1 Introduction

Increasing water scarcity is almost certain, as demand for all water uses continues to grow and water supply is pressured by a changing climate. One consequence is that the highest valued water uses will be increasingly emphasized at the expense of lower valued uses resulting in greater water productivity (WP) in remaining uses. An alternative narrative would emphasize conservation gains over time through increased efficiency of water use, thus making possible new or maintaining existing uses. This too is a story of increasing WP in the face of growing water scarcity.

Irrigated agriculture is the largest single use of the earth's freshwater, accounting for about 70% of all human water withdrawals (Wisser et al., 2008) and up to 90% of human consumptive use from water withdrawals (Rost et al., 2008). Despite the importance of irrigation in offstream water use, irrigation supplies support less than 10% of total cropland and grazing land evapotranspiration (Rost et al.). But irrigated production is in many regions particularly valuable from an economic perspective, with the value of irrigated crop production dominating that from nonirrigated land.

Most of irrigated agriculture will also face increasing water scarcity. Water demand will increase in most areas due to competing environmental and municipal demands, and from a growing human population with rising incomes demanding higher quality diets. Water supply will decrease as global temperature rises (e.g., Udall & Overpeck, 2017) and depletable groundwater stocks continue to be consumed (Döll et al., 2014). Again, higher valued irrigation will be increasingly favored with the consequence of greater WP. Over time, increases in nonwater inputs, including general and irrigation management will substitute for some water use resulting in water conservation at the farm level and increasing WP in typical irrigated agriculture.

Productivity statistics are widely used as indicators in economics: labor, land, and the productivity of capital address scarce production inputs and are widely reported and discussed. Similarly, WP is a single factor measure of output resulting from the use of the input. Averages over geographic regions and perhaps particular production activities are typically reported. While useful in understanding general economic development across regions and time, productivity indicators are rarely objectives in themselves, and provide little information on how to improve economic outcomes. Similarly, WP is a broadly informative indicator, and should not be seen as an objective itself or as a guide to a policy addressing water scarcity.

Changes in WP are instead the result of how scarcity is addressed through changes in water management and by economic adjustments brought about by rising water demands and falling supply. Much of this involves increasing nonwater inputs (Dawe, 2005) with changes in crop choice and reallocations between water users also part of the adjustment process. Scarcity is also the driver of innovation, and WP increases are the consequence of not only of substitution and conservation, but also of innovation. In contrast, perverse outcomes may be likely if physical measures of WP are viewed as a policy objective. For example, a number of cases can be identified where policies which directly increase WP lead in practice to increases in water consumption (Giordano et al., 2018; Perry, 2007; Ward & Pulido-Velazquez, 2008).

Much of the difficulty in using WP arises from water's confounding physical properties. Water is highly mobile and difficult to confine. In use, there are substantial private losses to evaporation, seepage, and drainage. But these private losses are often gains to downstream water users and must be considered in addressing basin water scarcity. Water supply is highly variable over time, making storage a necessity in many cases. But both storage and transport have high costs relative to value due to water's "bulkiness." Moreover, storage and transport costs may have strongly increasing returns to scale. Together these factors point to water being as much a public as a private good (Barker et al., 2003) leading to the conclusion that it often makes economic sense to manage water as a public good (Young & Haveman, 1985). These and other challenges to actual irrigation development and management were summarized almost 60 years ago by Boulding (1964):

Water is far from a simple commodity.
Water's a sociological oddity.
Water's a pasture for science to forage in,
Water's a mark of our dubious origin.
Water's a link with a distant futurity.
Water's a symbol of ritual purity.
Water is politics, water's religion.
Water is just about anyone's pigeon.
Water is frightening, water's endearing.
Water's a lot more than mere engineering.
Water is tragical, water is comical.
Water is far from the Pure Economical.
So studies of water, though free from aridity.
Are apt to produce a good deal of turbidity.

In this chapter I will start with the idealized Pure Economical before moving to the practical considerations in addressing scarcity given water's unique physical, social, and economic dimensions, and their implications for management across different spatial and time scales. An introduction to the relationship between scarcity and production in economics is given, with an application to water rights. Of greatest importance, contrasts in WP indices and meaning at field versus basin-wide levels are discussed, and alternative approaches to defining WP, including valuation of both outputs and inputs, are provided. Trends in estimated WP from both physical and economic perspectives are then given.

12.2 Productivity concepts as indicators

The general idea of productivity is widely used and understood in economics. It is both a concept at the core of production theory and a widely used statistical indicator. Most commonly, productivity is used in the context of labor, where it is the output resulting from a unit of labor. The numerator is typically measured in monetary terms when, for example, statistics for labor productivity are reported. If expressed in physical terms (e.g., loaves of bread/hour of a baker's work), this is termed the physical product.

The idea of WP is thus immediately understood as one of many single factor productivity (SFP) measures which could be considered in agricultural production. Most common amongst these is yield, or the physical productivity of land (e.g., kg of wheat ha^{-1}). The focus of yield is frequently on the physical measure, because crop prices, needed for monetary units (e.g., \$ha^{-1}), are highly variable and thus obscure interpretation of the underlying production process(es) of interest.

Similarly, WP is readily understood in economics as a SFP concept, whether as a metric for physical output ("crop per drop") or as one for the gross value of output per unit of water. Like any SFP measure, WP is primarily an indicator (Kumar & Dam, 2008) rather than an objective, and increasing WP would rarely be viewed in isolation as a goal in and of itself for achieving food security and sustainable water management.

Because WP is fundamentally a ratio of input to output, its use across different disciplines inevitably results in the use of specific units which alter interpretation and use. While each can be understood in context, there is little standardized terminology to distinguish these differing meanings and limitations of WP. Table 12.1 shows some of this complexity, with an emphasis on descriptive language as qualifiers, and typical units used in each definition. Limitations of different uses, particularly related to basin scale and water reuse and economic factors are also noted in Table 12.1.

In the economic perspective, the fundamental goal or objective is most often the improvement of human welfare. As a practical matter, this is typically quantified within a specific context as activities in which create benefits exceed costs. The significance is that there may be multiple

Table 12.1: Examples of nomenclature for use with water productivity.

Qualifier for water productivity (WP)	Examples	Example of units	Specific limitations and advantages
None	Multiple	Most general: input to output ratio	Complete interpretation needed
Hydrologic uses	Water productivity (WP)	kgm^{-3}; tons / acre-foot	Depletions versus withdrawals undefined; scale undefined
	Irrigation WP	kgm^{-3}	May refer to crop ET or total application
	Caloric WP	kcalories / m^3	energy focus
	Nutritional WP	Grams protein / m^3	nutritional focus
	Net physical WP	kgm^{-3}	Suggests reuse of recoverable seepage and return flows
Economic uses	WP or economic WP	€ m^{-3}; \$ / acre-foot	Gross or net undefined; no hydrologic detail
	Gross economic WP	Same	Neglects opportunity cost of inputs
	Net economic WP	Same	Includes opportunity costs; relevant to irrigator income
General	Net WP	kgm^{-3} or €m^{-3}	Suggests consideration of water reuse or value of inputs

approaches to increasing WP, but not all provide gains to human welfare. This is illustrated by Playán and Mateos (2006) who find that an emphasis on irrigation management leads to greater economic benefits than improvements to physical irrigation water structures. The resulting increases in WP may also not lead to reduced water consumption however, as the result may be higher ET and thus more water use per unit of irrigated land (Playán & Mateos, 2006; Grafton et al., 2018).

Increasingly equity is also explicitly considered in economics, though there is no single metric for expressing the distribution of outcomes across households or regions. And if equity across time (one meaning of sustainability) is the objective, then this could be addressed through considering the "generalized capacity to produce economic well-being" (Solow, 1993). Parallel to the limitations of physical measures of SFP, the physical connotations of most definitions of sustainability miss the main economic point: considerations of impacts on people and their welfare.

Economics also evaluates alternatives for meeting predetermined objectives, such as reducing water use (e.g., to meet an environmental objective), maintaining agricultural output under lower supply conditions (e.g., resulting from climate changes), or growing output under constant water supply. In these cases, the idea of *cost effectiveness* is typically used: identify the approach which achieves the objective at the lowest possible cost. This is presented in more detail in Section 12.2, and illustrated by Fig. 12.1. In contrast, while WP would be expected to increase in meeting each of these examples of objectives, there is no expectation that the most cost effective approaches would correspond to those with the greatest WP increases.

Because of the focus on human welfare as the underlying objective, a focus soley on physical measures of WP as a policy objective is suspect. Responses to rising scarcity are likely to include changes to labor and other inputs including land, energy, and private management. It is likely also that with rising demand for water or falling supply that complex changes in the availability and hence price of the many nonwater inputs will also change. The result is that the optimal input mix is likely to change over time, so that changes to WP may or may not be related to optimally producing agricultural output (in physical terms). And once output prices are added, the optimal mix of produced crops which maximize farm income (benefits minus costs) changes, with unpredictable results for WP for different crops.

These complications to the use of WP arise directly from standard economic production theory and are central to how productivity is considered in economics. While sometimes a useful indicator, particularly for macro level statistics (e.g., labor productivity at the national level), comparing productivity across regions or time with widely differing water scarcity and other conditions is problematic at best.

While the limitations to the use of WP and related efficiency metrics are well understood, they continue to be used as objectives in and of themselves. For example, the UNDP

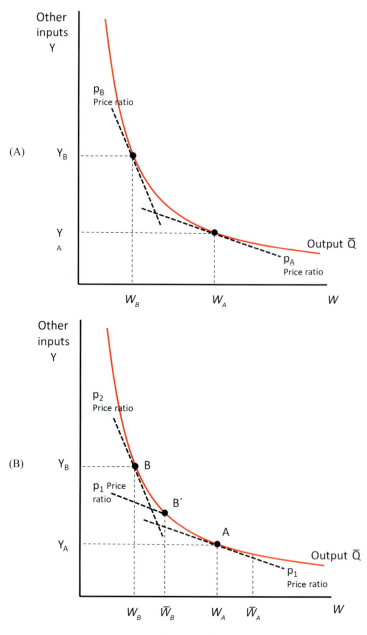

Figure 12.1
Cost minimization with increasing water scarcity and a variable water right. Flexible prices (A) and constant prices (B).

sustainable development goal SDG 6–4 (United Nations Department of Economic and Social Affairs, n.d.) calls for the "increase in water use efficiency [often identical in meaning to WP] across all sectors. Similarly, an FAO Conference (Working Group on

Sustainable Agricultural Water Use, 2020) appears to conflate the objective, human welfare outcomes, with the underlying physical reality of water scarcity in its title "Can Water Productivity improvements save us from Global Water Scarcity?" Individual panelists clearly understood the ambiguity, stating that "water productivity is not the core term to focus on" and should be a consequence of policies and not their objective. Similarly, there was clear recognition of the challenges of differing definitions, complexities introduced by scale, and the need for considering "economics and the value chain." Similar work at the state level in the United States discusses setting benchmarks for WP (Nebraska Water Productivity Report, 2019), potentially leading to perverse policy outcomes if scale and economic factors are not considered. The perspective of WP and efficiency as the objective also remains widespread in the technical literature. As just one example, Nair et al. (2013) provide a useful catalog of discipline specific meanings of water use efficiency (WUE), but fail to identify how these might be used or misused for policy purposes, concluding only that "dwindling water supplies make efficient use of irrigation water the prime priority." Seckler (1999) identifies the dilemma, starting his recapitulation on the uses of WP and water efficiency with the André Gide quote: "Everything that needs to be said has already been said. But since no one was listening, everything must be said again."

Bhaduri et al. (2016) note the importance of context, writing that "measurement of water efficiency should take into account relative water scarcity to assess the benefits of water saved from efficient usage," adding that such indicators "do not address affordability." Whether stated in terms of WP, WE, or WUE (often identical in meaning to WP), use of these indicators as policy objectives is limited because of their potential for divergence from human welfare objectives. And when hydrologic factors such as scale, and economic factors including value of inputs and the production output fail to be carefully defined these indicators face a significant likelihood of diverging from policies to address water scarcity at the lowest possible cost.

Some of these challenges are inevitable in the construction of macro indicators such as defining the water input. As noted in the FAO conference report [20], "it is important to have indicators that speak to actors to support the policy process." An example is the European Union (Organisation for economic co-operation & development OECD, 2021) calculation of WP across member states. Output prices are used to aggregate across crops, and the reported WP statistics could be interpreted in the same manner as, for example, labor productivity statistics.

Economic productivity statistics for a single production input will remain of general interest and address broad perspectives but are less helpful for more specific situations such as adjustments to crop mix and as the role of nonwater inputs. Lankford (2009) for example notes the importance of including human labor, noting that an apparently attractive policy intervention becomes immediately untenable when physical human labor requirements are considered. More generally, policy or management choices should be based on value added rather than gross values, and should include the value of inputs in the production process.

Additional complexity is introduced when broad interdependencies and cobenefits exist, as in the water-energy-food nexus (Conway et al., 2015) and embedded or virtual water used in production (Liu et al., 2018).

12.3 Production and scarcity in economics

Production is conceptualized in economics by a production function relating output to inputs. For example, production of a specific crop could be represented by

$$Q = f(W, \mathbf{Y}) \tag{12.1}$$

where Q is the crop output, W is the water input, and $\mathbf{Y}$ is a vector of the nonwater inputs such as land, labor, energy, and management. A single factor productivity measure is the marginal physical product, given for the water input by $\partial Q/\partial W$. For the simplified case of constant returns to scale and a single additional input given by $Q = W^\beta Y^{1-\beta}$ and $0 < \beta < 1$, this gives

$$\partial Q/\partial W = \beta (Y/W)^{1-\beta} \tag{2}$$

The marginal physical product of water (MPP_W) increases as water use decreases and as use of the nonwater inputs increase. An important qualification is the contradictory experimental finding from the irrigation literature that over many levels of water use the MPP_W actually decreases, making deficit irrigation typically an unattractive management choice. But in actual production environments with heterogeneous soil conditions and a time frame where planting choices can be made based on water availability, it becomes much more likely that the simple conclusions from Eq. (2) would hold.

While marginal productivities will vary with the levels of input use, more insight is gained by introducing prices. In economics, all inputs are scarce in the sense that there is an opportunity cost to their use. Land must be purchased or leased, labor hired, and a price for each is used to summarize land and labor scarcity. Irrigation water is in principle no different: it is costly to acquire and deliver it to the root zone of crops where it is needed. In this situation the costs of output (i.e., crop production) are minimized by a choice of inputs which is dependent on the relative scarcities, or prices of the inputs.

Fig. 12.1 illustrates the choice of water input, where all other inputs are represented by a composite. Panel (a) represents the neoclassical presentation, where there is a constant price (e.g., € unit^{-1}) for each input, and changes in relative scarcity of the inputs is reflected in changes to the price ratios between inputs. In Panel (a) the P_1 price ratio $= P_{W1}/P_{Y1}$ and represents the first scarcity scenario, where P_W is the cost to the irrigator for delivery and application of irrigation water W, and P_Y is a composite of the cost of all other inputs Y. The lowest total cost to the irrigator to produce the crop output $\overline{Q}$ is shown as A. Water use W_A and other input use Y_A is chosen. All other combinations of inputs would result either in higher total costs or lower output (or both). WP at the margin is now the partial derivative of the production function Q at this cost minimizing choice of inputs given in Eq. (2).

In an economy where prices reflect scarcity, increasing water scarcity (water scenario 2) would be presented to the irrigator as a high price for irrigation water, or $P_{W2} > P_{W1}$. In this second scenario the price ratio thus increases to P_2, and the cost minimizing solution for the irrigator to continue to produce an output $\overline{Q}$ is given by the point B, using inputs W_B and Y_B. This is the neoclassical substitution of inputs, driven by price changes in the face of changes in relative scarcity. Marginal WP is inversely related to the water use, and thus the cost minimizing input choice leads to higher WP. But this outcome is costly to the irrigator. Assuming constant prices for nonwater inputs Y, the price increase for the water input W must lead to higher total production costs to the irrigator.

While this explanation of input choice and adjustment to scarcity corresponds to many of the incentives facing irrigators, it neglects the role of water rights and the limited role played by water prices in most situations globally. In practice, the price of raw water to the irrigator is likely to be administratively set at a very low level, but with a quantity constraint (i.e., the quantity of the water right). Formally, a water right is assumed to include a fixed maximum level of water withdrawal and an administratively set price. The irrigator also faces additional costs for delivery and application of irrigation water included in P_{W1}. Unlike in panel (a), it is assumed that the sum of these water prices is not responsive to increasing water scarcity.

In this second scenario of a fixed water price, the irrigator chooses not the input combination B, but rather the combination B' representing input choices W_B' and Y_B'. That is, although output can be maintained given the production function illustrated in Fig. 12.1, not as much substitution occurs in (b) as in (a). The lowest cost solution to the irrigator represents a "corner solution" and they continue to use low cost water right up to the point of the water right. With less substitution of other inputs occurring, adjustment to scarcity is less complete than in panel (a), more irrigation water is used, and thus marginal WP is lower. Note that producer incomes are higher with solution B' than B because the lower water price is a cost saving to irrigators. But incomes are lower than with the larger water right $\overline{W}_A$ because the producer's costs are increased compared to the solution A. The result is that, given an assumption of constant crop output, the greatest increase in WP occurs when prices change to reflect scarcity. Water prices which do not adjust better preserve income, but WP increases and water savings are not as great.

12.4 Constructing water productivity indicators: physical and economic considerations

The practical considerations in constructing WP indicators due to hydrologic complexity are widely understood, particularly in capturing the role of return flows and water reuse at the basin scale (Barker et al., 2003; Ritzema, 2014; Seckler, 1996). To be useful as indicators

Table 12.2: Factors in estimating input and output in water productivity estimation.

Component	Category		Notes
Output	Physical	Moisture content	Standardized moisture contents are typically needed for comparisons
		Quality of product	Caloric or nutritional (e.g., protein) content across crops is variable
	Economic	Crop value	Commodity price and specific crop quality
		Risk	Variability of output across time and space
		Economies of scale	Per unit production costs likely increase as scale decreases
		Economies of scope	Multiple outputs are common (e.g., integrated feed and livestock)
		Regional benefits of agglomeration	Output markets improve as output and crop variety increase
Input	Physical	Timing	Highly variable input over time; output affected
		Other inputs	Includes management
		Scale and sequential use	Relates to *mobility*
		Recoverable fraction	Seepage and return flows available for reuse
		Bulkiness	Economies of scale in limiting losses from leakage and evaporation
	Economic	Value of inputs	Input mix, requirements, and costs
		Unpriced inputs	Externalities associated with use of inputs
		Bulkiness	Transport costs are high relative to value
Integrated		Measurement	Measurement gaps: cost is high relative to value due to *mobility* and *bulkiness*
		Complementary uses	At scale, complementary benefits are common (e.g., storage, recreation, and hydropower)
		Quality	Initial quality and following use affects production
		Innovation and learning curves	Knowledge diffusion over time can increase output and lower cost of production
		Cultural and social values	Nature of collection, distribution, and use affects perceived legitimacy

for policy purposes, WPs should also be constructed to include economic factors in addition to the relevant hydrologic considerations described by Young and Haveman (1985). In particular, the human value gained from crop production and the costs of nonwater inputs (including labor) should be used. Table 12.2 summarizes a range of considerations in constructing and interpreting meaningful WP indices.

12.4.1 Physical considerations

In the hydrologic and irrigation literature there is substantial attention to considering distinctions between applied versus depleted water, location of use, timing of water use, and scale of measurement. These distinctions directly relate to the denominator of WP, the measure of the water input. Perhaps most importantly, Seckler (1996) identified the need to consider the full basin scale for identifying true water savings in any measure of WP or water efficiency. Numerous authors identify the need to distinguish scale and beneficial versus nonbeneficial uses (e.g., Molden, 1997; Pereira et al., 2012). Perry (2007) provides an integrated approach to the wide variety of hydrologic settings, recommending use of the clarifying terms withdrawals, changes in storage, and consumed and nonconsumed fraction, to describe water use in a technically consistent way. Systems frameworks for water uses typically incorporate these distinctions by attempting to include all significant interactions between water users and the broader hydrologic environment.

In practice applied water is substantially greater than consumptive use in most irrigation settings. The result is that the unused irrigation portion is potentially available to support other uses (human and environmental) through return flows and groundwater seepage. If WP is measured as water applied to crops rather than the net measure of water consumption, then moves to higher efficiency irrigation technologies (e.g., from flood to sprinkler) will overestimate benefits of the new technologies. In many cases of widespread adoption of such higher efficiency technologies the result has been, despite increased WP, not decreased water use, but rather unsustainable increases in total consumptive water use (Giordano et al., 2018; Grafton et al., 2018; Ward & Pulido-Velazquez, 2008). This result is from both "water spreading," where water use practices and rights are implemented largely by diversion quantities, and roughly constant diversions are used across increasing land areas as new irrigation technologies are introduced. In such cases, net WP is likely to show no significant change, and the increase in net water use can be understood as a simple increase in the irrigated land base.

The more subtle case is when better control of irrigation applications using new technology results in net WP increases. Much of the adoption in practice of drip irrigation, for example, appears to be the direct result not of attention to a water "saving" technology, but to one which increases yield and simple WP (Taylor & Zilberman, 2017; Ward & Pulido-Velazquez, 2008). In such cases it is also common to find that net water consumption per hectare increases along with simple WP.

12.4.2 Economic considerations

A broad literature in economics addresses WP from a number several perspectives, generally predating introduction of the specific term WP by Molden (1997). In general, economic approaches use distinct terminology, including efficiency (see Scheierling et al., 2016), costs and benefits, income or net revenue, and economic welfare. The central distinction from typical agricultural and hydrologic approaches to WP are in the definition of the numerator of WP, or the output resulting from the water input. In economic approaches, typically a monetary or value measure of the physical output from water usage is used as a production metric.

For example, in a region producing both almonds and maize, estimates of productivity of water for each crop would rely on the price received (i.e., their per unit value) in order to allow comparisons based on either total or net value produced per unit of water. In the simplest, though incomplete comparison, WP would be the revenue per unit of water. But this measure uses water as the only production input. This is clearly a simplification, and the preferred measure would include other inputs by subtracting the monetary value of all production costs to give a net revenue per unit of water measure of WP (e.g., Jiang et al., 2016). It is common in economics to term this the "economic value of water" (Young, 2005). The concept and estimate is identical to the idea of net WP, where the output measure includes and adjustment for input use. In practice, included cost estimates often also incorporate external impacts such as the use of return flows, near term environmental effects, and depletion of aquifers.

Conceptually the "economic value of water" incorporates a holistic perspective to WP, not unlike the systems frameworks which can underly some hydrologic approaches to WP. The central idea (Young, 2005) is to include "all" significant opportunity costs (or benefits) in the measure of the net value produced from water use. In practice only a limited number of such impacts are included in typical economic work resulting in estimates of the "economic value of water" or measures of WP net of costs.

12.4.3 Net water productivity

Measures of WP should either incorporate net WP concepts or be presented in a way that definitions and limitations are clear. To the extent that WP is used as an approach to identifying production practices, making regional comparisons, or identifying trends, particular attention to identifying confounding changes from an "everything else equal" across practices, space, or time should be made.

Two broad categories are essential here. First, control for the endlessly complex characteristics of water itself must be made. Complexity results from the difficulty of containing and capturing water as it moves through soils, aquifers, surface water, plants, and into the atmosphere. In each case water quality, including temperature is affected by each environment. And in every location measurement is problematic, and changing conditions over time are the rule.

Second, the characteristics of the output and required inputs differ between crops and varieties, and also over time and space. Accounting across these differences is challenging when using economic values to compare incommensurate physical output (e.g., almonds and alfalfa) and inputs (e.g., fertilizer, transportation to markets, human labor). Attempts to control for such differences are more difficult or impossible when only physical units are used. Meaningful control may be reasonable in comparing outputs by using measures such as calories or nutrient content. Controlling across multiple and differing inputs presents additional challenges. One approach would be to use embedded energy of alternative inputs (Conway et al., 2015).

12.5 Trends in water productivity

An increasingly deep literature which specifically considers changes in WP (e.g., Kumar, 2021) is complemented by related work in economics providing empirical estimates of net WP in agriculture as measured by farm income, and by hydroeconomic modeling which implicitly deduces net WP in hydrologically complex systems as measured by shadow prices for water under alternative water policies and management.

12.5.1 Physical and economic water productivity trends

Reported measures of physical WP typically show growth over time which is explained primarily by growth in crop yields (i.e., land productivity). For example (Dawe, 2005) reports increasing WP from Asian rice based systems in the 1980s and 90s, primarily due to increasing crop yields. The reported WPs do not appear to include water use downstream of the study area. Dawe emphasizes also that the growth in WP "does not necessarily result in increased benefits to society" because the value of scarce land, capital, and labor are not considered. Dawe further notes that comparisons across geographically dissimilar regions are likely to be uninformative, just as yield comparisons across locations are not necessarily meaningful. For example, the Nebraska Water Productivity Report (2019) reports higher maize yields in eastern versus western NE due to altitude caused differences in the length of the growing season. Growth in estimated maize and soybean WPs over time are also largely the result of growth in yield. Despite the clear influence of a variety of inputs on crop production, the report asserts that addressing "WP gaps" can help reduce pressure on water resources. But note the paradox in this prescription for coping with increasing water scarcity: it is precisely in the western and more arid regions with lower yields due to the shorter growing season that the lower WPs are found. Moving to, for example, a shorter season crop would likely also increase seasonal ET, potentially exacerbating existing regional water scarcity.

El-Marsafawy et al. (2018) estimate WP of seasonal ET for a wide array of irrigated crops across the Nile Delta. They appear to use fixed crop coefficients together with a standardized reference ET calculated from observed climate variation to estimate crop ET over time. They find that WP increases across most crops due to "considerable increase[s] in crop yield."

Mainuddin and Kirby (2009), similarly find generally increasing WP of seasonal ET for lowland rice across SE Asian nations for 1992–2004 with ET estimates based on generalized crop coefficients. If WP is alternatively estimated in terms of gross water input, for which they use seasonal precipitation, there is much less evidence of WP growth. Mainuddin and Kirby also use crop prices to produce estimates of gross economic value per unit of ET. They find substantial and growing variation in gross economic productivity across countries and regions, and only modest productivity growth over time. Cai and Rosegrant (2003) predict substantial increases in WP in developing countries, largely resulting from increases in crop yield.

WP is now also reported systematically by the Organisation for economic co-operation and development OECD (2021) for European Union nations using the gross value of output per unit of water input. Care should be taken with interpretation, however, given the estimation and assumptions underlying the water input.

12.5.2 Economic net water productivity trends: value added, optimization, and hydroeconomic models

Water economics commonly makes use of methods which are deductive in nature, providing an ability to evaluate and make predictions of future hypothetical conditions. For example, in estimating future costs and benefits to support project evaluation or alternative policy reforms, such ex ante analysis is required. Underlying deductive methods are behavioral assumptions of profit or utility maximization (Young, 2005) and the normative assumption of an objective to maximize net social benefits. Typical development and interpretation of deductive models make the further normative assumption that monetary metrics are reasonable proxies for human welfare opportunity costs and benefits. One direct outcome of deductive method approaches are estimates of the contribution of irrigation and other water resources to producer incomes. Implicit in these outcomes are various measures of WP, while measures of technical efficiency are implied in model data on crop and farm budgets and of allocative efficiency are implied in net benefit estimates under alternative exogenous constraints.

A typical application is to allow for direct representation of a diminishing returns net benefits objective function. This allows model formulation using downward sloping derived irrigation water demand functions for the crop and/or farm level. Mass balance conditions representing system hydrology are generally included as linear constraints.

In many cases, the scarce water resource must be allocated between competing uses. A central feature to the solution of the programming problem is generation of the shadow price of the water input. The price reflects the simple opportunity cost of foregoing water use for a specific purpose. A central condition of the typical optimal solution (a so-called interior solution) is that this opportunity cost is the same for each purpose. That is, the water value or net WP is identical across uses. This must be so, because if a greater value were produced in one purpose versus another given a marginal reallocation, then that reallocation would increase total net benefits. A very useful feature of the shadow prices generated by programming problems is that they are a measure of the level of water scarcity. Higher shadow prices indicate a regime of greater scarcity, and lower shadow prices indicate lesser water scarcity. Note that the water shadow price has no direct relationship to water delivery or storage costs. In the typical case where water markets are not present and water is a nontraded good, the shadow price cannot be directly observed.

Valuation of the nonpriced crop production inputs is often challenging. A particular difficulty is valuing owned inputs, including land and management. While local rental markets often provide reasonable measures for irrigated and nonirrigated land, the value of owner provided management may be difficult to quantify. This is particularly problematic because benefits of management are potentially large (as is clear from the technical efficiency literature) compared to net income from crop production. Because the specific management input is potentially transferrable to other economic activity, it should be valued at its opportunity cost in the highest valued alternative. In practice, this alternative is rarely observable.

This difficulty in quantifying the value of management, a proportionally large but nonpriced opportunity cost, is one aspect of the broader challenge facing all deductive methods which ultimately rely on residual imputation. Estimates of opportunity costs and benefits are inherently imprecise. Thus net income, the difference between benefits and costs, will inevitably carry a larger uncertainty than estimates of either benefits or costs.

As economic optimization models of regional or producer behavior incorporate increasing levels of hydrologic detail they have come to be termed hydroeconomic models. The key distinguishing feature may be the presence of an implicit or explicit systems perspective. Importantly, distinctions and clarity between withdrawals, consumptive use, return flows, and groundwater impacts are almost always clear. A review article by Harou et al. (2009) addresses work which illustrates the impact of these physical externalities on other water users, including other agricultural producers. By explicitly incorporating such hydrological complexity ambiguities in physical meanings of WP are greatly reduced. One model outcome is the shadow price of water, or net economic WP.

Hydroeconomic models are particularly useful for addressing regional water development and allocation questions where the use of regional water resources is best treated as an integrated set. This is critical in understanding, for example, the impacts of increasing irrigation efficiency through new, more capital-intensive irrigation technology. While water withdrawals are likely to decline, consumptive use and hence regional water depletion may actually increase, as demonstrated in theory (Huffaker & Whittlesey, 2003) and in mathematical programming models (Brinegar & Ward, 2009; Scheierling et al., 2006) More generally, hydroeconomic models seek to incorporate the notion that regional water analysis typically includes a variety of interconnections (e.g., one user's return flows are the next user's supply), including those between ground and surface water. For example, Knapp and Olson (1995) demonstrate intertemporal opportunities for improved irrigation outcomes where producers can draw from both ground and surface water and have artificial recharge opportunities. A review of approaches based on deductive methods which link hydroeconomic models to policy is provided by Booker et al. (2012).

As computational resources have increased several modular approaches which couple sophisticated hydrologic and related modeling systems to economic optimization models have been realized (e.g., Draper et al., 2003; Maneta et al., 2009). Linkages to geographic information systems (GIS) are also being introduced (McKinney & Cai, 2002).

In addition to the ability to include the relevant hydrologic systems, one strength of hydroeconomic models is the explicit representation of policy alternatives. For example, Booker (1995) shows that Colorado River (United States) water users under severe and sustained drought can substantially reduce welfare impacts through water reallocations at the state level. Including external impacts of water use and storage including water quality, hydropower, and recreation results in substantially greater benefits of reallocations which cross state lines. Inclusion of these additional impacts of water use is consistent with Perry (2007), who notes that questions of impacts to water quality are typically not binary, but rather continuous.

California (United States) experienced severe droughts through several years in the early 21st century. Looking at 2015 in particular, (Howitt et al., 2015) use reported delivery shortfalls and a complex hydroeconomic model to estimate the economic responses of producers, including additional groundwater pumping, to estimate direct drought damages in agriculture of $1.8 billion. Such a drought damage estimate gives a measure of net economic WP when coupled with reductions in water depletions. Using their estimate of a net water shortage of 2.7 million acre-feet leads to average lost net WP of $680 per acre.

Much of the concern related to sustainability of irrigation water use is related to declining groundwater levels. (Hrozencik et al., 2017) construct a hydroeconomic model which addresses sustainability of groundwater use and farm income in a complex irrigated system. They show that policy interventions to curtail groundwater use have heterogeneous impacts across producers on both physical and net economic WP.

12.5.3 Other economic approaches to water productivity and efficiency

An alternative approach to understanding economic impacts of changes in water availability employs statistical methods using observations across time and space, or both. When both are used, these are termed panel methods, and are an increasingly important approach across economic economics to combine responses which occur as a time series and also can be observed cross-sectionally among participants in a sample. Dell et al. (2014) provides a broad summary addressing economic impacts of climate changes, and report on studies considering producer income, crop yield (i.e., land productivity), and labor productivity as independent variables regressed against water availability and numerous other control variables. While typically not focused on WP, land and labor productivity changes with climate are repeatedly reported. The greatest barrier to understanding WP in irrigated crop production through these methods is quantifying the irrigation water input for each observation. Schlenker et al. (2005) provide one crude approach, and are able to report separately on physical WP, finding evidence of declines in WP with one plausible climate change scenario.

While productivity in economics is clearly identified as a ratio of inputs to outputs, efficiency in economics is typically a dimensionless quantity measured as the ratio of an actual to a "best case" outcome termed the frontier (Latruffe, 2010). Most commonly, the frontier is estimated either statistically or using mathematical programming. In each case observations of individual producer behavior are the basis for developing the frontier and hence efficiency estimates for typical producers. Building upon this approach, Njuki and Bravo-Ureta (2019) controls for the use of nonwater inputs to conclude that that WP from irrigation, across most US states, has risen modestly over recent decades with technological progress and increased use of nonwater inputs the primary cause.

An alternative approach in production economics is to construct indices of "total factor productivity" (TFP) over time, and thus evaluate the contributions from all inputs over time. Growth in TFP is attributed to technological progress under the assumption that producers are perfectly efficient. Data limitations on appropriate prices for irrigation and other water inputs are a substantial barrier to use of the method (Alston & Pardey, 2014). Fuglie et al. (2020) overcome this limitation for Nile River irrigation in Egypt by constructing TFP indices from both private (zero cost to producers for public irrigation water) and social perspectives (full economic cost of providing irrigation water). They find that over the period 1961–2016 technological progress and efficiency gains in the use of irrigation water contributed significantly more to agricultural growth than the growth of irrigated area or water use over this period. Their conclusion is that under limited physical water supply, long-term growth in the value of irrigated agricultural production is possible when supported by both private and public investment.

12.6 Conclusions

WP is widely used and understood as an indicator similar in construction and function to crop yield and labor productivity. Its use across multiple disciplines and levels of specialty can be explained by the ease of visualizing the idea of "crop per drop." The result is that

WP, in principle, suffers from ambiguity in both meaning and implications which can point policy makers in perverse directions.

WP is also an attractive indicator because of its apparent (though misleading) ease of measurement. Unfortunately, it is exceptionally difficult to routinely measure the consumed fraction of water use, and as with many economic indicators, standard formulations of WP provide little actionable information in the absence of additional context. With hydrological and economic detail added so that the net output of irrigated production (net revenue or income) and the appropriate input (the nonreusable, or consumed fraction) for irrigation water is used, WP is an important economic indicator. It is in this case in fact the same meaning as the economic value of water for a particular use.

Following this approach, the water economics literature shows a rich set of adaptations to rising water scarcity which reduce economic and financial impacts in irrigated agriculture. The development literature emphasizes the role of rising crop yield over time (i.e., land productivity) in explaining increases in physical measures of WP. Economic interpretations of rising yield identify substitution of nonwater inputs (including management skill) as key factors in physical WP increases. Economics also emphasizes the importance of crop value to farm income, and demonstrates the role of changes in crop choice as water scarcity increases. The possibility of input and output adjustments leads to multiple avenues for adaptation which have as a consequence increases in net economic WP.

References

Alston, J. M., & Pardey, P. G. (2014). Agriculture in the global economy. *Journal of Economic Perspectives*, 28 (1), 121–146. Available from https://doi.org/10.1257/jep.28.1.121.

Barker, R., Dawe, D., & Inocencio, A. (2003). Economics of water productivity in managing water for agriculture. In J. W. Kijne, R. Barker, & D. Molden (Eds.), *Water productivity in agriculture: Limits and opportunities for improvement* (pp. 19–35). CAB International.

Bhaduri, A., Bogardi, J., Siddiqi, A., Voigt, H., Vörösmarty, C., Pahl-Wostl, C., Bunn, S. E., Shrivastava, P., Lawford, R., Foster, S., Kremer, H., Renaud, F. G., Bruns, A., & Osuna, V. R. (2016). Achieving sustainable development goals from a water perspective. In Frontiers in environmental science (4, Issue OCT). Frontiers Media S.A. <https://doi.org/10.3389/fenvs.2016.00064>.

Booker, J. F. (1995). Hydrologic and economic impacts of drought under alternative policy responses. *Journal of the American Water Resources Association*, 31(5), 889–906. Available from https://doi.org/10.1111/j.1752-1688.1995.tb03409.x.

Booker, J. F., Howitt, R. E., Michelsen, A. M., & Young, R. A. (2012). Economics and the modeling of water resources and policies. *Natural Resource Modeling*, 25(1), 168–218. Available from https://doi.org/10.1111/j.1939-7445.2011.00105.x.

Boulding, K. E. (1964). The economist and the engineer. In E. N. Castle, & S. C. Smith (Eds.), *Economics and public policy in water resource development* (pp. 156–166). Iowa State University Press.

Brinegar, H. R., & Ward, F. A. (2009). Basin impacts of irrigation water conservation policy. *Ecological Economics*, 69(2), 414–426.

Cai, X., & Rosegrant, M. W. (2003). World water productivity: current situation and future options. Ch. 10 in Water productivity in agriculture: limits and opportunities for improvement. 1, 163–178.

Conway, D., van Garderen, E. A., Deryng, D., Dorling, S., Krueger, T., Landman, W., Lankford, B., Lebek, K., Osborn, T., Ringler, C., Thurlow, J., Zhu, T., & Dalin, C. (2015). Climate and southern Africa's water-energy-food nexus. *Nature climate change* (5, 9, pp. 837–846). Nature Publishing Group. <https://doi.org/10.1038/nclimate2735>.

Dawe, D. (2005). Increasing water productivity in rice-based systems in asia-past trends, current problems, and future prospects. *Plant Production Science*, *8*(3), 221–230. Available from https://doi.org/10.1626/pps.8.221.

Dell, M., Jones, B., & Olken, B. (2014). What do we learn from the weather? The new climate-economy literature. *Journal of Economic Literature*, *52*(3), 740–798. Available from https://doi.org/10.1257/jel.52.3.740.

Döll, P., Schmied, H. M., Schuh, C., Portmann, F. T., & Eicker, A. (2014). Global-scale assessment of groundwater depletion and related groundwater abstractions: Combining hydrological modeling with information from well observations and GRACE satellites. *Water Resources Research*, *50*(7), 5698–5720. Available from https://doi.org/10.1002/2014WR015595.

Draper, A. J., Jenkins, M. W., Kirby, K. W., Lund, J. R., & Howitt, R. E. (2003). Economic-engineering optimization for california water management. *Journal of Water Resources Planning and Management*. Available from http://ascelibrary.org/doi/abs/10.1061/(ASCE)0733-9496(2003)129:3(155).

El-Marsafawy, S. M., Swelam, A., & Ghanem, A. (2018). Evolution of crop water productivity in the Nile Delta over three decades (1985-2015). *Water (Switzerland)*, *10*(9). Available from https://doi.org/10.3390/w10091168.

Fuglie, K., Dhehibi, B., & Shahat, A. el (2020). Water, policy, and productivity in Egyptian agriculture. *American Journal of Agricultural Economics*, *103*(4), 1378–1397. Available from https://doi.org/10.1111/ajae.12148.

Giordano, M., Turral, H., Scheierling, S. M., Tréguer, D. O., & Mccornick, P. G. (2018). *Beyond "More Crop per Drop": Evolving thinking on agricultural water productivity*. IWMI Research Report 169, International Water Management Report. <http://www.iwmi.org>.

Grafton, R. Q., Williams, J., Perry, C. J., Molle, F., Ringler, C., Steduto, P., Udall, B., Wheeler, S. A., Wang, Y., Garrick, D., & Allen, R. G. (2018). The paradox of irrigation efficiency. *Science (New York, N.Y.)*, *361* (6404), 748–750. Available from https://doi.org/10.1126/science.aat9314.

Harou, J. J., Pulido-Velazquez, M., Rosenberg, D. E., Medellín-Azuara, J., Lund, J. R., & Howitt, R. E. (2009). Hydro-economic models: Concepts, design, applications, and future prospects. *Journal of Hydrology*, *375* (3–4), 627–643. Available from https://doi.org/10.1016/j.jhydrol.2009.06.037.

Howitt, R., Macewan, D., Medellín-Azuara, J., Lund, J., & Sumner, D. (2015). *Economic analysis of the 2015 drought for California agriculture*.

Hrozencik, R. A., Manning, D. T., Suter, J. F., Goemans, C., & Bailey, R. T. (2017). The heterogeneous impacts of groundwater management policies in the Republican River Basin of Colorado. *Water Resources Research*, *53*(12), 10757–10778. Available from https://doi.org/10.1002/2017WR020927.

Huffaker, R., & Whittlesey, N. (2003). A theoretical analysis of economic incentive policies encouraging agricultural water conservation. *International Journal of Water Resources Development*, *19*(1), 37–53. Available from https://doi.org/10.1080/713672724.

Jiang, Y., Xu, X., Huang, Q., Huo, Z., & Huang, G. (2016). Optimizing regional irrigation water use by integrating a two-level optimization model and an agro-hydrological model. *Agricultural Water Management*, *178*, 76–88. Available from https://doi.org/10.1016/j.agwat.2016.08.035.

Knapp, K. C., & Olson, L. J. (1995). The economics of conjunctive groundwater management with stochastic surface supplies. *Journal of Environmental Economics and Management*, *28*(3), 340–356. Available from https://doi.org/10.1006/jeem.1995.1022.

Kumar, D. M. (2021). (1st ed.). *Water productivity and food security global trends and regional patterns*, (3). Elsevier.

Kumar, M., & Dam, J. van. (2008). *Improving water productivity in agriculture in developing economies:* In Search of New Avenues (M. Kumar, Ed.). Proceedings of the seventh annual partners meet, IWMI TATA Water Policy Research Program, ICRISAT, Patancheru, Hyderabad, India, 2–4 April 2008. Vol.1. International Water Management Institute (IWMI), South Asia Sub Regional Office. <https://core.ac.uk/download/pdf/6764778.pdf>.

Lankford, B. (2009). Viewpoint-the right irrigation? Policy directions for agricultural water management in Sub-Saharan Africa. *Water Alternatives*, 2(3), 476–480. Available from http://www.water-alternatives.org.

Latruffe, L. (2010). Competitiveness, productivity and efficiency in the agricultural and agri-food sectors. <https://doi.org/10.1787/5km91nkdt6d6-en>.

Liu, J., Zhao, X., Yang, H., Liu, Q., Xiao, H., & Cheng, G. (2018). Assessing China's "developing a water-saving society" policy at a river basin level: A structural decomposition analysis approach. *Journal of Cleaner Production*, 190, 799–808. Available from https://doi.org/10.1016/j.jclepro.2018.04.194.

Mainuddin, M., & Kirby, M. (2009). Spatial and temporal trends of water productivity in the lower Mekong River Basin. *Agricultural Water Management*, 96(11), 1567–1578. Available from https://doi.org/10.1016/j.agwat.2009.06.013.

Maneta, M., Torres, M., Vosti, S. A., Wallender, W. W., Allen, S., Bassoi, L. H., Bennett, L., Howitt, R., Rodrigues, L., & Young, J. (2009). Assessing agriculture–water links at the basin scale: Hydrologic and economic models of the São Francisco River Basin, Brazil. *Water International*, 34(1), 88–103. Available from https://doi.org/10.1080/02508060802669496.

McKinney, D. C., & Cai, X. (2002). Linking GIS and water resources management models: an object-oriented method. *Environmental Modelling & Software*, 17(5), 413–425. Available from https://doi.org/10.1016/S1364-8152(02)00015-4.

Molden, D. J. (1997). *Accounting for water use and productivity*. <https://www.researchgate.net/publication/42765708>.

Nair, S., Johnson, J. W., & Wang, C. (2013). Efficiency of irrigation water use: A review from the perspectives of multiple disciplines. *Agronomy Journal*. Available from https://doi.org/10.2134/agronj2012.0421.

Nebraska Water Productivity Report 2019. (2019).

Njuki, E., & Bravo-Ureta, B. E. (2019). Examining irrigation productivity in U.S. agriculture using a single-factor approach. *Journal of Productivity Analysis*, 51, 125–136. Available from https://doi.org/10.1007/s11123-019-00552-x.

Organisation for economic co-operation and development (OECD). (2021, October 5). *Water productivity*. Eurostat Database.

Pereira, L. S., Cordery, I., & Iacovides, I. (2012). Improved indicators of water use performance and productivity for sustainable water conservation and saving. *Agricultural Water Management*, 108, 39–51. Available from https://doi.org/10.1016/j.agwat.2011.08.022.

Perry, C. (2007). Efficient irrigation; Inefficient communication; flawed recommendations. *Irrigation and Drainage*, 56(4), 367–378. Available from https://doi.org/10.1002/ird.323.

Playán, E., & Mateos, L. (2006). Modernization and optimization of irrigation systems to increase water productivity. *Agricultural Water Management*, 80(1-3 SPEC. ISS.), 100–116. Available from https://doi.org/10.1016/j.agwat.2005.07.007.

Ritzema, R. S. (2014). Aqueous productivity: An enhanced productivity indicator for water. *Journal of Hydrology*, 517, 628–642. Available from https://doi.org/10.1016/j.jhydrol.2014.05.066.

Rost, S., Gerten, D., Bondeau, A., Lucht, W., Rohwer, J., & Schaphoff, S. (2008). Agricultural green and blue water consumption and its influence on the global water system. *Water Resources Research*, 44(9), 9405. Available from https://doi.org/10.1029/2007WR006331.

Scheierling, S. M., Loomis, J. B., & Young, R. A. (2006). Irrigation water demand: A *meta*-analysis of price elasticities. *Water Resources Research*, 42(1). Available from https://doi.org/10.1029/2005WR004009, n/a-n/a.

Scheierling, S., Treguer, D. O., & Booker, J. F. (2016). Water productivity in agriculture: Looking for water in the agricultural productivity and efficiency literature. *Water Economics and Policy*, 2(3). Available from https://doi.org/10.1142/S2382624X16500077.

Schlenker, W., Hanemann, W. M., & Fisher, A. C. (2005). Will U.S. agriculture really benefit from global warming? *Accounting for Irrigation in the Hedonic Approach*, 95(1).

Seckler, D. W. (1996). *The new era of water resources management: From "dry" to "wet" water savings*. International Irrigation Management Institute.

Seckler, D. (1999). Revisiting the "IWMI paradigm:" Increasing the efficiency and productivity of water use. IWMI Water Brief.

Solow, R. M. (1993). An almost practical step toward sustainability. *Resources Policy*, *19*(3), 162–172. Available from https://EconPapers.repec.org/RePEc:eee:jrpoli:v:19:y:1993:i:3:p:162-172.

Taylor, R., & Zilberman, D. (2017). Diffusion of drip irrigation: The case of California. *Applied Economic Perspectives and Policy*, *39*(1), 16–40. Available from https://doi.org/10.1093/aepp/ppw026.

Udall, B., & Overpeck, J. (2017). The twenty-first century Colorado River hot drought and implications for the future. *Water Resources Research*, *53*(3), 2404–2418. Available from https://doi.org/10.1002/2016WR019638.

United Nations Department of Economic and Social Affairs. (n.d.). Sustainable development goal 6. sustainable development goals. Retrieved October 3, 2021, <https://sdgs.un.org/goals/goal6>.

Ward, F. A., & Pulido-Velazquez, M. (2008). Water conservation in irrigation can increase water use Economic Sciences Engineering. *PNAS*, *25*, 18215–18220. Available from http://www.pnas.org/cgi/content/full/0805554105/DCSupplemental.http://www.pnas.orgcgidoi10.1073pnas.0805554105.

Wisser, D., Frolking, S., Douglas, E. M., Fekete, B. M., Vörösmarty, C. J., & Schumann, A. H. (2008). Global irrigation water demand: Variability and uncertainties arising from agricultural and climate data sets. *Geophysical Research Letters*, *35*(24). Available from https://doi.org/10.1029/2008GL035296.

Working Group on Sustainable Agricultural Water Use, *Can water productivity improvements save us from global water scarcity?* Workshop Report, 25 – 27 February 2020 | Ciheam-Bari, Valenzano, Italy. Food and Agriculture Organization of the United Nations.

Young, R. A. (2005). Determining the economic value of water. Resources for the future. <http://agris.fao.org/agris-search/search.do?recordID = US201300102646>.

Young, R. A., & Haveman, R. H. (1985). Economics of water resources: A survey. In A. V. Kneese, & J. L. Sweeney (Eds.), Handbook of natural resource and energy economics volume ii economics of water resources: A survey *(II)*. Elsevier Science Publishers B. V.

CHAPTER 13

Potential of municipal wastewater for resource recovery and reuse

Manzoor Qadir

United Nations University Institute for Water, Environment and Health (UNU-INWEH), Hamilton, ON, Canada

Chapter Outline

13.1 Introduction 263
13.2 Wastewater as a water source 264
13.3 Wastewater as a nutrient source 265
13.4 Wastewater as an energy source 268
13.5 Conclusions and prospects 269
References 270

13.1 Introduction

While impacting freshwater allocations and water quality, urbanization—coupled with population increase, improved living standards, industrialization, and economic development—continues to drive increased volumes of municipal wastewater (WWAP, 2017). Once branded as a waste and an environmental burden, wastewater is increasingly acknowledged and positioned as a valuable source of water, nutrients, organic carbon and organic matter, precious metals, and energy. With such an increasing value proposition recognition of wastewater by relevant institutions and professionals, the wastewater treatment facilities, which treat different waste streams, are viewed as resource recovery enterprises rather than wastewater management units (Drechsel et al., 2015).

Wastewater produced by homes, businesses, industries, and institutions mostly ends up in wastewater collection and conveyance channels, also called sewers, open depressions, or other water bodies directly in untreated form for dilution. Such forms of wastewater are collectively termed municipal wastewater (Sato et al., 2013). Wastewater irrigation contributes to agricultural production and provides support to the livelihoods of millions of urban and peri-urban smallholder farmers in several areas of the world. In addition, productive use of wastewater has been demonstrated for growing fish (aquaculture), supporting trees and shrubs (agroforestry), moving deep through soil profile to supplement groundwater (aquifer recharge), and contributing to making human life both possible and worth living (ecosystem services).

Most wastewater produced in developing countries is discharged into streams, ditches, or culverts with little or no treatment. Many farmers divert irrigation water from those sources and, thus utilize wastewater to produce crops for sale in local markets (WWAP, 2017). The farmers earn income that enhances their economic opportunities, yet the undesirable constituents in wastewater pose risks to farm households, field workers, and consumers (Grangier et al., 2012). It is essential to consider the public health and environmental implications of using wastewater for irrigation in such settings. In addition, there is a need to develop a better understanding of the cultural, social, and economic dimensions and policy and institutional settings relevant to water recycling and reuse.

Although recognized as valuable water, nutrients, and energy source, the scale of planned resource recovery from wastewater is highly variable across countries and regions (Otoo and Drechsel, 2017). In developing countries, the potential to transform wastewater from waste into a resource remains limited (Schuster-Wallace et al., 2015). A range of technical, social, policy, and economic roadblocks restrict the wastewater treatment plants from achieving the full potential of resource recovery from wastewater.

Based on a recent assessment of water, energy, and macronutrients embedded in municipal wastewater (Qadir et al., 2020), this chapter analyzes and presents global and regional trends highlighting the potential of wastewater as a valuable resource. While recognizing the challenges associated with achieving the potential of resource recovery from wastewater, this chapter provides insights into a range of technical, social, policy, and economic response options that can be implemented for effective resource recovery from wastewater.

13.2 Wastewater as a water source

Global assessments of the resources embedded in wastewater, based on the aggregation of the 2015 country-level data, suggest wastewater production across the world in the range of 360 to 380 km^3, that is, 360 to 380 billion m^3 or 360 to 380 trillion L (Qadir et al., 2020; Jones et al., 2021). Based on the 2015 data and its simulation for 2020, current wastewater production stands at 411 km^3, which is expected to reach 470 km^3 in 2030 and 574 km^3 in 2050, that is, 14% and 40% increase over 2020 level, respectively (Qadir et al., 2020). The major drivers for the increasing volumes of wastewater production are population increase, urbanization, improving living conditions, and economic development—all requiring more water for use to produce wastewater subsequently. The increasing trends of wastewater production give us a heads up for concerted efforts to manage large volumes of sewage in days to come safely. At the same time, it also showcases a prospect to address water scarcity in arid and semiarid countries and communities through effective management and treatment of wastewater and the use of treated wastewater for multiple opportunities.

In terms of use as a water source for agricultural irrigation, the current volume of wastewater at 411 km^3 (without further dilution) provides enough water to irrigate around 34 million ha, considering two crops per annum with a water requirement approximately 12,000 m^3 per ha for both crops. Such volume of wastewater can be used to bring new areas under irrigated agriculture or to swap with freshwater in areas where irrigated agriculture relies on freshwater supplies, which are at risk of meeting irrigation needs. In certain arid and semiarid regions, the substitution of freshwater with wastewater for irrigation has already been the case as farmers benefit from wastewater use in agriculture. This trend of wastewater use needs to persist. It is expected to increase further as wastewater volumes increase and offer an opportunity for its safe and productive use in agriculture and other sectors.

Regional analysis of wastewater production suggests Asia as the largest producer of wastewater with 176 km^3 (2020), followed by North America (70 km^3), Europe (69 km^3), Middle East and North Africa (37 km^3), Latin America and the Caribbeans (35 km^3), Sub-Saharan Africa (21 km^3), and Oceania, which produces small volumes of wastewater (3 km^3) compared with other global regions (Fig. 13.1). The considerable variation in wastewater production across global regions is largely attributed to human population and urbanization trends.

There is disparity across global regions in terms of wastewater production per capita where North America is the leading wastewater producer per capita (231 m^3) followed by Europe (124 m^3), the Middle East and North Africa (114 m^3), Oceania (88 m^3), Asia (82 m^3), Latin America and the Caribbean (65 m^3), and Sub-Saharan Africa (46 m^3), which produces one-fifth of the volume of wastewater produced per capita in North America. The global average sewage production stands at 95 m^3 per capita (Qadir et al., 2020).

13.3 Wastewater as a nutrient source

Global assessment of macronutrients in wastewater suggests 18.0 tera-gram (Tg = million metric tons) of nitrogen embedded in wastewater produced in 2020 (411 km^3) across the world annually. Phosphorus in wastewater stands at 3.2 Tg while potassium at 6.8 Tg of potassium (Fig. 13.2). Collectively, 28.0 Tg of nitrogen, phosphorus, and potassium occur in wastewater produced worldwide.

Considering the global demand for fertilizer nutrients, estimates reveal that nitrogen embedded in wastewater can offset about 15% of the global nitrogen demand as fertilizer nutrient. In comparison, phosphorus can supplement 7% and potassium 19% of the

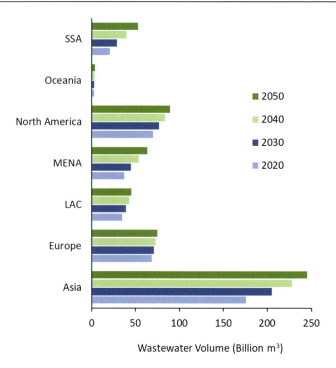

Figure 13.1
Wastewater production across global regions in 2020 and projections for 2030, 2040, and 2050. MENA stands for the Middle East and North Africa, SSA for Sub-Saharan Africa, and LAC for Latin America and the Caribbean regions. Source: Based on the data from Qadir, M., P. Drechsel, B.J. Cisneros, Y. Kim, A. Pramanik, P. Mehta, and O. Olaniyan. 2020. Global and regional potential of wastewater as a water, nutrient, and energy source. Natural Resources Forum 44 (1): 40–51.

worldwide fertilizer demand for these nutrients. Based on 2017 data, the need for these macronutrients for agriculture worldwide remains at 193 Tg (IFA, 2017). In terms of potential economic gains, the recovery of these nutrients from wastewater could generate a revenue of $14 billion globally (Qadir et al., 2020). These estimates assume that the nutrients recovered from wastewater have the same quality and market acceptance as industrial fertilizer.

Aside from the recovery of valuable nutrients and associated economic gains, there are vital environmental benefits to treating wastewater, such as minimizing pollution of surface and groundwater bodies (Otoo and Drechsel, 2017). It is crucial to remove or transform certain nitrogen and phosphorus compounds embedded in sewage due to their potential to trigger eutrophication or aquatic toxicity from nitrogen (ammonia). As some of the most polluted waters due to nutrient runoff are found in water-stressed regions (Damania et al., 2017), nutrient recovery in these regions from waste streams can help reduce water pollution resulting in environmental and economic gains.

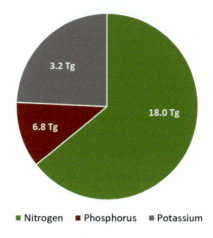

Figure 13.2
Nitrogen, phosphorus, and potassium present in wastewater at the global level. Calculated as a function of global volumes of wastewater produced in 2020 and average concentrations of these macronutrients in 53 wastewater samples representing different parts of the world. Source: Based on the data from Qadir, M., P. Drechsel, B.J. Cisneros, Y. Kim, A. Pramanik, P. Mehta, and O. Olaniyan. 2020. Global and regional potential of wastewater as a water, nutrient, and energy source. Natural Resources Forum 44 (1): 40–51.

Among the nutrients in wastewater, phosphorus recovery is crucial, as phosphorus is a nonrenewable resource obtained from finite deposits of rock phosphate confined to some countries. These rock phosphate resources, therefore, need to be used strategically. In addition, recent spikes in the price of mineral phosphorus are pushing phosphorus recovery from wastewater as an increasingly attractive option for economic reasons. Notably, between 50% and 80% of the total phosphorus compounds in wastewater are phosphates. These compounds leave the wastewater treatment facility in the treated effluent from where they can be recovered (Metcalf and Eddy, 2003).

For nitrogen, its compounds can be generated from atmospheric nitrogen and are not considered a limited resource, but their production is energy-intensive. Considering the energy costs of producing nitrogen fertilizers and that most nitrogen present in wastewater is soluble and in inorganic forms, nitrogen recovery from wastewater is also becoming more appealing (Puchongkawarin et al., 2015) and lowering the levels of eutrophication.

Potassium is abundant in wastewaters. Irrigation with wastewater may result in potassium availability that can fulfill crop nutrient requirements or even more. The chances of potassium accumulation in wastewater-irrigated soil are relatively high because potassium does not significantly leach through the soil (Arienzo et al., 2009). Potassium ions not taken up by plants are adsorbed at the soil's cation exchange sites, which minimize potassium leaching. Thus removing potassium from wastewater during its treatment or its release into treated effluent are attractive options.

13.4 Wastewater as an energy source

While energy is embedded in wastewater, the wastewater treatment process itself is energy-intensive. However, estimates suggest that wastewater contains more energy than is required for its treatment (Frinjs et al., 2013). In current practices, the energy potential of wastewater is yet to be fully exploited. However, innovations in specific anaerobic wastewater treatment systems have extended the range and scale of anaerobic treatment applications (Lew et al., 2009) to enhance dissolved methane recovery from wastewater.

Based on the anaerobic conversion of organic carbon to methane, the caloric value of methane (Verstraete et al., 2009), assuming full energy recovery from wastewater, and the electricity needs of a household on average 3350 kWh (World Energy Council, 2016), the energy value in 411 km^3 of wastewater is enough to provide electricity to 171 million households (Fig. 13.3) or 685 million people while assuming an average of four persons per household. As the volumes of wastewater are expected to increase over time, the energy embedded in wastewater produced during 2030 would be enough to fulfill the energy needs

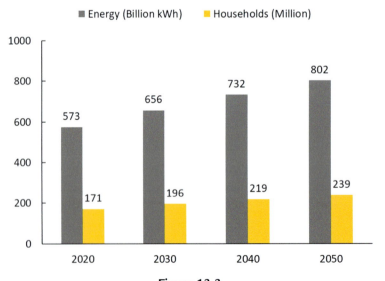

Figure 13.3

Potential for energy production from wastewater based on the 2020 wastewater production data and scenarios over coming decades for the years 2030, 2040, and 2050 based on the expected increase in wastewater volumes. *Source: Based on the data from Qadir, M., P. Drechsel, B.J. Cisneros, Y. Kim, A. Pramanik, P. Mehta, and O. Olaniyan. 2020. Global and regional potential of wastewater as a water, nutrient, and energy source. Natural Resources Forum 44 (1): 40–51.*

of 196 million households (783 million people). In 2040, the number of families potentially benefiting from energy generated by wastewater treatment plants could be in the order of 219 households (875 million people); in 2050, the number of homes using energy stemming from wastewater may reach 239 million households (958 people). Such scenarios stem from the assumption that considers maximum energy recovery without any technical and economic limitations in the processes addressing energy recovery from wastewater.

As it stands, the potential of recovering energy from wastewater is far from its full achievement (Otoo and Drechsel, 2017). The emerging innovations in research and practice on energy recovery from waste streams intensify energy recovery efforts from wastewater. Such measures may end up in an energy-neutral wastewater cycle through onsite renewable energy production together with improvements in energy efficiency in wastewater treatment facilities (Tarallo et al., 2015). Wastewater treatment plants can produce the energy needed to treat wastewater and contribute to the energy needs of cities and municipalities that generate waste streams. Implementing current best management practices on sustainable energy recovery provides a viable resource recovery opportunity in countries where wastewater treatment is relatively low (Lackey and Fillmore, 2017).

13.5 Conclusions and prospects

The importance of resources embedded in municipal wastewater and the benefits from the recovery of water, nutrients, and energy from wastewater are well documented and demonstrated at wastewater treatment plants. Despite such recognition, the potential of recovering such valuable resources from wastewater has not yet been achieved. However, a shift in research and practice is in progress to investigate further and support strategies that aim at safely managed wastewater.

The good news is that resource recovery innovations are being effectively implemented in high-income and upper-middle-income countries. Such innovations continue to be refined to bridge the gap between current resource recovery levels from waste streams and their resource recovery potential. The challenge remains in low-income and lower-middle-income countries, which need to expedite response options based on pertinent innovations as wastewater treatment in these countries continues to be low and resource recovery likewise. These countries need to embark on effective programs to collect and treat wastewater. The water supply and sanitation sector investments need to be secured and used effectively to harness the potential of resource recovery from wastewater.

The following response options regarding effective resource recovery from wastewater should be implemented:

1. Develop mechanisms and institutional strength to establish monitoring systems for the collection and reporting of data on different aspects of waste streams, such as

wastewater volumes generated by different sectors and its quality, wastewater collection and treatment status, quality of treated wastewater, resource recovery options used, if any, and quality of the recovered resources and their use, among others.

2. Establish interim measures to use different forms of wastewater until full treatment capacity and resource recovery potential are achieved in an anticipated time frame by transforming wastewater treatment plants into water resource recovery facilities, thereby valuing and realizing resource recovery from wastewater.
3. Develop wastewater treatment and resource recovery framework with time-bound targets and milestones for implementation at different scales—local, district, river basin—in partnership with relevant stakeholders while aiming at effective and environmentally acceptable management of wastewater and maximizing the benefits from improving resource recovery efficiency and achieving the potential of resource recovery and reuse.
4. Investigate and implement innovative financial models and sustainable business solutions that leverage the potential extra revenue streams from the recovery of resources at the wastewater treatment plants for use in different sectors.
5. Establish close and effective collaboration and linkages between the public sector responsible for wastewater management with the private sector to play an essential role in the initiatives on resource recovery from waste streams.
6. Make evidence-based decisions to implement pertinent policy, institutional, and regulatory frameworks to trigger action on the ground in transforming wastewater into a valuable resource rather than a waste looking for disposal.
7. Develop institutional and human capacity to deal with wastewater management's complex issues and develop business plans or evaluate existing schemes related to resource recovery from wastewater and reuse of recovered resources.

References

Arienzo, M., Christen, E. W., & Quayle, W. C. (2009). Phytotoxicity testing of winery wastewater for constructed wetland treatment. *Journal of Hazardous Materials, 169,* 94–99.

Damania, R., Desbureaux, S., Hyland, M., Islam, A., Moore, S., Rodella, A., Russ, J., & Zaveri, E. (2017). Uncharted waters: The new economics of water scarcity and variability. *The World Bank Group,* Washington, DC. Available from https://openknowledge.worldbank.org/handle/10986/28096.

Drechsel, P., Qadir, M., & Wichelns, D. (2015). *Wastewater: Economic asset in an urbanizing world* (p. 282) Dordrecht, the Netherlands: Springer Science + Business Media.

Frinjs, J., Hofman, J., & Nederlof, N. (2013). The potential of (waste)water as energy carrier. *Energy Conversion and Management, 65,* 357–363.

Grangier, C., Qadir, M., & Singh, M. (2012). Health implications for children in wastewater-irrigated peri-urban Aleppo, Syria. *Water Quality, Exposure and Health, 4*(4), 187–195.

IFA. (2017). Fertilizer Outlook 2017–2021. Production and international trade and agriculture services. International Fertilizer Association (IFA) annual conference, 22–24 May 2017, Marrakech, Morocco. Available at: https://www.fertilizer.org/images/Library_Downloads/2017_IFA_Annual_Conference_Marrakech_PIT_AG_Fertilizer_Outlook.pdf.

Jones, E. R., van Vliet, M. T. H., Qadir, M., & Bierkens, M. F. P. (2021). Country-level and gridded estimates of global wastewater production, collection, treatment, and reuse. *Earth System Science Data, 13*, 237−254.

Lackey K, Fillmore L (2017). Energy management for water utilities in Latin America and the Caribbean: Exploring energy efficiency and energy recovery potential in wastewater treatment plants. The World Bank Group, Washington, DC; and Water Environment Research Foundation, Alexandria, VA.

Lew, B., Tarre, S., Beliavski, M., Dosoretz, C., & Green, M. (2009). Anaerobic membrane bioreactor (AnMBR) for domestic wastewater treatment. *Desalination, 243*, 251−257.

Metcalf, & Eddy (Eds.), (2003). *Wastewater engineering—Treatment, and reuse* (3rd ed.). New York: McGraw Hill.

Otoo, M., & Drechsel, P. (Eds.), (2017). *Resource recovery from waste: Business models for energy, nutrient and water reuse in low- and middle-income countries*. London: Taylor & Francis Ltd.

Puchongkawarin, C., Gomez-Mont, C., Stuckey, D. C., & Chachuat, B. (2015). Optimization-based methodology for the development of wastewater facilities for energy and nutrient recovery. *Chemosphere, 140*, 150−158.

Qadir, M., Drechsel, P., Cisneros, B. J., Kim, Y., Pramanik, A., Mehta, P., & Olaniyan, O. (2020). Global and regional potential of wastewater as a water, nutrient, and energy source. *Natural Resources Forum, 44*(1), 40−51.

Sato, T., Qadir, M., Yamamoto, S., Endo, T., & Zahoor, A. (2013). Global, regional, and country-level need for data on wastewater generation, treatment, and reuse. *Agricultural Water Management, 130*, 1−13.

Schuster-Wallace, C. J., Qadir, M., Adeel, Z., Renaud, F., & Dickin, S. K. (2015). *Putting water and energy at the heart of sustainable development* (p. 27) United Nations University (UNU). Available from http://inweh.unu.edu.

Tarallo S, Shaw A, Kohl P, Eschborn R (2015). A guide to net-zero energy solutions for water resource recovery facilities (ENER1C12). Water Environment Research Foundation (WERF) Alexandria, VA.

Verstraete, W., Van de Caveye, P., & Diamantis, V. (2009). Maximum use of resources present in domestic "used water." *Bioresource Technology, 100*(23), 5537−5545.

World Energy Council. (2016). *Energy efficiency indicators: Average electricity consumption per electrified household*. Available from http://www.wec-indicators.enerdata.eu/household-electricity-use.html.

WWAP (United Nations World Water Assessment Programme). (2017). The United Nations world water development report 2017. Wastewater: The Untapped Resource. Paris, UNESCO. Available at: http://www.unesco.org/new/en/natural-sciences/environment/water/wwap/wwdr/2017-wastewater-the-untapped-resource/.

CHAPTER 14

Sustainable freshwater management—the South African approach

Oghenekaro Nelson Odume

Unilever Centre for Environmental Water Quality, Institute for Water Research, Rhodes University, Makhanda, South Africa

Chapter Outline

14.1 Introduction 273
14.2 The classification system and classification of water resources 275
14.3 The determination of the reserve and resource quality objectives 277
14.4 Reflections on the resource directed measure components using the Vaal Barrage catchment in South Africa as a case study 278
14.5 Source directed controls 287
14.6 Linking resource directed measure and source directed control instruments for sustainable freshwater resources management 287
14.7 Conclusions 290
Acknowledgments 290
References 290

14.1 Introduction

Freshwater resources are scarce globally and with a growing human population and expanding socioeconomic activities, particularly in developing countries, competition among users for the scarce water resources is likely to be on the rise in the next foreseeable future (DWA, 2013; Reid et al., 2019). Climate-related pressure, pollution, high rate of water abstraction, biodiversity loss and water quality and quantity deterioration present formidable challenges for sustainable freshwater resources management across scales (Bunn, 2016). At a global scale, critical freshwater planetary boundaries are either about to or already outstripped (Rockstrom et al., 2009). This calls for concerted efforts across jurisdictions for measures aimed at realizing the objective of sustainable utilization of freshwater resources, which is here interpreted as the balance between freshwater resource use and protection. Drawing on experiences from South Africa, this chapter presents an analysis of freshwater resources management. The intention is not to present a global or exhaustive picture of how freshwater resources are or should be managed, but to horn on the South African case study to demonstrate the concept of freshwater resources management and lessons that can be learned and transfer to other jurisdictions. By focusing on a specific example, the chapter provides an opportunity for comparative analysis and reflections between freshwater resource management in South Africa and elsewhere.

South Africa is an acclaimed international leader with its progressive and forward-looking water legislation, the National Water Act (NWA, Act No. 36 of 1998) (RSA, 1998). Equity, sustainability, and efficiency are the foundational principles of the Act, and are invoked when managing, utilizing, developing, protecting, and conserving water resources. The NWA differed radically from the 1956 Act by delinking land ownership from right to water. The Act concentrates ownership of all water resources in the hands of the national government. At the same time, the NWA provides for two legally binding rights to water, that for basic human needs, that is, the Human Need Reserve (HNR) and that for sustaining ecological functioning of the aquatic ecosystems, that is, the Ecological Reserve (ER). Once the two Reserves have been determined, all remaining water resources are administratively allocated for other uses, taking account of the three foundational principles of equity, sustainability, and efficiency. The implication is that by law water resources are to be used and protected in ways that ensure their sustainability.

Two complementary strategies have been developed to give effect to water resource use and protection in South Africa. These are the resource directed measures (RDM) and the source directed controls (SDC) (RSA, 1998; DWA, 2013). The RDM are measures directed at water resources to ensure their protection and they include the national water resource classification system (WRCS), the classification of every significant water resource, determination of the Reserve and setting of resource quality objectives (RQOs). On the other hand, the SDCs are instruments imposed on users to restrict and control the use of water resources not only in terms of ensuring water resource protection, but also ensuring that water resources are equitably allocated and use efficiently (DWAF, 2003). General authorization (GA), special permit, existing lawful uses, and water use licenses (WULs) are examples of SDC instruments (DWAF, 2006a). Used together, the two strategies (i.e., RDMs and SDCs) enable the balancing of water resource use and protection.

The entire RDM process begins with a catchment visioning exercise with the aim of bringing society together to collectively envision the desired future conditions of their catchments, considering a range of desired ecosystem services (DWAF, 2006b). The catchment visioning exercise is designed to give effect to the envisaged participatory approach to water resource management captured in the NWA (No. 36 of 1998; RSA, 1998). Catchment visioning is critical because in a diverse and pluralistic society, people tend to have diverse interests, values, aspirations, and choices, and in many cases, opting for one sets of choices may preclude the realization of other sets. For this reason, catchment visioning is meant to bring together a wide range of affected and interested stakeholders within a catchment as the starting point of the RDM process. It is critical to note that catchment visioning is a complex and time-consuming exercise that require a range of skills not only facilitation skills, but ethics skills that would ensure that the voices of the less powerful and often marginalized groups within the catchment are considered and taken forward in the decision-making process about the desired future state of the catchment.

The approach to catchment visioning is that it is fully participatory, inclusive, and context-sensitive in seeking to arrive at consensus among competing interests. During catchment visioning exercise the three foundational values of equity, sustainability and efficiency in the NWA serve as guide, beginning with a process of sensitizing all catchment communities, water resource users, including interested and affected parties who may not live within the catchment (DWAF, 2006a). Overall, the outcome of the catchment visioning exercise should reflect the collective objectives and aspirations for the catchment, which then give directions to other RDM processes that are to follow. In a way, the process of catchment visioning considers all dimensions of equity—procedural, distributive, contextual and recognitional (McDermott et al., 2013; Loft et al., 2017; Leach et al., 2018)—so that equity imperatives are taken seriously in the collective outcomes of the catchment visioning process.

14.2 The classification system and classification of water resources

Section 12 of the NWA provides for the establishment of a national WRCS for the classification of every significant water resource. The WRCS (DWAF, 2008a, 2008b) provides a set of guidance and procedures to follow when undertaking the classification process. The classification system allows water resources to be classified in a manner that reflects the desired protection and utility from a water resource. The classification system thus aims to balance water resource protection and use in a sustainable, equitable and efficient manner. The classification process considers ecological water requirements (EWRs) and user water requirements. The classification process is undertaken following a seven-step process (1) delineate the units of analysis and describe the status quo of the water resource, (2) link the value and condition of the water resource, (3) quantify the EWRs and changes in nonwater quality ecosystem goods, services and attributes, (4) determine an ecologically sustainable base configuration scenario and establish the starter configuration scenarios, (5) evaluate scenarios within the integrated water resource management (IWRM) process, (6) evaluate the scenarios with stakeholders, and (7) gazette the class configuration.

Depending on the desired protection and use a water resource is to be accorded, a resource is classified into one of three management classes (MCs): Class I (a resource with no noticeable or minimal human impacts); Class II (a resource moderately used and impacted by human activities with moderate deviation from natural/predevelopment conditions); and Class III (a resource that is heavily used and impacted with significant deviation from natural/predevelopment conditions) (DWAF, 2008a). It is important to note that the classification accorded a water resource has social, economic, and ecological implications regarding risks and development. For example, if a water resource for whatever reason is classified as Class I, the implication is the risks of activities likely to impact on the resources are kept at minimum, and this, on the other hand, can constrain further socioeconomic development within the catchment. Overall, the final MC of a water resource is a combination of the EWR and the requirements of other user sectors, for example, domestic, agriculture, industry, recreation. The EWRs are determined by assessing the present ecological state (PES), which are represented from categories A−F, and recommended ecological category (REC), represented from categories A−D. Table 14.1 shows descriptions of the categories as used during the determination of the MC of a water resource.

The second requirements captured in the MC are those of other users within the catchment. Unlike EWR reflected in categories A−F, users' requirements are reflected using the fitness for use categories of Ideal, Acceptable, Tolerable, and Unacceptable (DWAF, 2006b, Table 14.2). It is important to note that fitness for use is described in terms of the specific user sector requirements, for example, domestic, irrigation, recreation, and industrial.

The user requirements are mapped onto the ecological categories and configured into the final MC as follows (Fig. 14.1).

Table 14.1: Ecological categories and description applicable to water resources in South Africa (Kleynhans & Louw, 2008).

Ecological category name	Description	Explanation
A	Pristine, natural/unmodified	Natural
B	Largely natural	A small change in the natural functioning, processes, biota, and characteristics of the ecosystem
C	Moderately modified	Some loss of natural biota and habitats, but basic ecosystem functioning, processes and structure remain largely unchanged
D	Largely modified	A large loss of biota, habitats and impact on basic ecosystem function has occurred and are large
E	Seriously modified	Impact on ecosystem functioning is extensive, significant loss of biota, habitats, and impact on physical, chemical, and biological properties of the ecosystem is serious
F	Critically modified	Extreme modification of the system, including loss of habitats and biota. Ecosystem functioning is extremely compromised, and changes are irreversible

Table 14.2: Water user requirement categories.

Fitness for use category	Description
Ideal	100% always fit for use, water condition desirable for the intended use
Acceptable	Slight problem encountered, on few occasions for the intended use
Tolerable	Moderate to severe problems are encountered for the intended use, for a limited period
Unacceptable	Water is unacceptable for its intended use at all time

Source: *Adapted from DWAF (Department of Water Affairs and Forestry). (2006b). Guideline for determining resource water quality objectives (RWQOs), allocatable water quality and the stress of the water resource. Pretoria: Department of Water Affairs and Forestry.*

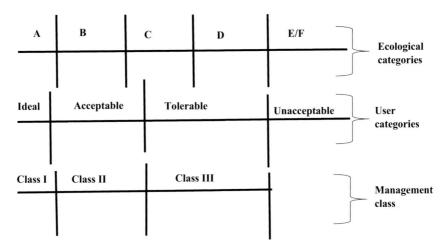

Figure 14.1
Relationship between the ecological and user categories and the management class (DWAF, 2006b).

14.3 The determination of the reserve and resource quality objectives

The Reserve and the RQOs, together with the classification of all significant water resources are the components of the RDM. The outcome of the classification process is the MC, the Reserve and the RQOs. The Reserve has two components, the Basic Human Need Reserve (BHNR) and the ER. The BHNR provides for the essential needs of individuals such as water for drinking, preparing food and for personal hygiene, whereas the ER is the quality, quantity, and assurance of supply of water needed to protect and maintain aquatic ecosystem functionality (King & Pienaar, 2011). The RQOs are descriptive and quantitative measures characterizing the desired level of protection and use of a water resource as defined by its MC and the RECs (DWA, 2013; DWS, 2016). The RQOs are thus the measurable goals that are assessed to determine whether progress is being made toward achieving the designated desired

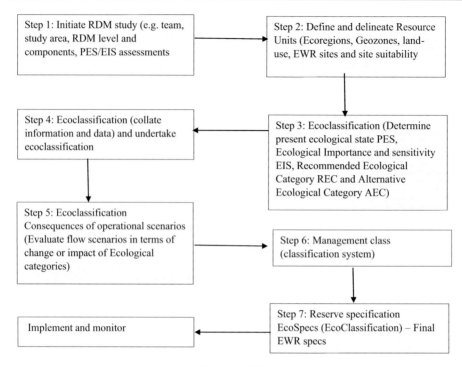

Figure 14.2
The seven steps for determining the Ecological Reserve. Source: *Adapted from DWAF (Department of Water Affairs and Forestry). (2008b). Methods for determining the water quality component of the ecological reserve for rivers. Pretoria: Department of Water Affairs and Forestry. Second Draft.*

future condition as captured in the REC. Overall, the RQOs translate the MC and the ecological needs determined in the Reserve into measurable objectives for tracking progress toward the desired future state (King & Pienaar, 2011). The seven steps of determining the ER and the RQOs are summarized in Figs. 14.2 and 14.3, respectively.

14.4 Reflections on the resource directed measure components using the Vaal Barrage catchment in South Africa as a case study

The Vaal River forms the boundary between Free State and Gauteng, Mpumalanga and North-West of South Africa. The Vaal Barrage within the Vaal River system was completed in 1923 following a project to secure a supply of potable water for Johannesburg and the Witwatersrand region following the growth of gold mining and financial activity in the area (Tempelhof, 2009). Within a few years of its construction, it became largely redundant as the upstream Vaal Dam was completed to act as an irrigation water source for farming, to further secure water for development in the Witwatersrand-Vaal Triangle region, and, in the short term, to offer employment during a worldwide depression

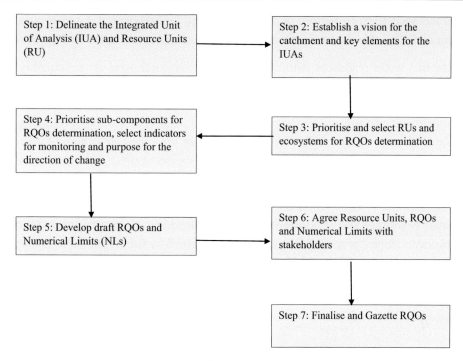

Figure 14.3
The seven steps for determining the resource quality objectives (DWA, 2011a).

(Tempelhof, 2009). Upon the completion of the Vaal Dam, the Vaal Barrage became a secondary water storage facility, and water from this region supported growing coal and gold mining, power generation and steel production (Turton et al., 2006; Tempelhof, 2009).

The Vaal River system has highly regulated and modified flow regimes. Flow along the Vaal River is modified by the Vaal Dam and Vaal Barrage in particular (DWAF, 2009; Tempelhof, 2009). Several interbasin transfer schemes including the Lesotho Highlands Water Project and the Heyshope, Zaaihoek and Tugela Transfer Schemes introduce water to the catchment, and water is also transferred out of the system to support Eskom and Sasol operations (Turton et al., 2006; DWAF, 2009). The river and its impoundments are also the largest water source for the densely populated Johannesburg and Vaal Triangle regions. Flow modification is known to have a serious impact on aquatic ecosystems (Bunn & Arthington, 2000).

A consequence of such development, initially without regulation, is a legacy of water quality issues. Fecal coliform counts in the area remain high (Tempelhof, 2009). Phosphate levels, which lead to eutrophication, are consistently high (Rand Water 2017 data). Salinity levels also remain high, and these are often accompanied by high levels of sulfate, indicating an origin in acid mine drainage from gold or coal mines (Rand Water 2017 data).

A map of the Vaal Barrage region, including the Klip, Suikerbosrand, Leeuspruit and Taaibosspruit Rives, which are the focus of this case study is presented in Fig. 14.4.

In ensuring that water resources within the Vaal Barrage catchment are sustainably utilized, the South African Department of Water and Sanitation gazetted the MC for the system, after the determination of their PES (DWS, 2016). The PES, RECs, and MC for the Vaal Barrage catchment and the associated rivers, that is, Leeuspruit, Taaibosspruit, Klip, Blesbokspruit, Rietspruit and Suikerbosrand Rivers are presented in Tables 14.3 and 14.4. Here, reflections are provided on the ecological and socioeconomic implications of the PES, REC, and MC (Tables 14.3 and 14.4).

Through the ER and HNR water is set aside for the functioning of ecosystems and basic human needs before the rest is administratively allocated to other user sectors. In the Vaal catchment for example, proposed Reserves were gazetted by the South African Department of Water and Sanitation (DWS, 2018). The Reserve takes the form of water quantity, quality, and reliability of supply to sustain vital ecosystem function. In Tables 14.5—14.8,

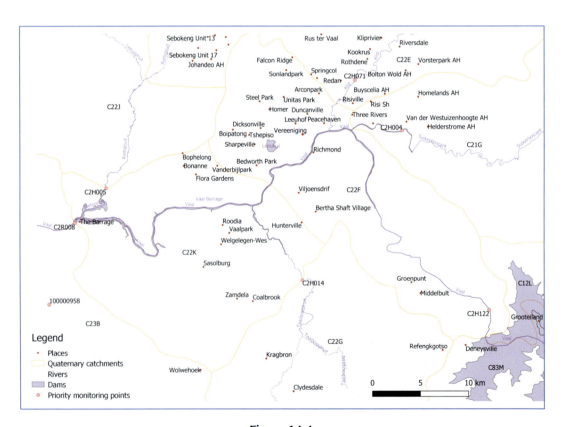

Figure 14.4

Vaal Barrage region showing place names, quaternary catchments, rivers, dams, and the South African Department of Water and Sanitation priority water quality monitoring points.

Table 14.3: Reflection on the management class (MC) and recommended ecological categories (REC) for the Suikerbosrand and Klip River system within the Vaal Barrage catchment.

IUA	IUA class	Node name	Major river	Tributary	PES	REC	Reflections
Suikerbosrand	II	UH.1	Vaal	Suikerbosrand	B/C	B	The system has been classified as a moderately used and impacted, with present ecological state being a transitional category from largely natural to moderately modified (UH.1). However, the desired state is to revert this trajectory to largely natural (REC: B). The socioeconomic implications are that the regulators would have to collaboratively work with other catchment stakeholders on measures to restrict water use and this may involve stricter license conditions or other measures that restrict water use. Similar implications hold for EWR 9 where the current state had to be reverted to B/C. Further, because the resource is classified as a category II, it is important to note that developments likely to have significant demand on water resources may not be allowed in this catchment or would have to be imposed with strict regulations to limit/control impacts.
		EWR 9	Vaal	Suikerbosrand	C	B/C	
Klip River (Gauteng)	III	UI.1	Klip	Rietspruit	E	D	Water resources within these catchments are heavily used and impacted as reflected in the MC III. The impacts on the systems are extensive, with significant loss of biota, habitat, and alteration of the physicochemical and hydrological properties of the systems, as currently indicated in PES E for UI.1, UI.2 and UI.3. Large impacts are associated with UI.4 and EWR 10 and 11. However, the regulator and society have agreed to accept significant risk in terms of protecting water resources within the catchment as reflected in a REC: D for all biophysical nodes except for EWR 10, for which a REC is C/D, which is still indicative of acceptance of high risk with regard to maintaining ecological integrity. Nevertheless, despite acceptance of higher risk in relation to ecological health of the systems, the regulator is likely to impose stricter water use license conditions in order to slightly improve the system and to halt further deterioration. Sensitive water users are likely to be vulnerable and at risk given that REC for most of the system is a D. However, given the strategic importance of these systems for economic development, a D is justifiable especially because of the economic costs that may be incurred to improve the systems to a C or C/B categories.
		UI.2	Vaal	Klip	E	D	
		UI.3	Vaal	Klip	E	D	
		UI.4	Vaal	Rietspruit	D/E	D	
		EWR 10	Vaal	Suikerbosrand	C/D	C/D	
		EWR 11	Suikerbosrand	Blesbokspruit	D	D	

EWR, Ecological water requirement; IUA, Integrated Unit of Analysis; PES, present ecological state.

Table 14.4: Reflection on the management class (MC) and recommended ecological categories (REC) for the Taaibosspruit and Vaal River reach from Vaal Dam to C23L.

IUA	IUA class	Node name	Major river	Tributary	PES	REC	Reflections
Taaibosspruit (UJ)	III	UJ.1	Vaal	Taaibosspruit	D	D	Water resources in the Taaibosspruit are currently heavily utilized and the system is largely modified with a large loss of biota, habitats, and alteration of the physicochemical and hydrological conditions. However, the regulator and society have agreed to keep and maintain the status quo as indicated by a REC D. The implication is that development that may significantly deviate the system from the current conditions may be prohibited or imposed with stricter conditions. From a water quality perspective, effluent quality ought to, as a minimum, meet historical standards that have kept the system within an ecological category D. Thus, development that may significantly further impact on the system may be subject to very strict licensing conditions as the minimum acceptable REC is D, and an E or F is unacceptable.
Vaal River reach from Vaal Dam to C23L (UM)	III	EWR 4	Vaal		C	B/C	The Vaal River reach from the Vaal Dam to C23L(UM) is heavily used supporting extensive socioeconomic activities as reflected in the MC III. EWR 4 is currently moderately modified as indicated by a PES of C and EWR 5 is in a transition state between being moderately modified and largely modified. However, for both biophysical nodes, the regulator and society have agreed to improve the system to one level higher than the current condition. The implication is that water users within the catchment may experience some sort of stricter license conditions and other measures to be able to improve the system as envisaged. For EWR 4, the regulator's appetite for risk in relation to ecological health is low as indicated by the B/C REC compared to EWR 5 with a REC of a C.
		EWR 5	Vaal		C/D	C	

EWR, Ecological water requirement; IUA, Integrated Unit of Analysis; PES, present ecological state.

Table 14.5: Reflections on the water quality component of the proposed Ecological Reserve for ecological water requirement (EWR) 9 Suikerbosrand River as published in government gazette no. 42127 (DWS, 2018).

WQ class	Variable	Specification	Reflection
EWR 9: Suikerbosrand River (UH): TEC: B/C; EIS			
Inorganic salts	$MgSO_4$	The 95th percentile of the data must be <37 mg L^{-1}	The Suikerbosrand River at EWR 9 is regarded as an ecologically important and sensitive system, and has thus been given a relatively high TEC of a B/C. The system is sensitive to input of salts, nutrients, and toxics and as well as activities that are likely to alter physical variables such as turbidity and dissolved oxygen. For these reasons, the WQ specifications for these variables are much stricter compared to EWR 10 and 11. The implication is that the regulator would impose much stricter requirements on water users within this subcatchment to meet the ecological Reserve as set out, thus potentially limiting future, large scale industrial developments and other activities likely to significantly impact on the ecological health of the system.
	Na_2SO_4	The 95th percentile of the data must be <51 mg L^{-1}	
	$MgCl_2$	The 95th percentile of the data must be <30 mg L^{-1}	
	$CaCl_2$	The 95th percentile of the data must be <57 mg L^{-1}	
	NaCl	The 95th percentile of the data must be <45 mg L^{-1}	
	$CaSO_4$	The 95th percentile of the data must be <351 mg L^{-1}	
Physical variables	EC	The 95th percentile of the data must be <55 mS m^{-1}	
	pH	The 5th percentile of the data must be 6.5–8.0 and the 95th percentile 8.0–8.8	
	Temperature	Small deviation from the natural temperature range	
	DO	The 5th percentile of the data must be ≥8 mg L^{-1}	
	Turbidity	Vary by a small amount from the natural turbidity range, minor silting of instream habitats acceptable	
Nutrients	TIN	The 50th percentile of the data must be <0.7 mg L^{-1}	
	PO_4-P	The 50th percentile of the date must be <0.020 mg L^{-1}	
Response variables	Chl-a phytoplankton	The 50th percentile of the data must be <20 mg L^{-1}	
	Chl-a periphyton	The 50th percentile of the data must be <21 mg m^{-2}	
Toxics	Ammonia	The 95th percentile of the data must be ≤0.073 mg L^{-1}	
	Fluoride	The 95th percentile of the data must be ≤1.5 mg L^{-1}	

DO, Dissolved oxygen; *EIS*, ecological importance and sensitivity; *TEC*, target ecological category; *WQ*, water quality.

Table 14.6: Reflections on the water quality (WQ) component of the proposed Ecological Reserve for ecological water requirement (EWR) 10 Suikerbosrand River as published in government gazette no. 42127 (DWS, 2018).

EWR 10: Suikerbosrand downstream: TEC: C/D; EIS: moderate

WQ class	Variables	Specifications	Reflections
Inorganic salts	$MgSO_4$	The 95th percentile of the data must be <37 mg L^{-1}	The Suikerbosrand River downstream at EWR 10 is regarded as being moderate in terms of EIS, and has thus been given a moderate TEC of a C/D. The system is sensitive to input of salts, nutrients, and toxics, although less so compared with the EWR 9. Its present ecological state of C/D implies that the regulator and potentially water users within the catchment would minimize future developments likely to change the status quo, if the water quality component of the Reserve is to be met. Nevertheless, the specifications at this EWR site are less stringent compared with those of EWR 9.
	Na_2SO_4	The 95th percentile of the data must be <51 mg L^{-1}	
	$MgCl_2$	The 95th percentile of the data must be <51 mg L^{-1}	
	$CaCl_2$	The 95th percentile of the data must be <105 mg L^{-1}	
	NaCl	The 95th percentile of the data must be <191 mg L^{-1}	
	$CaSO_4$	The 95th percentile of the data must be <351 mg L^{-1}	
Physical variables	EC	The 95th percentile of the data must be <85 mS m^{-1}	
	pH	The 5th percentile of the data must be 6.5–8.0 and the 95th percentile 8.0–8.8	
	Temperature	Small deviation from the natural temperature range	
	DO	The 5th percentile of the data must be ≥7 mg L^{-1}	
Nutrients	TIN	The 50th percentile of the data must be <0.7 mg L^{-1}	
	PO_4-P	The 50th percentile of the data must be <0.125 mg L^{-1}	
Response variables	Chl-a phytoplankton	The 50th percentile of the data must be <30 μg L^{-1}	
	Chl-a periphyton	The 50th percentile of the data must be <21 mg m^{-2}	
Toxics	Ammonia	The 95th percentile of the data must be ≤0.100 mg L^{-1}	
	Fluoride	The 95th percentile of the data must be ≤1.5 mg L^{-1}	

DO, Dissolved oxygen; EC, electrical conductivity; EIS, ecological importance and sensitivity; TEC, target ecological category; TIN, total inorganic nitrogen.

Table 14.7: Reflections on the water quality component of the proposed ecological reserve for Vaal River at De Neys as published in government gazette no. 42127 (DWS, 2018).

EWR 4: Vaal River at De Neys TEC: B/C; EIS

Class	Variables	Specifications	Reflections
Inorganic salts	$MgSO_4$	The 95th percentile of the data must be ≤ 37 mg L^{-1}	The Vaal River system at De Neys is regarded as being highly ecologically important and sensitive with a TEC of B/C. Compared with the PES of a C, a slight improvement is expected to achieve the TEC B/C. Given the sensitivity of the system, relatively stricter specifications are foreseen to achieve the determined water quality components of the Reserve. The implication is that while existing water users may not experience very strict imposition of water quality specifications in their water use licenses, measures are likely to be taken by the regulators in collaboration with catchment stakeholders to improve the water quality conditions of the system in order to achieve the proposed TEC, and to achieve the determined Reserved.
	Na_2SO_4	The 95th percentile of the data must be ≤ 33 mg L^{-1}	
	$MgCl_2$	The 95th percentile of the data must be ≤ 30 mg L^{-1}	
	$CaCl_2$	The 95th percentile of the data must be ≤ 57 mg L^{-1}	
	NaCl	The 95th percentile of the data must be ≤ 191 mg L^{-1}	
	$CaSO_4$	The 95th percentile of the data must be ≤ 351 mg L^{-1}	
Physical variables	EC	The 95th percentile of the data must be ≤ 30 mS m^{-1}	
	pH	The 5th percentile of the data must be 6.5–8.0 and the 95th percentile 8.0–8.8	
Nutrients	DO	The 5th percentile of the data must be ≥ 7 mg L^{-1}	
	TIN	The 50th percentile of the data must be <0.70 mg L^{-1}	
	PO_4-P	The 50th percentile of the data must be <0.125 mg L^{-1}	
Response variables	Chl-a phytoplankton	The 50th percentile of the data must be <10 µg L^{-1}	
	Chl-a periphyton	The 50th percentile of the data must be ≤ 1.7 mg m^{-2}	
Toxics	Ammonia	The 95th percentile of the data must be ≤ 0.1 mg L^{-1}	
	Fluoride	The 95th percentile of the data must be ≤ 1.5 mg L^{-1}	

DO, Dissolved oxygen; EIS, ecological importance and sensitivity; EWR, ecological water requirement; TEC, target ecological category; TIN, total inorganic nitrogen.

Table 14.8: Reflections on the water quality component of the proposed ecological Reserve for Vaal River at Scandinavia as published in government gazette no. 42127 (DWS, 2018).

EWR 5: Vaal River at Scandinavia, TEC: C; EIS			
Class	Variables	Specifications	Reflections
Inorganic salts	$MgSO_4$	The 95th percentile of the data must be ≤ 37 mg L^{-1}	The Vaal River system at EWR 5 is regarded as being highly ecologically important and sensitive, and is thus accorded a high EIS designation with a C TEC. The system is sensitive to further inputs of inorganic salts, nutrients, toxic substances, and activities that would impact on system variables such as dissolved oxygen and electrical conductivity. However, given that its present ecological state (PES) of C/D, the Reserve has been determined for C REC, implying that very serious measures restricting current water use are not envisaged but water users are likely to be imposed with reasonable license requirements to be able to afford the slight improvement from a C/D to a C. These conditions are envisaged to be more stringent, if for example, the system was to be improved from C/D to a B/C. Nevertheless, license conditions and/or other SDC instruments would still need to be imposed to achieve the Reserve as determined.
	Na_2SO_4	The 95th percentile of the data must be ≤ 51 mg L^{-1}	
	$MgCl_2$	The 95th percentile of the data must be ≤ 36 mg L^{-1}	
	$CaCl_2$	The 95th percentile of the data must be ≤ 105 mg L^{-1}	
	NaCl	The 95th percentile of the data must be ≤ 191 mg L^{-1}	
	$CaSO_4$	The 95th percentile of the data must be < 351 mg L^{-1}	
Physical variables	EC	The 95th percentile of the data must be < 85 mS m^{-1}	
	pH	The 5th percentile of the data must be 6.5–8.0 and the 95th percentile 8.8–9.2	
	Temperature	Temperatures should be close to natural range	
	DO	The 5th percentile of the data must be ≥ 6.0 mg L^{-1}	
Nutrients	TIN	The 50th percentile of the data must be ≤ 1 mg L^{-1}	
	PO_4-P	The 50th percentile of the data must be ≤ 0.025 mg L^{-1}	
Response variables	Chl-a phytoplankton	The 50th percentile of the data must be ≤ 20 μg L^{-1}	
	Chl-a periphyton	The 50th percentile of the data must be ≤ 21 mg m^{-2}	
Toxics	Ammonia	The 95th percentile of the data must be ≤ 0.1 mg L^{-1}	
	Fluoride	The 95th percentile of the data must be ≤ 1.5 mg L^{-1}	
	Atrazine	The 95th percentile of the data must be ≤ 100 μg L^{-1}	
Inorganic ions	Sulfate	The 95th percentile of the data must be ≤ 200 mg L^{-1}	

DO, Dissolved oxygen; EIS, ecological importance and sensitivity; EWR, ecological water requirement; SDCs, source directed controls; TEC, target ecological category; TIN, total inorganic nitrogen.

reflections are provided on the water quality components of the proposed Reserve for part of the Vaal catchment system, shedding light on how the ER provides a basis for ensuring sustainable utilization of water resources in South Africa.

14.5 Source directed controls

To control and minimize the impact of developments on water resources, Chapter 4 of the NWA provides for SDCs (DWAF, 2003). The SDCs are measures that ensure that activities likely to impact on water resources are controlled and minimized. Thus, the SDCs are the instruments used to achieve the objective of resource-based protection captured in the RDMs. A variety of instruments fall under SDCs, including GA, waste discharge charge system, compulsory licensing, special permit, economic incentives, self-regulation, environmental impact assessments, local municipal bye-laws and the precautionary principles (Scherman and Palmer, 2013). The NWA under Section 21 defines water use broadly including taking water from a resource, storing water, impeding, or diverting water flow, stream flow reduction activities, discharging waste/effluent directly into a water resource, recreation, altering water course etc. Nevertheless, the NWA also provides for permissible water use for which no licensing is required and such uses are captured in Section 22, regarded as Schedule 1 uses. Other permissible uses are those that are a continuation of existing lawful use or use that is authorized under a GA, provided the use complies with the specification in the GA. It is important to note that apart from Schedule 1 use, the regulator is moving toward compulsory licensing, and moving away from GA. Table 14.9 provides a summary of SDC-based tools currently being used to regulate and control impacts of developments on water resources. Some of these tools are outside the function of the national government. The local municipalities play significant roles in regulating developments on the environment through municipal bye-laws.

14.6 Linking resource directed measure and source directed control instruments for sustainable freshwater resources management

The link between RDM and SDC instruments is better conceptualized as an adaptive management cycle. Beginning with catchment visioning, which considers PES, desired future state, including the Reserve, a MC is determined for the water resource. The MC reflects the social, economic, and ecological contexts of the catchment. Specific goals in the form of RQOs are set to monitor trends toward or away from the desired goals while at the same time determining the available allocatable water — both in terms of quality and quantity. In heavily stressed river systems, allocatable water may not be available in which case new license application for water use may be declined (Fig. 14.5). In cases where allocatable water is available, the WUL application is assessed considering the RQOs as well as the legal provision in Section 27 of the NWA. WULs are then issued with

Table 14.9: Source directed control tools currently being used to regulate, control, and minimize impact of developments on water resources in South Africa.

Tools	Brief description
GA	GA is an instrument recognized in the Act that facilitates the use of water by a section of society/large group of people for identified uses without the need for specific individual licenses. Users of water under GA must register their use and adhere strictly with the provision of the GA, which is revised from time to time and published under government gazette.
Existing lawful use	This refers to water being lawfully use under the previous Act prior to the enactment of the NWA.
Water use license	Licenses are instruments use to control the use of water that falls outside the GA and Schedule 1 uses as well as those within the confine of lawful use. The Act makes provision for a range of condition that can be written into and specify in a license, depending on the type, nature and extent of water use and potential impact, risk and severity. Users issued a license are required to adhere strictly to the conditions of the license. They are obligated to use water only for the specified intended use(s) in their licenses. Compulsory licenses are important in catchments considered stressed and/or with competing users. Usually, water users under a license are required to monitor their uses, and the effect of their uses as measure toward compliance monitoring and assessment.
Water quality standards	These are water quality standards written into water use licenses that forms part of the license conditions. These standards are legally binding and enforceable by the regulators.
Water use charges	Waste discharge charge system and water resource management charge are the two instruments in this category. The objective is to enforce the polluter pay principle as well as the user pay principle in the case of water resource management charges. In the case of waste discharge charge system, the polluter pays according to the waste load released and potential impacts of the waste on the receiving environment.
Self-regulatory instruments, e.g., ISOs	Water users are encouraged to invoke the principle of self-regulation by meeting identified criteria. ISO 14001 certification is a widely used instrument in this category.
Incentive programs	Users of water resources can be encouraged through various incentives to minimize their impact on water resources. In South Africa for example, the Green Drop Certification program (DWA, 2011b) is an example of incentive-driven system targeting mainly the wastewater treatment sector. The program encourages excellence by providing incentives that promote good standards and behavior in the sector while improving effluent quality.
EIA	EIA is globally recognized a practice that is intended to better understand impact of developments on the environment, and to devise means to mitigate such impact, while promoting responsible social-economic developments. In South Africa for example, EIA is provided for under NEMA (Act No 107 of 1998). In various regulations flowing from the Act, certain developments trigger EIA process, with the objective of seeking ways to minimize the impact of such developments on the environment, including water resources.

EIA, Environmental Impact Assessment; *GA*, general authorization; *NEMA*, National Environmental Management Act; *NWA*, National Water Act.

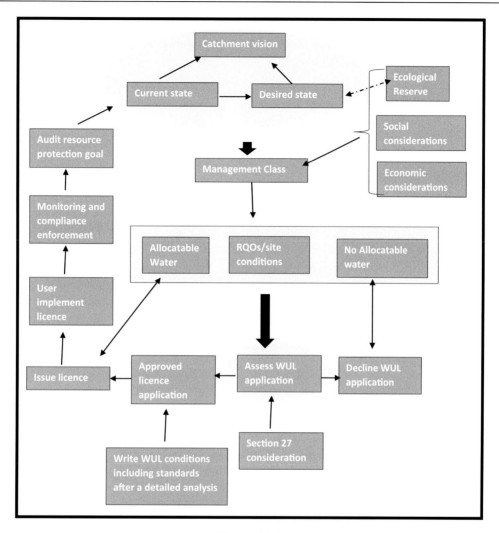

Figure 14.5
Adaptive management cycle showing the link between resource directed measures and source directed control (e.g., water use licenses) instrument. Source: *Adapted from DWAF (Department of Water Affairs and Forestry). (2006a). Guidelines for setting license conditions for resource directed management of water quality (1st ed.). Pretoria: Department of Water Affairs and Forestry.*

conditions often including water quality and quantity standards that are align to the objective of resource protection. It is the responsibility of the water resource users to implement and monitor the conditions specified in their licenses, while the regulator ensures that monitoring, compliance, and enforcement do happen. As the goal is to balance the use and protection of freshwater resources, the regulator is mandated to continuously audit the resource protection goals, whether they are being met through the instrumentality of water use licensing. This audit then informs another process of determining the current ecological state, desired future state, and catchment visioning, completing the adaptive management cycle as indicated in Fig. 14.5.

14.7 Conclusions

In the face of escalating urbanization, a growing human population and increase demand for freshwater resources, it is increasingly becoming critical to balance the use and protection of scarce water resources. This chapter presents a review and reflection on the strategies adopted to realize sustainable utilization of freshwater resources in South Africa. The RDMs are targeted toward resource protection while those of the SDCs are targeting toward controlling the activities of water resource users in ways that minimize impact on the resources. Although implementation of these measures remains a challenge in South Africa, they are sufficiently robust to ensure sustainable freshwater resources utilization. The challenges to the implementation of these measures are mainly related to governance failure across scale, a review of which is beyond the scope of this chapter. Nevertheless, it is believed that lessons can be derived from the South African strategies reviewed in this chapter, which can find applications in other jurisdictions.

Acknowledgments

The Water Research Commission of South Africa is acknowledged for funding the project (Project No. K5/2910/1&2) upon which this chapter is based.

References

Bunn, S. E. (2016). Grand challenge for the future of freshwater ecosystems. *Frontiers in Environmental Science*, 4(21).
Bunn, S. E., & Arthington, A. H. (2000). Basic principles and ecological consequences of altered flow regimes for aquatic biodiversity. *Environmental Management*, 30(4), 492–507.
DWA (Department of Water Affairs). (2011a). *Procedures to develop and implement resource quality objectives*. Pretoria: Department of Water Affairs.
DWA (Department of Water Affairs). (2011b). *Green drop handbook. Version 1*. Pretoria: Department of Water Affairs.
DWA (Department of Water Affairs). (2013). *National water resource strategy* (2nd ed.). Pretoria: Department of Water Affairs.
DWAF (Department of Water Affairs and Forestry). (2003). *Source management in South Africa* (1st ed.). Pretoria: Department of Water Affairs and Forestry.
DWAF (Department of Water Affairs and Forestry). (2006a). *Guidelines for setting license conditions for resource directed management of water quality* (1st ed.). Pretoria: Department of Water Affairs and Forestry.
DWAF (Department of Water Affairs and Forestry). (2006b). *Guideline for determining resource water quality objectives (RWQOs), allocatable water quality and the stress of the water resource*. Pretoria: Department of Water Affairs and Forestry.

DWAF (Department of Water Affairs and Forestry). (2008a). *Draft regulations for the establishment of a water resource classification system*. Government gazette no. 31417. Pretoria.

DWAF (Department of Water Affairs and Forestry). (2008b). *Methods for determining the water quality component of the ecological reserve for rivers*. Pretoria: Department of Water Affairs and Forestry. Second Draft.

DWAF (Department of Water Affairs and Forestry). (2009). *Integrated water quality management plan for the Vaal River system: Task 8: Water Quality Management Strategy for the Vaal River system*. Report no. P RSA C000/00/2305/7. Pretoria: Directorate National Water Resource Planning, Department of Water Affairs and Forestry.

DWS (Department of Water and Sanitation). (2016). *Classes and resource quality objectives of water resources for catchments of the Upper Vaal*. Government gazette no 39943, 22 April 2016. Pretoria.

DWS (Department of Water and Sanitation). (2018). *Proposed reserve determination of water resources for the Vaal catchment*. Government gazette no. 42127. Pretoria.

King, J., & Pienaar, H. (2011). *Sustainable use of South Africa's inland waters*. WRC report no. TT 491/11. Pretoria: Water Research Commission.

Kleynhans, C. J., & Louw, M. D. (2008). *Module A: EcoClassification and EcoStatus determination in River EcoClassification: Manual for EcoStatus Determination (version 2)*. Joint Water Research Commission and Department of Water Affairs and Forestry report. WRC report no. TT 329/08.

Leach, M., Reyers, B., Bai, X., Brondizio, E. S., Cook, C., Diaz, S., Espindola, G., Scobie, M., Stafford-Smith, M., & Subramanian, S. M. (2018). Equity and sustainability in the Anthropocene: A social-ecological systems perspective on their intertwined futures. *Global Sustainability*, *1*(e13), 1–13.

Loft, L., Le, D. N., Pham, T. T., Yang, A. L., Tjajadi, J. S., & Wong, G. Y. (2017). Whose equity matters? National to local equity perceptions in Vietnam's payments for forest ecosystem services scheme. *Ecological Economicsi*, *135*, 164–175.

McDermott, M., Mahanty, S., & Schreckenberg, K. (2013). Examining equity: A multidimensional framework for assessing equity in payments for ecosystem services. *Environmental Science and Policy*, *33*, 416–427.

Reid, A. J., Carlson, A. K., Creed, I. F., Eliason, E. J., Gell, P. A., Johnson, P. T. J., Kidd, K. A., MacCormack, T. J., Olden, J. D., Ormerod, S. J., Smol, J. P., Taylor, W. W., Tockner, K., Vermaire, J. C., Dudgeon, D., & Cooke, S. J. (2019). Emerging threats and persistent conversation challenges for freshwater biodiversity. *Biological Reviews*, *94*, 849–873.

Rockstrom, J., Steffen, W., Noone, K., Persson, A., Chapin, F. S., Lambin, E., Lenton, T. M., Scheffer, M., Folke, C., Schellnhuber, H. J., Nykvist, B., de Wit, C. A., Hughes, T., van der Leeu, S., Rhode, H., Sorlin, S., Snder, P. K., Constanza, R., Svedin, U., . . . Foley, J. (2009). Planetary boundaries: Exploring the safe operating space for humanity. *Ecology and Society*, *14*(2), 32.

RSA (Republic of South Africa). (1998). *National Water Act (Act No 36 of 1998)*. Government gazette no. 19182. Cape Town.

Scherman, P. A., & Palmer, C. G. (2013). *Critical analysis of environmental water quality in South Africa: Historical and current trends*. WRC deliverable, project no. K5/2184.

Tempelhof, J. W. N. (2009). Civil society and sanitation hydropolitics: A case study of South Africa's Vaal River Barrage. *Physics and Chemistry of the Earth*, *34*, 164–175.

Turton, A., Schultz, C., Buckle, H., Kgomongoe, M., Malungani, T., & Drackner, M. (2006). Gold, scorched earth and water: The hydropolitics of Johannesburg. *International Journal of Water Resources Development*, *22*(2), 313–335.

CHAPTER 15

Sustainable water management with a focus on climate change

Thomas Shahady

University of Lynchburg, Lynchburg, VA, United States

Chapter Outline

15.1 Introduction and background 294
    15.1.1 What is sustainable water management 294
    15.1.2 The central premise of climate change impacting water resources 294

15.2 Impacts from climate change 295
    15.2.1 Changes to the water cycle 295
    15.2.2 Changes in water storage 296
    15.2.3 Changes in precipitation patterns 297
    15.2.4 Changes to evapotranspiration rates 298
    15.2.5 Changes in hydraulic retention time 298
    15.2.6 Changes in pollutant loading and processing 298

15.3 Managing worsening concerns 299
    15.3.1 Eutrophication 300
    15.3.2 Dead zones 300
    15.3.3 Red tides 300
    15.3.4 Flooding 301
    15.3.5 Drought 302
    15.3.6 Pollutants of emerging concern 302

15.4 Sustainable water management: a need for change 304
    15.4.1 Challenges to effective management 304
    15.4.2 Stream restoration 305
    15.4.3 Green infrastructure 306
    15.4.4 Dam removal and wetland creation 307
    15.4.5 Stormwater management 308
    15.4.6 Sanitation 308
    15.4.7 Sociopolitics and economics 309
    15.4.8 Mitigation, protection, and ecological services 310

15.5 A sustainable water future 310

References 311

15.1 Introduction and background

15.1.1 What is sustainable water management

In its most basic sense, sustainability is the use of resources in a manner that is mindful of future generations. Concerning sustainable water management, a more robust definition is needed that includes supply, use and sanitation. In this sense, the system must produce high quality water reliability and show resilience from extensive use and a variety of stressors. Water management systems should perform "as built" providing those whom depend upon it a sense of reliability and function. The performance of the system needs to be adequate under extreme or unexpected conditions—show durability. And to be truly sustainable, persists through intergenerational time periods while maintaining basic integrity.

Considering the importance of water, there are a surprising number of limitations and knowledge gaps about key aspects of this resource such as storage, flux, water quality, and water infrastructure (Garrick et al., 2017). To build sustainable systems without an understanding of these critical components is difficult. Yet, water sustainability is possible when the idea is to build a system not to overcome failure but to manage it (Butler et al., 2017). Within an ever-changing climate and demand for use, future water management needs this approach. While our current system appears inadequate to sustainably manage water, it is not practical to completely rebuild the current system to achieve that goal. We need to manage the problem, adequately define success and properly manage the challenges ahead.

15.1.2 The central premise of climate change impacting water resources

Climate change resulting from elevated atmospheric CO_2 (eCO_2), will cause shifts in temperature and precipitation. Thus this will inevitably change the basic hydrological cycle. The key variables for consideration will be how precipitation is intercepted, surface water collection and groundwater recharge. Key storage components in streams, lakes, reservoirs and aquifers will demand smart management and adequate use rather than indiscriminate allocation simply based on need. Climate change models are effective in guiding our understanding of temperature change but inadequate in hydrology predictions due to land-atmosphere interactions with atmospheric water vapor content (Turral et al., 2014). So, while the planet is warming and predictably so, the water environment response and our ability to manage it is less certain.

More fundamental are principles imbedded in civil engineering and agriculture. Our management approach continually assumes patterns of precipitation represented by well-developed probability curves will continue into the future. But evidence suggests, from groundwater infiltration to surface runoff that historical patterns no longer adequately represent the cycle. How climate change impacts each of these areas, our anthropogenic response to these impacts and our collective management of water use will plot our sustainable future path.

Finally, former predictive principles must be abandoned. And for many, this is difficult. For example, retention and detention basins, the standard of stormwater management, are designed based upon sets of standard precipitation curves (Schmadel et al., 2019). These curves derived from historical precipitation and storm patterns are becoming obsolete. The assumption that precipitation follows some form of predictability or can be engineered deterministically is no longer a valid approach (Milly et al., 2008; Döll et al., 2015). Previous designs were adequate because precipitation was predictable within errors coefficients small enough to allow good design. As these coefficients continually expand, the resilience of our built systems will be tested again and again. Future water managers instead of managing water resources based a set of expectations will need to consider a broad range of possible future scenarios translated into risk and developed as an expectation or reasonable probability. We must expand from our known and expected into the probable and then minimize the reasonable risk.

As climate change unfolds through the upcoming years, decades and centuries, we will be well versed to learn from history. Application of crisis management theory is needed. A crisis is never planned for but only reacted to and managed. The problem needs worked to generate solutions and not hyperbole generating a crisis. Each obstacle should be evaluated and solved within the bounds of risk. The epic proclamation that "Failure is not an Option" from those on the Apollo 13 mission should guide our effort to find adequate solutions. While impact from the current climate change crisis is minimal, worsening changes are expected and will demand our very best leadership.

15.2 Impacts from climate change

15.2.1 Changes to the water cycle

Stream degradation is now better understood to be driven by two key parameters—hydrologic flow and morphological change (Gregory, 2019; Anim & Banahene, 2021). Evidence suggests it is a hydrological problem rather than driven by chemical or biological changes. These response parameters certainly respond, but it is usually due to the changing hydrological regime. The ecosystems surrounding these water resources have been altered in the past and continue to change under a nonrelenting pattern of development. Erosion, sedimentation, various pollutants, channelization, and flooding continually plague our waterways worldwide (Maklin & Lewin, 2019; Fletcher et al., 2014).

The now understood changes to our water cycle from a somewhat predictable and deterministic system toward that of an unpredictable and stochastic system will be the legacy of climate change. All expectations suggest this cycle will accelerate as increasing temperatures drive evaporation and accelerate water transfer from the sea to the land (Turral et al., 2014). In this scenario, dry areas of the world become hotter and dryer while areas with high rates of precipitation become significantly wetter. We know variability around these predictions with increase resulting in extreme drought and floods and unpredictable runoff rates. Each compartment of this cycle, and in particular those human populations depend upon for survival, will be altered.

15.2.2 Changes in water storage

Water storage is likely to progress through some profound changes as we look at how we manage water and changes from climate stress. Storage generally occurs in aquifers replenished by groundwater recharge and surface water storage. Multiple trends suggest these traditional storage systems may become inadequate to meet increasing water demand.

Agricultural water needs are expected to increase globally and water used to fuel this demand will need to be augmented from groundwater aquifers (Cotterman et al., 2018). In the most important aquifers in the world such as the Ogalala, California Central Valley, North China Plain, Northwest India and Pakistan and the Tigris-Euphrates, dwindling supply and overuse will render these supplies unsustainable (Scanlon et al., 2012; Famiglietti et al., 2011; Cao et al., 2013; Rodell et al., 2009; Voss et al., 2013). To adjust, crop type and irrigation practice will need to change to slow the rate of depletion (Haacker et al., 2019). Yet these changes may exacerbate aquifer recharge as grasses and shrubs, a likely shift from crop production, will further lower the rate of infiltration (Riley et al., 2019). Loss of productive cropland and aquifer depletion could create significant food shortages in the future.

Another concern is storage capacity. Over 50% of river systems in the world are regulated by dams with a storage capacity of more than 75% of mean annual runoff (Biemans et al., 2011). These dams increase surface water storage sevenfold mitigating flooding, supplementing irrigation and reducing sea level rise from climate change (Vörösmarty, 1997; Chao et al., 2008). Removal has become the desired means of dam management primarily driven for ecological uplift but with understandably indeterminant consequences in other aspects of watershed function (Shahady & Cleary, 2021). Many of these removals are fueled by concern about failing infrastructure's ability to properly perform during floods and future intense storms (Khan et al., 2015). These removals will deplete needed surface storage capacity and further stress already depleted aquifers. Loss of dam storage capacity, overuse of remaining surface water and a push toward increasing reliance on nonrenewable

Figure 15.1
Hydropower dam representing water storage and electricity generation both vulnerable to impacts of climate change.

groundwater will further challenge development toward a sustainable water future (Wada & Bierkens, 2014) (Fig. 15.1).

15.2.3 Changes in precipitation patterns

As climate change intensifies, extreme rainfall events are more and more like to occur exceeding current return periods by 17% through 2030 up to 50% by 2080 (Wang et al., 2014). These extreme events are very problematic as peak rainfall intensity rate extends the impact of the event itself by an order of magnitude (Dunkerley, 2008). When much of the rainfall in any rainstorm occurs in bursts, precipitation interacts with the land surface differently than when an event is uniform throughout. Rainfall bursts increase runoff intensity by 85%–570% with the greatest intensity occurring when bursts happen at the end of the event (Dunkerley, 2012).

The impact from these changes will be tremendous. It is estimated changing precipitation patterns will reduce groundwater recharge up to 70%, increasing flooding from 23%–29%, and contribute to a 25% loss of hydropower generation while increasing pathogen loading (Kundzewicz et al., 2008). It is difficult to imagine how these changes will be managed. Many streams are already stressed and concerns over dropping water levels of major aquifers is already concerning. Many parts of the world heavily depend on aquifers to supply drinking water, purify waste, and produce crops and livestock. Losses to key aquifers coupled with changing stream flow and water storage will move to the forefront of emerging problems as climate change accelerates.

15.2.4 Changes to evapotranspiration rates

In temperate humid climates, evapotranspiration (ET) returns more than 50% of precipitation to the atmosphere (Zhang et al., 2016). With that much moisture transferred in this manner, ET is clearly a significant component impacting stream flow and groundwater recharge. Two components must be considered here—temperature and plant composition. Both of these components will change as the planet warms.

Cover type, particularly changes between native vegetations and planted crops, will influence ET. Observations suggest this change can so profoundly impact stream flow they can change from perennial to ephemeral (Brown et al., 2013). However, clear patterns are hard to discern. Hamilton et al. (2018) suggests that groundwater is still the biggest contributor to annual stream flow while ET was consistently ~60% despite changes in plant composition. But, the balance between ET, groundwater and stream flow must be managed. Changes in plant growth rates, flowering, land distributions and adaptation will contribute to a changing water cycle and must be part of water management (Parmesan & Hanley, 2015).

15.2.5 Changes in hydraulic retention time

Implicit in the water cycle is the time water takes to flow from land interception, through each of the various compartments and then eventual return to the oceans. Research suggests hydraulic retention time is a key parameter that controls the structure, water quality, processing of exogenous inputs, and organization of aquatic ecosystems (Francisco et al., 2006). This includes residence times of groundwater flow, the dynamics of water recharge, and differential flow of water between wet and dry precipitation patterns (Rodriguez et al., 2018). As climate change proceeds, and residence times are impacted, we will see concurrent changes to aquatic ecosystems and the biogeochemical processing within each compartment. Our engineered systems need to be retrofitted to meet this change. While this is a simple concept, it fundamentally challenges the assumptions used throughout the civil engineering profession. This may be one of the most difficult problems to solve.

15.2.6 Changes in pollutant loading and processing

Water quality parameters such as temperature, dissolved organic matter (DOM), nutrients; and pollutants such as plastics, sediment, pathogens, and toxins will be influenced by climate change. The greater the precipitation event, the more severe the impact. These events will increase pollution loading through direct transport, will overwhelm water infrastructure, cause malfunction of critical treatment systems and destroy property (Kistemann et al., 2002). These storms will increase land and stream bank erosion unable to process excessive amounts precipitation. These events will sharply diminish chemical and microbial water quality in receiving waters by increasing the abundance of protists such as *Cryptosporidium* and *Giardia* and various microbiological pathogens associated with *Escherichia coli* indicators (Khan et al., 2015).

Further problems will be associated with drought. Low river flows and higher temperatures will bring about low concentrations of dissolved oxygen, greater availability of nutrients, higher pH, concentrations of metals and other pollutants and increased risk for toxic algae blooms (Delpla et al., 2009). Extraction of water for irrigation, drinking and other human use from these declining sources will introduce additional levels of risk. These problems will prove to be very challenging as poor water quality adversely impacts both livestock and crop production as these areas impacted by drought from climate change water will be in high demand.

The interplay between wet and dry years driven by climate change may further complicate these predictions. Deviations from mean annual precipitation in wet climates differed between 30%–40% from current predictions while moisture in arid climates changed between 60%–150% (Knapp et al., 2015). Dry climates will see much greater swings in moisture regime, while wet climates will see increased probability of extreme rain events and increased days between moisture events. These changes along with warming trends will create further management challenges.

15.3 Managing worsening concerns

Many water quality problems currently exist in the absence of any exacerbation from climate change. Europe has put forth directives to control water pollutants among all of the member states (CEC, 1991). The United States Environmental Protection Agency (EPA) uses various forms of the National Pollutant Discharge Elimination Permit (NPDES) program to control water pollutants entering receiving streams (USEPA, 2020). In other parts of the world, management of pollutants are varied. In China, worsening water quality has forced the central government to generate national pollution discharge standards to control pollution (Li et al., 2012). In other parts of the world, pollution control is less centrally regulated. While Latin America has concentrated central governments, water protection may be spread throughout differing agencies (Shahady & Boniface, 2018). Significant challenges current exist for sustainable water management and need addressed now before the impacts from climate change generate even greater challenges.

15.3.1 Eutrophication

Eutrophication due to agricultural expansion, growing populations fossil fuel consumption and rising temperatures is already a global problem. Associated with eutrophication is the concurrent increase in CH_4 emissions from lakes and reservoirs. These methane emissions, themselves a potent greenhouse gas, could impact climate change to such an extent as to negate reductions achieved from fossil fuel reductions (Beaulieau et al., 2019). Even more alarming, if eutrophication continues to increase, methane production from lakes, wetlands and reservoirs will counterbalance all marine and terrestrial carbon burial in the global carbon budget.

Examining the problem economically, costs from eutrophication will far exceed costs from other aspects of climate change (Downing et al., 2021). This problem rooted in water resource management is not garnering the attention it needs. We need to elevate increasing eutrophication into the climate change narrative to realize gains in other sectors such as fossil fuel reduction and CO_2 management, eutrophication transcends sustainable water management impacting the phenomenon of climate change itself.

15.3.2 Dead zones

The proliferation of dead zones along many of the world's largest and most sensitive marine environments has arisen rapidly. A dead zone describes the development of hypoxic (low oxygen) and anoxic (no oxygen) areas in water. This problem is an ancillary impact from eutrophication generated through human activities. Ecosystems severely affected by this phenomenon may be at the threshold of collapse where fisheries, food webs, and biodiversity will be irreparably harmed (Diaz, 2001).

The coverage of dead zones worldwide continues to expand (Diaz & Rosenberg, 2008). Continued transport of dissolved organic carbon and nutrients will fuel this expansion. Most concerning is the exacerbation of this problem with climate change. Climate change is likely the original cause of these dead zones, so known changes such as rising temperatures, ocean acidification, sea level rise, precipitation, wind, and storm patterns will only exacerbate it (Altieri & Gedan, 2015). Climate change may make our coastal areas uninhabitable and very difficult to manage.

15.3.3 Red tides

Harmful algal blooms (HABs) have developed into a significant water quality problem worldwide. This problem is generated by various species of algae and phytoplankton (most prominently dinoflagellates) and is known to plague freshwater and coastal areas of all sizes worldwide (Hallegraeff et al., 2021). Human health toxins, fish and shellfish poisonings, livestock impact, and coral bleaching are all problems associated with this phenomenon. As with the other eutrophication problems, nutrient transport likely exacerbated by climate change will likely increase the problem.

15.3.4 Flooding

Flooding is now considered one of the most pressing water sustainability problems because of the damage and economic costs flooding causes. The idea that precipitation will become more intense as climate change progresses has been studied and is well documented (Groisman et al., 2005; Trenberth, 2011). Many studies have suggested flooding will become a major problem (Semadeni-Davies et al., 2008) and in the United Kingdom has required planners to include a 20% increase in all flood designs moving forward (Prudhomme et al., 2013). The role of impervious surfaces is critical as we examine issues related to climate change. There is debate between the importance of impervious surface

Figure 15.2
Destruction in the wake of Tropical Storm Nate that hit Costa Rica with flooding and damage unprecedented in its history. The storm caused 11 deaths, major landsides, loss of property, drinking water and electricity.

cover and the importance of connectivity on how our urban centers will respond to changing precipitation patterns (Zahmatkesh et al., 2015).

Moving forward, it is likely flood risk will increase by 30-fold with climate change (Yao et al., 2016). What is unknown and hardly predicable is how our constructed infrastructure and overall built environment will respond. In Mexico City, a majority of disasters (10 × greater) have occurred from flooding when compared to other meteorological events such as drought, wildfires, wind storms or hail (Lankao, 2010). Certainly, how urban areas are interconnected in stormwater infrastructure will impact how water runoff moves through these systems and flood risk. The costs, in terms of physical damage, toll on human lives and loss of ecological integrity will be tremendous. While changing the inevitable increase in flooding is not possible, the adaptation to this increasing trend is necessary.

15.3.5 Drought

While some areas of the globe will experience increased precipitation, others will begin to dry. Climate change will likely increase the severity, duration, and variability of dry weather in already arid areas and likely create new areas of water stress (Dai, 2013; Cook et al., 2015). The severity of drought will be a function of characteristics of the catchment as a function of changing rates of precipitation (Van Loon & Laaha, 2015). Moving forward, it is expected that many types of arid ecosystems are expand (Zarch et al., 2017).

The risk of drought and movement toward arid conditions is likely to increase. In California, drought conditions coupled with warming is now an expectation while during the past century it was more of an anomaly (Diffenbaugh et al., 2015). This will cause many changes in the management of both land and water. Needed release from surface storage will be intensely desired and debate on wise use of these resources will intensify. Irrigation from groundwater will naturally increase and concerns over aquifer depletion will ensue. Facing the loss of agriculture, which is an important component of these economies, water will be exploited until it reaches levels of unsustainability.

The combined impact of flood and drought risk will create unprecedented need for good water management. The historical development of any region from a patchwork of infrastructure will be tested in the face of unprecedented challenges ahead. As new patterns of water flow evolve, so will the necessary management.

15.3.6 Pollutants of emerging concern

Plastic pollution has become a global environmental crisis. Plastics are known to threaten aquatic life in both the streams where this pollutant originates and in ocean environments where streams flow terminates (van Emmerik & Schwarz, 2020). Plastics can cause

Figure 15.3
Plastics carried by water flow through an ephemeral channel.

entanglement of wildlife and as these plastics break down, microplastics are created that are ingested. In addition to wildlife impacts, plastics have created negative consequences to tourism, fishing, shorelines and vessels. It is safe to say, plastics represent a severe threat to freshwater, estuarine and marine ecosystems worldwide.

Greater flooding will increase plastic transport from rivers into marine ecosystems. Worldwide, a tenfold increase in plastic mobilization is likely from just a low severity flood with a 10-year return period (Roebroek et al., 2021). As discussed, increasing storm intensity will increase flooding and in response the magnitude of pollution from plastics. Production of plastics and life cycle in the use must be modified as part of a sustainable water future.

Another concern is hormones and human use products. These pollutants are emerging contaminants (ECs) that include pharmaceuticals and personal care products that often bind to sediment or plastic particles creating and additional pollutant concern (Wilkinson et al., 2017). These (ECs) will be discharged from storage sinks in rivers along with plastics and sediment directly during storm events. On the other hand, low flow and drought conditions have the potential of concentrating these pollutants when flow is diminished (Bunke et al., 2019).

Regardless, the fate of these ECs once introduced into the aquatic environment remains relatively unresolved.

Finally, a concern over the increase in bacteria, protozoa and viruses currently plagues water quality due to increase with climate change. This response will be dependent on river flow land use ability to filter contaminated water (Bussi et al., 2017). Riparian buffer, the preservation of vegetation along stream banks, can control this problem. Uncontrolled, it is expected that increased flooding will increase the disease burden (Sterk et al., 2013).

15.4 Sustainable water management: a need for change

15.4.1 Challenges to effective management

The identification and study of climate change threats currently supersedes our combined effort to repair and manage them. If we are going to manage water sustainably, we must push forward meaningful improvement programs. Significant research already exists to suggest watersheds are stressed by over development and reduction of key ecological services necessary to maintain the integrity of the system (Paul & Meyer, 2001; Chin, 2006). While improvement to these systems is underway, the greater problem is documenting such improvements to gain acceptance among stakeholders to continue to move the process forward. Funding continues to be allocated for watershed restoration without coordinated measurement of the outcomes.

Improvement metrics must be adequately designed to measure effective restoration followed by effective communication of these improvements (Karr, 2006). So, while change is occurring while watersheds are under repair or improvement, the verification cannot be theoretical. Under the guise of climate change, we need real solutions rather than benchtop exercises. Infrastructure must be adapted to the coming changes in precipitation patterns and to make the proper decisions good data must be followed. Good metrics must be developed and used to track successful management.

Secondly, research needs to get in sync with management (Harmsworth et al., 2011). The scientific community demands novelty in research projects and not basic monitoring or hotspot identifications. Management requires an understanding of what is and is not working, meeting societal needs whether resources expended are effective. Managers often rely on circular logic to proclaim success since resources were allocated or improvement structures installed the project must be successful. We need investment into continued monitoring, verification of watershed improvement and a system of reward for researchers involved. A system where research projects are funded and published based on novelty and university work environments that demand conformity, creative management will suffer. Finally, connectivity and political boundaries must be overcome. As watersheds are natural divides of landscapes, political boundaries are not. Effective management through one jurisdiction can be counteracted by another. Watershed boundaries may be the next frontier we must understand to effectively manage water resources sustainably.

15.4.2 Stream restoration

Stream restoration is currently a desired methodology to improve water resources and the ecological integrity of damaged stream ecosystems. The idea is to return these systems to some theoretical state that will improve water quality (from the current degraded state) and improve water health and sustainable management of water resources. In these designs, hydrology, sediment transport and watershed processes are reworked to develop a renewed channel design believed to be the best representation of what the stream should approximate (Niezgoda & Johnson, 2005). In these designs, streambanks are stabilized, the stream channel is reconnected to the floodplain, and riparian vegetation restored.

The results from these restorations have performed within expected results or expenditures. Certainly, in some instances channel restoration will contribute to watershed improvement downstream (Merriam & Petty, 2019). But, bioavailable phosphorus from stream banks is usually less than ¼ of all phosphorus and denitrification rates are proportional to nitrate concentrations. These reductions while touted as the direct benefit of restoration appear to be minimal. Restoration of riparian buffers and floodplains are much more efficient than stream bank repair for nutrient management. A combination of restoration strategies that reduce nutrient inputs to streams, re-establish riparian functions, provide balanced water and sediment regimes, and increase in-stream nutrient processing and retention will likely be most effective for improved water quality (Lammers & Bledsoe, 2017). Riparian protection, enhancement, and hyporheic reconnection appear to provide greater improvements over stream bank restoration.

Research suggests stream restoration alone may be insufficient for reducing nutrient concentrations to desired levels if no upland best management practices (BMP) are incorporated (Selvakumar et al., 2010). Riparian restoration had the highest success in increasing nutrient uptake rates and reducing fluxes (88%), followed by in-stream structure installation (63%), wetland creation (25%), and then channel reconstruction (14%) (Palmer et al., 2014). Riparian protection resulted in 3—4 times the nutrient reduction rates by comparison to stream bank restoration (Miller et al., 2014).

In Lynchburg Virginia, United States, stream restoration was instigated to improve water quality. In one section of stream, over 2 million dollars were spent to restore 2744 linear feet (Fig. 15.2). The stream was incised and impacted due to over 65% impervious watershed draining 5.1 hectares from a network of concrete pipe and drainage ditches (Fig. 15.3). While the stream was restored, the concerns with these projects from a sustainability perspective is the lack of problem solving. This stream will incise again due

Figure 15.4
Urbanized ephemeral channel draining a 65% impervious watershed.

to the impervious drainage network and increasing force of water that will flow into this project. Design specifications used were from NOAAs Hydrometeorological Data based on storm statistics without modification for climate change. While communities are spending time and resources on stream repair, concern over the infrastructure that has caused the problem and working with the problem is now our greatest need (Figs. 15.4 and 15.5).

15.4.3 Green infrastructure

Green infrastructure is the design of systems to intercept stormwater for storage and treatment before release from urban designs. These systems invoke various BMP and low impact development (LID) designs such as bioretention, grass swales, green roofs, wetlands, ponds and other infiltration devices. Research suggests that these designs improve water quality under low or minimal stormwater flow (Liu et al., 2015; Aulenbach et al., 2017). They are also improvements over traditional "gray infrastructure" that has minimal ability to improve water quality (Gaffin et al., 2012).

Figure 15.5
Restored ephemeral channel receiving stormwater flow from Fig. 15.4. While restoration has improved this channel, climate change is likely to return it to its previous impacted condition.

While LID is a good sustainability practice, it has limited effectiveness in mitigating impacts from climate change expected from extreme and increased precipitation events (Pour et al., 2020; Demuzere et al., 2014). But, this practice has promise to help mitigate some concerns. Water storage is increased and in drought-stricken areas will provide an additional source of water. These practices also provide cooling through insulation lower energy (Ramyar et al., 2021).

15.4.4 Dam removal and wetland creation

The multitude of dams throughout watersheds creates a unique management concern with impending climate change. With a continued desire for ecological uplift and concerns over aging infrastructure and overtopping from extreme precipitation events, removal has gained increased popularity (Shahady & Cleary, 2021). However, the reservoirs behind these dams have well-developed fringe wetlands, are excellent trapping mechanisms for nutrients, sediments and other pollutants and provide recreation and other community recreational opportunities (Selvakumar et al., 2010; Gold et al., 2016). Under climate change, the resilience of these dams should be evaluated and then determine whether removal is prudent.

Eutrophication management related to greenhouse gases and carbon sequestration must also be incorporated. The creation of wetlands to mitigate CH_4 slowing the rates of eutrophication is possible in these systems (Bartolucci et al., 2021). They act as filtration mitigating the downstream flow into estuaries, lakes and reservoirs. This approach works when water level in wetlands managed appropriately.

15.4.5 Stormwater management

Stormwater management encompasses the network of pipes, impervious surfaces, storage ponds or other devices to mitigate peak discharge. This involves a variety of BMP with a goal to infiltrate stormwater rather than direct surface discharge. Much of the current infrastructure was built to standards of a 10-year, 24-hour storm on probability curves developed in the mid-20th century. It is now clear that precipitation intensity and amounts are increasing, which suggests that infrastructure will be inadequate moving forward (Rosenberg et al., 2010; Horton et al., 2010). There is considerable variability in these predictions, yet it is reasonable to assume current stormwater infrastructure must be retrofit or management scenarios adjusted to understand this reality.

Efforts to deal with this problem are underway. Because this infrastructure is already in place, and costs to completely replace these systems prohibitive, an adaptive strategy may be the best approach. With this strategy, the resilience of the system is tested, adaptive tipping points (ATP) identified, and suggestions on improvements to the system quantified and recommended (Gersonius et al., 2012). This is a good approach to the problem but does have some concerns as policy makers often wait until the tipping points are reached before acting. The ability for a given stormwater system to assimilate increasing storm frequency and volume is related to the degree of build-out and closure of the system. Older and more developed areas with a high degree of impervious surface and connectivity will not be very adaptable to climate change (Moore et al., 2016). In any scenario, the addition of wetlands and riparian corridors may be the best way to mitigate impending problems.

15.4.6 Sanitation

Sanitation must be properly managed in a sustainable water future. Providing for basic sanitation needs, protection from the adverse health effects from untreated waste and elimination of open defecation are understood as proper sustainability goals (Mara & Evans, 2017). Adequately removing waste must occur to maintain water supply while protecting aquatic life and human health.

Concerns over an ability to properly manage sanitation with climate change are extensive (Dickin et al., 2020). Increases in the variability of the water cycle will lead to uncertain supply impacting treatment. Extreme storms and flooding will destroy sanitation infrastructure, flood existing systems and limit transportation and energy needed to maintain these systems. Drought conditions will increase pollutant load as discharge from waste water treatment may be the only flow in rivers. All aspects of climate change have the potential to increase the spread of diarrheal diseases. As conditions worsens, people in drought-stricken areas will begin to utilize recycled water that is full of the ECs and other pollutants (Rodriguez-Narvaez et al., 2017). Hardening sanitation infrastructure against climate change will be a challenge. This will involve improving construction quality of facilities to enhance disaster resilience; using dry toilets in water insecure areas; raising substructure height and/or using sealed substructure options to avoid groundwater contamination in rainy seasons or following flooding; raised community toilet facilities for use during floods; using composted fecal waste as fertilizer to combat increasing climatic strain on agriculture (Gordon & Hueso, 2021).

15.4.7 Sociopolitics and economics

Sociopolitical and economic concerns always drive the extent of sustainability success. Without political will and support along with appropriate funding to create a sustainable water infrastructure, water needs and availability will be problematic. Sustainability must become the first societal goal then followed by adaptation and resiliency to withstand the variability expected from climate change.

Many types of activities will move us toward a more sustainable future. Activities such as green infrastructure will create community among stakeholders opening avenues to discuss needed change (Olsson & Head, 2015). Good policy while understanding the natural sciences must incorporate the social sciences and economic policy to generate good governance of our water resources (Hong & Chang, 2020). Also, stakeholders must be engaged in the process and continually informed over success. People generally understand the need for sustainability solutions but become frustrated when money, time and effort are widely dispersed without understandable results. When success or management is clearly explained such ideas are accepted and supported (Balasubramanya & Stifel, 2020). Community acceptance is an integral part of any sustainability program and should not be neglected.

Planners need to study and inventory water infrastructure to determine a best course of action for community support. Divisions must be recognized and dealt with not ignored. In the agricultural sector, irrigation must be supported but then translated into a community to not alienate those who have other desires for the resource. As climate change intensifies, concerns with water use and poverty will command greater understanding and analysis if we are to manage water sustainably (Brody et al., 2013).

Connectivity will become an even more pressing problem. When examining the resiliency of infrastructure to flooding, high-intensity, clustered development performed much better than low-intensity, more sprawling development patterns (Zhu, 2012). Encouraging more clustered development will also generate coincidental benefits, such as clean air and water, critical habitat protection and protection of open space. Unfortunately, this division coincides with political divides along with poverty and overcrowding. The inability of urban areas to assimilate migrants successfully will lead to squalor and intensifying poverty along with greater social stratification (Zhu & Simarmata, 2015). This social stratification will further divide public versus private governance and availability and use of commons such as water (Junk et al., 2013).

15.4.8 Mitigation, protection, and ecological services

Programs to protect, enhance, and conserve wetlands, streams, lakes and reservoirs will improve water quality and sequester carbon to help mitigate climate change. It is clear that wetlands are significant carbon, nutrient and pollutant sinks (Yasarer & Sturm, 2016). Protection and enhancement of these systems will be needed to ensure water purification services as climate change progresses in the future. Streams, lakes and reservoirs will provide similar services moving forward. These systems will supply hydropower, flood risk reduction, recreation, nutrient attenuation, aquatic ecosystem support, and water supply for municipal, industrial, and agricultural uses (Tzilivakis et al., 2019). Programs that allow for the mitigation of land disturbance for preservation of these ecosystems will become more valuable in the future.

Another way to enhance sustainability is to pay for forest protection, greening of agricultural land or other approaches. Ecological Focus Areas (EFAs) allow farmers to adapt and manage farmland that meets requirements of European biodiversity enhancement (Chan et al., 2017). Individuals that meet the requirements of the program are paid based upon land and biodiversity enhancements. A similar program used in other parts of the world are payments for ecosystem services (PES). These are preservation programs generally funded by a tax on other goods (such as portion of a gasoline tax) that are transferred to individuals that put land into a preservation program (Garrick et al., 2017). While problems with this program occur, these may be some of the best opportunities to protect water quality through engagement socially and economically with natural resource protection.

15.5 A sustainable water future

Four steps toward sustainable development of water resources include: good measurement of the resource; valuation at all relevant scales including environmental, social, political and economic; good inclusive decision making; and good responsive governance (Sohn et al., 2019). We must recognize the future of water resource management must move from previous designs towards a model of ever-changing climate (Smith & Katz, 2013). A world where the hydrologic cycle is predictable based upon previous patterns cannot drive a sustainable water future.

We know there is an increasing in frequency of billion-dollar water related disasters (Bathurst et al., 2018). This is not sustainable water management and suggests decreasing success toward a sustainable future. While expected changes may be embedded in our climate future, our response is not. This change can start through an understanding of what systems need respected and where our enhancements are needed.

References

Altieri, A., & Gedan, K. (2015). Climate change and dead zones. *Global Change Biology*, *21*, 1395−1406.

Anim, D., & Banahene, P. (2021). Urbanization and stream ecosystems: The role of flow hydraulics towards an improved understanding in addressing urban stream degradation. *Environmental Reviews*, *29*, 401−414.

Aulenbach, B., Landers, M., Musser, J., & Painter, J. (2017). Effects of impervious area and BMP implementation and design on storm runoff and water quality in eight small watersheds. *The Journal of the American Water Resources Association*, *53*, 382−399.

Balasubramanya, S., & Stifel, D. (2020). Viewpoint: Water, agriculture & poverty in an era of climate change: Why do we know so little? *Food Policy*, *93*, 101905.

Bartolucci, N., Anderson, T., & Ballantine, K. (2021). Restoration of retired agricultural land to wetland mitigates greenhouse gas emissions. *Restoration Ecology*, *29*, e13314.

Bathurst, J., Birkinshaw, S., Johnson, H., Kenny, A., Napier, A., Raven, S., Robinson, J., & Stroud, R. (2018). Runoff, flood peaks and proportional response in a combined nested and paired forest plantation/peat grassland catchment. *Journal of Hydrology*, *564*, 916−927.

Beaulieau, J., DelSontro, T., & Downing, J. (2019). Eutrophication will increase methane emissions from lakes and impoundments during the 21st century. *Nature Communications*, *10*, 1375.

Biemans, H., Haddeland, I., Kabat, P., Ludwig, F., Hutjes, R. W., Heinke, J., von Bloh, W., & Gerten, D. (2011). Impact of reservoirs on river discharge and irrigation water supply during the 20th century. *Water Resources Research*, *47*, W03509.

Brody, S., Kim, H., & Gunn, J. (2013). Examining the impact of development patterns on flooding on the Gulf of Mexico coast. *Urban Studies (Edinburgh, Scotland)*, *50*, 789−806.

Brown, A., Westem, A., McMahon, T., & Zhang, L. (2013). Impact of forest cover changes on annual streamflow and flow duration curves. *Journal of Hydrology*, *483*, 39−50.

Bunke, D., Moritz, S., Brack, W., Herraez, D., Posthuma, L., & Nuss, M. (2019). Developments in society and implications for emerging pollutants in the aquatic environment. *Environmental Sciences Europe*, *31*.

Bussi, G., Whitehead, P., Thomas, A., Masante, D., Jones, L., Cosby, J., Emmett, B., Malham, S., Prudhomme, C., & Prosser, H. (2017). Climate and land-use change impact on faecal indicator bacteria in a temperate maritime catchment (the River Conwy, Wales). *Journal of Hydrology*, *553*, 248−26.

Butler, D., Ward, S., Sweetapple, C., Astaraie-Imani, M., Diao, K., Farmani, R., & Fu, G. (2017). Reliable, resilient and sustainable water management: The safe & sure approach: Reliable, resilient, and sustainable water management. *Global Challenges*, *1*, 63−77.

Cao, G., Zheng, C., Scanlon, B., Liu, J., & Li, W. (2013). Use of flow modeling to assess sustainability of groundwater resources in the North China Plain. *Water Resources Research*, *49*, 159−175.

CEC (Council of the European Community). Council Directive concerning urban wastewater treatment. (91/271/EC). (1991); Official Journal L135/40.

Chan, K., Anderson, E., Chapman, M., Jespersen, K., & Olmsted, P. (2017). Payments for ecosystem services: Rife with problems and potential—For transformation towards sustainability. *Ecological Economics*, *140*, 110–122.

Chao, B., Wu, Y., & Li, Y. (2008). Impact of artificial reservoir water impoundment on global sea level. *Science (New York, N.Y.)*, *320*, 212–214.

Chin, A. (2006). Urban transformation of river landscapes in a global context. *Geomorphology*, *79*, 460–487.

Cook, B., Ault, T., & Smerdon, J. (2015). Unprecedented 21st century drought risk in the American Southwest and Central Plains. *Science Advances*, *1*, e1400082.

Cotterman, K., Kendall, A., Basso, B., & Hyndman, D. (2018). Groundwater depletion and climate change: Future prospects of crop production in the Central High Plains Aquifer. *Climatic Change*, *146*, 187–200.

Dai, A. (2013). Increasing drought under global warming in observations and models. *Nature Climate Change*, *3*, 52–58.

Delpla, I., Jung, A., Baures, E., Clement, M., & Thomas, O. (2009). Impacts of climate change on surface water quality in relation to drinking water production. *Environment International*, *8*, 1225–1233.

Demuzere, M., Orru, K., Heidrich, O., Olazabal, E., Geneletti, D., Orru, H., Behave, A., Mittal, N., Feliu, E., & Faehnle, M. (2014). Mitigating and adapting to climate change: Multi-functional and multi-scale assessment of green urban infrastructure. *Journal of Environmental Management*, *146*, 107–115.

Diaz, R. (2001). Overview of hypoxia around the world. *Journal of Environmental Quality*, *30*, 275–281.

Diaz, R., & Rosenberg, R. (2008). Spreading dead zones and consequences for marine eco- systems. *Science (New York, N.Y.)*, *321*, 926–929.

Dickin, S., Bayoumi, M., Giné, R., Andersson, K., & Jimenez, A. (2020). Sustainable sanitation and gaps in global climate policy and financing. *NPJ Clean Water*, *3*, 24.

Diffenbaugh, N., Swain, D., & Touma, D. (2015). Anthropogenic warming has increased drought risk in California. *Proceedings of the National Academy of Sciences*, *112*, 3931–3936.

Downing, J., Polasky, S., Olmstead, S., & Newbold, S. (2021). Protecting local water quality has global benefits. *Nature Communications*, *12*, 2709.

Dunkerley, D. (2008). Rain event properties in nature and in rainfall simulation experiments: A comparative review with recommendations for increasingly systematic study and reporting. *Hydrological Processes*, *22*, 4415–4435.

Dunkerley, D. (2012). Effects of rainfall intensity fluctuations on infiltration and runoff: Rainfall simulation on dryland soils, Fowlers Gap, Australia. *Hydrological Processes*, *26*, 2211–2224.

Döll, P., Jiménez-Cisneros, B., Oki, T., Arnell, N., Benito, G., Cogley, J., Jiang, T., Kundzewicz, Z., Mwakalila, S., & Nishijima, A. (2015). Integrating risks of climate change into water management. *Hydrological Sciences Journal*, *60*, 4–13.

Famiglietti, J., Lo, M., Ho, S., Bethune, J., Anderson, K., Syed, T., Swenson, S., de Linage, C., & Rodell, M. (2011). Satellites measure recent rates of groundwater depletion in California's Central Valley. *Geophysical Research Letters*, *38*, L03403.

Fletcher, T., Vietz, G., & Walsh, C. (2014). Protection of stream ecosystems from urban stormwater runoff: The multiple benefits of an ecohydrological approach. *Progress in Physical Geography*, *38*, 543–555.

Francisco, R., Moreno-Ostos, E., & Armengol, J. (2006). The residence time of river water in reservoirs. *Ecological Modelling*, *191*, 260–274.

Gaffin, S., Rosenzweig, C., & Kong, A. (2012). Adapting to climate change through urban green infrastructure. *Nature Climate Change*, *2*, 704.

Garrick, D., Hall, J., Dobson, A., Damania, R., Quentin Grafton, R., Hope, R., Hepburn, C., Bark, R., Boltz, F., De Stefano, L., O'Donnell, E., Matthews, N., & Money, A. (2017). Valuing water for sustainable development. *Science (New York, N.Y.)*, *358*, 1003.

Gersonius, B., Nasruddin, F., Ashley, R., Jeuken, A., Pathirana, A., & Zevenbergen, C. (2012). Developing the evidence base for mainstreaming adaptation of stormwater systems to climate change. *Water Research*, *46*, 6824–6835.

Gold, A., Addy, K., Morrison, A., & Simpson, M. (2016). Will dam removal increase nitrogen flux to estuaries? *Water*, *8*, 522.

Gordon, T., & Hueso, A. (2021). Integrating sanitation and climate change adaptation: Lessons learned from case studies of WaterAid's work in four countries. *Waterlines*, *40*, 107−114.

Gregory, G. (2019). Human influence on the morphological adjustment of river channels: The evolution of pertinent concepts in river science. *River Research and Applications*, *8*, 1097−1106.

Groisman, P., Knight, R., Easterling, D., Karl, T., Hegerl, G., & Razuvaev, V. (2005). Trends in intense precipitation in the climate record. *Journal of Climate*, *18*, 1326−1350.

Haacker, E., Cotterman, K., Smidt, S., Kendall, A., & Hyndman, D. (2019). Effects of management areas, drought, and commodity prices on groundwater decline patterns across the High Plains Aquifer, Agric. *Water Management*, *218*, 259−273.

Hallegraeff, G., Enevoldsen, H., & Zingone, A. (2021). Global harmful algal bloom status reporting. *Harmful Algae*, *102*, 101992.

Hamilton, S., Robertson, G., Hussain, M., Lowrie, C., & Basso, B. (2018). Evapotranspiration is resilient in the face of land cover and climate change in a humid temperate catchment. *Hydrological Processes*, *32*, 655−663.

Harmsworth, G., Young, R., Walker, D., Clapcott, J., & James, T. (2011). Linkages between cultural and scientific indicators of river and stream health. *New Zealand Journal of Marine and Freshwater Research*, *45*, 423−436.

Hong, C., & Chang, H. (2020). Residents' perception of flood risk and urban stream restoration using multi-criteria decision analysis. *River Research and Applications*, *36*, 2078−2088.

Horton, R., Rosenzweig, C., Gornitz, V., Bader, D., & O'Grady, M. (2010). Climate risk information: Climate change scenarios and implications for NYC infrastructure. *New York City Panel on Climate Change. Ann NY Academy Sci*, *1196*, 147−228.

Junk, W., An, S., Finlayson, C., Gopal, B., Kvet, J., Mitchell, S., Mitsch, W., & Robarts, R. (2013). Current state of knowledge regarding the world's wetlands and their future under global climate change: A synthesis. *Aquatic Sciences*, *75*, 151−167.

Karr, J. (2006). Seven foundations of biological monitoring and assessment. *Biologia Ambientale*, *20*, 7−18.

Khan, S., Deere, D., Leusch, F., Humpage, A., Jenkins, M., & Cunliffe, D. (2015). Extreme weather events: Should drinking water quality management systems adapt to changing risk profiles? *Water Research*, *85*, 124−136.

Kistemann, T., Classen, T., Koch, C., Dagendorf, F., Fischeder, R., Gebel, J., Vacata, V., & Exner, M. (2002). Microbial load of drinking water reservoirs tributaries during extreme rainfall and runoff. *Applied and Environmental Microbiology*, *68*, 2188−2197.

Knapp, A., Hoover, D., Wilcox, K., Avolio, M., Koerner, S., La Pierre, K., Loik, M., Yiqi, L., Sala, O., & Smith, M. (2015). Characterizing differences in precipitation regimes of extreme wet and dry years: Implications for climate change experiments. *Global Change Biology*, *21*, 2624−2633.

Kundzewicz, Z., Mata, L., Arnell, N., Döll, P., Jimenez, B., Miller, K., Oki, T., Sen, Z., & Shiklomanov, I. (2008). The implications of projected climate change for freshwater resources and their management. *Hydrological Sciences Journal*, *53*, 3−10.

Lammers, R., & Bledsoe, B. (2017). What role does stream restoration play in nutrient management? *Critical reviews in Environmental Science and Technology*, 1−37.

Lankao, P. (2010). Water in Mexico City: What will climate change bring to its history of water-related hazards and vulnerabilities? *Environment & Urbanization*, *22*, 157−178.

Li, W., Sheng, G., Zeng, R., Liu, X., & Yu, H. (2012). China's wastewater discharge standards in urbanization evolution, challenges and implications. *Environmental Science and Pollution Research*, *19*, 1422−1431.

Liu, Y., Bralts, V., & Engel, B. (2015). Evaluating the effectiveness of management practices on hydrology and water quality at watershed scale with a rainfall-runoff model. *The Science of the Total Environment*, *511*, 298−308.

Maklin, M., & Lewin, J. (2019). River stresses in anthropogenic times: Large-scale global patterns and extended environmental timelines. *Progress in Physical Geography*, *43*, 3−23.

Mara, D., & Evans, B. (2017). The sanitation and hygiene targets of the sustainable development goals: Scope and challenges. *Journal of Water Sanitation and Hygiene for Development*, 8, 1–16.

Merriam, E., & Petty, J. (2019). Stream channel restoration increases climate resiliency in a thermally vulnerable Appalachian river. *Restoration Ecology*, 27, 1420–1428.

Miller, R., Fox, G., Penn, C., Wilson, S., Parnell, A., Purvis, R., & Criswell, K. (2014). Estimating sediment and phosphorus loads from streambanks with and without riparian protection. *Agriculture, Ecosystems & Environment*, 189, 70–81.

Milly, P., Betancourt, J., Falkenmark, M., Hirsch, R., Kundzewicz, Z., Lettenmaier, D., & Stouffer, R. (2008). Stationarity is dead: Whither water management? *Science (New York, N.Y.)*, 319, 573–574.

Moore, T., Gulliver, J., Stack, L., & Simpson, M. (2016). Stormwater management and climate change: Vulnerability and capacity for adaptation in urban and suburban contexts. *Climatic Change*, 138, 491–504.

Niezgoda, S., & Johnson, P. (2005). Improving the urban stream restoration effort: Identifying critical form and processes relationships. *Environmental Management*, 35, 579–592.

Olsson, L., & Head, B. (2015). Urban water governance in times of multiple stressors: An editorial. *Ecology and Society*, 20, 27.

Palmer, M., Hondula, K., & Koch, B. (2014). Ecological restoration of streams and rivers: Shifting strategies and shifting goals. *Annual Review of Ecology, Evolution, and Systematics*, 45, 247–272.

Parmesan, C., & Hanley, M. (2015). Plants and climate change: Complexities and surprises. *Annals of Botany*, 116, 849–864.

Paul, M., & Meyer, J. (2001). Streams in the urban landscape. *Annual Review of Ecology and Systematics*, 32, 333–365.

Pour, S., Wahab, A., Shahid, S., Asaduzzaman, M., & Dewan, A. (2020). Low impact development techniques to mitigate the impacts of climate-change-induced urban floods: Current trends, issues and challenges. *Sustainable Cities and Society*, 62, 102373.

Prudhomme, C., Crooks, S., Kay, A., & Reynard, N. (2013). Climate change and river flooding: Part 1 classifying the sensitivity of British catchments. *Climatic Change*, 119, 933–948.

Ramyar, R., Ackerman, A., & Johnston, D. (2021). Adapting cities for climate change through urban green infrastructure planning. *Cities (London, England)*, 117, 103316.

Riley, D., Mieno, T., Schoengold, K., & Brozović, N. (2019). The impact of land cover on groundwater recharge in the High Plains: An application to the Conservation Reserve Program. *The Science of the Total Environment*, 696, 133871.

Rodell, M., Velicogna, I., & Famiglietti, J. (2009). Satellite-based estimates of groundwater depletion in India. *Nature*, 460, 999–1002.

Rodriguez, N., McGuire, K., & Klaus, J. (2018). Time-varying storage – Water age relationships in a catchment with a Mediterranean climate. *Water Resources Research*, 54, 3988–4008.

Rodriguez-Narvaez, O., Peralta-Hernandez, J., Goonetilleke, A., & Bandala, E. (2017). Treatment technologies for emerging contaminants in water: A review. *Chemical Engineering Journal*, 323, 361–380.

Roebroek, C., Harrigan, S., van Emmerik, T., Baugh, C., Eilander, D., Prudhomme, C., & Pappenberger, F. (2021). Plastics in global rivers: Are floods making it worse? *Environmental Research Letters*, 16, 025003.

Rosenberg, E., Keys, P., Booth, D., Hartley, D., Burkey, J., Steinemann, A., & Lettenmaier, D. (2010). Precipitation extremes and the impacts of climate change on stormwater infrastructure in Washington state. *Climate Change*, 102, 319–349.

Scanlon, B., Faunt, C., Longuevergne, L., Reedy, R., Alley, W., McGuire, V., & McMahon, P. (2012). Groundwater depletion and sustainability of irrigation in the US high plains and Central Valley. *Proceedings of the National Academy of Sciences of the United States of America*, 109, 9320–9325.

Schmadel, N., Harvey, G., Schwarz, R., Alexander, J., Gomez-Velez, D., Scott, D., & Ator, S. (2019). Small ponds in headwater catchments are a dominant influence on regional nutrient and sediment budgets. *Geophysical Research Letters*, 46, 9669–9677.

Selvakumar, A., O'Connor, T., & Struck, S. (2010). Role of stream restoration on improving benthic macroinvertebrates and in-stream water quality in an urban watershed: Case study. *Journal of Environmental Engineering, 136*, 127–139.

Semadeni-Davies, A., Hernebring, C., Svensson, G., & Gustafsson, L. (2008). The impacts of climate change and urbanization on drainage in Helsingborg, Sweden: Combined sewer system. *Journal of Hydrology, 350*, 100–113.

Shahady, T., & Boniface, H. (2018). Water quality management thorough community engagement in Costa Rica. *Journal of Environmental Science and Studies, 8*, 488–502.

Shahady, T., & Cleary, W. (2021). Influence of a low-head dam on water quality of an urban river system. *Journal of Environmental Management, 297*, 113334.

Smith, A. B., & Katz, R. W. (2013). US billion-dollar weather and climate disasters: Data sources, trends, accuracy and biases. *Natural Hazard, 67*, 387–410.

Sohn, W., Kim, J., Li, M., & Brown, R. (2019). The influence of climate on the effectiveness of low impact development: A systematic review. *Journal of Environmental Management, 236*, 365–379.

Sterk, A., Schijven, J., de Nijs, T., & de Roda, A. (2013). Husman, direct and indirect effects of climate change on the risk of infection by water-transmitted pathogens. *Environmental Science & Technology, 47*, 12648–12660.

Trenberth, K. (2011). Changes in precipitation with climate change. *Climate Research, 47*, 123–138.

Turral, H., Burke, J., & Faurès, J.-M. (2014). Climate change, water and food security (Rome: FAO) 200. *Water Report, 174*.

Tzilivakis, J., Warner, D., Green, A., & Lewis, K. (2019). Spatial analysis of the benefits and burdens of ecological focus areas for water-related ecosystem services vulnerable to climate change in Europe. *Mitigation and Adaptation Strategies for Global Change, 24*, 205–233.

USEPA (United States Environmental Protection Agency). National Pollutant Discharge Elimination System. (2020).

van Emmerik, T., & Schwarz, A. (2020). Plastic debris in rivers. *Water, 7*, 1–24.

Van Loon, A., & Laaha, G. (2015). Hydrological drought severity explained by climate and catchment characteristics. *Journal of Hydrology, 526*, 3–14.

Voss, K., Famiglietti, J., Lo, M., de Linage, C., Rodell, M., & Swenson, S. (2013). Groundwater depletion in the middle East from GRACE with implications for transboundary water management in the Tigris–Euphrates-Western Iran region. *Water Resources Research, 49*, 904–914.

Vörösmarty, C. (1997). The storage and aging of continental runoff in large reservoir systems of the world. *Ambio, 26*(1997), 210–219.

Wada, Y., & Bierkens, M. (2014). Sustainability of global water use: Past reconstruction and future projections. *Environmental Research Letters, 9*, 104003.

Wang, X., Huang, G., & Liu, J. (2014). Projected increases in intensity and frequency of rainfall extremes through a regional climate modeling approach. *Journal of Geophysical Research Atmospheres, 119*, 213–271.

Wilkinson, J., Hooda, P., Barker, J., Barton, S., & Swinden, J. (2017). Occurrence, fate and transformation of emerging contaminates in water: An overarching review of the field. *Environmental Pollution, 231*, 954–970.

Yao, L., Wei, W., & Chen, L. (2016). How does imperviousness impact the urban rainfall-runoff process under various storm cases? *Ecological Indicators, 60*, 893–905.

Yasarer, L., & Sturm, B. (2016). Potential impacts of climate change on reservoir services and management approaches. *Lake and Reservoir Management, 32*, 13–26.

Zahmatkesh, Z., Burian, S., Karamouz, M., Tavakol-Davani, H., & Goharian, E. (2015). Low impact development practices to mitigate climate change effects on urban stormwater runoff: Case study of New York City. *Journal of Irrigation and Drainage Engineering, 141*, 04014043.

Zarch, M., Sivakumar, B., Malekinezhad, H., & Sharma, A. (2017). Future aridity under conditions of global climate change. *Journal of Hydrology, 554*, 451–469.

Zhang, Y., Peña-Arancibia, J., McVicar, T., Chiew, F., Vaze, J., Liu, C., & Pan, M. (2016). Multi-decadal trends in global terrestrial evapotranspiration and its components. *Scientific Reports, 6*, 19124.

Zhu, J. (2012). Development of sustainable urban forms for high-density low-income Asian countries: The case of Vietnam: The institutional hindrance of the commons and anticommons. *Cities (London, England), 2*, 77–87.

Zhu, J., & Simarmata, H. (2015). Formal land rights vs informal land rights: Governance for sustainable urbanization in the Jakarta metropolitan region, Indonesia. *Land Use Policy, 43*, 63–73.

CHAPTER 16

Food-energy-water nexus and assessment models

Anju Vijayan Nair and Veera Gnaneswar Gude

Richard A Rula School of Civil and Environmental Engineering, Mississippi State University, Mississippi State, Starkville, MS, United States

Chapter Outline

16.1 Introduction 317
16.2 Assessment of food-energy-water nexus 318
16.3 Food-energy-water nexus model development 319
    16.3.1 WEF Nexus Tool 2.0 319
    16.3.2 CLEWS 320
    16.3.3 WEF Nexus Rapid Appraisal Tool 320
    16.3.4 MuSIASEM 320
    16.3.5 Foreseer 320
    16.3.6 WEAP-LEAP 321
16.4 Comparison of different models 322
16.5 Applications of F-E-W nexus models 322
16.6 Future considerations 328
References 329

16.1 Introduction

Food, water, and energy can be considered as the basic necessities for sustaining life and for promoting social and economic development. The assessment of food, energy, water, and its nexus is gaining importance as they are the central factors for attaining sustainable development. The Food-Energy-Water (F-E-W) nexus refers to the complex interconnections (synergies and trade-offs) that exist between water, energy, and food sectors, in order to achieve a balanced and sustainable development (D'Odorico et al., 2018). Demand for these resources is increasing significantly due to population growth and migration, economic development, international trade, urbanization, diversifying diets, cultural and technological changes, as well as climate variability and changes (D'Odorico et al., 2018; Scanlon et al., 2017). The water, energy, and food components are interlinked in many ways, depending upon its field of application. For example, in the context of agricultural applications, water and energy requirements and availability will directly affect food production. Also, depending upon the type of crop produced, there will be a considerable variation in the water and energy demands. In order to attain overall development, the issues and concerns with respect to food, energy, and water need to be addressed collectively in a sustainable manner. The different elements of the F-E-W nexus in the context of agriculture are illustrated using Fig. 16.1.

The F-E-W nexus can be considered as a compilation of three individual nexuses: food-energy nexus, food-water nexus, and energy-water nexus (Gold & Webber, 2015; Serrano-Tovar et al., 2019). Once the interactions and interlinkages between individual nexuses are known, the overall F-E-W nexus can be easily accomplished. For this, evaluation and assessment of food, energy, water, and its nexus are inevitable. This chapter will present currently used modeling platforms for understanding the F-E-W nexus in different contexts and at different scales. First, a list of modeling tools will be presented, then a comparison of different models and detailed discussion of their applications in different contexts will be made followed by discussion of future possibilities for improving F-E-W nexus models.

16.2 Assessment of food-energy-water nexus

F-E-W nexus is studied through the development of various assessment techniques and mathematical models (Dargin et al., 2019; Davies & Simonovic, 2010; Givens et al., 2018;

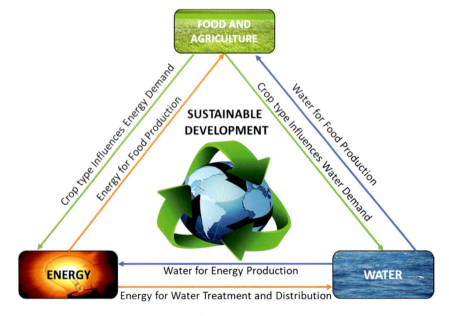

Figure 16.1
The elements of Food-Energy-Water nexus in the context of agriculture.

Gold & Webber, 2015; Guan et al., 2020; Gunda & Tidwell, 2019; Karnib, 2017; Kraucunas et al., 2015; Mabrey & Vittorio, 2018; Martinez-Hernandez et al., 2017; Serrano-Tovar et al., 2019; Tan et al., 2017; Wa'el et al., 2017; Walker et al., 2014; White et al., 2017; Wicaksono & Jeong, 2017). The development of models depends on the overall objective of the scenario under investigation (Serrano-Tovar et al., 2019). Some models are developed to evaluate the F-E-W nexus at the household level (White et al., 2017) while others are developed to understand the nexus at much larger scale, including cities, regions and countries (Givens et al., 2018; Kraucunas et al., 2015; Tan et al., 2017; Wa'el et al., 2017). The complexity of F-E-W nexus and the models depends on the problem at hand and the number of variables that need consideration and the end deliverables. Due to the need for inclusion of factors related to environment, economics, and socioecological and policy related disciplines, models are often designed to function under layer-by-layer approach (Gunda & Tidwell, 2019). A variety of tools should be connected or gathered to derive the desired solutions. Depending on the nature and objective of the study, assess to large input data may be required which include local, regional, or national level water, energy and food portfolios. Some models have been developed to evaluate the impact of climate change or contribution of renewable energy advancement for sustainable energy supplies, and others focus on land use optimization, and societal and ecosystem balances.

16.3 Food-energy-water nexus model development

While there are many models developed to understand F-E-W nexus in different contexts, a few models have gained traction and are used more widely than others. These include WEF Nexus Tool, CLEWS (Climate, Land use, Energy-Water Strategies), WEF nexus rapid appraisal tool, MuSIASEM (Multiscale Integrated Analysis of Societal and Ecosystem Metabolism), WEAP (Water Evaluation and Planning System) and LEAP (Long Range Alternatives Planning System), Foreseer, and Diagnostic tools for investment in water for agriculture and energy (Gold & Webber, 2015; Serrano-Tovar et al., 2019). A few highlights of some of the models are described below.

16.3.1 WEF Nexus Tool 2.0

- The WEF Nexus Tool 2.0 is an input-output model developed by Daher and Mohtar (2015) for the purpose of analyzing the national resource requirements associated with different food self-sufficiency scenarios.
- Users of the tool identify data inputs that provide a localized, contextual basis to the model: local food profile, national water and energy portfolios, agricultural conditions, and food import-export portfolio.
- As a result, the tool specifies the total water, land, and energy requirements, carbon footprint, financial costs, and sustainability of the user-defined food efficiency scenario.
- The tool is web-accessible and open-access for users.

16.3.2 CLEWS

- CLEWS is a framework for a cross-sectoral systems approach to nexus challenges developed by Howells (2013).
- The framework is focused on identifying feedbacks across these systems and uses the interconnections to determine how changes in one sector influence others.
- CLEWS has been applied to various case-studies across Africa, small island developing states, and European transboundary basins with emphasis on context-specific nexus issues, such as (but not limited to) links between water availability, hydropower production, ecosystem services, and agricultural intensification.

16.3.3 WEF Nexus Rapid Appraisal Tool

- Developed by the Food and Agricultural Organization of the United Nations, the WEF Nexus Rapid Appraisal Tool is an online version the Nexus Assessment framework (Flammini et al., 2014).
- It provides a quick method for assessing specific policy and technology interventions with respect to bioeconomic pressures at national scale.
- FAO specifies the tool's intended use is for communication and awareness raising purposes.
- It supplies the user with a set of 10 "nexus context assessment" indicators and 30 "nexus intervention assessment" indicators and allows the user to adjust weights applied to each index according to their relative importance.
- The intervention scenarios currently look at power irrigation, a bioenergy, hydropower and water desalination interventions from the perspective of water, energy, food, labor, and cost components.

16.3.4 MuSIASEM

- MuSIASEM (Giampietro & Mayumi, 2000) is a framework that builds on concepts from bioeconomics and the flow-fund model.
- The framework analyzes the "metabolic pattern of energy, food, and water" in relation to land use changes, population dynamics, greenhouse gas emissions at both national and subnational scales.
- It has been used for both diagnostic purposes as well to simulate scenarios defined by the user simulation purposes.

16.3.5 Foreseer

- The Foreseer Tool uses Sankey diagrams to visualize and trace energy, water and land resources from source to service, under various socioeconomic and climate change scenarios.

- Through the online interface, users create customized water and energy policy scenarios while environmental impacts are computed as an output of the tool.
- Foreseer developers at the University of Cambridge have published one global version of Foreseer, and three national and local scale models applicable to China, the United Kingdom, and the state of California (Allwood et al., 2016).

16.3.6 WEAP-LEAP

- WEAP and LEAP are two software models developed by the Stockholm Environment Institute.
- Individually, the tools have been applied worldwide to support alternative policy measures in water resources and energy challenges. The models were integrated in 2014, becoming "WEAP-LEAP."
- The model works by exchanging parameters and outputs, such as hydropower generated or cooling water requirements. Together, they can represent evolving conditions in both water and energy systems (Heaps, 2008).

Other models which have been developed include ANEMI (Davies & Simonovic, 2010; Wicaksono & Jeong, 2017), Q-nexus (Karnib, 2017), and NexSym (Martinez-Hernandez et al., 2017). The first-generation ANEMI model (named after the Greek God) was developed as an integrated assessment model to model and simulate all involved elements, such as climate−carbon cycle, economy, land use, population, natural hydrological cycle, water demand, and water quality system. The first model did not simulate the individual sectors in detail. The second-generation model (ANEMI-2) was developed in 2011 to improve the capability of the previous model by including food production and enhancing the capability of optimizing the energy-economy element (Davies & Simonovic, 2010; Wicaksono & Jeong, 2017). Both models were based on a system dynamics model that is capable of explicitly modeling the feedback between components. The model was developed using the Ventana Simulation Environment (Vensim) software that has the capability to conceptualize, simulate, analyze, and optimize the system dynamics model.

The Q-Nexus Model is a practical and mathematically-based quantitative WEF nexus assessment framework that serves as platform to quantify, plan, simulate, and optimize water, energy and food as an interconnected system of resources (Karnib, 2017). It is built on input-output theory and is based on the quantitative balance of the WEF total quantities. The framework is based on the balance of the water, energy and food total quantities through two main quantitative conceptual elements: (1) the intersectoral use quantities matrix and (2) the final demand quantities vector. The sum of these two components gives the total resources quantities vector. Water, energy and food inflows are identified. Based on these inflows, nine intersectoral relationships between water, energy, and food are considered: (1) water for water; (2) water for energy; (3) water for food; (4) energy for water; (5) energy for energy; (6) energy for food; (7) food for water; (8) food for energy and (9) food for food.

NexSym tool provides a dynamic modeling of local technoecological interactions relevant to WEF operations (Martinez-Hernandez et al., 2017). The modular tool integrates models for ecosystems, WEF production and consumption components and allows the user to build, simulate and analyze a "flowsheet" of a local system. This enables elucidation of critical interactions and gaining knowledge and understanding that supports innovative solutions by balancing resource supply and demand and increasing synergies between components, while maintaining ecosystems. NexSym allows to explore not only how parts of the nexus are affected by a change in another part, but also to evaluate key interactions that could be developed into synergistic integrations.

16.4 Comparison of different models

The applicability and benefits of each of these models vary widely. Table 16.1 presents a summary of comparative elements of different models. There are also knowledge gaps and limitations in each of these models as shown in Table 16.1.

16.5 Applications of F-E-W nexus models

Regional level F-E-W nexus: Few research groups have made attempts to understand the F-E-W nexus at city or regional or river basin scales. A tool called Platform for Regional Integrated Modeling and Analysis (PRIMA) was developed by the Pacific Northwest National Laboratory simulates interactions between natural and human systems at scales relevant to decision-making (Kraucunas et al., 2015). A number of individual sector models were put together to develop this tool. Central to this model development are the stakeholder decision support needs are used to select the appropriate model components, information exchanges, and uncertainty characterization approach needed for a particular application of the platform. Key attributes of this model are:

- Stakeholder engagement: stakeholders include members of the scientific community and policy & decision makers from the public and private sectors
- Uncertainty characterization: focuses on the uncertainties relevant to a particular stakeholder decision
- Flexible coupling approach: a flexible software architecture is developed to facilitate coupling of relevant models and outputs depending on stakeholder decisions
- Consistency with global boundary conditions
- Portable and modular: PRIMA can be applied to any region in the world, as long as appropriate data can be obtained to parameterize, initialize, and evaluate the relevant component models

Table 16.1: A summary of comparative elements of different F-E-W models (Gold & Webber, 2015; Serrano-Tovar et al., 2019).

Model name	Primary uses	Advantages	Gaps	Nexus components
WEF Nexus Tool 2.0	For analyzing the national resource requirements associated with different food self-sufficiency scenarios.	• The tool specifies total water, land, and energy requirements, carbon footprint, financial costs, and sustainability of the user-defined food efficiency scenario. • The tool is web-accessible and open-access for users.	Application of model is limited to Qatar only.	Water, Energy, Food
Water Evaluation and Planning Model (WEAP)	• For conducting integrated water resources planning assessments. • It estimates water demand, supply, runoff, infiltration, crop requirements, flows and storage and pollution generation, treatment, discharge and instream water quality under varying hydrologic and policy scenarios.	• It is a scalable resource planning tool and allows to compare water supplies and demands and provides capabilities for forecasting demand. • Free for nonprofit organizations, universities, government and academia in developing countries.	Main applications are limited to distribution of water resources.	Water, Biomass, and Climate
Multiscale Integrated Analysis of Societal and Ecosystem Metabolism Framework (MuSIASEM)	For examining patterns of metabolism of socioeconomic systems by providing a characterization at different levels and scales of socioeconomic activities and ecological constraints.	Allows for context-specific flexibility in constructing multilevel socioeconomic structures.	• The holistic approach generates a large output of information, which is not easy to interpret and summarize. • Does not address factors which determine equitable or inequitable sharing of resources within sectors.	Food, Energy, Water
Climate, Land, Energy and Water Strategies (CLEWS)	For quantifying resource use, greenhouse gas emissions and costs associated with meeting energy, water and food security goals.	Applied to various case-studies across Africa, small island developing states, and European transboundary basins with emphasis on context-specific nexus issues.	CLEWS models help integrated planning at the local and national levels, but they cannot foretell the future.	Water, Energy, Land use, Climate

(*Continued*)

Table 16.1: (Continued)

Model name	Primary uses	Advantages	Gaps	Nexus components
Water–energy–food Nexus Framework	To help decision makers better understand risks associated with managing resources ahead of time and respond proactively in times of crisis.	• Recognizes that the water–energy–food nexus is subject to external factors like global governance failure, economic disparity and geopolitical conflict. • Shows the connection between energy and economic securities in the context of energy shortages and potential impact of social stability.	• Does not address the fluctuation of resource supply over time. • Most of the inputs and outputs are static and do not reflect potential changes over time and space.	Food, Energy, Water
Diagnostic Tool For Investment In Water For Agriculture And Energy	Provides estimates of ongoing planned investment in water resources for food and energy production projects.	• Analyzes the impact of hydropower and irrigation projects on three aspects: (1) human, social and environment (2) poverty and food security (3) health and nutrition. • Facilitates the identification of practical solutions, which reflect the institutional, legal and policy realities of a country.	Does not provide any mechanism for spatial and temporal analysis.	Water, Energy, Food
Foreseer	Uses Sankey diagrams to visualize and trace energy, water and land resources from source to service, under various socioeconomic and climate change scenarios.	One global version of Foreseer, and three national and local scale models applicable to China, the United Kingdom, and the state of California are available online.	Applications limited to agricultural uses (crop production).	Energy, Water, Land

- Open source: each component of the PRIMA framework was developed from an existing community model or built specifically for the PRIMA initiative using an open-source approach

Another study evaluating F-E-W nexus in Phoenix, Arizona is also based on stakeholder need analysis (White et al., 2017). The case study was developed based on data collection and interviews with different stakeholders. There are many issues involved in developing such large scale analysis which include resilience, security and sustainability of these coupled systems. Not only how these interconnected systems are physically connected should be understood but also how the systems are institutionally linked should be considered. Stakeholder involvement has revealed that nexus governance could be improved through awareness and education, consensus and collaboration, transparency, economic incentives, working across scales, and incremental reforms.

Theoretical and methodological approaches relating to processes, metrics, or modeling are developed to address the F-E-W resilience and sustainability in Yakima River Basin (Givens et al., 2018). The challenges or complexities that need consideration are defining resilience; social justice; a process to achieve end-goals of resilience; conceptualizing and adapting changes; and boundaries and scale of the system under consideration. This study developed conceptual maps which are useful tools for representing complex system dynamics, including social aspects. Conceptual maps have been used to bring more explicit attention to social, community, and ecosystem service factors and to highlight composition and competition of communities.

A study focusing on F-E-W nexus for Blue Nile River Basin developed submodels (water submodel—comprised of the dam, its hydrological inflow, its release rate decision, evaporation, spillage; Agricultural Submodel-includes irrigation water demand; Energy Submodel: includes hydropower energy production and releasing rules) (Tan et al., 2017). The main objectives of the reservoir optimization model are to minimize the irrigation and municipal water supply deficit and maximize hydropower generation. The decision variables in this model are the operating rule decisions, which involve storage volume and release rates. Three types of possible scenarios are simulated and optimized: Power Priority (store the water to maximize hydropower generation), Water Priority (dam will only release the water according to the demand of downstream users and will not store the water for hydropower benefit), and Equal Priority (to improve the system by finding an optimal solution that generates the best performance out of the reservoir considering the balance between the water and power priority scenarios). The outcome is a solution that reduces water deficit, and, at the same time, increases power production, which would be in the best interest of all countries in the Blue Nile basin. The evaluation results of this study indicate that the development of the dam provides Ethiopia with a greater control of the water to generate energy.

Desalination and renewable energy integration in agriculture: A novel approach to resource nexus assessment is based on MuSIASEM (Serrano-Tovar et al., 2019). The approach is illustrated with a case study of desalination plant in the Canary Islands, Spain. The application of processor for different hierarchical levels of the system is explained using Fig. 16.1. Three different processes considered in the system are: (1) Production of electricity by the wind farm (local wind electricity supply); (2) Production of fresh water by the desalination plant (local desalted water supply); and (3) Production of crops on the local farm (local crop supply). Electricity (produced from wind farm) and the desalted water (from desalination plant) are used for irrigating the local farm thereby establishing F-E-W nexus. MuSIASEM is used as an approach to study the sustainability of integrated WEF systems. Questions answered through the third study are: How much net energy, water and food are being produced? Is this F-E-W system self-sufficient? What are the possible external and internal constraints? What is the economic viability of the system? Is it feasible to extrapolate this system to other areas? Overall, MuSIASEM is shown as an effective tool which utilizes the concept of biological metabolism applied to socioecological systems. It can be applied to different levels of system analysis.

A water treatment model was developed to estimate the power requirement for brackish groundwater reverse osmosis (BWRO) desalination plant supported by wind- and/or solar-generated electricity (Gold & Webber, 2015). Three models were used as the basis: (1) water treatment model; (2) energetic model; (3) integrated optimization model. The energy model was designed to (1) size a wind farm based on this power requirement and (2) size a solar farm to preheat water before reverse osmosis treatment. Further, the integrated model combines results from the water treatment and energy models. The integrated model optimized the performances of the proposed facility to maximize daily operational profits. Using these models, four different scenarios are analyzed to compare desalination powered by different energy sources and a combination of these sources: Scenario A models a desalination facility integrated with solar power that can either use solar-generated electricity for water treatment or sell solar-generated electricity to the grid; Scenario B models a desalination facility integrated with wind power that can either use wind-generated electricity for water treatment or sell wind-generated electricity to the grid; Scenario C models a desalination facility integrated with a wind farm and colocated with a solar farm. Wind-generated electricity is used to power the water treatment process while the solar panels are used to reduce the energetic intensity of desalination by preheating brackish water; and Scenario D models the traditional approach of a desalination facility that is powered by electricity purchased from the grid. Results indicate that integrated facility can reduce grid-purchased electricity costs by 88% during summer months and 89% during winter when compared to a stand-alone desalination plant.

Urban development and metabolism: A Multisectoral Systems Analysis (MSA), a tool for research and for supporting decision-making for policy and investment was illustrated in the context of Greater London, with three objectives: (1) estimating resource fluxes (nutrients, water and energy) entering, leaving and circulating within the city-watershed system; (2) revealing the synergies and antagonisms resulting from various combinations of water-sector innovations; and (3) estimating the economic benefits associated with implementing these technologies, from the point of view of production of fertilizer and energy, and the reduction of greenhouse gases (Walker et al., 2014). The MSA framework is built upon three components. The first component is the methodology of Substance Flow Analysis, the second involves the definition of metabolic performance metrics based on material and energy flows, and the third component relies on a Regionalized Sensitivity Analysis procedure.

Another study developed an accounting method by using material and energy flow analysis (MEFA) to quantify the urban F-E-W nexus. System boundaries and processes of (1) food system, (2) energy system, (3) water system, and (4) F-E-W nexus of a simplified urban system are used to illustrate the simplified interactions of urban F-E-W systems (Liang et al., 2018). The study developed a method to quantify the urban F-E-W flows and stocks and their interactions based on MEFA, which quantifies flows and stocks of a substance, a group of substances, bulk materials, or energy within a system during a given period of time. An Excel-based template streamlines the data compilation process, for easy adoption by other cities. The method was validated using a case study on Detroit Metropolitan Area. Data from the US Department of Agriculture (USDA), US Geological Survey (USGS), USDA Economics Research Service, Detroit Water and Sewerage Department (DWSD). By properly adjusting certain data estimation methods and parameters, the proposed methods and templates can also be used in F-E-W nexus analyses of other countries.

WEAP is also used in metropolitan scale water-sector evaluation studies. WEAP is used to simulate the allocation of water from different sources to the demand sectors, which include power plants and agriculture that quantify food and energy resources, respectively. WEAP simulates water fluxes from the different sources to all demand sectors. Future scenarios of water and energy demand and supply, as well as agricultural production (ellipsoidal elements) are applied to explore future impacts on the F-E-W nexus (Wa'el et al., 2017).

Household scale evaluation: An integrated model, capturing the interactions between F-E-W at end-use level at a household scale was developed based on a survey of 419 households over winter and summer months for the city of Duhok, Iraq (White et al., 2017). The model estimates F-E-W demand and the generated organic waste and wastewater quantities. It also investigates the impact of change in user behavior, diet, income, family size and climate. A bottom-up approach is used to develop the model, comprising the interactions between water, energy and food at end-use level. This model is coded using SIMILE modeling environment. SIMILE is a system dynamics modeling software that is used for modeling the interactions between various system components and capturing the changes in this system behavior over time. Based on the survey conducted in the selected city for case study, values for input parameters are entered into the model. The sensitivity of the model output to the input parameters and the model validity is tested using uncertainty assessment analysis. The results of the developed model are compared against the available historical data which are published in reports. The results show that the estimated values of the WEF model are close to the measured historical data.

16.6 Future considerations

F-E-W Nexus models enables a systematic evaluation of complex scenarios at household, municipal, regional and national scales. Many modeling approaches have been developed to assess and determine the interlinks between the food, energy and water elements. While these models are useful in addressing the issues at the F-E-W nexus, each of these models come with limitations that need proper consideration. For instance, most of the models are centered around optimization of resources for various uses based on end-user needs. However, generation and utilization of these resources will result in waste production which is not included in the models (Gunda & Tidwell, 2019). While these models are developed to address the most pressing challenges such as global climate change, environmental pollution and sustainable development, without inclusion of the by-products or wastes in the modeling loop, the analysis will be incomplete. Subsequent waste management will require additional costs and may lead to new environmental burdens. F-E-W nexus is also linked to climate change, land use patterns, and waste management practices. Therefore, models should be developed to incorporate these dynamic linkages. Future models should have the capability to simulate the feedback between elements in a single modeling framework social and political dimensions of these individual sectors are also important in the model development which is lacking in many current models. While models predict the required mass and energy balances, implementation of these strategies will not be possible without stakeholders engagement. The following are some of the recommendations that should be considered for future model development.

- Future models should be developed at the national level (which are currently lacking) and they have the capability to simulate the feedback between elements in a single modeling framework rather than performing in different modeling platforms.
- Various contributions of the agricultural sector should be considered in future applications to better understand the energy and water nexus.
- The challenge for future simulation models is to include climate-related changes, such as variations in temperature and rainfall pattern or intensity, in nexus simulation.
- The future models should identify options to optimize the allocation of resources in an economic and sustainable way.

- Visualizing the simulation results will promote more involvement of stakeholders in the nexus dialog and will make the decision process more transparent to nontechnical parties.
- Inclusion of the economic sector in the model development for creating sustainable development scenarios is also important for technological feasibility studies.
- Cost-benefit analysis of the alternatives at household, municipal, regional and national scales should be included in nexus models.
- Models should be developed to encourage use of sustainable materials that are locally available and are environment-friendly in potential F-E-W nexus solutions.
- Self-sufficiency and resiliency aspects of the integrated systems should be considered as a whole rather than individual elements or sectors.

References

Allwood, J., Konadu, D., Mourao, Z., Lupton, R., Richards, K., Fenner, R., Skelton, S., & McMahon, R. (2016). Integrated land-water-energy assessment using the foreseer tool. InEGU General Assembly Conference Abstracts 2016 Apr (pp. EPSC2016–17130).

Daher, B. T., & Mohtar, R. H. (2015). Water–energy–food (WEF) Nexus Tool 2.0: Guiding integrative resource planning and decision-making assessment tools. *Water International, 40*(5–6), 748–771. Available from https://doi.org/10.1080/02508060.2015.1074148.

Dargin, J., Daher, B., & Mohtar, R. H. (2019). Complexity vs simplicity in water energy food nexus (WEF) assessment tools. *Science of the Total Environment, 650*, 1566–1575.

Davies, E. G., & Simonovic, S. P. (2010). ANEMI: A new model for integrated assessment of global change. *Interdisciplinary Environmental Review, 11*(2–3), 127–161.

D'Odorico, P., Davis, K. F., Rosa, L., Carr, J. A., Chiarelli, D., Dell'Angelo, J., Gephart, J., MacDonald, G. K., Seekell, D. A., Suweis, S., & Rulli, M. C. (2018). The global food-energy-water nexus. *Reviews of Geophysics, 56*(3), 456–531.

Flammini, A., Puri, M., Pluschke, L., & Dubois, O. (2014). Walking the nexus talk: Assessing the water-energy-food nexus in the context of the sustainable energy for all initiative. Climate, Energy and Tenure Division (NRC), Food and Agriculture Organization of the United Nations.

Giampietro, M., & Mayumi, K. (2000). Multiple-scale integrated assessments of societal metabolism: Integrating Biophysical and Economic Representations Across Scales. *Population and Environment, 22*(2), 155–210. Available from https://doi.org/10.1023/A:1026643707370.

Givens, J. E., Padowski, J., Guzman, C. D., Malek, K., Witinok-Huber, R., Cosens, B., Briscoe, M., Boll, J., & Adam, J. (2018). Incorporating social system dynamics in the Columbia River Basin: Food-energy-water resilience and sustainability modeling in the Yakima River Basin. *Frontiers in Environmental Science, 6*, 104.

Gold, G. M., & Webber, M. E. (2015). The energy-water nexus: An analysis and comparison of various configurations integrating desalination with renewable power. *Resources, 4*(2), 227–276.

Guan, X., Mascaro, G., Sampson, D., & Maciejewski, R. (2020). A metropolitan scale water management analysis of the food-energy-water nexus. *Science of the Total Environment, 701*, 134478.

Gunda, T., & Tidwell, V. C. (2019). A uniform practice for conceptualizing and communicating food-energy-water nexus studies. *Earth's Future, 7*(5), 504–515.

Heaps, C. (2008). *An introduction to LEAP*. Stockholm Environment Institute, pp. 1–6.

Howells, M., Hermann, S., Welsch, M., Bazilian, M., Segerström, R., Alfstad, T., Gielen, D., Rogner, R., Fischer, G., van Velthuizen, H., Wiberg, D., Young, C., Roehrl, R. A., Mueller, A., Steduto, P., & Ramma, I. (2013). Integrated analysis of climate change, land-use, energy and water strategies. *Nature Climate Change, 3*(7), 621–626. Available from https://doi.org/10.1038/nclimate1789.

Karnib, A. (2017). *Water, energy and food nexus: The Q-Nexus model*. In: 10th World Congress on Water Resources and Environment 2017.

Howells, R. (2013). It's not (just) "the environment, stupid!" Values, motivations, and routes to engagement of people adopting lower-carbon lifestyles. *Global Environmental Change*, 23, 281–290.

Kraucunas, I., Clarke, L., Dirks, J., Hathaway, J., Hejazi, M., Hibbard, K., Huang, M., Jin, C., Kintner-Meyer, M., van Dam, K. K., & Leung, R. (2015). Investigating the nexus of climate, energy, water, and land at decision-relevant scales: The Platform for Regional Integrated Modeling and Analysis (PRIMA). *Climatic Change*, 129(3), 573–588.

Liang, S., Qu, S., Zhao, Q., Zhang, X., Daigger, G. T., Newell, J. P., Miller, S. A., Johnson, J. X., Love, N. G., Zhang, L., & Yang, Z. (2018). Quantifying the urban food–energy–water nexus: The case of the detroit metropolitan area. *Environmental Science & Technology*, 53(2), 779–788.

Mabrey, D., & Vittorio, M. (2018). Moving from theory to practice in the water–energy–food nexus: An evaluation of existing models and frameworks. *Water-Energy Nexus*, 1(1), 17–25.

Martinez-Hernandez, E., Leach, M., & Yang, A. (2017). Understanding water-energy-food and ecosystem interactions using the nexus simulation tool NexSym. *Applied Energy*, 206, 1009–1021.

Scanlon, B. R., Ruddell, B. L., Reed, P. M., Hook, R. I., Zheng, C., Tidwell, V. C., & Siebert, S. (2017). The food-energy-water nexus: Transforming science for society. *Water Resources Research*, 53(5), 3550–3556.

Serrano-Tovar, T., Suárez, B. P., Musicki, A., Juan, A., Cabello, V., & Giampietro, M. (2019). Structuring an integrated water-energy-food nexus assessment of a local wind energy desalination system for irrigation. *Science of the Total Environment*, 689, 945–957.

Tan, C. C., Erfani, T., & Erfani, R. (2017). Water for energy and food: A system modelling approach for Blue Nile River basin. *Environments*, 4(1), 15.

Wa'el, A. H., Memon, F. A., & Savic, D. A. (2017). An integrated model to evaluate water-energy-food nexus at a household scale. *Environmental Modelling & Software*, 93, 366–380.

Walker, R. V., Beck, M. B., Hall, J. W., Dawson, R. J., & Heidrich, O. (2014). The energy-water-food nexus: Strategic analysis of technologies for transforming the urban metabolism. *Journal of Environmental Management*, 141, 104–115.

White, D. D., Jones, J. L., Maciejewski, R., Aggarwal, R., & Mascaro, G. (2017). Stakeholder analysis for the food-energy-water nexus in Phoenix, Arizona: Implications for nexus governance. *Sustainability*, 9(12), 2204.

Wicaksono, A., Jeong, G., & Kang, D. (2017). Water, energy, and food nexus: Review of global implementation and simulation model development. *Water Policy*, 19(3), 440–462.

CHAPTER 17

Emerging water pollutants

Daniel A. Vallero
Pratt School of Engineering, Duke University, Durham, NC, United States

Chapter Outline
17.1 Introduction 331
17.2 Dose-response relationships 332
17.3 Uncertainty 337
17.4 Conclusions 339
References 340

17.1 Introduction

The government agencies who regulate water quality around the world must not only address existing pollutants but must also plan for potential problems and risks. These concerns fall into several categories, including newly identified compounds that heretofore were not detected with earlier technologies, substances that were not considered inordinately toxic or hazardous, but which recent studies indicate otherwise, or substances that had been suspected of toxicity and hazard, but which recent data support this suspicion.

Once a substance reaches a level of concern from the public and governmental agencies, studies usually become more focused on the type and severity of toxicity and hazard. Governments and academic researchers conduct in vivo, in vitro, in silico, and epidemiological studies to characterize the threats and also to identify threshold concentrations at which actions must be taken, for example, bans and controls. Human exposure to emerging contaminants must be mapped against toxicological findings, since risk is a function of both hazard and exposure. Even when biological studies indicate thresholds for toxicity and harm, exposure studies often must rely on more indirect methods. For example, exposure to substances such as per- and polyfluoroalkyl substances (PFAS) has been estimated by inserting secondary data from a large range of studies conducted into pharmacokinetic (PK) models, thereby predicting their concentrations in blood serum (East et al., 2021).

Of the thousands of PFAS componds, perfluorooctanoic acid (PFOA) and perfluorooctanoate sulfonate (PFAS) have been produced in the largest amounts within the

United States (United States Environmental Protection Agency, 2017). Since the production and use of certain compounds change with time, especially as their hazards become better documented given new controls and public awareness, and because additional concentration data become available, studies of emerging contaminants must be updated frequently (Egeghy et al., 2016; Malloy et al., 2016; Roca et al., 2017; Vallero & Gunsch, 2020; Vallero, 2020). In a risk-based approach, the threshold concentrations of a pollutant are designed to protect humans and ecosystems. For emerging contaminants, these concentrations are first directed toward actions to protect human health. For some contaminants, ecological concerns may precede those of human health, for example, gene flow of genetic material when genetically modified crops share genetically modified DNA with adjacent habitats.

Like any pollutant, the level of protection as indicated by environmental quality criteria for a pollutant is based on the potential of an unacceptable outcome, that is, its hazard. Often, there are two approaches when addressing a hazardous pollutant in water, that is, effluent limits and water quality limits. Effluent limits apply to any source that releases the contaminant to a waterbody, for example, 100 mg L^{-1} of a chemical compound. Of course, the actual level of concern for emerging contaminants vary considerably given the large number of candidates of chemicals of concern. For example, the US publishes lists of contaminants that are not yet regulated, but are known or anticipated to be found in the environment. If, for instance, a chemical or biological agent of concern is expected to reach public water systems and may need to be regulated under the Safe Drinking Water Act (SDWA), is placed on the Contaminant Candidate List. The most recent draft (CCL5) includes 66 chemicals, 3 chemical groups (PFAS, cyanotoxins, and disinfection byproducts) and 12 microbial contaminants (See Table 17.1) (US Environmental Protection Agency, 2021).

17.2 Dose-response relationships

The first consideration of chemical concern is often the dose expected to elicit an adverse response, that is, the mass or volume that comes into contact with an organism. Dose is expressed in many way, such as the applied dose, absorbed dose, or the biologically effective dose, which is the amount that reaches a particular organ. For example, a neurotoxin harms the central nervous system (CNS), which means only the mass or volume of that hepatotoxin that reaches the brain and nerve cells will be the biological effective dose. The remaining absorbed dose is metabolized, stored or, in a best-case scenario, eliminated from the body before causing any harm.

Increasing the concentration of a chemical or pathogen that contacts an organism will increase the biologically effective dose and the probability that a response will occur, that is, the expected adverse outcome. This gradient is known as "dose-response" curve (See Fig. 17.1). Increasing the amount of the substance coming into contact with the receptor increases the absorbed and biologically effective dose and, consequently, the

Table 17.1: Drinking water contaminant candidate list 5-Draft.

Candidate substance	CAS number[a]	DTXSID number[b]
1,2,3-Trichloropropane	96−18-4	DTXSID9021390
1,4-Dioxane	123−91-1	DTXSID4020533
17-Alpha ethynyl estradiol	57−63-6	DTXSID5020576
2,4-Dinitrophenol	51−28-5	DTXSID0020523
2-Aminotoluene	95−53-4	DTXSID1026164
2-Hydroxyatrazine	2163−68-0	DTXSID6037807
4-Nonylphenol (all isomers)	25154−52-3	DTXSID3021857
6-Chloro-1,3,5-triazine-2,4-diamine	3397−62-4	DTXSID1037806
Acephate	30560−19-1	DTXSID8023846
Acrolein	107−02-8	DTXSID5020023
Alpha-Hexachlorocyclohexane (alpha-HCH)	319−84-6	DTXSID2020684
Anthraquinone	84−65-1	DTXSID3020095
Bensulide	741−58-2	DTXSID9032329
Bisphenol A	80−05-7	DTXSID7020182
Boron	7440−42-8	DTXSID3023922
Bromoxynil	1689−84-5	DTXSID3022162
Carbaryl	63−25-2	DTXSID9020247
Carbendazim (MBC)	10605−21-7	DTXSID4024729
Chlordecone (Kepone)	143−50-0	DTXSID1020770
Chlorpyrifos	2921−88-2	DTXSID4020458
Cobalt	7440−48-4	DTXSID1031040
Cyanotoxins[c]	Multiple	Multiple
Deethylatrazine	6190−65-4	DTXSID5037494
Desisopropyl atrazine	1007−28-9	DTXSID0037495
Desvenlafaxine	93413−62-8	DTXSID40869118
Diazinon	333−41-5	DTXSID9020407
Dicrotophos	141−66-2	DTXSID9023914
Dieldrin	60−57-1	DTXSID9020453
Dimethoate	60−51-5	DTXSID7020479
Disinfection byproducts (DBPs)[d]	Multiple	Multiple
Diuron	330−54-1	DTXSID0020446
Ethalfluralin	55283−68-6	DTXSID8032386
Ethoprop	13194−48-4	DTXSID4032611
Fipronil	120068−37-3	DTXSID4034609
Fluconazole	86386−73-4	DTXSID3020627
Flufenacet	142459−58-3	DTXSID2032552
Fluometuron	2164−17-2	DTXSID8020628
Iprodione	36734−19-7	DTXSID3024154
Lithium	7439−93-2	DTXSID5036761
Malathion	121−75-5	DTXSID4020791
Manganese	7439−96-5	DTXSID2024169
Methomyl	16752−77-5	DTXSID1022267
Methyl tert-butyl ether (MTBE)	1634−04-4	DTXSID3020833
Methylmercury	22967−92-6	DTXSID9024198
Molybdenum	7439−98-7	DTXSID1024207
Norflurazon	27314−13-2	DTXSID8024234
Oxyfluorfen	42874−03-3	DTXSID7024241

(Continued)

Table 17.1: (Continued)

Candidate substance	CAS number[a]	DTXSID number[b]
Per- and polyfluoroalkyl substances (PFAS)[e]	Multiple	Multiple
Permethrin	52645–53-1	DTXSID8022292
Phorate	298–02-2	DTXSID4032459
Phosmet	732–11-6	DTXSID5024261
Phostebupirim	96182–53-5	DTXSID1032482
Profenofos	41198–08-7	DTXSID3032464
Propachlor	1918–16-7	DTXSID4024274
Propanil	709–98-8	DTXSID8022111
Propargite	2312–35-8	DTXSID4024276
Propazine	139–40-2	DTXSID3021196
Propoxur	114–26-1	DTXSID7021948
Quinoline	91–22-5	DTXSID1021798
Tebuconazole	107534–96-3	DTXSID9032113
Terbufos	13071–79-9	DTXSID2022254
Thiamethoxam	153719–23-4	DTXSID2034962
Tri-allate	2303–17-5	DTXSID5024344
Tribufos	78–48-8	DTXSID1024174
Tributyl phosphate	126–73-8	DTXSID3021986
Trimethylbenzene (1,2,4-)	95–63-6	DTXSID6021402
Tris(2-chloroethyl) phosphate (TCEP)	115–96-8	DTXSID5021411
Tungsten	7440–33-7	DTXSID8052481
Vanadium	7440–62-2	DTXSID2040282

[a]Chemical Abstracts Service Registry Number (CASRN) is a unique identifier assigned by the Chemical Abstracts Service (a division of the American Chemical Society) to every chemical substance (organic and inorganic compounds, polymers, elements, nuclear particles, etc.) in the open scientific literature. It contains up to 10 digits, separated by hyphens into three parts.
[b]Distributed Structure Searchable Toxicity Substance Identifiers (DTXSID) is a unique substance identifier used in EPA's CompTox Chemicals database (Williams et al., 2017), where a substance can be any single chemical, mixture or polymer.
[c]Toxins naturally produced and released by some species of cyanobacteria (previously known as "blue-green algae"). The group of cyanotoxins includes, but is not limited to: Anatoxin-a, cylindrospermopsin, microcystins, and saxitoxin.
[d]This group includes 23 unregulated DBPs.
[e]This group is inclusive of any PFAS (except for PFOA and PFOS). For the purposes of this document, the structural definition of PFAS includes per- and polyfluorinated substances that structurally contain the unit R-(CF2)-C(F)(R′)R″. Both the CF2 and CF moieties are saturated carbons and none of the R groups (R, R′ or R″) can be hydrogen (Toxic Substances Control Act, 2021).
Source: From U.S. Environmental Protection Agency, "Drinking Water Contaminant Candidate List 5-Draft: A Proposed Rule by the Environmental Protection Agency on 07/19/2021," Federal Register, vol. FR Doc. 2021−15121, Filed 7−16-21; 8:45 am, 2021. [Online]. Available: https://www.federalregister.gov/documents/2021/07/19/2021−15121/drinking-water-contaminant-candidate-list-5-draft.

probability of an adverse outcome (National Research Council, 2012; Vallero, 2015; Vallero, 2021).

Constructing a dose-response assessment follows a sequence of steps (EPA, 2015). The dose-response data from animal and human studies are fitted with a mathematical model. From this model, the upper confidence limit (e.g., 95%) line equation is expressed and the confidence limit line is then extrapolated to a response point immediately below the lowest

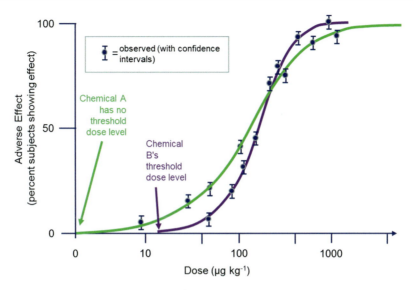

Figure 17.1
Examples of dose-response curves for emerging contaminants. Curve A represents the "no-threshold" curve, which predicts an adverse response (e.g., cancer) even if exposed to a single molecule ("one-hit model"). Here, the low dose end of the curve, that is, below which experimental data are available, is linear. Thus Curve A represents a linearized multistage model. Curve B represents toxicity above a certain threshold, that is, the no observable adverse effect level (NOAEL) or, the no observable effect concentration (NOAEC) if inhaled. Below these thresholds, no response has been observed in the experimental data. The no observable effect concentration (NOEC) is the highest concentration where no effect is on survival is observed ($NOEC_{survival}$) or where no effect on growth or reproduction is observed ($NOEC_{growth}$). Both curves are sigmoidal in shape due to the saturation effect at high dose, that is, less response with increasing dose. Source: *Adapted from Vallero D., Environmental contaminants: Assessment and control. Academic Press, 2010.*

measured response point from the experimental data, that is, the "point of departure" (POD). For carcinogens, the response is assumed to be a linear function of dose from the POD to zero response at zero dose. The last step is to calculate the dose on the line that is estimated to produce the response.

The sigmoidal-shaped curves in Fig. 17.1 are common for toxic chemicals (Schulte et al., 2013). An essential compound curve or substances like vitamins and certain metallic compounds is U-shaped (See Fig. 17.2). The left-hand side represents the low dose region or deficient region. The right-hand side is the high dose region, that is, toxicity. The optimal region lies between these two adverse responses (Calabrese & Baldwin, 2001).

Dose-response curves are derived from animal studies (in vivo toxicological studies) and from studies of human populations, that is, epidemiology. "Petri dish" (i.e., in vitro) studies, such as mutagenicity studies (Flückiger-Isler et al., 2004; Kamber et al., 2009) can be a first

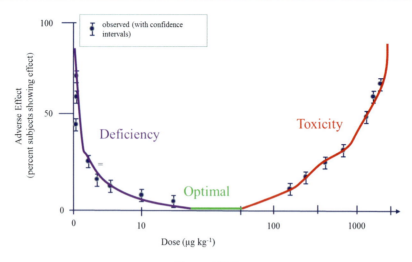

Figure 17.2
U-shaped curve with dose regions of deficiency and toxicity. Between them is the optimal range, that is, sufficiently high to avoid deficiency and sufficiently low to avoid toxicity. This applies to nutrients and vitamins, as well as some metals, for example, trivalent chromium. Source: *Adapted from Vallero D. A., Translating diverse environmental data into reliable information: How to coordinate evidence from different sources. Academic Press, 2017.*

screen and may complement dose-response assessments. For example, the "Ames *Salmonella*/microsome mutagenicity assay" displays the short-term reverse mutation in histidine dependent *Salmonella* strains of bacteria (Ames et al., 1975; McCann et al., 1975). The culture allows only those bacteria whose genes revert to histidine interdependence to form colonies. As the mutagen population increases in the culture, a biological gradient is ascertained. The greater the amount of the chemical added, the greater the number of mutagens and the larger the size of colonies on the plate. The test can be used as a first screen for a substance's potential carcinogenicity, since cancer is a genetic disease and a manifestation of mutations (Kamber et al., 2009).

The toxicity criteria include both acute and chronic effects and include both human and ecosystem effects. For example, a manufacturer of a new chemical may have to show that there are no toxic effects in particular amphibian species exposed to concentrations below 1 $\mu g\ L^{-1}$. If the amphibians show effects at 0.9 $\mu g\ L^{-1}$, the new chemical would be considered to be toxic.

If a substance can cause damage with only a few doses, it is considered to be acutely toxic. Conversely, chronic toxicity occurs when a human or ecological receptor is exposed over a protracted period of time, with recurring exposures. A "no-threshold" model assumes that there is no safe level of exposure to that substance, so the curve intercepts the *x*-axis at 0 (Calabrese, 2019).

The steepness of dose-response curves represents the contaminant's toxic potency or severity. Curve B is steeper than Curve A in Fig. 17.1 because the response elicited by chemical in Curve B is more potent than that of the chemical in Curve A. Potency is but one factor in the risk. A chemical that is very potent in its ability to elicit a rather innocuous effect, like a headache, is less worrisome than another chemical with a rather gentle slope for birth defects or cancer. Thus it is wise to compare curves of different agents for the same outcome.

17.3 Uncertainty

Emerging contaminants all share a high degree of uncertainty about outcomes, timing, and effectiveness of proposed actions, that is, the uncertainty on the ability to meet an objective or desired outcome. Thus the "effect" is the deviation from the "expected" outcome (Solomon, 2017). The expectations of parties vary. Policy makers and the public may care most about the availability of affordable products than the protection of the environment and public health (Vallero, 2019).

Uncertainty about an emerging contaminant can also stem from temporal, spatial and operational variability and the lack of knowledge (Van Asselt, 2000). Uncertainties in risk assessment vary by many factors, for example, types of habitats, sources of pollution, diseases in populations, geographic diversity, and relevance to better studied contaminants, for example, applying quantitative structural activity relationships (QSARs) from known chemicals to emerging contaminants (Card et al., 2017; Kimura et al., 1996; Lagunin et al., 2011; (Q)Sar, 2007; US Environmental Protection Agency, 2015; Vilar et al., 2008). For example, within a square meter of soil, the microbial populations, soil texture, organic matter and chemical makeup can be highly diverse, and may differ substantially from a nearby square meter of soil. Extrapolations to the effect of dredging near a 10-hectare wetland or trying to determine the impact of a leaking underground gasoline tank propagates even greater amounts and types of uncertainty. Uncertainty in the real world compared to highly controlled experiments is compounded by the measurement imprecision, voids and gaps in observations, practical obstacles (e.g., no access to private property and physical barriers), lack of consistency in measured and modeled results, and simply not being able to understand what the information means, that is, structural uncertainty (Vallero, 2017).

Safety thresholds for chemical hazards are known as reference doses (RfD), usually based on oral exposure data. For inhalation, the safe dose is known as the reference concentration (RfC), which is calculated in the same manner as the RfD, using units that apply to air (e.g., $\mu g \ m^{-3}$). Animal studies and epidemiology provide the knowledgebases for these values. Assays, that is, in vitro studies, can also inform the calculations of RfDs and RfCs. The main reasons for uncertainty include variability among the animals and human populations being tested, as well as differences in response to the compound by different species (e.g., one species may have decreased kidney function, while another may show neurological effects). These uncertainties can also be part of microbial data. For example, certain immunocompromised subpopulations and children may respond differently to microbial exposures that are below thresholds for the general population.

Lacking long-term human studies, the hazards and risks will be extrapolated from acute or subchronic studies of the same or similar agents, usually from secondary data sources that likely had different research objectives than those needed for an emerging contaminants (Cohen Hubal et al., 2020; DeLuca et al., 2021; Evans et al., 2014; Hubal et al., 2020; Mitchell et al., 2013a; 2013b; Wambaugh et al., 2013; Weitekamp et al., 2021; Whaley et al., 2021). The routes and pathways of exposure used to administer the agent to subjects may differ from those applications in the real world. For example, if the dose in a research study is administered orally, but the pollutant is more likely to be inhaled, this route-to-route extrapolation adds inconsistency, which translates into uncertainty (Vallero, 2021).

The factors underlying the uncertainties are quantified as specific uncertainty factors (UFs). The uncertainties in the RfD are largely due to the differences between results found in animal testing and expected outcomes in human population. As in other bioengineering operations, a factor of safety must be added to calculations to account for UFs. So, for environmental risk analyses and assessments, the safe level is expressed in the RfD, or in air the RfC. This is the dose or concentration below which regulatory agencies do not expect a specific unacceptable outcome. Thus all the uncertainty factors adjust the actual measured levels of no effect (i.e., the threshold values, NOAELs and LOAELs) in the direction of a zero concentration. This is calculated as:

$$\text{RfD} = \frac{\text{NOAEL}}{\text{UF}_{\text{inter}} \times \text{UF}_{\text{intra}} \times \text{UF}_{\text{iother}}} \quad (17.1)$$

The calculation of the RfC is identical:

$$\text{RfC} = \frac{\text{NOAEC}}{\text{U}_{\text{inter}} \times \text{U}_{\text{intra}} \times \text{U}_{\text{other}}} \quad (17.2)$$

The increase in the size of the denominator, that is, total uncertainty, means that the safe level of exposure continues to drop, so that the RfC and RfD become increasingly small, that is, assumed high toxicity given the high levels of uncertainty.

The first of the three types of uncertainty are those resulting from the difference in the species tested and that of *Homo sapiens* (UF_{inter}). Humans may differ from other species in their sensitivity to substances, so their response to be exposed to the compound will differ from that of the tested species, that is, UF_{inter}. The second uncertainty factor results from certain subpopulations being more sensitive to the effects of a compound than the general human population; known as intraspecies uncertainty factors (UF_{intra}). The third type of uncertainties (UF_{other}) results when the available data and science is lacking, such as when a lowest observed adverse effect is used rather than a NOAEL. This happens when data show a dose at which an effect is observed, but the "no effect" threshold must be extrapolated. Since the UFs are in the denominator, the greater the uncertainties, the closer the safe level (i.e., the RfD) is to zero, that is, the threshold is divided by these factors. The UFs are usually multiples of 10, although the UF_{other} can range from 2 to 10.

Errors are a major source of uncertainty. When an occurrence or event is not identified during an observation or measurement, but in fact the occurrence or event exists, this is a false negative. If sampling and analysis are not sufficiently sensitive and lacks resolution it can lead to a false negative (Adams and Kanaroglou, 2016). Other problems, for example, level of detection (LoD) being too high, chain-of-custody with missing or erroneous documentation, improper interpretation of results, and poor documentation and data entry can also be sources of false negative findings.

The other major type of error is one that leads to a false positive, that is, an occurrence or event is found but n fact does not exist (Boffetta et al., 2008). This can be a major challenge for emerging contaminants. For example, consider the situation where previous evidence used by an agency to list a compound as a potential endocrine disruptor, but has subsequently found that a wealth of new information is now showing that it has no such effect (Furman et al., 2012). The decision makers must decide whether the new information is sufficient to remove protections that were only required based on the compound's apparently false positive endocrine disruption data. Sometimes, the false positive results from scientific shortfalls, for example, based upon faulty models, or models that only work well for lower organisms (i.e. interspecies uncertainty). Often, this is rectified by scientific advancement, for example, models that are subsequently developed consider newly available data that account for physical, chemical, and biological complexities of higher-level organisms, including humans.

False positives may also lead to overemphasis on less severe problems that cause public health officials to devote inordinate amounts of time and resources address too many problems (Psaltopoulos et al., 2017). False positives also may discourage people to make use of potentially beneficial products. False positives, especially when they occur frequently, create credibility gaps between scientists and the public.

17.4 Conclusions

Addressing emerging contaminants requires high quality, scientifically sound data, and proper applications of models and other decision tools. Whether the substance of concern is chemical or biological, whether it is a water pollutant, air pollutant or greenhouse gas, it must be addressed sooner rather than later. This means that scientists and decision makers must begin to consider potential hazards, exposures, and risks before they become obvious and to consider actions needed to reduce these hazards, exposures, and risks.

References

Adams, M. D., & Kanaroglou, P. S. (2016). A criticality index for air pollution monitors. *Atmospheric Pollution Research*, *7*(3), 482–487. Available from https://doi.org/10.1016/j.apr.2015.11.004.

Ames, B. N., McCann, J., & Yamasaki, E. (1975). Methods for detecting carcinogens and mutagens with the *Salmonella*/mammalian-microsome mutagenicity test. *Mutation Research/Environmental Mutagenesis and Related Subjects*, *31*(6), 347–363.

Boffetta, P., McLaughlin, J. K., La Vecchia, C., Tarone, R. E., Lipworth, L., & Blot, W. J. (2008). False-positive results in cancer epidemiology: A plea for epistemological modesty. *Journal of the National Cancer Institute*, *100*(14), 988–995.

Calabrese, E. J. (2019). The linear no-threshold (LNT) dose response model: A comprehensive assessment of its historical and scientific foundations. *Chemico-Biological Interactions*, *301*, 6–25.

Calabrese, E. J., & Baldwin, L. A. (2001). The frequency of U-shaped dose responses in the toxicological literature. *Toxicological Sciences*, *62*(2), 330–338.

Card, M. L., et al. (2017). History of EPI Suite[trade mark sign] and future perspectives on chemical property estimation in US Toxic Substances Control Act new chemical risk assessments. *Environmental Science: Processes & Impacts*, *19*(3), 203–212. Available from https://doi.org/10.1039/C7EM00064B.

Cohen Hubal, E. A., Reif, D. M., Slover, R., Mullikin, A., & Little, J. C. (2020). Children's environmental health: A systems approach for anticipating impacts from chemicals. *International Journal of Environmental Research and Public Health*, *17*(22), 8337.

DeLuca, N. M., Angrish, M., Wilkins, A., Thayer, K., & Hubal, E. A. C. (2021). Human exposure pathways to poly-and perfluoroalkyl substances (PFAS) from indoor media: A systematic review protocol. *Environment International*, *146*, 106308.

East, A., Egeghy, P. P., Hubal, E. A. C., Slover, R., & Vallero, D. A. (2021). Computational estimates of daily aggregate exposure to PFOA/PFOS from 2011 to 2017 using a basic intake model. *Journal of Exposure Science & Environmental Epidemiology*, 1–13.

Egeghy, P. P., et al. (2016). "Computational exposure science: An emerging discipline to support 21st-century risk assessment,". *Environmental Health Perspectives (Online)*, *124*(6), 697.

EPA (2015). *Proposed guidelines for carcinogen risk assessment.* <http://cfpub.epa.gov/ncea/cfm/recordisplay.cfm?deid = 2833>.

Evans, A. M., Rice, G. E., Wright, J. M., & Teuschler, L. K. (2014). Exploratory cumulative risk assessment (CRA) approaches using secondary data. *Human and Ecological Risk Assessment: An International Journal*, *20*(3), 704–723.

Flückiger-Isler, S., et al. (2004). Assessment of the performance of the Ames II™ assay: A collaborative study with 19 coded compounds. *Mutation Research/Genetic Toxicology and Environmental Mutagenesis*, *558*(1), 181–197.

Furman, J. L., Jensen, K., & Murray, F. (2012). Governing knowledge in the scientific community: Exploring the role of retractions in biomedicine. *Research Policy*, *41*(2), 276–290.

Hubal, E. A. C., et al. (2020). Advancing systematic-review methodology in exposure science for environmental health decision making. *Journal of Exposure Science & Environmental Epidemiology*, *30*(6), 906–916.

Kamber, M., Flückiger-Isler, S., Engelhardt, G., Jaeckh, R., & Zeiger, E. (2009). Comparison of the Ames II and traditional Ames test responses with respect to mutagenicity, strain specificities, need for metabolism and correlation with rodent carcinogenicity. *Mutagenesis*, gep017.

Kimura, T., Miyashita, Y., Funatsu, K., & Sasaki, S.-I. (1996). Quantitative structure − activity relationships of the synthetic substrates for elastase enzyme using nonlinear partial least squares regression. *Journal of Chemical Information and Computer Sciences*, *36*(2), 185–189.

Lagunin, A., Zakharov, A., Filimonov, D., & Poroikov, V. (2011). QSAR modelling of rat acute toxicity on the basis of PASS prediction. *Molecular Informatics*, *30*(2-3), 241–250.

Malloy, T., Trump, B. D., & Linkov, I. (2016). *Risk-based and prevention-based governance for emerging materials*. ACS Publications.

McCann, J., Choi, E., Yamasaki, E., & Ames, B. N. (1975). Detection of carcinogens as mutagens in the Salmonella/microsome test: Assay of 300 chemicals. *Proceedings of the National Academy of Sciences, 72*(12), 5135–5139.

Mitchell, J., et al. (2013a). A decision analytic approach to exposure-based chemical prioritization. *PLoS One, 8*(8), e70911. Available from https://doi.org/10.1371/journal.pone.0070911.

Mitchell, J., et al. (2013b). Comparison of modeling approaches to prioritize chemicals based on estimates of exposure and exposure potential. *Science of the Total Environment, 458*, 555–567.

National Research Council. (2012). *Exposure science in the 21st century: A vision and a strategy* (p. 196) Washington, DC: The National Academies Press (in English).

Psaltopoulos, D., Wade, A. J., Skuras, D., Kernan, M., Tyllianakis, E., & Erlandsson, M. (2017). False positive and false negative errors in the design and implementation of agri-environmental policies: A case study on water quality and agricultural nutrients. *Science of The Total Environment, 575*, 1087–1099.

(Q)Sar (2007). *69, Guidance document on the validation of (quantitative) structure-activity relationship [(Q)Sar] models*.

Roca, J. B., Vaishnav, P., Morgan, M. G., Mendonça, J., & Fuchs, E. (2017). When risks cannot be seen: Regulating uncertainty in emerging technologies. *Research Policy, 46*(7), 1215–1233.

Schulte, P. A., et al. (2013). Occupational safety and health, green chemistry, and sustainability: A review of areas of convergence. *Environmental Health, 12*(1), 1.

Solomon J. D., (2017). *Communicating reliability, risk and resiliency to decision makers*. Duke University, October 12.

Toxic Subtances Control Act, Proposed Rule. 86 FR 33926, (2021).

U.S. Environmental Protection Agency. (2015). Quantitative structure activity relationship. <http://www.epa.gov/nrmrl/std/qsar/qsar.html> Accessed 19.03.21.

U.S. Environmental Protection Agency. (2021). *Drinking water contaminant candidate list 5-draft: A proposed rule by the Environmental Protection Agency on 07/19/2021," Federal Register*, vol. FR Doc. 2021–15121, Filed 7–16-21; 8:45 am,. [Online]. Available: <https://www.federalregister.gov/documents/2021/07/19/2021-15121/drinking-water-contaminant-candidate-list-5-draft>.

United States Environmental Protection Agency. (2017). Technical fact sheet—perfluorooctane sulfonate (PFOS) and perfluorooctanoic acid (PFOA).

Vallero, D. (2010). *Environmental contaminants: Assessment and control*. Academic Press.

Vallero, D. (2015). *Environmental biotechnology: A biosystems approach*. Elsevier Science.

Vallero, D. A. (2017). *Translating diverse environmental data into reliable information: How to coordinate evidence from different sources*. Academic Press.

Vallero, D. A. (2019). *Air pollution calculations: Quantifying pollutant formation, transport, transformation, fate and risks*. Elsevier.

Vallero, D. A. (2020). *Estimating and predicting exposure to products from emerging technologies. Synthetic biology 2020: Frontiers in risk analysis and governance* (pp. 107–142). Springer.

Vallero, D. A. (2021). *Environmental systems science: Theory and practical applications* (1st ed.). Elsevier Academic Press.

Vallero, D. A., & Gunsch, C. K. (2020). Applications and implications of emerging biotechnologies in environmental engineering. *Journal of Environmental Engineering, 146*(6), 03120005.

Van Asselt, M. B. (2000). *Perspectives on uncertainty and risk. Perspectives on uncertainty and risk* (pp. 407–417). Springer.

Vilar, S., Cozza, G., & Moro, S. (2008). Medicinal chemistry and the molecular operating environment (MOE): Application of QSAR and molecular docking to drug discovery. *Current Topics in Medicinal Chemistry, 8*(18), 1555–1572.

Wambaugh, J. F., et al. (2013). High-throughput models for exposure-based chemical prioritization in the ExpoCast project. *Environmental Science & Technology, 47*(15), 8479–8488.

Weitekamp, C. A., Phillips, L. J., Carlson, L. M., DeLuca, N. M., Hubal, E. A. C., & Lehmann, G. M. (2021). A state-of-the-science review of polychlorinated biphenyl exposures at background levels: Relative contributions of exposure routes. *Science of The Total Environment*, 145912.

Whaley, P., et al. (2021). Improving the quality of toxicology and environmental health systematic reviews: What journal editors can do. *ALTEX-Alternatives to Animal Experimentation*, *38*(3), 513–522.

Williams, A. J., et al. (2017). The CompTox chemistry dashboard: A community data resource for environmental chemistry. *Journal of Cheminformatics*, *9*(1), 61.

CHAPTER 18

Local representations of a changing climate

Juan Baztan[1], Scott Bremer[2], Charlotte da Cunha[1], Anne De Rudder[3], Lionel Jaffrès[4], Bethany Jorgensen[5], Werner Krauß[6], Benedikt Marschütz[7], Didier Peeters[8], Elisabeth Schøyen Jensen[2], Jean-Paul Vanderlinden[1], Arjan Wardekker[2] and Zhiwei Zhu[1]

[1]University of Versailles Saint-Quentin-en-Yvelines, UPS-CEARC, Guyancourt, France [2]Centre for the Study of the Sciences and the Humanities, University of Bergen, Bergen Norway [3]Institut Royal d'Aéronomie Spatiale de Belgique, Bruxelles, Belgium [4]Theatre du Grain, Brest, France [5]Civic Ecology Lab, Cornell University, Ithaca, NY, United States [6]University of Bremen, Bremen, Germany [7]Klima und Energiefonds, Vienna, Austria [8]Université Libre de Bruxelles, Bruxelles, Belgium

Chapter Outline

18.1 Introduction 344
18.2 Local conditions of a changing climate 344
    18.2.1 Bergen, Norway 344
    18.2.2 Brest, Kerourien, France 345
    18.2.3 Dordrecht, the Netherlands 346
    18.2.4 Gulf of Morbihan, France 347
    18.2.5 Jade Bay, Germany 348
18.3 Art and science local representation processes by site and related challenges 349
    18.3.1 Bergen, Norway 349
    18.3.2 Brest, Kerourien, France 349
    18.3.3 Dordrecht, the Netherlands 351
    18.3.4 Gulf of Morbihan, France 352
    18.3.5 Jade Bay, Germany 354
18.4 Metadata and dynamic mapping perspectives for local representations 356
18.5 Lessons learned and final conclusions 358
    18.5.1 Conclusions on local representations for codeveloping climate services 359
    18.5.2 Final conclusions 360
Acknowledgments 361
References 361

18.1 Introduction

Here, we share local representations of changing climate from five sites in four countries across Europe: Bergen (Norway); Brest, Kerourien (France); Dordrecht (Netherlands); Gulf of Morbihan (France), and the Jade Bay (Germany). These case studies represent opportunities for local consciousness-raising on climate-related questions, along with forthcoming implementation of climate services at local, regional, national, and international levels. We explore novel ways to transform climate science into action-oriented place-based climate services that engage, enable, and empower local communities, knowledge-brokers, and scientists to act locally and identify future information needs and the nature of the climate science needed to address the local communities' concerns, aspirations, and goals in light of climate variability and climate change. Our experiences working with these five sites also provide a glimpse into the diversity of climate contexts in Europe. This chapter is one of the outcomes from concluding a 4-year initiative codeveloping place-based climate services for action: CoCliServ; this chapter is a brief version of the WP4 project deliverables, available in full via cc-by-sa-nc doi:353039226 and doi:350386446.

18.2 Local conditions of a changing climate

18.2.1 Bergen, Norway

In Bergen, codeveloping place-based climate services, CoCliServ, is based on a municipal climate adaptation context and engaged in citizen science projects. Bergen is a harbor city in the fjords of western Norway. King Olav Kyrre founded the city in 1070, and its name *Bergen* means "the green meadow among the mountains." The historic center is on the flat land wrapped around the sheltered "Vågen" harbor, and surrounded by seven low mountains. The harbor itself is adjacent to a fjord – "Byfjorden" – which is sheltered from the North Sea by a chain of islands including the islands of Sotra and Askøy. Today Bergen is the administrative center of Hordaland County and comprises eight boroughs extending over an area of 465 square kilometers, with a population in 2016 of 278,121 inhabitants. Here, the CoCliServ study focused on the historic center of Bergen and the immediately surrounding suburbs that fall within the Bergenhus borough. Bergenhus contains most of the historic sites of the city and is the most densely urbanized, with shops, offices, apartments, and houses. Bergen has historically been an administrative and trading center. Once founded, it became the capital and administrative center for Norway until the late 13th Century when King Håkon V moved these functions to Oslo, though Bergen has remained an administrative center for western Norway. Bergen has always had a strong focus on trade through the harbor, historically tied to the trade of fish and dried cod in particular, which the city was granted a monopoly to trade in the 13th Century. By the mid-14th Century, a commercial and defensive confederation of German merchant guilds—the Hanseatic League—established a "kontor" in Bryggen, alongside Vågen harbor. Bergen became one of the four most important Hanseatic trading centers up until the mid-18th Century when the Germans left the kontor to Norwegians. Over this time, Bergen was the center of trade in Norway, which also saw it become the largest city in Norway up until the 1830s when it was overtaken by Oslo. It has always been a prominent international city, with influences to its culture and language through trade ties to England, the Netherlands, Germany, and France.

18.2.2 Brest, Kerourien, France

In Kerourien, CoCliServ used climate change as a medium to connect the past, present, and future of social minorities in a collaborative effort with local artist groups. Kerourien is a neighborhood in the urbanized area of Brest, France, and is mostly structured around postwar housing projects. During World War II, the city of Brest was one of the worst damaged areas on France's west coast. From 1940 to 1944, it was the target of 165 bombings and 480 alerts, which resulted in 965 dead and 740 seriously wounded. The Kerourian farming area was also greatly impacted. The Kerourien reconstruction story begins January 8th, 1964. Albert Cortellari designed the site plan for the first Habitations à Loyer Modéré, (rent dwellings), HLM tower project in the western part of the city of Brest. The project included 500 apartments to be built on seven hectares of land the city purchased for 20 million francs using expropriation procedures that are recorded in the municipal archives. This housing project marked a clear turning point in the progressive transformation of Kerourien from a rural to a peri urban area. According to the 2013 census, Kerourien has 1200 inhabitants. It is a priority area within St. Pierre, as indicated in city policy statements starting in 2014. The most salient aspect of Kerourien is its diverse population. Rooted in a place with fragile economic conditions, residents face the challenge of unemployment. Thirty-two percent of residents between the ages of 15 and 64 are unemployed. For those between the ages of 15 and 24, the rate jumps to 46%. Thirty-two percent of women are unemployed. Only 35% of young adults ages 18−24 are enrolled in universities or other academic institutions. The most recent results from L'Hévéder's work in 2016 show the seasonal sea surface temperature (SST) of the ocean bordering the city using high-resolution satellite data. Coastal seas, well separated from offshore waters by intense frontal structures, show colder SST by $1°C-2°C$ in summer. A significant warming trend is observed in the autumn season. This positive trend is stronger offshore, with an annual mean SST increase of $0.32°C$ decade^{-1}, but weaker in coastal waters ($0.23°C$ decade^{-1}), where strong vertical mixing induced by tides and winds acts to reduce surface warming. In the Iroise Sea the increase in annual mean SST in CMIP5 future scenario simulations ranges from $0.5°C$ (RCP2.6) to $2.5°C$ (RCP8.5) by year 2100, with a seasonal modulation leading to a more intense warming in summer than in winter. This increase in

SST may strongly affect marine biology, particularly phytoplankton phenology, macro-algae biomass and benthic fauna, including exploited shellfish. Governmental representations of climate converge mainly at the regional level and are summarized in the "Climate Plan: Energy & Territory 2014—2019" from the Regional Council of Brittany. This document establishes a framework rooted in the IPCC's efforts and identifies energy, transport, agriculture, fisheries, and infrastructure as the main sectors to focus on for improving sustainability through training, economics, planning, and environmental and international actions. The climate change representations embedded in Kerourien residents present a challenge, since they are manifested implicitly and explicitly, rooted in residents' past, present, and anticipated future life conditions. This opens the question: what can we learn from these embedded representations about the connections between knowledge and action?

18.2.3 Dordrecht, the Netherlands

In Dordrecht, CoCliServ joined municipal climate adaptation projects in addressing the needs of citizens and questions of social inequality. Dordrecht is located in the Rhine-Meuse-Scheldt Delta, one of the large European deltas. It is built on former peat swampland, and the soil consists mainly of peat, riverine clay, and to a lesser extent, sand. The presence of peat leads to ground compaction and consequently soil subsidence. The riverine clay, which is water impenetrable, reduces the uptake of water into the soil, and resulting in challenges with discharging rainwater. With receding water levels in the area, people built polders (drained land), with the first one dating back to 1603. From that date onward, the city developed significantly in size until the latest polder forms were built in 1926 around the Biesbosch on the island of Dordrecht. Whereas several severe floods happened before people started building the first polders in Dordrecht, much less is known about them, potentially due to their less severe effect on the already destroyed landscape around Dordrecht. Nienhuis (2008) refers to the big Allerheiligen Flood in 1570, which affected large parts of the North Sea coast due to a large storm that made landfall and was followed by several floods from rivers affecting the whole Delta region including Dordrecht. Accounts of this event on Dordrecht are not known in detail. Despite all the new dikes, after the St. Elisabeth's Flood (1421 AD), the city was completely surrounded by rivers and continues to be an "island on a river crossroad." The St. Elisabeth flood in 1421 had, by far, the most severe impact on the island, though many more floods occurred in the centuries to come. Interviews on the historical embeddedness of climate revealed, together with flood-stones in the city of Dordrecht, more recent historical floods in 1901, 1906, 1916, 1928, 1936, 1953, and 1954 (Marschütz et al., 2020). Recently, the municipality concluded that during flood events there will be too little time to evacuate its citizens, and that urban flood risk management will need to be rethought using concepts such as urban flood resilience and vertical evacuation. The municipality aims to actively collaborate with citizens on climate adaptation and water management. Their work is backed by the National Water Plan, which states the need for adapting to climate change, specifically in relation to water. The province of South Holland has also considered the need to adapt to a changing climate and its effects, with this province being particularly at risk as large parts of South Holland are situated below sea level. Dordrecht in particular has been developing the concept of multilayer safety to bring flood resilience to a new level and incorporating sustainable urban planning as a primary defense to limit effects of flooding, as well as making the city and the island more self-reliable considering only 15% of the population can be evacuated. We have focused on the Reeland district of Dordrecht, with a specific interest in the Vogelbuurt neighborhood. The area has been affected by flooding through heavy precipitation events in recent years. The municipality and neighborhood are exploring on how to cope with weather-related issues and climate change through adaptation, with much local energy and with active local organizations. Furthermore, large-scale restructuring and maintenance (e.g., replacement of social housing estates), sewer replacements, and redesign of public green spaces and sporting facilities are planned. This provides a window of opportunity to explicitly take citizens' desires and climate change concerns into account when redesigning the area.

18.2.4 Gulf of Morbihan, France

In the Gulf of Morbihan, we worked together with nongovernmental organizations that have a strong focus on raising climate awareness. Located within the southern fringe of Brittany (at 47° 36′ North, 2° 48′ West), the Gulf is an attractive location for many reasons: geography and geology, history (and prehistory), environment and biodiversity, economy and tourism, and climate. These characteristics, in part antagonistic, justified the creation of the PNR in 2014. It works to protect and enhance the natural, cultural and human heritage of its territory by implementing an innovative policy of land use planning including mitigation and adaptation actions, and economic, social and cultural development, especially of a tourism respectful of the environment. The name "Gulf of Morbihan" is derived from two characteristics and one fallacy. It is a gulf of about 25 km in diameter (surrounded by a coastal trail of more than 400 km length, due to the numerous capes and bays), several tens of islands, from 6 km to a few hundred meters in length, and with a narrow entrance to the Atlantic Ocean (about 1 km wide). Morbihan, the name of the administrative department, means "little sea" in the Breton language. This little sea does not refer to the modern Gulf, but to the coastal sea which borders it to the west, limited by the "Pointe de Quiberon" to the northwest, the Belle Isle, Houat and Hoedic islands to the west and the "La Vilaine" river estuary to the south, a vast valley about 20 m deep, inundated after the end of the last deglaciation about 10−5 kyr ago. The Gulf itself was a small estuary joining two little rivers, which has been progressively inundated over these last 2000 years. The submersion derives from a large-scale tilt of Southwestern Brittany and its associated continental shelf, along the fault following the western boundary of the Hercynian western mountains (submersion speed of about 1 mm per year) confirmed by 14C dating of submerged oak roots and megalithic menhirs of about 5000−7000 years BP, and Gallo-roman houses and roads. Climatically, the Gulf of Morbihan is within the general zone of conflicting influence between the mid-latitude Atlantic Ocean system and its ocean atmosphere interactions (with seasonal and interannual oscillations of the dry Açores tropical anticyclone system, and, at its north, the chain of temperate depressions and associated rain and westerly winds). Furthermore, seasonally the winter climate involves expansions of the continental polar anticyclone systems, together with cold north-easterly winds and the summer experiences expansion of the warm subtropical continental anticyclone. Local people are sensitive to climate change, with the winter rains more frequent, and dryer summers. Hay and wheat harvests are typically one month earlier nowadays than 50 years ago.

18.2.5 Jade Bay, Germany

The Jade Bay (Jadebusen) area is part of the UNESCO World heritage site Lower Saxony Wadden Sea, and it is situated between the river Ems and the river Weser. This North Sea bight reaches far into inland, and is the result of many centuries of interaction between humans and the sea. Especially between the 15th and the 19th centuries, devastating storm floods had extended the sea far inland. Since the end of the 19th century, a dike line of 55 km almost fully encircles the Bay and ensures the maintenance of the deep-water trenches, the so-called inner and outer Jade (named after the river Jade), which form Germany's only deep-water port in Wilhelmshaven. The Jade bight resembles a giant bathtub, surrounded by dikes and with a neck to the North Sea, which serves as the entry and exit of tidal waters. The bathtub is filled twice a day with water; with each tide, 450 million m^3 water from the sea floods an area of 164 km^2, while during low tide, only 44 km^2 are covered with water. The Jade Bay is surrounded by two districts, Wesermarsch in the east (the Butjadingen marshlands, between Weser Delta and Jade Bay) and Friesland in the west, with its main towns Jever and Varel. Water remains the crucial threat and challenge in this area: from the seaside, storm floods threaten the land, and on the land, highly complex drainage and pump systems keep the rainfall water out of this low-lying landscape and pump it into the sea. The marshlands, moors and the *Geest* (land formed by alluvial sediments from the ice age) are mostly used for pastures, cattle, and energy production, such as crop for bio-gas tanks and wind turbines. As a result, climate change in the Jade Bay area is mostly addressed by the fields of coastal climate adaptation, land use and energy transition.

18.3 Art and science local representation processes by site and related challenges

While the diversity of each local site determined the local representation processes, each local representation process aimed at three main representation forms: art and science integration, metadata, and geo-referencing. While the art and science components have their roots in each site, the metadata and geo-referencing aspects have been connected throughout the representation processes in ongoing discussions and explorations. However, they are less salient in the stabilized final representation forms. We discuss this further in the section dedicated to lessons learned.

18.3.1 Bergen, Norway

On three different occasions, Bergen's partners introduced groups of children and adults to the historic Norwegian calendar and invited them to create their own primstav (ancient Norwegian stick) for their own lives, and through this engaged them in thinking and conversation about the rhythms of their year, seasonal adaptation, climate adaptation, and climate change.

Around 150 primstavs were made through this approach. While one might argue that the representations made through this exercise are seasonal clichés, the main objective of the exercise was to spark reflections on (1) how we do different things at different parts of the year, (2) that this has been different at different times in history, for example, the time of the historic primstav and now, and (3) that this might change again in the future, for instance, due to climate change. Based on the conversations we had with the children and adults during the exercise and the primstavs they created, we think the exercise worked well. We find the primstav to be a powerful image, a concrete object articulating what living with seasonal rhythms can be, suggestive in the context of climate change and climate adaptation; an object "good to think with" (Levi-Strauss, 1964). The adults and children were fascinated by the primstavs and excited about the task of making one for their own life, and they found the suggested connection of this to climate change and climate adaptation to be interesting.

We found that, as the exercise is currently designed, children younger than seven have some trouble responding to it, though they also seemed to enjoy the drawing. On the other hand, we found that most adults enjoyed it and responded very well to the primstav and climate adaptation analogy, and to the drawing and woodwork.

18.3.2 Brest, Kerourien, France

For the Kerourien, Brest, France partners, the coproduction work focused on collecting narratives in an ad hoc process that evolved along the way. The initial step of our coproduction process allowed us to anchor our actions in local stories and relate directly to our partner community and its values. This allowed us to free the coproduction team from dominant (and technocratic) climate change and adaptation discourses. Rather than adopting the pervading culture represented in the climate literature available to the community, we adopted coproduction work with narratives associated with everyday life, hardships, the joys and pain of migration, and engagement for greater justice.

Paying attention to local stories and the role of weather and climate within these stories led us to the realization that locally place-based climate service coproduction may actually entail working with multiple locations and associated issues. Coproduction challenged our routines. It pushed us to reconceptualize "place" as extending beyond the circumscribed location where our coproduction partners were living at the time. In the course of our work, place became a relational concept, the definition of which belonged to the members of the coproducing community—what mattered was their sense of place (see Stedman, 2003). Sense of place is an integrative concept (Saarinen et al., 1982), and it carries the characteristics of the physical environment and of the individual or group perceiving it. Sense of place plays an important role in place attachment, and others have shown, as we also observed, how memory is critical for migrant populations' relationships to places (Rishbeth & Powell, 2013). By adopting this extended concept of place, we recognized knowledge transcends national boundaries, and that time scales may relate to individual trajectories of past, present, and hoped-for futures. In their work analyzing the interplay of occupation, place, and identity, Huot and Rudman (2010) propose that individuals perform their identity in relation to place and occupation. This resonates with our results and the dynamic nature of the judgment individuals expressed of the place where they live and of the (now imagined) place that they once left, and to which many long to return. The status of an individual's perception of place shifts through time, as a manifestation of changes in their context, occupation and identity. Place-based coproduced climate services in such situations need to be reinvented to offer information that is dynamic, reconfigurable, and multilayered. This is another central challenge for the climate service coproduction research agenda.

Through the lens of priority setting, climate service coproduction has much to learn from participatory research and participatory planning. For instance, one aspect we did not address explicitly in this experiment in coproduction was that of gender. In the realm of participatory research there are many analyses showing that one should be explicit about gender and other identity dynamics at play—the "Whose voices? Whose choices?" questions that need to be answered (Cornwall, 2003). In the case of climate service coproduction, the dominant discourse may totally blind coproducers with its technocratic, pseudo-neutral, scientific stance; it seems too often to consider gender, race, class, and other social categories as not necessarily part of what deserves attention. Within the realm of climate change, our results point to "the importance of (re)politicizing co-production by allowing for pluralism and for the contestation of knowledge" (Turnhout et al., 2020). As Krauss (2020) writes, "a focus on narratives shifts the attention from the impact of climate on society to the myriad of entanglements between human and non-human actors in a changing climate." This shift in focus would allow us to ground further steps of climate service coproduction in the priorities of those most vulnerable to the vagaries of the world.

18.3.3 Dordrecht, the Netherlands

The Dutch partners worked with "knowledge directors" and creatives, such as graphical designers, applying their skills to climate change and adaptation to codevelop a visual workshop to help participants develop their vision and scenario design. Designing future visions and scenarios for timescales up to 30 years from the present can be very challenging, for experts and nonexperts. A clear advantage of this approach is that it helped facilitate the research process and local design of climate adaptation visions, options, and scenarios. It helped bridge the disparate worlds of different scientific disciplines and different stakeholder groups, including policymakers, residents, and researchers. This type of art approach also involved a very tactile, physical mediation, which stimulated active discussion and creative work during the workshops and put all participants on an even playing field. The latter is particularly important, for instance if there are tensions, or perceived barriers or differences in expertise between participants. The use of line art and uncluttered white space enticed people to add their own contributions. Visualizations of hypothetical neighborhood streets, with elements connected to the scenarios, and a timeline were printed on large cardboard wall posters, which allowed multiple participants to work on them. The collaboration between the research team and the graphic designers in developing the artwork also went very smoothly. One might speculate that graphic designers are used to working with a wide range of clients for purpose-driven assignments, compared to more art-driven processes such as painting or theater. Of course, the specific situation in the case study and the Dutch team also helped: the Dutch research team and project leader are very interdisciplinary, and Studio Lakmoes is specialized in knowledge visualization. Overall, this led to a very useful and productive process.

We encountered several challenges and limitations as well. Graphic designs and line art worked well to represent physical elements that are easily visualized into recognizable objects, such as houses, people, clouds, etc. However, the narratives also involved more nebulous notions such as identity, history, sense of community, and emotions such as hope and fear. Combining the visual art with narrative handouts (short descriptions with quotes per vision) worked well to bridge this potential gap. It may also be useful to combine this approach with other art forms. The approach was successful in combining the natural science and social science dimensions of adaptation, but in line with the previous point, it may be challenging to integrate with humanities-related aspects unless combined with other forms. Regarding the human dimensions of graphic design in art-science collaboration, while this approach worked very well, one could also argue that it is fairly goal-directed and instrumental. It helped elicit the "insider perspective" of the local communities. Other art forms might be better at providing an "outsider perspective"; for example, an artist observing the proceedings and providing their own unique perspective on it.

18.3.4 Gulf of Morbihan, France

For Morbihan's partners the aim was to engage local stakeholders through several forms and approaches, always aiming to achieve greater collective construction. This brought its challenges and pleasant surprises. The main outputs are: a collaboration with a designer to produce creative tools to help facilitate workshops; a long-term exhibition; and a small traveling exhibition with climate data panels and a comprehensive storytelling exercise connecting them to the metadata and dynamic mapping tool. The artistic process is embedded in extracts of interviews, led by the social sciences research team, and in the results of the workshop. All the information shared during the workshop was seeded by the presentation of local climate data and services.

In the Morbihan case, five artists came together to work on the projects along with a designer and the scientific team, resulting in some misunderstandings and possible frustrations. After a strong beginning, thanks to coordination work done by the designer and the local partner, Clim'action, it became clear with time that the will and expectations of some individuals in the group had given contradictory information on the possibilities of the project. Realigning everyone to share a common understanding of the project's objective and constraints took time but once the long-term objective was clarified, it became a strong foundation to build on.

The main objective was to develop artistic work on the basis of material collected through the interviews and workshop. In the process, we noticed the weak presence of some high-stake topics such as the loss of biodiversity, acidification of oceans, and changes in the rain regimes. This led to debate about the mission of the art and science project: (1) to only illustrate the concerns raised by the interviewed stakeholders, considering those will be the main ones of interest for the population of Vannes; and/or (2) to inject topics identified as key by the scientific and artistic teams to extend the horizons of the visitors.

Developing artistic work has been found to be an effective way to convey future narratives. We used maps and creative tools to help workshop participants engaged in the discussion with other approaches. The development of the Cataravane (the term used for one of the art forms is a neologism combining the words for "caravan" and "catamaran") and the storytelling exercise led to discussions among the artistic team and created the possibility to discuss our visions for the future. In order to push those exchanges beyond the boundaries of the team, multiple animation activities were planned to accompany the art and science process. The objective was to organize animation at the end of the exhibition, seizing the potential benefits from the immersive experience to further discuss spectators' perception of climate change and will to act or take decisions. Unfortunately, the difficulties of setting up the exhibition both administrative (due to COVID) and human did not allow us to carry out these animations.

The coastal path turned out to have a fundamental role as an apt location for the artistic exhibition. Propositions such as the "Pathway of possibles" allow us to argue that artistic work can also become chronotopes. Here, chronotopes refers to points in the geography of a community where time and space intersect and fuse. Time takes on flesh and becomes visible for human contemplation; likewise, space becomes charged and responsive to the movements of time and history and the enduring character of a people (...) Thus chronotopes stand as monuments to the community itself, as symbols of it, as forces operation to shape its members' images of themselves (Bakhtin, 1981).

Such works would result from interpretations of collectively built narratives of past, present, and future change converted into physical elements to allow people to follow ongoing changes in the territory. Scientific researchers and artists would analyze scientific information collectively to capture the spatial and temporal dimensions of current transformations, which would then be communicated through artworks acting as markers, for instance, of expected sea level rise or estimated coastal erosion in the future. This "sneak-peak" into the future, as well as the ability to observe the speed of these changes through these new chronotopes, could inspire community-led transformative practices.

The team confronted various difficulties in conducting a collective process on the artistic aspect of the project. Artistic media is the result of a complex combination of its creator's perception, personal artistic interpretation, and the message the artist wants to convey. Most of the artists engaged in the Morbihan project produce in situ creations. The uncertainty of the location for the long-term and short-term exhibitions was perceived as a difficulty for the artists, and perhaps originated from a misunderstanding or mismatch between the expectations of the project's research team and the creative processes of the artists. We also encountered the difficulty of creating a horizontal management dynamic with individuals from different backgrounds and sensitivities. It was not always clear who was in charge of the final decisions and how should those decisions should be taken. This was only exacerbated by the constraints created through the COVID-19 context of only being able to meet virtually. We speculate that the group dynamic would have taken another turn if we could have met in person.

The COVID-19 situation prevented in-person meetings on-site and between the artists. The horizontal management that was implemented was constrained by the budget discussion, which was in the hands of the scientific team or under the responsibility of the local partner. These two elements would usually be easily discussed in person through formal and informal conversations. The distance created by the virtual meetings emphasized difficulties of the inclusive process. Moreover, the artistic process mobilized individuals with different sensitivities and expectations for the project and its development. At various moments, the artists felt excluded from the process; we believe in-person meetings would have helped to avoid such situations.

The project benefited from the contacts and existing networks of artists at the site, involved in climate/environmental awareness art performance. However, it is important to note that in parallel to the official communication within the artistic committee, alternative communication and dynamics within the preexisting network were creating other narratives about the project and its aims.

18.3.5 Jade Bay, Germany

The initial intention of the art and science cooperation was to document ethnographic work in the coastal village of Dangast, the most prominent tourist location and one of the oldest sea bathing areas in this section of North Sea coast. The environmental photographer Werner Rudhart was invited, as was done for previous works on the Swiss Alps (Krauß, 2019), for two stays in Dangast; unfortunately, the second stay had to be canceled due to Covid-19 (Krauß, 2021a). The results of the first stay—in total several hundreds of photos were taken—served to be useful for documenting the coastal landscape. Many of the photos have been used in articles and presentations at conferences, and an e-book coauthored by Werner Krauß and Werner Rudhart about climate change in this coastal area is in preparation, with the working title "Blind spots – climate change in a coastal area." The e-book serves to demonstrate that climate and its changes are not abstract phenomena calculated by scientists, but visible in the work and maintenance of the dikes and the land.

The artist working for Jade Bay lives and works in rhythms which are different from those of their academic counterparts. "Hiring" an artist for a project involves complicated time- and financial-management. Furthermore, it is impossible to predict in advance if the collaboration will be successful. Art is a process dependent upon intuition and serendipity, as is ethnography to a certain degree. In practice, the photo documentation of the ethnographic fieldwork was an interdisciplinary and continuous dialog between the photographer, the human and nonhuman motifs, and the anthropologist. A documentary is not just a depiction of something that already exists; rather, it is an active process producing something new. Photography did not simply add something to the body of ethnographic work that already existed; instead, it turned into a collaborative activity with its own dynamic and new insights.

The art-and-anthropology collaboration is different from the collaboration with science. Climate science is exclusively about the production and exchange of data and information. From the anthropologist's perspective, there was a strong feeling of uneasiness about reducing complex field experiences and long conversations or interactions into "data" or "information," or to figure out whether local perceptions of climate change were "right" or "wrong." The photographer shared this feeling of unease from the beginning: what kind of data was he expected to produce? What kind of product was he expected to deliver? It is difficult to get rid of this kind of thinking, which is common in interdisciplinary projects. In our art-and-anthropology collaboration, we discussed this kind of pressure from the beginning and started to reset the default position. We both had already worked in different settings about landscapes, and we compared our methods: how do we engage with a landscape, what does it take to understand, to see and to feel the specific atmosphere of a territory, understood as an amalgam of the physical, the geological, the social and political atmospheres? Is it possible to "see" climate change, and what does it mean to live in the Anthropocene? Our explorations were focused on the blind spots; our goals were: (1) to visualize climate change; and (2) to show "the phantoms of the Anthropocene," as Tsing et al. (2017) put it, in a specific combination of words and pictures.

Dangast as a field site turned out to be instructive. The work of the famous painter Franz Radziwill, who spent most of his life in Dangast, served as important inspiration. In his paintings in the style of magic realism, he links different elements of reality in new ways and integrates the horrors of the World Wars, the lifeworld of modernity and environmental degradation (Krauß, 2021b). Most of all, he was fascinated by the permanently changing light in this coastal area. Photography is about light, too, and light is linked to weather and time. The light of the North Sea coast is one of the reasons people love to spend their time in this area, and it is integral part of the photos of Werner Rudhart.

In a first step, we absorbed local climate information in many ways: as daily weather, in talks about the weather, by deciphering historical landmarks or in interviews. People were curious about our work and easily understood our intention to depict climate change beyond statistics and numbers. Our project served as an entry point for debating what it means to live in a coastal landscape with a changing climate.

In a second step, we focused on the dikes that surround the Jade Bay. Our working hypothesis was that climate change sits in the dikes, and that the dikes are the result of work, maintenance, construction, and planning (Krauß, 2020). In documenting the work that it takes to protect the land, traces of climate change become visible.

Finally, the simultaneous existence of the UNESCO world heritage site Lower Saxony Wadden Sea and the highly industrialized landscape with Wilhemshaven on the horizon, invited us to document the phantoms of the Anthropocene. Unfortunately, Covid-19 prevented our finalizing the last step of the project, the portrayal of the main interviewees of the anthropological research as inhabitants of the Anthropocene. There is only a small group of portraits; Werner Rudhart arranged for each inhabitant to be photographed in specific settings where the anthropogenic landscape merged with the professions of the interviewees.

The art-and-anthropology cooperation turned out to be successful despite the interruption of the pandemic, and the results open up a new way to engage with climate change as a practice, as a place-based coastal phenomenon. The photos display dike maintenance, scenes of constructed land- and sea-scapes, moments of regional Fridays for Future demonstrations as much as the interaction of humans and nonhumans in unusual arrangements. Focusing on the blind spots in climate communication thus enables new forms of collaboration and place-based climate action.

18.4 Metadata and dynamic mapping perspectives for local representations

Constructing and using a metadata model (metadata refers to a set of data that describes and gives information about other data) and a metadata tool is a complex undertaking. The time-consuming aspects of preliminary tasks may mask the hypothetical future benefits. For example, encoding metadata for the whole corpus of narratives was too tedious a task for most partners, whether via the Excel spreadsheets provided or directly into the QGis-based tool we developed, even with templates to simplify the task. The future advantages that would potentially result from such efforts remained too unclear—and were indeed not demonstrated—to convince project partners to dive into this effort. Timing was also a challenge, since narrative collection and the metadata scheme proceeded and matured in parallel. Site leaders had already started to use their traditional methods of analysis when the possibilities that the metadata offered in this respect became clearer. However, the prospective of visual representations of results was an incentive.

In many respects, we opted for a resolutely open metadata model:

- no field was mandatory;
- new options could be added to drop-down menus;
- new metadata fields could be proposed;
- numerous fields accepted free text.

   The obvious advantages of such a model are its flexibility, the freedom allowed to the person entering the metadata, and the high degree of nuance that can be expressed. This approach also has strong drawbacks. The fact that the scheme expands as time goes by and more options or fields are added hinders the comparison of narratives for which metadata were recorded at different stages of its evolution. Empty fields may result in discarding a narrative from statistical studies. Free text cannot be automatically read and makes comparison between narratives arduous. To guarantee sensible information content, easy analysis and at least minimal comparability, a balance between flexibility and rigor must be achieved. Flexibility may be preferred as long as the scheme is under development but at some point, the model must be frozen in order to be usable.

Data of a subjective nature are essentially different from the data used in the physical sciences for which metadata schemes have been established for a long time. Even more crucially, the subjectivity of the researcher who documents a narrative influences the metadata content, which is less likely to be the case for metadata pertaining to physical measurements. Whether a number lies under a given threshold, for example, will not depend on the metadata provider, whereas an element of a narrative may be interpreted by one researcher as an allusion to a potential climate-related point, and not by one of their colleagues. In this sense, users must be aware of the relative nature of narrative metadata.

Due in part to the novelty of the approach and in part to the different styles of practice, concepts and vocabulary of the researchers in their respective fields, developing a metadata scheme for narratives involved substantial communication and lasted for the duration of the project. The scheme that was progressively completed to meet the partners' expectations was only tested on some examples, not numerous enough to conclude if the concept was adequate, even less to document all the collected narratives and serve as an analysis or dissemination tool. Yet, we hope the outcome of this pioneering exercise will be a starting point for future projects.

GIS (Geographic Information System) brings together various more- or less-sophisticated techniques, of which cartography is probably the best known. Cartography produces rich and relatively easy-to-understand graphics to enlighten a discourse. It also lays the groundwork for certain analyses by allowing the spatial dimension of studied problems to be taken into account. Geographers start by mapping a phenomenon and then use this spatialization to look for correlations with other aspects or other mapped phenomena. By attempting a hybridization between humanities, social, and natural sciences, the CoCliServ project explored various paths, including GIS.

The ambition of CoCliServ in relation to mapping was more than just to produce beautiful illustrative maps to brighten up reports or to support meeting discussions with local communities. That said, we did also make and use maps for these purposes, as maps are a common tool in scientific discussions about environmental issues, and are presented in other deliverables. Beyond this, we intended to explore the feasibility of producing a tool to dynamically share field observations from many types of sources and ensure a maximum preservation of the collected information. In the social sciences, not all information can be saved and standardized; there is always a limit, as a written text will fall short of capturing and conveying the depth and intensity of personal feelings or lived experience. The intention here was to develop a database model to push the limits of data compilation and digitization from informal social science sources. Our first step was to create a metadata scheme with the field investigators, the second was to produce a database model and the third was to make a user-friendly interface for different types of users, dealing with different types of information. These three theoretical steps were explored simultaneously with lots of back and forth during the project. "Whether such a dream may really be a nightmare is another topic" (Peuquet & Marble, 1990). We aimed to compose an IT tool for users who are not familiar with such technology in their usual activities (e.g., Caquard, 2013; Vivant et al., 2014). This proved a never-ending task, since the needs were always too specific and a versatile solution would require high computer literacy, as it would need to remain very abstract. The challenge is to find the right balance between solutions that are simple to grasp but unable to account for the nuances of human reality, and very sophisticated solutions that can accommodate such nuances but are too complicated to learn and use.

We have come to the conclusion that mixing these opposite constraints might be feasible if:
- users are ready to learn;
- there are a few users responsible for more difficult tasks;
- someone is there to manage the system on a technical level.

18.5 Lessons learned and final conclusions

Three papers directly linked with the sites' efforts were published in the past year (Baztan et al., 2020; da Cunha et al., 2020; Krauß & Bremer, 2020). These papers are the first arts and sciences papers published in the dedicated journal, "Climate Risk Management." These publications illustrate the first key point for our consortium and the broader research community: it is important to validate experimental local approaches with mainstream dynamics through peer-review publication.

The second lesson from our CoCliServ experiences is the room for improvement revealed through the peer-review validation process. This room for improvement expands in three main directions: the art and science processes and their methodologies; making social transformation intentions explicit; and linking local challenges with national and European Framework Directives related to climate services.

The third key lesson is to understand the power of using art forms to represent local climate information. Now we know how local stakeholders engage in the process and how important they are for engaging with local and regional actors on climate change; from here, we need to link arts and sciences with other modes of representation. Art-based intervention has been proposed in the context of climate change for quite some time (Lippard, 2007; Volpe, 2018). This is not without pitfalls such as instrumentalizing art and reproducing dominant categories and codes through art (Miles, 2010). To counter these risks, experiments have shown that through public participation and activism, art may be empowering, and may shift attention to issues that question dominant paradigms (Sommer & Klöckner, 2019).

Several dimensions have been identified for collaborations between art and science: new understandings and capacities within and across the arts and sciences involved (Gabrys & Yusoff, 2012); catalyzing explorations of the scientific context and critical re-imaginations of research practices (Rödder, 2017); helping to engage multiple senses and emphasizing social interaction within research practices; aiding participating researchers in thinking creatively (Jacobson et al., 2016); redesigning social relations to natural systems (Armstrong & Leimbach, 2019); rearticulating politics and knowledge (Latour, 2011); offering more effective approaches to engaging multiple publics in climate-compatible behavior change; and engaging explicitly with the underresearched issue of the role of place attachment and local, situated knowledge in mediating the influence of climate change communication (Burke et al., 2018).

Drawing from these observations, we developed the working hypothesis that iterative art and science approaches have the potential for instigating and sustaining community dialog through efforts to coproduce climate services. We see art as essential for making the concept of climate services more meaningful in a specific place, and our approach focused on narratives as an entry point for coconstruction.

Developing strong connections between art and science enables the re-articulation of the scientific description of the world (Latour, 2011). The teams at each CoCliServ site, in their diversity, explored polysemic concepts such as climate change and associated "services" through their own local arts and sciences approaches. Our investigation entailed examining the potential of art "gestures" (Citton, 2012) and the "practices of everyday life" (de Certeau, 1990) to facilitate cultural translation between different fields of knowledge and the associated diversity of priorities in bringing them to action. While the five sites each have their own particularities to their approaches, here we synthesize the main lessons learned from the sites along with the common points in the sites' processes, and the final conclusion.

18.5.1 Conclusions on local representations for codeveloping climate services

For the purpose of this research, we used Vaughan and Dessai's (2014) definition of climate services: "The aim of climate services is to provide people and organizations with timely, tailored climate-related knowledge and information that they can use to reduce climate-related losses and enhance benefits, including the protection of lives, livelihoods, and property." When considering this definition, we focused on providing people with climate tailored knowledge, "people" understood to be the members of local communities. The central challenge we wished to address was that of "tailoring" climate knowledge and information for communities at the margins, or at the core, who may or may not be aware of climate issues. Such tailoring of climate knowledge is closely associated with the ability to establish iteration (Dilling & Lemos, 2011) and dialog between scientists and nonscientists in the course of knowledge production and use. Climate knowledge coproduction, the "deliberate collaboration of different people to achieve a common goal" (Bremer & Meisch, 2017) has been proposed for quite some time (e.g., Lemos & Morehouse, 2005) to address challenges of initiating and maintaining such reiterations and dialogs. But what if this dialog needs to be established on grounds other than those of climate change?

Bremer and Meisch (2017) conducted an extensive mapping of the literature on climate change research coproduction. They identify a series of eight "conceptual lenses" and call for a "self-reflexive transparency when using coproduction concepts" to address the concepts' polysemy. Within their framework, CoCliServ work lies at the juncture of several objectives associated with these various lenses: we wanted to integrate nonscientists as coinvestigators (extended lens); we aimed to sustain interactions between climate science providers and users (iterative interaction lens); we pursued a goal of empowering local experience, and thus of local knowledge (empowerment lens); we recognized the need to facilitate social learning about climate issues (social learning lens); and we were embedded in a culturally rooted goal to improve public service through the joint engagement of government agencies and citizens in the production of new knowledge (public services lens).

All these objectives are associated with acknowledging the current uneven distribution of access to, and benefits from, climate services development. For instance, Harjanne (2017) surveyed institutions related to climate services to identify how they justify the need for climate services (as a departure from climate science), and identified the following: global and widespread nature of the climate challenge; specific industry needs; socio-economic value; technological potential; and deficient supply and demand. We envisioned the coproduction of climate services, not because we perceived coproduction as a "value in itself" (Voorberg et al., 2015), but because we understood coproduction as a means to create and nurture sustained interaction with communities while contributing to their empowerment. We aimed to explore means for correcting the inequitable distribution of climate change knowledge for action.

We hypothesized that part of this uneven distribution may reflect that not all communities are equal and some are facing such immediate challenges that climate change may be invisible to them. This hypothesis called for working to shift awareness to the actual or potential, current or future, connections between everyday nonclimate concerns and climate issues. Such a shift called for a practical intervention, centered on local culture. We chose to work hand-in-hand with artists to conduct such an intervention, as art is well identified as an approach to make visible the "invisible or almost-visible" phenomenon of climate change (Knebusch, 2008). Art is also identified as facilitating access to narratives in general, and climate narratives in particular (Roosen et al., 2018).

18.5.2 Final conclusions

Representations and their associated modes were at the very core of what the five sites were working toward: turning "matters of fact" into "matters of concern," and using narratives to transcend oppositions in representations of scientific and local knowledge, nature-culture, global-local, in order to coproduce climate services. The complexity of the starting point and the exigence of the learning process brought the partners through three key stages that we consider major contributions for the climate services community:

1. Climate services do not necessarily appear in the top priorities of the population. Our efforts demonstrate that the modes of representation and their associated processes contain the priorities of society as a whole. Standardizing across the five sites resulted in protocols to connect the populations' priorities with climate services, and from there climate services with the most pertinent and integrative local modes of representation.
2. In order to have access to and represent the complexity related to climate services and its connection with society as a whole, we found a way to create an integrative context for the three modes of representation: metadata, dynamic mapping, and art. In this we learned to be explicit in the related processes at play in the emergence of new forms of representation that link and integrate what was previously separate for reasons of disciplinary inertia.
3. Local particularities and constraints clearly appear in our efforts. Universal concepts related to climate services can be applied locally with long-term perspectives only if they are concordant with local values and "acceptable" processes.

The formal context for "modes of representation" enabled us to recognize the importance of making social transformation intentions that are explicitly linked with local challenges and values to connect with national and European Framework Directives related to climate services. We reiterate the importance of having local stakeholders engage in the climate services process in order to forge common commitments and incorporate value perspectives, even those that may be polarized, throughout society as a whole.

Acknowledgments

Project CoCliServ is part of ERA4CS, an ERA-NET initiated by JPI Climate, and funded by FORMAS (SE), BELSPO (BE), BMBF (DE), BMWFW (AT), IFD (DK), MINECO (ES), ANR (FR) with cofunding by the European Union (Grant 690462).

References

Armstrong, K., & Leimbach, T. (2019). Art-eco-science. Field collaborations. *Antennae: The Journal of Nature in Visual Culture* (48), 108–127.

Bakhtin, M. M. (1981). *The Dialogic Imagination: Four Essays. Translated by Caryl Emerson & Michael Holquist*. University of Texas Press.

Baztan, J., Vanderlinden, J.-P., Jaffrès, L., Jorgensen, B., & Zhu, Z. (2020). Facing climate injustices: Community trust-building for climate services through arts and sciences narrative co-production. *Climate Risk Management*, 30, 100253. Available from https://doi.org/10.1016/j.crm.2020.100253.

Bremer, S., & Meisch, S. (2017). Co-production in climate change research: Reviewing different perspectives. *Wiley Interdisciplinary Reviews: Climate Change*, 8(6), e482. Available from https://doi.org/10.1002/wcc.482.

Burke, M., Ockwell, D., & Whitmarsh, L. (2018). Participatory arts and affective engagement with climate change: The missing link in achieving climate compatible behaviour change? *Global Environmental Change*, 49, 95–105.

Caquard, S. (2013). Cartography I: Mapping narrative cartography. *Progress in Human Geography*, 37(1), 135–144. Available from https://doi.org/10.1177/0309132511423796.

Citton, Y. (2012). Gestes d'humanités: anthropologie sauvage de nos expériences esthétiques. *Armand Colin* (2012).

Cornwall, A. (2003). Whose voices? Whose choices? Reflections on gender and participatory development. *World Development*, 31(8), 1325–1342.

da Cunha, C., Rocha, A., Cardon, M., Breton, F., Labeyrie, L., & Vanderlinden, J.-P. (2020). Adaptation planning in France: Inputs from narratives of change in support of a community-led fore-sight process. *Climate Risk Management*, 30, 100243. Available from https://doi.org/10.1016/j.crm.2020.100243.

de Certeau, M. (1990). L'invention du quotidien. *Arts de faire*, 1, 57–63.

Dilling, L., & Lemos, M. C. (2011). Creating usable science: Opportunities and constraints for climate knowledge use and their implications for science policy. *Global Environmental Change*, 21(2), 680–689. Available from https://doi.org/10.1016/j.gloenvcha.2010.11.006.

Gabrys, J., & Yusoff, K. (2012). Arts, sciences and climate change: Practices and politics at the threshold. *Science as Culture*, 21(1), 1–24.

Harjanne, A. (2017). Servitizing climate science. Institutional analysis of climate services discourse and its implications. *Global Environmental Change*, 46, 1–16.

Huot, S., & Rudman, D. L. (2010). The performances and places of identity: Conceptualizing intersections of occupation, identity and place in the process of migration. *Journal of Occupational Science*, 17(2), 68–77.

Jacobson, S. K., Seavey, J. R., & Mueller, R. C. (2016). Integrated science and art education for creative climate change communication. *Ecology and Society*, 21(3).

Knebusch, J. (2008). Art and climate (change) perception: Outline of a phenomenology of climate. In S. Kagan, & V. Kirchberd (Eds.), *Sustainability: A new frontier for the arts and cultures* (pp. 242–261). Frankfurt: Verlag fur Akademische Schriften.

Krauß, W. (2019). Alpine landscapes in the Anthropocene: Alternative common futures. *Landscape Research*, 43(8), 1021–1031. Available from https://doi.org/10.1080/01426397.2018.1503242.

Krauß, W. (2020). Narratives of change and the co-development of climate services for action. *Climate Risk Management*. Available from https://doi.org/10.1016/j.crm.2020.100217.

Krauß, W. (2021a). Coastal atmospheres: The peninsula of the blessed and the art of noticing. *Maritime Studies*, 329–340. Available from https://doi.org/10.1007/s40152-021-00241-2.

Krauß, W. (2021b). 'Meat is stupid': Covid-19 and the co-development of climate activism. *Social Anthropology*, 29(1), 241–244. Available from https://doi.org/10.1111/1469-8676.12993.

Krauß, W., & Bremer, S. (2020). The role of place-based narratives of change in climate risk governance. *Climate Risk Management*, 28, 100221. Available from https://doi.org/10.1016/j.crm.2020.100221.

Latour, B., 2011. Waiting for Gaia. Composing the common world through art and politics, Lecture given for the launching of SPEAP. French Institute, London.

Lemos, M. C., & Morehouse, B. J. (2005). The co-production of science and policy in integrated climate assessments. *Global Environmental Change*, 15(1), 57–68. Available from https://doi.org/10.1016/j.gloenvcha.2004.09.004.

Levi-Strauss, C. (1964). *Totemism*. London: Merlin Press.

Lippard, L. (Ed.), (2007). *Weather Report: Art and Climate Change*. Boulder, CO: Boulder Museum of Contemporary Art.

Marschütz, B., Bremer, S., Runhaar, H., Hegger, D., Mees, H., Vervoort, J., & Wardekker, A. (2020). Local narratives of change as an entry point for building urban climate resilience. *Climate Risk Management*, 28, 100223.

Miles, M. (2010). Representing nature: Art and climate change. *Cultural Geographies*, *17*(1), 19–35.
Peuquet, D. J., & Marble, D. F. (Eds.), (1990). *Introductory readings in geographical information systems*. London: Taylor and Francis.
Rishbeth, C., & Powell, M. (2013). Place attachment and memory: Landscapes of belonging as experienced post-migration. *Landscape Research*, *38*(2), 160–178. Available from https://doi.org/10.1080/01426397.2011.642344.
Rödder, S. (2017). The climate of science-art and the art-science of the climate: Meeting points, boundary objects and boundary work. *Minerva*, *55*(1), 93–116.
Roosen, L. J., Klöckner, C. A., & Swim, J. K. (2018). Visual art as a way to communicate climate change: A psychological perspective on climate change–related art. *World Art*, *8*(1), 85–110.
Saarinen, T. F., Sell, J. L., & Husband, E. (1982). Environmental perception: International efforts. *Progress in Geography*, *6*(4), 515–546.
Sommer, L. K., & Klöckner, C. A. (2019). Does activist art have the capacity to raise awareness in audiences?—A study on climate change art at the ArtCOP21 event in Paris. *Psychology of Aesthetics, Creativity, and the Arts*.
Stedman, R. C. (2003). Is it really just a social construction?: The contribution of the physical environment to sense of place. *Society & Natural Resources*, *16*(8), 671–685.
Tsing, A. L., Swanson, H. A., Gan, E., & Bubandt, N. (Eds.), (2017). *Arts of living on a damaged planet. Ghosts and monsters of the Anthropocene*. Minnesota: The University of Minnesota Press.
Turnhout, E., Metze, T., Wyborn, C., Klenk, N., & Louder, E. (2020). The politics of co-production: participation, power, and transformation. *Current Opinion in Environmental Sustainability*, *42*, 15–21.
Vivant, A.-L., Garmyn, D., Gal, L., & Piveteau, P. (2014). The Agr communication system provides a benefit to the populations of Listeria monocytogenes in soil. *Frontiers in Cellular and Infection Microbiology*, *4*. Available from https://doi.org/10.3389/fcimb.2014.00160.
Volpe, C. (2018). Art and climate change: contemporary artists respond to global crisis. *Zygo*, *53*(2), 613–623.
Voorberg, W. H., Bekkers, V. J., & Tummers, L. G. (2015). A systematic review of co-creation and co-production: Embarking on the social innovation journey. *Public Management Review*, *17*(9), 1333–1357. Available from https://doi.org/10.1080/14719037.2014.930505.

CHAPTER 19

Agricultural water pollution

Thomas Shahady
University of Lynchburg, Lynchburg, VA United States

Chapter Outline

19.1 Introduction and background 365
    19.1.1 Water consumption and contamination tied to agriculture 365
    19.1.2 Pollutants 366
    19.1.3 Land Use 367
19.2 Pollution problems generated from agricultural practices 368
    19.2.1 Biosolids and contamination 368
    19.2.2 Fertilizers and eutrophication 369
    19.2.3 Toxic contamination from pesticides 370
    19.2.4 Bacterial contamination 370
19.3 Shifting practices and climate change 372
    19.3.1 Globalization and trade 372
    19.3.2 Climate change 372
    19.3.3 Land management (best management practices) 373
    19.3.4 Emerging concerns 373
19.4 Water pollution control 374
    19.4.1 Water quality 374
    19.4.2 Best management practices 375
    19.4.3 Land management practices 375
    19.4.4 Intensive management practices 376
19.5 Agriculture and a sustainable future 377
References 377

19.1 Introduction and background

19.1.1 Water consumption and contamination tied to agriculture

When considering global water use, Hoekstra and Chapagain (Hoekstra & Chapagain, 2007) suggest volume of water consumed is related to wealth, level of meat consumption, climatic conditions and agricultural efficiency. We know that global water use has increased sixfold over the previous 100 years and will continue to increase at roughly 1% each year forward (UN World Water Development Report, 2018).

If trending is accurate, by 2050, the global demand for crop production will increase 70%–110% (Alexandratos, 2009; Tilman et al., 2011) placing tremendous stress on water resources. Following this trend if wealth continues to increase will be greater consumption of meat and dairy products (Delgado, 2003). With this increase, the use and demand for pesticides, fertilizers and farmable land will continue (Cassidy et al., 2013; Metson et al., 2012). This will put an even greater burden on stressed water and ecological systems to process the waste and maintain some suitable stasis within ecological systems. In the upcoming decades, we may not be able to produce agricultural products in the same way as the past assuming the environment will substantially process the waste. New paradigms and processes will be necessary.

Agricultural Water Pollution is now considered an emerging and worsening pollution problem because of current and increasing food demands (Evans et al., 2019). As global population and consumption needs increase, more land use will be needed to meet demand, more water for irrigation, and greater consumption of agrochemicals, animal feeds and antibiotics (Mateo-Sagasta et al., 2018). This generates increasing changes in water flow patterns and greater concentrations of pollutants enter watercourses creating worsening water quality problems. Intensification of both crop and livestock production can cause these problems to become untenable. Agriculture already outproduces human populations in contaminated wastewater and excreta (Evans et al., 2019). Simply increasing agricultural output to meet increasing demand is not a solution.

19.1.2 Pollutants

Pollutants generated by agriculture are multifold. Nutrients such as phosphorus and nitrogen, sediment, pesticides, and microorganisms are the most prevalent and currently problematic (Manyi-Loh et al., 2016; Noe & Hupp, 2005; Tilman et al., 2011). Pollutants generated by various types of agricultural practices can now be generalized (Cuffney et al., 2000; Stone et al., 2005). One example is the cash crop industry. It has become an increasing phenomenon supporting economic progress globally but also degrading water quality (Su et al., 2016). Driven by productivity rather than sustainability (Munoz et al., 2008), growers use very high concentrations of nutrients to produce crops. This practice is highly polluting to surrounding water bodies. Livestock grazing has been implicated as the driver of impairment for most US stream kilometers (Hughes & Vadas, 2021) (Fig. 19.1).

Genetically modified crops (GMOs) have created a revolution in agriculture. Driven by ease of production and promise of greater profits, much of the soybean, corn, and cotton now grown is GMO (NAS National Academy of Sciences, 2016). Yet production using GMO crops has not increased productivity but required greater uses of pesticides and fertilizers (Quarles, 2017). Extensive application of glyphosate to soils is understood to damage mycorrhizae, nitrogen fixation and the microbiota lowering the natural production of nutrients (Duke et al., 2014; Wolmarans & Swart, 2014). Herbicide use has also increased

Figure 19.1
Coffee plants in Costa Rica. This is an example of cash crops for profit even on a small scale.

with the planting of GMO crops (Benbrook, 2016). The greater reliance on these crops will increase the pollution to waterways.

Concentrated animal feeding operations (CAFOs) are a type of agriculture where animals are confined to indoor stalls their entire lives at very high densities until they are taken for slaughter. These operations produce excessive amounts of waste, nutrient, and bacteriological pollution often many times greater than comparable human waste (Burkholder et al., 2007). These types of operations are also responsible for chronic pollution problems in streams in direct proximity to impacted areas (Mallin et al., 2015). The amount of livestock production produced in this fashion along with concurrent increases in nutrient and bacteria pollution associated with this practice is increasing (Raff & Meyer, 2021). This is a type of agriculture may be very problematic as it increases because of severe concentrations of waste and nutrient pollution.

19.1.3 Land Use

The direct impact from the conversion of forested watersheds into agricultural production is almost immediately followed by visible changes in stream water quality. Forest ecosystems are the natural buffer for precipitation infiltration into groundwater recharging the system and maintaining excellent water quality. As this system is altered, a steady trend of degradation occurs. This is not limited to only forested watersheds. Any alteration of the natural balance of land cover to precipitation directly influences streams that drain the landscape.

Many studies have documented these changes. Sobolewski (Sobolewski, 2016) found that lakes in catchments with greater than 60% agricultural land use contained much worse water quality than those with lesser amounts. Much of the problem is associated with export of nutrients, sediment, and microorganisms. The export of nitrogen, phosphorus, agrochemicals, and pathogens are significantly increased from land used in the production of livestock (Hooda et al., 2000). While the nutrients are valuable for soil health, waste in these operations is considered a disposal problem rather than a useful source of plant nutrition.

Overall impact of land use on receiving water quality may be specific to the type of agricultural activity rather than the cumulative total impact of agriculture (Wijesiri et al., 2018). They found that *type* of agriculture was much more significant than amount of agricultural land disturbance throughout a basin. Their results suggest irrigated agriculture and plantations while comprising significant increases in land area were weak contributors of pollutants. Concurrently, grazing and production forestry were the largest pollutant contributors. In my work in Costa Rica, while CAFO operations demonstrated significant impairment to receiving water quality, overall land use patterns were more predictive for metric indicators. The evaluation of activity type and translation into water quality metric evaluators is an important part in determining land use impacts on water quality (Fig. 19.2).

19.2 Pollution problems generated from agricultural practices

19.2.1 Biosolids and contamination

The application of biosolids (concentrated human sewage sludge) on agricultural lands is a management practice considered as a soil conditioner that improves the physical, chemical,

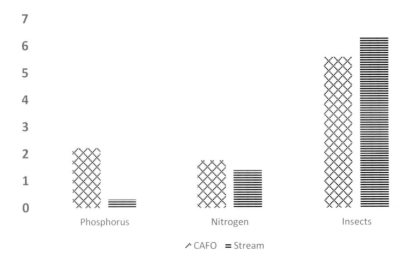

Figure 19.2
Phosphorus, nitrogen, and insect metrics comparing a CAFO impacted stream with adjacent receiving stream. The CAFO impacted stream displayed declining water quality but most apparent is the large dichotomy in phosphorus discharged based on the high density of pigs in the CAFO. *CAFO*, concentrated animal feeding operations.

and biological properties of soils while solving disposal problems for municipal sewage treatment plants (Lu et al., 2012). Application has been shown to be beneficial for many types of agricultural applications. Yet, concerns over the practice continue including persistence of pathogens, eutrophication from excessive phosphorus contained in the biosolids, improper land management and other types of pollution (Lu et al., 2012; Pourcher et al., 2007). The resistance of pathogens to antibiotics along with bacteria developing multidrug resistance in the environment has created an international health crisis (Kalavrouziotis, 2016).

Further, continuation of this practice creates a scientific conundrum when application of a recognized pollutant is allowed as an agricultural practice. In the process to purify wastewater, the contaminates are concentrated into a sewage sludge. As a result, the US National Academy of Sciences has issued a warning that standard strategies to manage the risk of land application do not protect public health (Kalavrouziotis, 2016; National Academy of Sciences, 2002). The collective impact of this problem is hard to quantify. Interestingly, while manure application contains the greatest risk due to bacterial infections, human waste from biosolids carries the greatest viral risk for infection (Oun et al., 2014). Research to date suggest COVID-19 is not a virus of concern in the spreading of biosolids (Jones et al., 2020).

19.2.2 Fertilizers and eutrophication

Agricultural nonpoint source pollution is considered a major source of stream and lake contamination and has been implicated as the reason many water quality goals have not been attained (Daniel et al., 1998; Steffen et al., 2015). Because the environmental problem of eutrophication is driven by nutrient pollution, and agriculture is driven by the application of nutrients to crops, nutrient management has become a central point of focus (Conley et al., 2009). Ever since the seminal work of limnologists to link the concentrations of phosphorus (a limiting nutrient) to worsening eutrophication (Schindler, 1977), efforts to limit this nutrient into freshwater ensued. Coupled with land use changes, application of manufactured fertilizers and loss of natural capacity for rivers to assimilate nutrients, hypoxia from major river systems has developed (Rabalais et al., 2002).

This pollution has developed from four basic areas due to agriculture: nutrient levels in livestock feed, soil management, manure management, and transport management. Moving into organic farming is a possibility. While lowering yields, it reduces the need for extensive cereal production as feed and nutrient management problems surrounding this issue (Gaudar et al., 2021). Altering nutrient composition of feeds is also a possibility. Soil management is of the utmost importance not only in the management of nutrients but in the entirety of life itself (Powlson et al., 2011). By controlling erosion, you will control nutrients. Finally, manure and transport into rivers necessitates control. An efficient manure management system will limit or prevent manure or its constituents from gaining undesirable access to the larger environment. Sound manure management contributes to health and environmental, economic and social benefits (Malomo et al., 2018).

19.2.3 Toxic contamination from pesticides

Beginning with the development and widespread use of DDT (Dichlorodiphenyltrichloroethane), and later fungicides, herbicides and pesticides, worldwide use of these chemicals in agriculture tops 4 billion tons worldwide (Langenbach et al., 2021). Impact of pesticides on both the environment (Hayes et al., 2006) and human health (Ye et al., 2017) is considerable. What makes the continued use of pesticides so concerning is the lack of adequate knowledge on proper use, lack of awareness among users of health and environmental impacts, lack of extension support to ensure safety, and inadequate regulation by the government (Bhandari, 2014). The continued use of these products to boost agricultural output at environmental expense will begin to take its toll on many essential ecosystem services.

Disposal of packaging in the field, lack of proper PPE (Personal Protective Equipment), improper cleaning of clothes, low understanding of environmental impacts, and low educational levels all plague this problem (Sánchez-Gervacio et al., 2021). The belief that water pollution is critical to the use of pesticides is important in the control of pesticide use (Diendéré et al., 2018). While other pollutants are more diffuse in impacts, pesticide toxicity is often directed at stream wildlife. All of these pollutants may have antagonistic effects. There is a critical need for more studies on the interactive impacts of these pollutants with other contaminants (Cornejo et al., 2019).

19.2.4 Bacterial contamination

Precipitation creating extensive erosion of agricultural lands then depositing this sediment rich in organic material and nutrients along stream beds creates the ideal environment for bacteria such as *Escherichia coli* to prosper. We know that sediment loading as a direct result of erosion of exposed land or streambank failures contains heavy loads of bacteria (Chen & Liu, 2017). These bacteria survive in this sediment much longer than in overlying waters (Pachepsky & Shelton, 2011). This sediment acts as a sink with the potential of continued bacterial contamination to overlying waters long after the contamination event (Mallin et al., 2017). Thus, agriculture is threatening aquatic resources through creation of extensive beds of bacteria, potential pathogens and other pollutants that will be resuspended continually as sediment moves through these systems.

Therefore agricultural land use in this instance is the driving mechanisms for elevated *E. coli* concentrations (Pandeya et al., 2012). More specifically, rainfall events directly after the application of manure or upon cattle-impacted lands can cause elevated bacteria in receiving streams (Guber et al., 2007; Soupir et al., 2006). Crop lands can be an additional source for elevated bacterial loadings (VanderZaag et al., 2010) depending on land management strategies. In watersheds predominately covered by forest, we find relatively good water quality while in watersheds increasingly plagued by human alteration and economic activities, we find greater concentrations of water pollutants (Baker, 2003; Buck et al., 2004; Tong & Chen, 2002).

While concern over land use and river contamination can be quantified, the translation to public health risk and disease is more elusive. Using climate and epidemiological records, Rose et al. (2000) found statistical evidence suggesting a correlation between storm events and disease outbreak. DeFlorio-Barker et al. (2018) estimated the cost of recreational waterborne illness on United States surface waters to be an estimated 4 billion annually. Surface water contamination results in an estimated 90 million illnesses nationwide with a cost of $2.2–$3.7 billion annually. Illnesses of moderate severity are responsible for over 65% of the economic burden while severe illnesses (hospitalization or death) were responsible for approximately 8% of the total economic burden.

We now understand that agricultural land runoff from intensive livestock farming practices increases risks to environmental and human health. This occurs by disseminating pollutants and overwhelming ecosystem processes reducing the ability of the natural processes to assimilate and reduce harmful bacteria and toxic chemicals (Beattie et al., 2020). The interactions of all of these components create a system of multiple stressors that is difficult to quantify (Ormerod et al., 2010). These are emerging problems that need study as we quantify agricultural water quality problems (Fig. 19.3).

Figure 19.3

Heavy sediment contamination in an agricultural dominated river system. *Escherichia coli* associated in this river often exceeds 2000 cfu mL^{-1} and is a health risk to livestock and people using this resource.

19.3 Shifting practices and climate change

19.3.1 Globalization and trade

The lack of water resources or the inability to produce enough food due to available land or water shortage can be remedied globally when food is traded between countries. Water and land used in food production is thus transferred virtually (defined as water and land use embedded in an agricultural good) exchanged from the exporter to the importer (Hoekstra & Chapagain, 2008). In this manner, agricultural use of water resources, land, irrigation and production of food are now interlinked economically throughout world. It is becoming understood that global trade is repositioning land and water between developed and undeveloped countries. Developed countries are usually net importers of these virtual resources while the undeveloped world is burdened with production and export (Chen et al., 2021). As this trend continues, these countries with limited environmental technologies will worsen and exacerbate the waste of environmental resources.

The imbalance between virtual water and land use import and export is not sustainable. The idea of virtual water is otherwise inherently flawed as water itself provides so many benefits beyond just agricultural production (Barnes, 2013). Thus the global economy and movement of agricultural goods while redistributing water is not curing the differential distribution of water. And further it is redistributing pollution problems. Here again, the wealthy or developed nations tend to export pollution through virtual water and land use trades while remain world absorbs these impacts (O'Bannon et al., 2014). Global trade and agriculture will drive many of these concerns in the future.

19.3.2 Climate change

Changes in water distribution from climate change may have a profound impact on agriculture. Exacerbating water scarcity into South Asia and Europe (Gosling & Arnell, 2016) will strongly intensify virtual water trade and tensions globally. Changing precipitation patterns along with differing runoff scenarios should increase areas of drought and flooding. In the case of China, the largest producer of agricultural crops globally, changing precipitation patterns will have profound impacts on rainfed crop production and global trade (Fulu et al., 2003). These areas if experiencing loss of productivity or failure will impact global supplies.

A dichotomy exists pertaining to agriculture and climate change. Concerns over precipitation patterns changing yields and other biophysical damage is real and continues to be concerning (Challinor et al., 2014). The variability surrounding predictions is forcing considerations on how to properly manage our agricultural future. But on the other hand, agriculture is responsible for considerable emissions that contribute to the problem (Vermeulen et al.,

2012). Agriculture will need to adjust to changing patterns in precipitation and warming and global production of food is likely to be quite different in the future.

19.3.3 Land management (best management practices)

The state of the art to control nonpoint source pollution from agriculture consists of various forms of land management, livestock exclusions, treatment structures, or reductions in use. Through estimation of pollutant loading and education about these problems, solutions can be derived. Management options range from fencing cattle and riparian buffer enhancement to complex treatment systems that treat all of the waste generated. Crop management includes understanding of the nutrient needs with various strategies from tillage to harvesting. In all, it will take a fundamental change in agricultural practices to meet increasing demand for food while reducing pollutant impacts.

Conservation agriculture (CA) is a collective term describing practices designed to do minimal damage to the environment. It describes various practices such as no tillage, minimal tillage or tillage with mulch all designed to minimize the impact from agriculture (Palm et al., 2014). Certainly, many benefits have been found from this practice including increasing organic matter in top soil along with less soil erosion thus increasing soil properties. The problem of increased nutrient runoff still exists as CA does reduce sediment loss but not necessarily precipitation runoff (Xiong et al., 2018) or the concentration of nutrients in this runoff (Uribe et al., 2018). However, because of the improvements to topsoil, research does suggest that pesticides are retained at much greater sorption rates than expected by soils (Alletto et al., 2010).

There is a strong suggestion that fencing and riparian enhancements (a width greater than 5–10 m) improves water quality by reducing nutrients and bacteria in receiving waters in temperate areas (Grudzinski et al., 2020). Greater study throughout multiple global biomes are needed to quantify the results of this practice. The direct treatment of manures can be achieved through use of various technologies such as aerated lagoons, activated sludge systems with anoxic tanks and sequencing batch reactors (Bernet & Beline, 2009). Lagoons are difficult to manage and produce less than satisfactory results. Built systems including activated sludge and SBRs (sequencing batch reactors) using single tank technologies seem to provide better treatments because they are technologically and economically better suited to farm use. Applications of these technologies will be essential as more meat production is needed to meet rising food demand (Fig. 19.4).

19.3.4 Emerging concerns

Emerging concerns (ECs) generate a disturbing new pattern of pollutants from agriculture entering the aquatic environment. The vast array of contaminants such as various

Figure 19.4
The fencing of cattle from streams is the most common and often most effective way to improve water quality. Here a cow is freely using the stream where it has a variety of negative impacts.

pharmaceuticals, personal care products and metabolites from biosolids applications, hormones, veterinary medicines and metabolites from manure spreading and nanomaterials used in crop production even in the smallest concentration are now concerning (Boxall, 2012). Synergistic multiplication in the environment compound this problem. These contaminants can bind to sediment or microplastics further complicating the concern, confounding and proving difficult for science to discern the impact, and creating alarm that the problem is greater than previously thought and will be persistent in the environment for an extended period of time (Wilkinson et al., 2017). Complexes with sediment and plastics further distributes these ECs great distances throughout the globe.

Major consequences and adverse effects of ECs on human health and the environment are prevalent (Rasheed et al., 2019). The proliferation through manufacture and now detection has far outpaced any comprehensive understanding on environmental and human health and treatment technologies for removal before introduction into the environment. The ability of these contaminants to be problematic at very low concentrations (nanograms per liter), easily dissolve is water, avoid removal through conventional wastewater treatment technologies and require phase change (liquid to gas or solid) for removal presents a tremendous future challenge (Rasheed et al., 2019). Fate, removal technologies and much needed ecotoxicology should be better understood before continued production and release of these ECs continues at its current pace.

19.4 Water pollution control

19.4.1 Water quality

Studies suggest agriculture is responsible for up to half of the water quality impairments to surface waters (USEPA, 2002). Further, stream health measured by the robustness of fish population metrics, metrics based on macroinvertebrates indices and overall chemical water quality metrics suggest as the catchment agriculture land use increases these measures show impairment (Chen & Olden, 2020; Martins et al., 2021; Silva et al., 2018). Bacterial contamination is another problem suggesting agricultural subcatchments tend to be more polluting in comparison to mixed use development and certainly forested subcatchments (Petersen & Hubbart, 2020). It is clear that agriculture is impacting water quality and needs to be considered if improvements are to be realized.

Needed improvements to stream habitat and riparian buffers creating reductions to water sediment and nutrient loading seem to be the most daunting of the problems. More recently, it is the streambanks that appear to significantly load streams with sediment and associated problems with up to 70% of the loading (Cashman et al., 2018; Gellis & Gorman, 2018). Further, these areas contain legacy nutrients, bacteria, and sediment that load into streams from the increased water force generated by land use changes (Fox et al., 2016). Agriculture and land disturbance is a significant contributor to this problem.

19.4.2 Best management practices

Best management practices (BMPs) include an entire suite of possibilities to minimize the impact of agriculture on receiving waters. Often, these watersheds are very large and diverse presenting challenges to formulate a plan that will be effective in controlling degrading water quality. In other instances, improvements take considerable amounts of time and may fatigue those working on them. These improvements can be simplified into simple land management or livestock exclusions up through more sophisticated water treatment systems and lagoons.

The effective planning and monitoring of BMPs remains a challenge. According to Kroll et al. Kroll et al. (2018), lack of holistic planning that does not include geographic context or considerations of scale, failure to tie monitoring to specific goals and predicted improvements and limited and parochial approach to monitoring taken by funding agencies all hinder effective water quality improvements based on BMP installation. Ecosystem response to agricultural BMPs is often not observed in short-term or in small-scale studies. The literature is sparse but more likely to show changes over longer term (greater than 25 years) than in the short term (less than 3 years) (Kroll & Oakland, 2018).

19.4.3 Land management practices

Multiple land management problems generate additional problems. In the production of crops, precision agriculture management is a promising technique to reduce pollutant loading to adjacent waterways (Davis & Pradolin, 2016; Diacono et al., 2013). This type of farming uses various types of technology during planting, growing, harvesting and shipping of farm products. With the advent of GPS guidance and remote sensing capabilities, everything from efficiently guiding farm equipment during harvest to understanding soil fertility, moisture and herbicide application is controlled through maps, controllers, computers and software. These techniques have demonstrated efficiency and environmental improvements for farmers.

While these techniques greatly improve efficiency of farming multiple problems still hamper good environmental improvement. Overgrazing and lack of adequate riparian buffer corridors are detrimental to water quality. Quite concerning is the idea that riparian buffers are extensively damaged and may be the worst in our history (USDI, 1994). Minimum widths are needed to offer adequate protections to water resources (Sweeney & Newbold, 2014). When examined more closely, deciduous trees provide the best nutrient, sediment and water flow reduction for riparian buffers in the critical 10–30 m width [the minimum width needed for water quality improvement (Zhang et al., 2017)]. The differential uptake rates by species (deciduous vs coniferous trees or grass) accounts for this disparity. In watersheds where allochthonous energy inputs are required for stream health deciduous trees add additional ecological benefits. Often, improvement of riparian buffer size and health is the most economical and widely used practice to increase stream health.

Overgrazing and livestock density need greater management. Cattle tend to graze in response to topography, distance to water, and plant composition (Raynor et al., 2021). This can create differential use of grazing land and congregate livestock into areas that increase the environmental impact. The rotation of livestock to prevent overgrazing is a good practice. Other methods to disperse livestock are needed as well.

Working with the landscape is another promising technique. Often, drainage ditches connect fields with each other and also adjacent waterways. Reworking ditches into a "two-stage" system where the features within the ditch are created to better mimic a natural stream then planted with vegetation designed for maximum uptake of nutrients and slowing of water flow allowing infiltration (Hodaj et al., 2017). Such systems can be a cost-effective addition to land management practices in agriculture (Castaldelli et al., 2015). Additional benefits to these vegetated buffer strips include the reduction in herbicide runoff due to sorption and degradation of these chemicals within the soil and vegetation (Krutz et al., 2005). In conjunction with various types of wetland systems (Wang et al., 2018), these systems when properly designed will significantly reduce or even eliminate the contaminant loading into adjacent waterways.

19.4.4 Intensive management practices

Traditional farming where crops were produced and residues then utilized by livestock has been replaced, due to rising production costs versus price of products, intensive forms of agriculture (Martinez et al., 2009). These operations reply on external feeds and energy, not well assimilated in farm production, to be released into the environment without treatment. These problems are intensifying in both industrialized and less developed countries worldwide.

There are many types of systems available to treat waste generated through farming practices most notably mixers, separators or reactors to reduce pollutant content of wastes (Burton & Turner, 2003). More problematic is the return on investment and the prohibitive costs. In large operations, some financial reward is possible through the generation of electricity from anerobic digestion but more likely any benefit gained by these technologies cannot be realized and only cost-sharing by governments or legislation will increase use (Martinez et al., 2009). So, while these technologies are beneficial in the reduction of water pollution, scant use, and prohibitive costs prevent the overriding application for agriculture worldwide.

19.5 Agriculture and a sustainable future

Protecting water quality needs to be our greatest global focus as the benefits of good water quality need to be reflected in our values, economics and lives (Downing et al., 2021). Debate over the use of cropping systems purely for human food production while managing livestock only on pastures may become more mainstream as pollution generated by climate change or increasing production becomes more pronounced (Martin et al., 2020). Future agricultural decision making may be influenced by social media, expected norms by society, and increased pressures from regulations (Liu et al., 2018). While the evolution of agriculture is unknown, pressures brought forward by expectations for sustainability and better water resource management may broadly influence what we see in our food supply.

Moving forward, the need for farmers and scientists to collaborate together to develop pathways for improving sustainability rather than studies of transfer of developed farming technologies should ensue (Collins et al., 2016). The future of agriculture should incorporate more sustainability practices, and we should concurrently see increases in water quality once impacted by agriculture. Appropriate practices must be incorporated if we intend to reduce water pollution.

References

Alexandratos, N. (2009). *Proceedings of a technical meeting of experts: How to feed the world in 2050. Rome* (pp. 1–32). Food and Agriculture Organiuzation of the United Nations.

Alletto, L., Coquet, Y., Benoit, P., Heddadj, D., & Barriuso, E. (2010). Tillage management effects on pesticide fate in soils: A review. *Agronomy for Sustainable Development, 30*, 367–400.

Baker, A. (2003). Land use and water quality. *Hydrological Processes, 17*, 2499–2501.

Barnes, J. (2013). Water, water everywhere but not a drop to drink: The false promise of virtual water. *Critique of Anthropology, 33*, 371–389.

Beattie, R., Bandla, A., Swarup, S., & Hristova, K. (2020). Freshwater sediment microbial communities are not resilient to disturbance from agricultural land runoff. *Frontiers in Microbiology, 11*, 539921.

Benbrook, C. (2016). Impacts of genetically engineered crops on pesticide use in the U.S.—The first sixteen years. *Environmental Sciences Europe, 24*, 24.

Bernet, N., & Beline, F. (2009). Challenges and innovations on biological treatment of livestock effluents. *Bioresource Technology, 100*, 5431–5436.

Bhandari, G. (2014). An overview of agrochemicals and their effects on. *Atomic Energy Education Society, 2*, 66–73.

Boxall, A. (2012). *New and Emerging water pollutants from agriculture. Research Report*. Organization for Economic Cooperation and Development.

Buck, O., Niyogi, D., & Townsend, C. (2004). Scale-dependence of land use effects on water quality of streams in agricultural catchments. *Environmental Pollution (Barking, Essex: 1987), 130*, 287–299.

Burkholder, J., Libra, B., Weyer, P., Heathcote, S., Kolpin, D., Thorne, P., & Wichman, M. (2007). Impacts of waste from concentrated animal feeding operations on water quality. *Environmental Health Perspectives, 115*, 308–312.

Burton, C., & Turner, C. (Eds.), (2003). *Manure management—Treatment strategies for sustainable agriculture* (second ed., p. 490). Wrest Park, Silsoe, Bedford, UK: Silsoe Research Institute.

Cashman, M., Gellis, A., Gorman Sanisaca, L., Noe, G., Cogliandro, V., & Baker, A. (2018). Bank-derived sediment dominates in a suburban Chesapeake Bay watershed, Upper Difficult Run, Virginia, USA. *River Research and Applications, 34*, 1032–1044.

Cassidy, E., West, P., Gerber, J., & Foley, J. (2013). Redefining agricultural yields; from tonnes to people nourished per hectare. *Environmental Research Letters, 8*, 034015, 8p.

Castaldelli, G., Soana, E., Racchetti, E., Vincenzi, F., Fano, E., & Bartoli, M. (2015). Vegetated canals mitigate nitrogen surplus in agricultural watersheds. *Agriculture, Ecosystems & Environment, 212*, 253–262.

Challinor, A., Watson, J., Lobell, D., Howden, S., Smith, D., & Chhetri, N. (2014). A *meta*-analysis of crop yield under climate change and adaptation. *Nature Climate Change, 4*, 287–291.

Chen, K., & Olden, J. (2020). Threshold responses of riverine fish communities to land use conversion across regions of the world. *Global Change Biology, 26*, 4952–4965.

Chen, W., Kang, J., & Han, M. (2021). Global environmental inequality: Evidence from embodied land and virtual water trade. *Science of the Total Environment, 783*, 146992.

Chen, W., & Liu, W. (2017). Investigating the fate and transport of fecal coliform contamination in a tidal estuarine system using a three-dimensional model. *Marine Pollution Bulletin, 116*, 365–384.

Collins, A., Zhang, Y., Winter, M., Inman, A., Jones, J., Johnes, P., Cleasby, W., Vrain, E., Lovett, A., & Noble, L. (2016). Tackling agricultural diffuse pollution: what might uptake of farmer-preferred measures deliver for emissions to water and air? *The Science of the Total Environment, 547*, 269–281.

Conley, D., Paerl, H., Howarth, R., Boesch, D., Seitzinger, S., Havens, K., Lancelot, C., & Likens, G. (2009). Controlling eutrophication: Nitrogen and phosphorus. *Science (New York, N.Y.), 323*, 1014–1015.

Cornejo, A., Tonin, A., Checa, B., Tuñon, A., Pérez, D., Coronado, E., González, S., Ríos, T., Macchi, P., Correa-Araneda, F., & Boyero, L. (2019). Effects of multiple stressors associated with agriculture on stream macroinvertebrate communities in a tropical catchment. *PLoS One, 14*, e0220528.

Cuffney, T., Meador, M., Porter, S., & Gurtz, M. (2000). Responses of physical, chemical, and biological indicators of water quality to a gradient of agricultural land use in the Yakima River Basin, Washington. *Environmental Monitoring and Assessment, 64*, 259–270.

Daniel, T., Sharpley, A., & Lemunyon, J. (1998). Agricultural phosphorus and eutrophication: A symposium overview. *Journal of Environmental Quality, 27*, 251–257.

Davis, A., & Pradolin, J. (2016). Precision herbicide application technologies to decrease herbicide losses in furrow irrigation outflows in a Northeastern Australian cropping system. *Journal of Agricultural and Food Chemistry, 64*, 4021.

DeFlorio-Barker, S., Wing, C., Jones, R., & Dorevitch, S. (2018). Estimate of incidence and cost of recreational waterborne illness on United States surface waters. *Environmental Health, 17*, 1−10.

Delgado, C. (2003). Rising consumption of meat and milk in developing countries has created a new food revolution. *The Journal of Nutrition, 133*, 3907S−3910SS.

Diacono, M., Rubino, P., & Montemurro, F. (2013). Precision nitrogen management of wheat. A review. *Agronomy for Sustainable Development, 33*, 39.

Diendéré, A., Nguyen, G., Del Corso, J., & Kephaliacos, C. (2018). Modeling the relationship between pesticide use and farmers' beliefs about water pollution in Burkina Faso. *Ecological Economics, 151*, 114−121.

Downing, J., Polasky, S., Olmstead, S., & Newbold, S. (2021). Protecting local water quality has global benefits. *Nature Communications, 12*, 2709.

Duke, S., Lydon, J., Koskinen, W., et al. (2014). Glyphosate effects on plant mineral nutrition, crop rhizosphere microbiota, and plant disease in glyphosate resistant crops. *Journal of Agricultural and Food Chemistry, 60*, 10375−10397.

Evans, A., Mateo-Sagasta, J., Qadir, M., Boelee, E., & Ippolito, A. (2019). Agricultural water pollution: Key knowledge gaps and research needs. *Current Opinion in Environmental Sustainability, 36*(1029), 20−27.

Fox, G., Garey, A., Rebecca, A., Purvis, A., & Penn, C. (2016). Streambanks: A net source of sediment and phosphorus to streams and rivers. *Journal of Environmental Management, 181*, 602−614.

Fulu, T., Yokozawa, M., Hayashi, Y., & Lin, E. (2003). Future climate change, the agricultural water cycle, and agricultural production in China. *Agriculture, Ecosystems & Environment, 95*(1), 203−215.

Gaudar, U., Pellerin, S., Benoit, M., Durand, G., Dumont, B., Barbieri, P., & Nesme, T. (2021). Comparing productivity and feed-use efficiency between organic and conventional livestock animals. *Environmental Research Letters, 16*, 024012.

Gellis, A., & Gorman, L. (2018). Sanisaca, sediment fingerprinting to delineate sources of sediment in the agricultural and forested Smith Creek watershed, Virginia, USA. *Journal of the American Water Resources Association, 54*, 1197−1221.

Gosling, S. N., & Arnell, N. W. (2016). A global assessment of the impact of climate change on water scarcity. *Climatic Change, 134*, 371−385.

Grudzinski, B., Frits, K., & Dodds, W. (2020). Does riparian fencing protect stream water quality in cattle-grazed lands? *Environmental Management, 66*, 121−135.

Guber, A., Pachepsky, Y., Shelton, D., & Yu, O. (2007). Effect of bovine manure on fecal coliform attachment to soil and soil particles of different sizes. *Applied and Environmental Microbiology, 73*, 3363−3370.

Hayes, T., Case, P., Chui, S., Chung, D., Haeffele, C., Haston, K., Lee, M., & Mai, V. (2006). Pesticide mixtures, endocrine disruption, and 1356 amphibian declines: are we underestimating the impact? *Environmental Health Perspectives, 114*, 40.

Hodaj, A., Bowling, L., Frankenberger, J., & Chaubey, I. (2017). Impact of two-stage ditch on channel water quality. *Agricultural Water Management, 192*, 126−137.

Hoekstra, A., & Chapagain, A. (2008). *Globalization of water: Sharing the planet's freshwater resources*. Oxford, UK: Blackwell Publishing.

Hoekstra, A., & Chapagain, A. (2007). Water footprints of nations: Water use by people as a function of their consumption pattern. *Water Resources Management, 21*, 3548.

Hooda, P., Edwards, A., Anderson, H., & Miller, A. (2000). A review of water quality concerns in livestock farming areas. *Science of the Total Environment, 250*, 143−167.

Hughes, R., & Vadas, R. (2021). Agricultural effects on streams and rivers: A western USA focus. *Water, 13*, 1901.

Jones, D., Quintela Baluja, M., Graham, D., Corbishley, A., McDonald, J., Malham, S., Hillary, L., Connor, T., Gaze, W., Moura, I., Wilcox, M., & Farkas, K. (2020). Shedding of SARS-CoV-2 in feces and urine and its potential role in person-to-person transmission and the environment-based spread of COVID-19. *Science of The Total Environment, 749*, 141364.

Kalavrouziotis, I. (2016). Long term wastewater and biosolids application in agriculture. *International Journal of Water and Wastewater Treatment, 2*.

Kroll, S., Horwitz, R., Keller, D., Sweeney, B., Jackson, J., & Perez, L. (2018). Large-scale protection and restoration programs aimed at protecting stream ecosystem integrity: The role of science-based goal-setting, monitoring, and data management. *Freshwater Science, 38*, 23–39.

Kroll, S., & Oakland, H. (2018). Review of studies documenting the effects of agricultural best management practices on stream ecosystem integrity. *Natural Areas Journal, 39*, 58–77.

Krutz, L., Lenseman, S., Zablotowicz, R., & Matocha, M. (2005). Reducing herbicide runoff from agricultural fields with vegetative filter strips: A review. *Weed Science, 53*, 353–367.

Langenbach, T., de Campos, T., & Caldas, L. (2021). , January 30th In Syed Abdul Rehman Khan (Ed.), *Why airborne pesticides are so dangerous, environmental sustainability—Preparing for tomorrow*. IntechOpen.

Liu, T., Bruins, R., & Heberling, M. (2018). Factors influencing farmers' adoption of best management practices: A review and synthesis. *Sustainability, 10*, 432, 84.

Lu, Q., He, Z., & Stoffella, P. (2012). Land application of biosolids in the USA: A review. *Applied and Environmental Soil Science, 2012*, 1–11.

Mallin, M., Cahoon, L., Toothman, B., Parsons, D., McIver, M., Ortwine, M., & Harrington, R. (2017). Impacts of a raw sewage spill on water and sediment quality in an urbanized estuary. *Marine Pollution Bulletin, 54*, 81–88.

Mallin, M., McIver, M., Robuck, A., & Dickens, A. (2015). Industrial swine and poultry production causes chronic nutrient and fecal microbial stream pollution. *Water, Air, and Soil Pollution, 226*, 1–13.

Malomo, G., Madugu, A., & Bolu, S. (2018). , August 29th In Anna Aladjadjiyan (Ed.), *Sustainable Animal Manure Management Strategies and Practices, Agricultural Waste and Residues*. IntechOpen.

Manyi-Loh, C., Mamphweli, S., Meyer, E., Makaka, G., Simon, M., & Okoh, A. (2016). An overview of the control of bacterial pathogens in cattle manure. *International Journal of Environmental Research and Public Health, 13*, 843.

Martin, G., Duru, M., Durand, J., Julier, B., Litrico, I., Louarn, G., Gastal, F., Novak, S., Médiène, S., Valentin-Morison, M., Paschalidou, F., & Jeuffroy, M. (2020). Role of ley pastures in tomorrow's cropping systems—A review. *Agronomy for Sustainable Development, 40*, 17.

Martinez, J., Dabert, P., Barrington, S., & Burton, C. (2009). Livestock waste treatment systems for environmental quality, food safety, and sustainability. *Bioresource Technology, 100*, 5527–5536.

Martins, I., Macedo, D., Hughes, R., & Callisto, M. (2021). Major risks to aquatic biotic condition in a Neotropical Savanna river basin. *River Research and Applications, 37*, 858–868.

Mateo-Sagasta, J., Marjani, S., & Turral, H. (2018). Global drivers of pollution from agriculture. In J. Mateo-Sagasta, S. Marjani, & H. Turral (Eds.), *More people, more food, worse water? A global review of water pollution from agriculture*. Food and Agricultural Organization and Rome, and International Waste Management Institute Colombo, (2018).

Metson, G., Bennett, E., & Elser, J. (2012). The role of diet in phosphorus demand. *Environmental Research Letters, 7*(2021), 044043, 10 p.

Munoz, P., Anton, A., Paranjpe, A., Arino, J., & Montero, J. (2008). High decrease in nitrate leaching by lower N input without reducing greenhouse tomato yield. *Agronomy for Sustainable Development, 28*, 489–495.

NAS (National Academy of Sciences). (2016). *Genetically engineered crops: Experiences and prospects* (p. 420) Washington, DC: National Academy Press.

National Academy of Sciences. (2002). *Biosolids applied to land; advancing standards and practices*. Washington D.C: National Academy Press.

Noe, G., & Hupp, C. (2005). Carbon, nitrogen, and phosphorus accumulation in floodplains of Atlantic Coastal Plain rivers, USA. *Ecological Applications: A Publication of the Ecological Society of America, 15*, 1178–1190.

O'Bannon, C., Carr, J., Seekell, D., & D'Odorico, P. (2014). Globalization of agricultural pollution due to international trade. *Hydrology and Earth System Sciences, 18*, 503–510.

Ormerod, S., Dobson, M., Hildrew, A., & Townsend, C. (2010). Multiple stressors in freshwater ecosystems. *Freshwater Biology, 55*, 1–4.

Oun, A., Kumar, A., Harrigan, T., Angelakis, A., & Xagoraraki, I. (2014). Effects of biosolids and manure application on microbial water quality in rural areas in the US. *Water, 6*, 3701–3723.

Pachepsky, Y., & Shelton, D. (2011). Escherichia coli and fecal coliforms in freshwater and estuarine sediments. *Critical Reviews in Environmental Science and Technology, 41*, 1067−1110.

Palm, C., Blanco-Canqui, H., DeClerck, F., Gatere, L., & Grace, P. (2014). Conservation agriculture and ecosystem services: An overview. *Agriculture, Ecosystems & Environment, 187*, 87−105.

Pandeya, P., Soupir, M., Haddad, M., & Rothwell, J. (2012). Assessing the impacts of watershed indexes and precipitation on spatial in-stream E. coli concentrations. *Ecological Indicators, 23*, 641−652.

Petersen, F., & Hubbart, J. (2020). Spatial and temporal characterization of *Escherichia coli*, suspended particulate matter and land use practice relationships in a mixed-land use contemporary watershed. *Water, 12*, 1228.

Pourcher, A., Françoise, P., Virginie, F., Agnieszka, G., Vasilica, S., & Gérard, M. (2007). Survival of fecal indicators and enteroviruses in soil after land-spreading of municipal sewage sludge. *Applied Soil Ecology, 35*, 473−479.

Powlson, D., Gregory, P., Whalley, W., Quinton, J., Hopkins, D., Whitmore, A., Hirsch, P., & Goulding, K. (2011). Soil management in relation to sustainable agriculture and ecosystem services. *Food Policy, 36*(1), S72−S87.

Quarles, W. (2017). Glyphosate, GMO soybean yields and environmental pollution. *IPM Practitioner, 35*, 1−7.

Rabalais, N., Turner, R., & Scavia, D. (2002). Beyond science into policy: Gulf of Mexico hypoxia and the Mississippi River. *Bioscience, 52*, 129−142.

Raff, Z., & Meyer, A. (2021). CAFOs and surface water quality: Evidence from Wisconsin. *American Journal of Agricultural Economics*, 1−29, 00.

Rasheed, T., Bilal, M., Nabeel, F., Adeel, M., & Iqbal, H. (2019). Environmentally-related contaminants of high concern: Potential sources and analytical modalities for detection, quantification, and treatment. *Environment International, 122*, 52−66.

Raynor, E., Gersie, S., Stephenson, M., Clark, P., Spiegal, S., Boughton, R., Bailey, D., Cibils, A., Smith, B., Derner, J., Estell, R., Nielson, R., & Augustine, D. (2021). Cattle grazing distribution patterns related to topography across diverse rangeland ecosystems of North America. *Rangeland Ecology & Management, 75*, 91−103.

Rose, J., Daeschner, D., Easterling, F., Curriero, S., Lele, S., & Patz, J. (2000). Climate and waterborne disease outbreaks. *Journal of the American Water Works Association, 92*, 77−87.

Sánchez-Gervacio, B., Bedolla-Solano, R., Rosas-Acevedo, J., Legorreta-Soberanis, J., Valencia-Quintana, R., & Juárez-López, A. (2021). Pesticide management by subsistence farmers in Mexico: Baseline of a pilot study to design an intervention program. *Human and Ecological Risk Assessment, 27*, 1112−1125.

Schindler, D. (1977). Evolution of phosphorus limitation in lakes. *Science (New York, N.Y.), 95*, 260−262.

Silva, D., Herlihy, A., Hughes, R., Macedo, D., & Callisto, M. (2018). Assessing the extent and relative risk of aquatic stressors on stream macroinvertebrate assemblages in the neotropical savanna. *Science of the Total Environment, 633*, 179−188.

Sobolewski, W. (2016). Effect of agricultural land use on the water quality of Polish Lakes: A regional approach. *Polish Journal of Environmental Studies, 25*, 2705−2710.

Soupir, M., Mostaghimi, S., Yagow, E., Hagedorn, C., & Vaughan, D. (2006). Transport of fecal bacteria from poultry litter and cattle manures applied to pastureland. E. coli and Enterococci in dairy cowpats. *Water, Air, & Soil Pollution, 169*, 125−136.

Steffen, W., Richardson, K., Rockstrom, J., Cornell, S., Fetzer, I., Bennett, E., Biggs, R., Carpenter, S., et al. (2015). Planetary boundaries: guiding human development on a changing planet. *Science (New York, N.Y.), 347*, 1259855.

Stone, M., Whiles, M., Webber, J., Williard, K., & Reeve, J. (2005). Macroinvertebrate communities in agriculturally impacted southern Illinois streams: patterns with riparian vegetation, water quality, and in-stream habitat quality. *Journal of Environmental Quality, 34*, 907−917.

Su, S., Zhou, X., Wan, C., Li, Y., & Kong, W. (2016). Land use changes to cash crop plantations: crop types, multilevel determinants and policy implications. *Land Use Policy, 50*, 379−389.

Sweeney, B., & Newbold, J. (2014). Streamside forest buffer width needed to protect stream water quality, habitat, and organisms: a literature review. *Journal of the American Water Resources Association, 50*, 560−584.

Tilman, D., Balzer, C., Hill, J., & Befort, B. (2011). Global food demand and the sustainable intensification of agriculture. *Proceedings of the National Academy of Science of the United States of America*, *108*(50), 20260–20264.

Tong, S., & Chen, W. (2002). Modeling the relationship between land use and surface water quality. *Journal of Environmental Management*, *66*, 377–393.

UN World Water Development Report. Nature-based Solutions for Water. (2018).

Uribe, N., Corzo, G., Quintero, M., van Griensven, A., & Solomatine, D. (2018). Impact of conservation tillage on nitrogen and pohosphorus runoff losses in a potato crop system in Fuquene watershed, Columbia. *Agricultural Water Management*, *209*, 62–72.

USDI (U.S. Department of the Interior). Rangeland Reform, Draft Environmental Impact Statement; Bureau of Land Management: Washington, DC, USA, (1994).

USEPA (U.S. Environmental Protection Agency). National Water Quality Inventory Report. EPA-841-F-02–003; (2002).

VanderZaag, A., Campbell, K., Jamieson, K., Sinclair, R., & Hynes, A. (2010). Survival of *Escherichia coli* in agricultural soil and presence in tile drainage and shallow groundwater. *Canadian Journal of Soil Science*, *90*, 495–505.

Vermeulen, S., Campbell, B., & Ingram, J. (2012). Climate change and food systems. *Annual Review of Environment and Resources*, *37*, 195–222.

Wang, M., Zhang, D., Dong, J., & Tan, S. (2018). Application of constructed wetlands for treating agricultural runoff and agro-industrial wastewater: A review. *Hydrobiologia*, *805*, 1–31.

Wijesiri, B., Deilami, K., & Goonetilleke, A. (2018). Evaluating the relationship between temporal changes in land use and resulting water quality. *Environmental Pollution*, *234*, 480–486.

Wilkinson, J., Hooda, P., Barker, J., Barton, S., & Swinden, J. (2017). Occurrence, fate and transformation of emerging contaminats in water: An overarching review of the field. *Environmental Pollution*, *231*, 954–970.

Wolmarans, K., & Swart, W. (2014). Influence of glyphosate, other herbicides and genetically modified herbicide resistant crops on soil microbiota: A review. *South African Journal of Plant and Soil*, *31*, 177–186.

Xiong, M., Sun, R., & Shen, L. (2018). Effects of soil conservation techniques on water erosion control: A global analysis. *The Science of the Total Environment*, *645*, 753–760.

Ye, M., Beach, J., Martin, J., & Senthilselvan, A. (2017). Pesticides exposures and respiratory health in general population. *Journal of Environmental Sciences*, *51*, 361–370.

Zhang, C., Li, S., Qi, J., Xing, Z., & Meng, F. (2017). Assessing impacts of riparian buffer zones on sediment and nutrient loadings into streams at watershed scale using an integrated REMM-SWAT model. *Hydrological Processes*, *31*, 916–924.

CHAPTER 20

Environmental impacts on global water resources and poverty, with a focus on climate change

Claudia Yazmín Ortega Montoya[1] and Juan Carlos Tejeda González[2]

[1]Escuela de Humanidades y Educación, Tecnologico de Monterrey, Torreón, Mexico,
[2]Facultad de Ingeniería Civil, Universidad de Colima, Colima, Mexico

Chapter Outline

20.1 Environmental impacts on national/regional water resources   384
20.2 Impacts of climate change on poverty   385
20.3 Exposure and vulnerability impacts   387
20.4 Slow-onset events   389
20.5 Future trends   393
References   394

Research into the impacts of climate change on water resources is commonly analyzed by coupling climate models and hydrological models (Tramblay et al., 2020), However, most of the analysis to date uses historical hydrological data that are irrelevant to the current context of continuous and significant changes to the Earth systems. This will have a significant future impact, because with no data series to feed the models, they will become obsolete, unless there is a breakthrough in the systems' modeling development.

The impacts of climate change on Global Water Resources (GWR) imply an additional burden for marginalized groups because multidimensional poverty involves higher levels of vulnerabilities to related extreme and slow-onset events. Such impacts tend to exacerbate poverty and inequalities through reproduction cycles of poverty−vulnerability/health detriment.

The aim of this chapter is to describe the challenges that climate change impacts on water resources presents to poverty reduction.

20.1 Environmental impacts on national/regional water resources

The problems related to GWR can be categorized into two strategic features: availability and quality. Many countries have undertaken research on the effects of climate change on their water resources with the objective of preparing their policies, infrastructure, and management for current and future impacts; for example, the USA developed the report "Water: The Potential Consequences of Climate Variability and Change for the Water Resources of the United States" in 2010, where significant impacts were identified including major changes in timing and magnitude of runoff, damage to coastal aquifers and water supplies, risk of flooding and drought with widely distributed effects, and potential damage to fish and wildlife species and habitats such wetlands (Insid. EPA's Water Policy Rep., 2001). More recently, research focused on the realistic representation of the hydrological cycle and related extreme events in Earth system models has been developed in order to have a better understanding of regional climate change impacts on water resources, for example, events such as mesoscale convective systems, which are responsible for over half of extreme daily rainfall events, and they are notoriously difficult to simulate in traditional global climate models. This research showed that it is important and necessary to analyze such mesoscale processes to effectively integrate them into the next-generation climate models, because global scale models do not always represent the reality on smaller scales (Feng et al., 2021).

In Central America, the impacts likely to occur in Nicaragua on the water supply and sanitation sector due to climate change are the frequency and intensity of floods; diminishing groundwater levels due to overexploitation, mainly during the dry season, and increasing contamination of water sources such as aquifers and springs. All of these impacts are significantly increased by the country's vulnerability to certain decision-making elements that only exacerbate the effects of climate change on their water resources (World Bank, 2013). In the case of Mexico, it is estimated that runoff will decrease and thus the aquifer's recharge will decrease too, and because of this shortage of water availability in combination with the expected population and economic growth, there will be significant water stress in the country; also, food production will be drastically affected due to the increase in evaporation (Austria & Gómez, 2012).

In Swaziland, Matondo and Msibi estimated the impacts of climate change on the country's hydrology will result in higher stream flows in summer months (resulting in flooding) and lower stream flows in winter months (causing drought-related problems), because of expected higher temperatures and evapotranspiration rates during the winter months (Matondo & Msibi, 2001). In another study, Juana, Mangadi, and Strzepek identified the socioeconomic impacts of climate change on water resources in South Africa, where the results suggested that with the reduction in water availability due to climate change, there will be a decline in the sectoral output in the country, focusing on the decline of the wages of unskilled and medium-skilled laborers and thus the deterioration in the welfare of the lowest-, low- and medium-income households (Juana et al., 2012).

In the case of the Netherlands, it is expected that climate change will bring three major consequences: more water in the winters (an increase in the total amount of rainfall, the intensity of showers, and river discharges), longer drier periods, and a decrease in river discharges in summers, and finally, the rising of the sea level (van Dijk, 2006). It is important to adopt a more holistic approach to identify mechanisms that will help to protect the available water resources without compromising the development of economic activities such as farming or tourism. For example, the concept of "ecological integrity" in EU's nature conservation law is a good example of a discrete body of rules that helps to sustain highly valued ecosystems, such as groundwater, an element widely affected by climate change variations (McIntyre, 2017).

All of the national/regional examples described above are aligned with the changes in climate related to water specified by the IPCC Technical Paper VI "Climate change and water" of 2008, both from observed and estimated impacts: increases in global average air and ocean temperatures, widespread melting of snow and ice and rising global sea level, and in terms of more specific impacts to the hydrological cycle and systems: changes in precipitation patterns, including the intensity of events and the presence of extreme events (both rainfall and drought), increase in atmospheric water vapor, and the increase in evaporation, among others (Bates et al., 2008).

However, even though many countries have developed studies to understand the effects of climate change on their water resources, and some institutions have performed research on a global scale based on the reports made by the Intergovernmental Panel on Climate Change (IPCC), there is a consensus that the historical data series will not be enough for future water management and new models must be based on multiple climate change scenarios (Austria & Gómez, 2012).

20.2 Impacts of climate change on poverty

When considering the causes of anthropogenic influence on climate change and their consequences, surprisingly, 20 of the 36 highest greenhouse gases (GHG)-emitting countries are among the least vulnerable to the negative impacts of present and future climate change, while 11 of the 17 countries with low or moderate GHG emissions are acutely vulnerable to the negative impacts of climate change (Althor et al., 2016). These contrasting outcomes are related to the exposure and vulnerabilities of societies and are proof that impacts on a global scale issue are not equitably distributed.

Poverty refers to the lack of access to satisfiers required to fulfill physical or psychological necessities. People living in poverty and disadvantaged groups are disproportionately exposed to climate change, because they live in rural areas with great dependency on agricultural, fishing, and other ecosystem-related incomes (UN DESA, 2020) which are the most highly affected environmental features of climate change. Another factor related to the disproportionate outcomes of climate change on the poverty of disadvantaged groups is related to the conditions that tend to increase their vulnerabilities to climate change.

Vulnerability is a feature of a person or group of people that influences their capacity to anticipate, deal with, resist, and/or recover from the impact of a hazard (Wisner, 2006). Multidimensional poverty accounts for the broad disadvantages in accessing satisfiers able to provide wellbeing to people. People living in poverty experience a lack of diverse wellbeing satisfiers including food, health, education, housing, etc. The 2020 UN Multidimensional Poverty report declared that rural areas have a concentrated population in poverty where poor people tend to experience lower vaccination rates, secondary school achievement, and insecure work, among other satisfiers (UNDP & OPHI, 2020). Climate change impacts on the environment tend to exacerbate existing vulnerabilities and increase the gap of human inequalities. Fig. 20.1 illustrates the cycle of poverty reproduction in extreme events and slow-onset events caused by climate change.

In the case of extreme and slow-onset events, the lack of resources caused by poverty generates vulnerabilities in socioeconomic systems exposed to such impacts. Vulnerability can adopt diverse forms such as social, physical, or environmental. In all cases, this variability causes a higher susceptibility to loss and incapability for recovery after extreme events. Negative outcomes of climate change generate barriers that inhibit people's ability

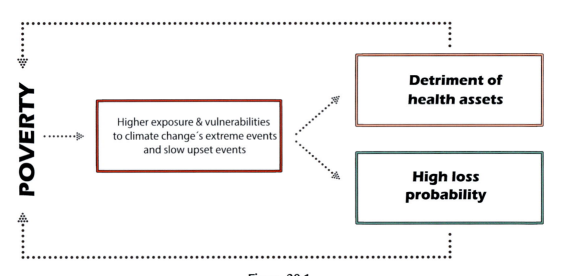

Figure 20.1
Cycle of poverty reproduction in extreme events and slow-onset events caused by climate change [own elaboration].

to overcome poverty (UN DESA, 2020). Climate change also manifests as a detriment to health assets from which wellbeing originates, such safe water, food security, and personal and social stability.

The most vulnerable human populations are those in the countryside that are directly linked to changes in water resources and that experience the highest levels of loss because of the short- and long-term effects of climate change. In all cases, the populations that already experience poverty and a higher level of vulnerability to water resources are the same ones that additionally experience the greatest harmful effects of climate change, this being a determining factor in preventing higher levels of wellbeing and human fulfillment.

The main elements associated with the relationship between climate change and its impact on water resources and poverty are described in the following section.

20.3 Exposure and vulnerability impacts

The occurrence of extreme events such as heatwaves, hurricanes, droughts, and wildfires are some of the main barriers to achieving effective sustainable development. Extreme events erode social wellbeing causing death, injuries, loss of fields, infrastructure damage, and loss of ecosystem services. Those events affect all societies but the wealth of those in poverty is more likely to be concentrated on protecting material forms, such as housing or livestock, because their assets are more fragile (UN DESA, 2020). Climate change is increasing the frequency and severity of extreme events; that, in conjunction with pandemic crises, is challenging the efforts of society to overcome poverty. Extreme weather events were responsible for 91% of disasters between 1998 and 2017, causing direct economic losses valued at US$ 2,245 billion (CRED & UNISDR, 2018).

New scientific methods and regional model simulations allow studies to relate extreme events with climate change. For instance, British Columbia's extreme wildfire in 2017 set a record of 1.2 million burned forest hectares and was attributable to extreme warm and dry conditions where anthropogenic factors were responsible with over 95% probability for the observed maximum temperature anomalies and a burned area increase factor of 7−11 (Kirchmeier-Young et al., 2019). A global effect also caused by climate change is drier dry seasons in extratropical latitudes including Europe, western North America, northern Asia, southern South America, Australia, and eastern Africa (Padrón et al., 2020).

Disasters are sometimes considered external shocks, but disaster risk results from the complex interaction between development processes generating conditions of exposure, vulnerability, and hazard (UNDRR, 2020). The priorities of the UN's Sendai Framework for Disaster Risk Reduction attempt to reduce vulnerability in communities by understanding and allocating resources to the four stages of emergency management: prevention, management, resistance, and recovery (Fig. 20.2).

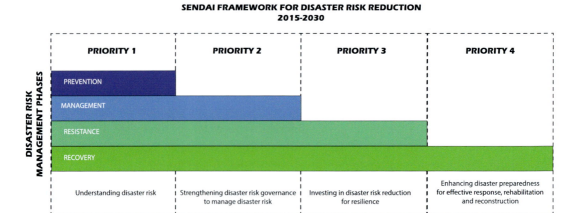

Figure 20.2
Priorities of the Sendai Framework for Disaster Risk Reduction and its relationship with the emergency management stages. Source: *Own elaboration from data (UN, Sendai Framework for Disaster Risk Reduction 2015-2030, 2015. [Online]. Available: https://www.preventionweb.net/files/43291_sendaiframeworkfordrren.pdf; UN, 2015).*

Climate change and its effects worsen existing poverty and when interacting with nonclimatic stressors and entrenched structural inequalities tend to shape vulnerabilities exacerbating inequalities (Olsson et al., 2014). The World Bank has calculated that disasters push 26 million people into poverty every year (The World Bank, 2017). In the South of the Sahara, flood shocks are associated with a 35% decrease in total and food per capita consumption and a 17% increase in extreme poverty, with smallholder farmers the most vulnerable to weather variability (Azzarri & Signorelli, 2020). In the same context, New Zealand projected an increase in damages caused by extreme climate events ranging between 7% and 8% for the period 2020–40, and between 9% and 25% for the period 2080 to 2100 (Pastor-Paz et al., 2020).

The social and economic deprivation that a person or groups experience shapes vulnerability and exposure. The multidimensional aspects of poverty are closed related to the success of disaster risk management:

- Health and health services deprivation. Good health and access to quality health services can play a critical role in resisting and recovering from an extreme event.
- House deprivation. Economic deprivation and the basic need for refuge often forces poor people to build irregular settlements in risk zones; the lack of historic memory of migrants is an important factor in this problem. Housing construction codes and materials play a key role in the resistance to extreme events.
- Income deprivation. The daily pursuit to meet basic needs centers poor people's focus on present challenges, but the prevention process requires actions and processes tended toward anticipating and arranging resources for future potential losses.

- Educational deprivation. Access to school promotes acquisition of skills and risk perception that may influence personal and social preparedness and resistance strategies. Communication and social interaction can also be favorable to social resistance and the recovery from climate-related events.
- Empowerment deprivation. Effectiveness of disaster risk management requires a participative approach to integrate all sectors of society (women, indigenous, migrants, elderly, etc.), into inclusive strategies and responses. Social cohesion can generate positive outcomes on the resistance and recovery stage, for instance, in collaborative reconstruction.

This close relationship between poverty and vulnerability clarifies the cycle of poverty reproduction that takes place when scarcity of economic and social resources originate in the most vulnerable systems that experience the most serious consequences of extreme events; such consequences exacerbate poverty, impair resilience and response ability, and bar progress toward opportunities. The United Nations Office for Disaster Risk Reduction has found that people exposed to natural hazards in the poorest nations were more than seven times more likely to die, and six times more likely to be injured, to lose their home, to be displaced or evacuated, or to require emergency assistance than equivalent populations in the richest nations (CRED & UNISDR, 2018).

Extreme events often damage medical facilities and support services hindering effective response actions. Such events can also magnify the consequences for vulnerable populations leading to hygiene deprivation and increasing the risk of outbreaks of infectious disease, for instance, floods contaminating freshwater supplies, which heightens the risk of water-borne diseases and disease-carrying insects.

Extreme events also cause a decrease in or loss of harvest increasing the risk of malnutrition and undernutrition causing 3.1 million deaths every year (WHO, 2021a). Economic pressures caused by extreme events transform social interactions, modify gender bargaining positions, and cause women to experience negative health outcomes after extreme events in mortality, health, education, and transmission of HIV (Carleton & Hsiang, 2016).

20.4 Slow-onset events

Slow-onset events related to climate change include detriment of ecosystem services, such as land erosion and degradation, air and water quality, biodiversity, and disease protection. These affect many of the social and environmental determinants of health related to GWR—safe drinking water, sufficient food, and secure shelter (WHO, 2021a). It is important to highlight that health is the basis of human capital, and together with other capitals govern the level of livelihoods (Ding et al., 2021).

Climate change can generate additional economic pressures on households, due to emergency expenditures caused by extreme events or the increase in energy consumption to provide indoor comfort. Health detriments also imply economic expenditures and limit peoples and groups' ability to work and to be capable of providing satisfiers or overcoming poverty. The detriment to health also causes economic vulnerabilities that promote marginalized groups to experience the most adverse consequences of climate change because of their lack of resources and human capital to adapt and recover from weather impacts. This can act as a reproduction cycle of poverty—vulnerability reproduction just like the case of extreme events.

Children and elderly people are the most susceptible groups of populations that can experience the most severe consequences of climate change in health due to their biological susceptibility related to their life state. Developing countries are also expected to be the most vulnerable because of their lack of medical health facilities and proper infrastructure. According to a World Health Organization (WHO) assessment, climate change is expected to cause approximately 250,000 additional deaths per year between 2030 and 2050; 38,000 due to heat exposure in elderly people, 48,000 due to diarrhea, 60,000 due to malaria, and 95,000 due to childhood undernutrition (WHO, 2021a).

Safe drinking water

Climate change, in concurrence with anthropogenic pressure on global resources, is a critical cause of water insecurity, defined as the inability to access and benefit from affordable, adequate, reliable, and safe water (Jepson et al., 2017). Due to the dependence of human life on water, water insecurity carries negative health, economic, social, and demographic implications (Carleton & Hsiang, 2016). The degree of negative consequences depends on the level of water vulnerability and local adaptive capacities related to economic and social factors.

The UN General Assembly stated in 2010 that it "recognizes the right to safe and clean drinking water and sanitation as a human right that is essential for the full enjoyment of life and all human rights" (United Nations General Assembly, 2010), so it is arguably significant to consider the GWR as the key element for the development of our present and future, as "the impacts of climate change on hydrology and water resources management has been addressed by various research programs and global, regional, and national assessments" (Muttiah & Wurbs, 2002). Failing to mitigate and adapt to the effects of climate change in GWR management will develop sociopolitical impacts far more significant than the biophysical impacts of the water resources themselves.

The social experience of water scarcity can originate from physical scarcity when water demand outstrips supply of the resource; meanwhile, in the case of economic scarcity the lack of infrastructure or institutional capability prevents communities from accessing the resource even when it is available. According to the Food and Agriculture

Organization (FAO), by 2025, 1800 million people are expected to be living in countries or regions with "absolute" water scarcity (<500 m^3 per year per capita), and two-thirds of the world's population could be under "stress" conditions (between 500 and 1000 m^3 per year per capita) (FAO, 2012).

The presence of extreme events, changes in precipitation, and evapotranspiration can affect the availability and quality of drinking water. Pressures over water supplies can force the use of poor or unsuitable water with drastic repercussions for industry, human health, and the associated costs of health care (Duran-Encalada et al., 2017). The main factors affecting the quality of water supplies are natural and anthropogenic pollutants and water-borne diseases.

Surface water bodies face threats related to urbanization including receiving an excess of external nutrients (eutrophication), microbial pathogens, and pollutants. Eutrophication contributes to the spread of gastrointestinal and dermatological diseases and conjunctivitis (Zhidkova et al., 2020), while pathogens present in water supplies can transmit diseases such diarrhea, cholera, dysentery, and typhoid (WHO, 2021b).

Naturally-occurring fungal and algal toxins are another pollutant variation in surface water supplies. The importance of the development of new toxins is their capability for adverse effects on ecosystems and for reaching the human food chain. Botana (2016) reported cylindrospermopsin and microcystin in freshwater, and vector related toxins including: tetrodotoxin in gastropods in the South of Europe, ciguatoxin in fish in the South of Spain, palytoxin in mussels in the Mediterranean Sea, pinnatoxin in Europe, and okadaic acid in the south of the USA. African elephants' mortality has also been attributed to hypertrophic water's biotoxins produced by cyanobacteria that caused the death of more than 330 specimen (Wang et al., 2021).

Impacts from climate change are also expected to increase the transmission of vector-borne diseases such as malaria, dengue, chikungunya fever, and zika virus fever. Climatic events influence insect vectors' reproduction and survival rates, habitat, prosperity, and transmission; similarly, they affect the proliferation and survival rates of the infectious agent inside the vector (Rayan et al., 2021), facilitating the reemergence or expansion of endemic vector-borne diseases and promoting the migration of these diseases to new regions (Noyes et al., 2009).

Altering the climate pattern affects GWR, and a strong correlation between climate change and seasonal variation on arsenic content in groundwater levels has been reported (Aribam et al., 2021). Jarsjö et al. (2020) studied alterations in contaminant loading of groundwater and surface water systems by performing hydrogeological−geochemical model projections with climate change scenarios. They have found that a moderate increase (0.2 m) in average levels of shallow groundwater systems can increase mass flows of metals (As and Pb) through groundwater by a factor of 2−10 as well as their mobilization in regions subject to increased amplitude of groundwater levels fluctuations.

Variations in climate and associated changes in weather patterns as well as nonclimate-related drivers are anticipated to affect various points in the source-to-receptor pathways, modifying how and to what extent humans are exposed to toxic chemicals and how they respond to harmful effects from those exposures (Noyes et al., 2009).

Food security

Food production is key for human health and is in close relationship with water resource management and the commitment to the UN Sustainable Development Goals of no hunger and poverty. The adverse consequences of extreme events on food security are largely studied: droughts, floods, and irregular rainfall exacerbate poverty with droughts having the greatest impact on farmers' welfare loss, followed by floods (Maganga et al., 2021). For this reason, climate change has the potential to worsen poverty and aggravate conditions of agriculture-dependent groups and the poorest countries.

As the global food system includes all activities and stakeholders in food production, transport, manufacturing, retailing, and consumption, as well as waste, impacts, nutrition, health, and wellbeing, the impacts of climate change on food security can be through many routes and will extend to all activities in the food system (Brown et al., 2015; Mbow et al., 2019). The main climate drivers affecting food security are temperature and precipitation, thus extreme climate events resulting in inland and coastal flooding, will have the significant impact of reducing people's ability to obtain or prepare food (Mbow et al., 2019).

Khanal et al. (2021) explained that climate change is related to two long term factors that are likely to further lower agricultural productivity: (1) Negative crop yield due to increases in temperature and (2) Extreme weather events that cause pests, reduce water and nutrient use efficiency, and increase yield variability.

Abeldaño Zuñiga et al. (2021) identified that soil degradation, desertification, rising sea levels, rising temperatures, ocean acidification, and retreat of glaciers are the main factors associated with food security and poverty in rural communities where land is the primary mean of production. The rise of temperatures will also create changes in the actual trade networks because high temperatures increase spoilage, so a more efficient management of organic product with high energy expenditure is expected to increase food cost.

As climate change will affect the availability, access, use, and stability of the food chains worldwide, it is imperative that people and agricultural activities change their usual performance and start to adapt to new growing crops techniques and consumer behavior to mitigate the impacts of climate change on food security (Mbow et al., 2019).

Secure shelter and social stability

The adverse consequences of climate change on food safety, water security, and extreme events affect households' wellbeing and threaten safe and secure shelter. The presence of extreme events forces people to move from their homes and become climate refugees. The United Nations High Commissioner for Refugees

(UNHCR) found that in the past decade, weather-related events caused an average of 21.5 million new displacements each year, more than any conflict worldwide (UNHCR, 2021).

In addition to extreme events, climate change can exacerbate political tensions and prompt latent conflicts, mainly those associated with resources. Conflicts cause social instability and that poses an additional risk of displacement. Even when climate change does not cause conflict directly, any resource conflict may be a secondary or tertiary effect of climate change (Anthonia et al., 2021).

Displaced communities that are settled in refugee camps experience a lack of secure shelter similar to marginalized communities that inhabit slums and physically vulnerable rural or urban settlements that are unprotected from inclement weather and do not have the basic services to live, such as safe water and sanitation (Selendy, 2016). Vulnerable settlements also expose people to other climate-related risk such as vector-transmitted diseases and infectious diseases.

Extreme events do not only impact social wellbeing by forcing people to move, Stoler et al. (2020) also established a framework linking water insecurity to perceived stress and socioeconomic disruption. They identified that water is related to three wellbeing determinants: (1) physical health and hygiene, (2) psychosocial health, and (3) nonagricultural livelihood; consequently, slow-onset events linked to climate change can cause an increase in perceived stress and disruption on the social and economic levels closely linked to migration.

20.5 Future trends

The management of water resources urgently requires a broader approach so the resources can ensure the different uses related to human wellbeing considering not only quantity but quality of water to contribute to reduce poverty and vulnerabilities (Balasubramanya & Stifel, 2020) in a time marked by climate change.

As changes in socioeconomic systems are experienced during extreme events and slow-onset events, societies need to invest greater resources to study, monitor, and control these risks, given the extreme danger they cause to inequality reduction and the permanence of ecosystem services that sustain human life on Earth. Adaptive strategies include the use of geographic information systems technologies for monitoring vulnerable areas, diversification of cultures, adoption of agroecological practices, safety nets on farm households, reduction in the gender gap in land governance, and implementation of educational plans (Abeldaño Zuñiga et al., 2021; Maganga et al., 2021).

The recognition of the Anthropocene as a stage in which human actions have caused a great deal of change in ecosystems places humanity at a decisive moment in our history since the fate of human life on Earth depends almost entirely on the actions and measures taken today. In this situation, it is necessary to identify the adverse and irreversible effects that the current economic model has caused for the planet, and we need to imagine new paths and routes of responsible action where all human beings and the various forms of life are included.

References

Abeldaño Zuñiga, R. A., Lima, G. N., & González Villoria, A. M. (2021). Impact of slow-onset events related to Climate Change on food security in Latin America and the Caribbean. *Current Opinion in Environmental Sustainability, 50*, 215–224. Available from http://10.0.3.248/j.cosust.2021.04.011.

Althor, G., Watson, J. E. M., & Fuller, R. A. (2016). Global mismatch between greenhouse gas emissions and the burden of climate change. *Scientific Reports, 6*(1), 20281. Available from https://doi.org/10.1038/srep20281.

Anthonia, A. N., Bello, Y. A., Oliatan, O. M., & Macaulay, I. A. (2021). Reviewing the links between climate change and resource conflict. *Global Journal of Pure and Applied Sciences, 27*(2), 243–252.

Aribam, B., Alam, W., & Thokchom, B. (2021). Chapter 8—Water, arsenic, and climate change. In: B. Thokchom, P. Qiu, P. Singh, & P. K. B. T.-W. C. in the E. of G. C. C. Iyer, (Eds.), Elsevier, pp. 167–190.

Austria, P. F. M., & Gómez, C. P. (2012). Effects of climate change on water availability in Mexico. *Tecnol. y ciencias del agua, 3*(1), 5–20.

Azzarri, C., & Signorelli, S. (2020). Climate and poverty in Africa South of the Sahara. *World Development, 125*, 104691. Available from https://doi.org/10.1016/j.worlddev.2019.104691.

Balasubramanya, S., & Stifel, D. (2020). Viewpoint: Water, agriculture & poverty in an era of climate change: Why do we know so little?

Bates, B. C., Kundzewicz, Z. W., Wu, S., & Palutikof, J. (Eds.). (2008). *Climate change and water. Technical paper of the intergovernmental panel on climate change*. Geneva.

Botana, L. M. (2016). Toxicological perspective on climate change: Aquatic toxins. *Chemical Research in Toxicology, 29*(4), 619–625.

Brown, M. E., et al. (2015). *Climate change, global food security, and the U.S. Food System*. U.S. Department of Agriculture.

Carleton, T. A., & Hsiang, S. M. (2016). Social and economic impacts of climate. *Science (80-.), 353*(6304), aad9837-1–aad9837-15. Available from http://10.0.4.102/science.aad9837.

CRED and UNISDR. (2018). Economic losses, poverty & disasters: 1998-2017. Centre for Research on the Epidemiology of Disasters, United Nations Office for Disaster Risk Reduction.

Ding, Y.-J., et al. (2021). An overview of climate change impacts on the society in China. *Advances in Climate Change Research, 12*(2), 210–223. Available from https://doi.org/10.1016/j.accre.2021.03.002.

Duran-Encalada, J. A., Paucar-Caceres, A., Bandala, E. R., & Wright, G. H. (2017). The impact of global climate change on water quantity and quality: A system dynamics approach to the US–Mexican transborder region. *European Journal of Operational Research, 256*(2), 567–581. Available from https://doi.org/10.1016/j.ejor.2016.06.016.

FAO. (2012). Coping with water scarcity. An action framework for agriculture and food security. Available from http://www.fao.org/land-water/water/water-scarcity/en/ (accessed 25.07.21.).

Feng, Z., Song, F., Sakaguchi, K., & Leung, L. R. (2021). Evaluation of mesoscale convective systems in climate simulations: Methodological development and results from MPAS-CAM over the United States. *Journal of Climate, 34*(7), 2611–2633. Available from https://doi.org/10.1175/JCLI-D-20-0136.1.

Insid. EPA's Water Policy Rep. (2001). Report warns of climate change impacts on water resources. *Insid. EPA's Water Policy Reports, 10*(1), 12–13. Available from https://www.jstor.org/stable/26821574.

Jarsjö, J., et al. (2020). Projecting impacts of climate change on metal mobilization at contaminated sites: Controls by the groundwater level. *The Science of the Total Environment, 712*, 135560. Available from https://doi.org/10.1016/j.scitotenv.2019.135560.

Jepson, W. E., Wutich, A., Colllins, S. M., Boateng, G. O., & Young, S. L. (2017). Progress in household water insecurity metrics: A cross-disciplinary approach. *WIREs Water, 4*(3), e1214. Available from https://doi.org/10.1002/wat2.1214.

Juana, J. S., Mangadi, K. T., & Strzepek, K. M. (2012). The socio-economic impacts of climate change on water resources in South Africa. *Water International, 37*(3), 265.

Khanal, U., Wilson, C., Rahman, S., Lee, B. L., & Hoang, V.-N. (2021). Smallholder farmers' adaptation to climate change and its potential contribution to UN's sustainable development goals of zero hunger and no poverty. *Journal of Cleaner Production, 281*. Available from http://10.0.3.248/j.jclepro.2020.124999.

Kirchmeier-Young, M. C., Gillett, N. P., Zwiers, F. W., Cannon, A. J., & Anslow, F. S. (2019). Attribution of the influence of human-induced climate change on an extreme fire season. *Earth's Futur., 7*(1), 2–10.

Maganga, A. M., Chiwaula, L., & Kambewa, P. (2021). Climate induced vulnerability to poverty among smallholder farmers: Evidence from Malawi. *World Development Perspectives, 21*, 100273. Available from https://doi.org/10.1016/j.wdp.2020.100273.

Matondo, J. I., & Msibi, K. M. (2001). Estimation of the impact of climate change on hydrology and water resources in Swaziland. *Water International, 26*(3), 425.

Mbow, C. et al. (2019). Food security. In: P. R. Shukla, J. Skea, E. C. Buendia, V. Masson-Delmotte, H.-O. Pörtner, D. C. Roberts, P. Zhai, R. Slade, S. Connors0, R. van Diemen, M. Ferrat, E. Haughey, S. Luz, S. Neogi, M. Pathak, J. Petzold, J. P. Pereira, P. Vyas, E. Huntley, K. Kissick, M. Belkacemi, and J. Malley (Eds.), *Climate change and land: An IPCC special report on climate change, desertification, land degradation, sustainable land management, food security, and greenhouse gas fluxes in terrestrial ecosystems* (pp. 437–560).

McIntyre, O. (2017). EU legal protection for ecologically significant groundwater in the context of climate change vulnerability. *Water International, 42*(6), 709.

Muttiah, R. S., & Wurbs, R. A. (2002). Modeling the impacts of climate change on water supply reliabilities. *Water International, 27*(3), 407.

Noyes, P. D., et al. (2009). The toxicology of climate change: Environmental contaminants in a warming world. *Environment International, 35*(6), 971.

Olsson, L., et al. (2014). Livelihoods and poverty. In C. B. Field, V. R. Barros, D. J. Dokken, K. J. Mach, M. D. Mastrandrea, T. E. Bilir, M. Chatterjee, K. L. Ebi, Y. O. Estrada, R. C. Genova, B. Girma, E. S. Kissel, A. N. Levy, S. MacCracken, P. R. Mastrandrea, & L. L. White (Eds.), *Climate change 2014: Impacts, adaptation, and vulnerability. Part A: Global and sectoral aspects. Contribution of working group II to the fifth assessment report of the intergovernmental panel on climate change* (pp. 793–832). Cambridge, UK and New York: Cambridge University Press.

Padrón, R. S., et al. (2020). Observed changes in dry-season water availability attributed to human-induced climate change. *Nature Geoscience, 13*(7), 477. Available from https://doi.org/10.1038/s41561-020-0594-1.

Pastor-Paz, J., Noy, I., Sin, I., Sood, A., Fleming-Munoz, D., & Owen, S. (2020). Projecting the effect of climate change on residential property damages caused by extreme weather events. *Journal of Environmental Management, 276*, 111012. Available from https://doi.org/10.1016/j.jenvman.2020.111012.

Rayan, R. A., Choudhury, M., Deb, M., Chakravorty, A., Devi, R. M., & Mehta, J. (2021). Chapter 10 - Climate change: Impact on waterborne infectious diseases. In: B. Thokchom, P. Qiu, P. Singh, & P. K. B. T.-W. C. in the E. of G. C. C. Iyer, (Eds.), Elsevier, pp. 213–228.

Selendy, J. M. H. (2016). *Water, food and shelter security in natural hazards*. Oxford University Press. Available from http://doi.org/10.1093/acrefore/9780199389407.013.61.

Stoler, J., et al. (2020). Cash water expenditures are associated with household water insecurity, food insecurity, and perceived stress in study sites across 20 low- and middle-income countries. *The Science of the Total Environment, 716*, 135881. Available from https://doi.org/10.1016/j.scitotenv.2019.135881.

The World Bank. (2017). Results brief - Climate insurance. https://www.worldbank.org/en/results/2017/12/01/climate-insurance (accessed 03.07.21.).

Tramblay, Y., Llasat, M. C., Randin, C., & Coppola, E. (2020). Climate change impacts on water resources in the Mediterranean. *Regional Environmental Change, 20*(3). Available from https://doi.org/10.1007/s10113-020-01665-y.

UN. (2015). Sendai framework for disaster risk reduction 2015-2030. Available from https://www.preventionweb.net/files/43291_sendaiframeworkfordrren.pdf.

UN DESA. (2020). World social report 2020 inequality in a rapidly changing world. [Online]. Available: https://www.un.org/development/desa/dspd/wp-content/uploads/sites/22/2020/01/World-Social-Report-2020-FullReport.pdf.

UNDP and OPHI. (2020). Global multidimensional poverty index 2020 – Charting pathways out of multidimensional poverty: Achieving the SDGs report. Available from https://ophi.org.uk/wp-content/uploads/G-MPI_Report_2020_Charting_Pathways.pdf.

UNDRR. (2020). Disaster risk. *Understanging Disaster Risk*. https://www.preventionweb.net/understanding-disaster-risk/component-risk/disaster-risk (accessed 18.07.21.).

UNHCR. (2021). Displaced on the frontlines of the climate emergency. *Multiplying risks*.

United Nations General Assembly. (2010). The human right to water and sanitation. *Resolution adopted by the General Assembly on 28 July 2010, A/RES/64/292*. https://documents-dds-ny.un.org/doc/UNDOC/GEN/N09/479/35/PDF/N0947935.pdf?OpenElement (accessed 20.08.21.).

van Dijk, J. M. (2006). Water assessment in the Netherlands. *Impact Assessment and Project Appraisal, 24*(3), 199.

Wang, H., et al. (2021). From unusual suspect to serial killer: Cyanotoxins boosted by climate change may jeopardize megafauna. *Innovation, 2*(2), 100092. Available from https://doi.org/10.1016/j.xinn.2021.100092.

WHO. (2021a). Climate change and health. https://www.who.int/news-room/fact-sheets/detail/climate-change-and-health (accessed 07.07.21.).

WHO. (2021b). Drinking-water. https://www.who.int/news-room/fact-sheets/detail/drinking-water (accessed 10.08.21.).

Wisner, B. (2006). Risk reduction indicators... social vulnerability. *Annex B-6, TRIAMS Work. Paper-Risk Reduction Indicators*.

World Bank. (2013). Climate change impacts on water resources and adaptation in the rural water supply and sanitation sector in Nicaragua. Latin America and Caribbean Region Environment and Water Resources occasional paper series. Washington. Available from https://openknowledge.worldbank.org/handle/10986/16586.

Zhidkova, A. Y., Podberesnij, V. V., Zarubina, R. V., & Kononova, O. A. (2020). The effect of eutrophication on human health on the example of the Gulf of Taganrog of the Sea of Azov. *IOP Conference Series: Earth and Environmental Science, 548*, 52053. Available from https://doi.org/10.1088/1755-1315/548/5/052053.

SECTION C

Policy issues

CHAPTER 21

Climate change and water justice

M. Mills-Novoa[1,2], R. Boelens[3,4,5] and J. Hoogesteger[3]

[1]Department of Environmental Science, Policy and Management, University of California: Berkeley, Berkeley, CA, United States, [2]Energy and Resources Group, University of California: Berkeley, Berkeley, CA, United States, [3]Water Resources Management Group, Department of Environmental Sciences, Wageningen University, Wageningen, The Netherlands, [4]Centre for Latin American Research and Documentation (CEDLA), University of Amsterdam, Amsterdam, The Netherlands, [5]Faculty of Agronomy, Central University of Ecuador, Quito, Ecuador

Chapter Outline

21.1 Introduction 399
21.2 What is justice and for whom? Examples from the water sector 401
21.3 Defining water justice 403
21.4 Water justice and climate change vulnerability 404
21.5 Water justice and climate change's proposed solutions 406
    21.5.1 Mitigation 407
    21.5.2 Adaptation 408
21.6 Naturalizing climate change 410
21.7 Struggle(s) for water and climate justice 411
21.8 Conclusions 412
References 412

21.1 Introduction

The impact of climate change on water has never been more apparent. In August 2021, the first official declaration of a water shortage was issued for the Colorado River basin, which supports 40 million people. Just weeks later, Hurricane Ida caused widespread and highly destructive flooding in large swaths of the Southeast and Northeast United States, affecting urban and rural communities alike. Meanwhile, wildfires like Caldor Fire in California raged across the Western United States. These mega wildfires compromise water quality in the short time by depositing ash in vital waterways and create long term water quality challenges by increasing erosion and sediment deposition across a watershed. These examples were drawn from just one country, over a few weeks' time. In the same period, devastating floods inundated entire regions in China, the Philippines, Belgium, Germany, and the Netherlands, destroying villages and taking numerous lives. Severe drought caused famine across southern Madagascar. Wildfires raged across Greece, Canada, and Russia, and numerous hurricanes and cyclones brought damage and flooding to Haiti, Mexico, Barbados, Ghana, Burkina Faso, India, Nigeria, and others. Climate change is here and affecting the availability, distribution, and quality of water around the world.

The recent release of the sixth assessment report of the Intergovernmental Panel on Climate Change (IPCC) underscores the impact of climate change on water resources. As the climate warms, the global water cycle will continue to intensify, creating more variability across regions, between seasons, and from year to year (IPCC, 2021). We can also expect more severe droughts and flooding as dry and wet events become more extreme. Monsoon precipitation is expected to increase, creating the potential for devastating flooding across much of South and Southeast Asia and Western Africa (IPCC, 2021). This intensifying water cycle holds real consequences for human communities and ecosystems.

Despite these escalating impacts, it is important to acknowledge that climate change is not happening in a vacuum. Climate change impacts are layered onto and exacerbate pre-existing inequalities and injustices. The world's poorest are most vulnerable to these climate change impacts and have the lowest capacity to adapt (Cardona et al., 2012). Climate change is unfolding in a world where nearly 1 in 3 people do not have access to safe drinking water (UNICEF & WHO, 2019) and climate change is poised to erode many of the development gains of recent decades (Roy et al., 2018). Additionally, the proposed solutions to climate change, through both cutting greenhouse gas emissions (mitigation) and responding to impacts (adaptation) can bolster or undermine water justice.

In this chapter, we examine ways in which climate change is and will continue to (re)shape water (in)justice. We identify three important areas of intersection between water justice and climate change:

1. Water injustice produces climate change vulnerability and climate change entrenches water injustices
2. Proposed solutions to climate change can and will have implications for water justice
3. Water justice and climate justice struggles can and should build greater unity.

By exploring these important intersections between climate change and water justice, we argue that water justice and climate justice struggles and scholarship would benefit considerably from one another.

In this chapter, we first present the multifaceted concept of justice with a particular focus on how this is reflected in the water sector. Second, we explore the intersection of water injustice and climate change vulnerability. Third, we present emblematic climate change mitigation and adaptation strategies and how they can affect water justice. Fourth, we briefly discuss the pitfalls of the over attribution of climate change as the driver of water injustice. Fifth, we examine the tensions and potential synergies between climate and water justice struggles.

21.2 What is justice and for whom? Examples from the water sector

Central to understanding water justice, is clarifying what justice means. It is not possible to understand water justice, or water governance for that matter, without a theory of justice (Perreault, 2014). Justice is not a universal truth. Rather, there are varied notions of justice that lead to different decisions, goals, and processes. Therefore it is important to explicitly consider what concepts of justice are being advanced (Boelens et al., 2018). Justice encompasses not only distributive and procedural justice but also recognitional justice with struggles aligning around one or more of these facets (Fraser, 2000; Zwarteveen & Boelens, 2014). Furthermore, there is a difference between formally accredited justice, which might be gained through state action, and socially perceived justice. This difference depends on how a group constructs "fairness" and shapes the nature of struggles for justice (Boelens, 2009). Taken together, concepts of justice are highly place-based and situational.

To examine the underlying theories of justice, we present some key concepts of justice that have mobilized water governance models. These theories of justice are not meant to be exhaustive, and we ground each theory in a relevant example.

First, justice can be considered as the greatest good for the greatest number. Based on the work of utilitarian Jeremy Bentham, justice is understood here as the impact of a policy or decision on the aggregate social welfare of a society (Bentham, 1988). An inherent issue with this approach, which is applied in many mainstream water policies and governance models around the world, is that there is not a consideration of how that welfare is distributed across society.

In the context of water governance, a utilitarian approach to justice is exemplified by rural to urban water transfers. Globally, urban freshwater demand and changing water availability has driven the development of 103 rural to urban mega water transfers serving 69 cities, moving approximately 16 billion m^3 of water per year over almost 13,000 km (Garrick et al., 2016). These transfers privilege urban water uses and users over rural communities, evoking a Benthamian rationale around the distribution of water resources across landscapes and societies (Hommes & Boelens, 2018). The "greatest number" (often consisting of more visible and dominant citizen groups) legitimizes water-based suffering and water depletion of less visible and politically/economically marginalized rural populations, territories, and ecological systems.

Second, Rawls' theory of justice, which explicitly critiques utilitarianism, defines justice as fairness. Rawls takes issue with the aggregational approach of Bentham, arguing for a greater consideration of the differences between people (Rawls, 1971). Instead, Rawls argues that citizens have equal claim to equal basic rights and liberties. In doing this, he focuses on the "least advantaged members of society" meaning that a policy is only desirable if it benefits the most marginalized (Rawls, 2005, p. 6). Rawls stresses procedural fairness and ethics-based autonomous decision making to achieve this. In this conception of justice, decisions are made behind an abstract "veil of ignorance" where decisionmakers do not know the impact of their decisions on themselves or potential biasing factors such as the class, gender, race, or other axes of societal difference for the people involved. A critique of Rawlsian conceptualizations of justice is that universalistic distributive models and processes do not consider the historical conditions that created uneven distributions (Schlosberg, 2004). Moreover, the theoretical "veil of ignorance" often turns out to be discriminatory for peasant, Indigenous, and female water user groups in many countries, where justice plays out in accordance with gender, ethnicity, and class-based power structures. Rather than unrealistic theoretical and universalist suppositions about justice that are often made in policy decision-making spaces, water users claim and advance water justice and climate adaptation measures based on their local day-to-day experiences and contexts in profoundly unjust sociohydrological systems.

Over the last decade, the push to establish water as a basic human right reflects aspects of Rawlsian justice wherein this right is universal across a society and the state is responsible for ensuring this right. In 2010, the United Nations passed a resolution on the Human Right to Water and Sanitation, which states, "The right to safe and clean drinking water and sanitation is a human right that is essential for the full enjoyment of life and all human rights" (UN, 2010). While this resolution was not ratified by all countries, it was not opposed and thus was adopted though large gaps in interpretation and implementation remain (Harris et al., 2013; Sultana & Loftus, 2015; Sultana, 2018).

Third, libertarianism defines justice by the greatest individual freedom (Friedman, 1963; Nozick, 1974). Like Rawlsianism, libertarianism emphasizes rights and liberties, but these rights and liberties are more strictly defined. In libertarianism, a core right is the right to own property and emphasis is placed on negative freedom where rights and liberties are constrained by the actions of others or the state. In orthodox libertarianism, government intervention is only warranted to protect personal liberty and private property. Under this theory of libertarian entitlement, there is an outright rejection of redistributive policy through taxes or other mechanisms.

Libertarianism has been applied to water governance through the creation of water markets. From Australia to Chile to the American West, water markets have been created, privatizing and commodifying water as a tradeable good that can be bought and sold on an open market. The impact of these markets on the environment, social equity, and economic development is hotly contested (Bauer, 2015; Cardoso & Pacheco-Pizarro, 2021; Hasselman & Stoker, 2017; Höhl et al., 2021; Prieto, 2016).

Fourth, the capabilities approach to justice merges the aggregational view of utilitarianism with the focus on individual circumstances of Rawlsianism and libertarianism. Amartya Sen, who proposed the capabilities approach, defines capability as an individual's capacity to satisfy basic needs in society, which is shaped by institutional structures (Sen, 2009). Unlike libertarianism, Sen emphasizes positive freedoms like the right, capacity, or ability to achieve something. A state is thus judged against its citizens' actual capabilities as opposed to idealized rights accorded to them in law. Under a capabilities approach, justice is measured by the fulfillment of human potential through both the material satisfaction of human needs and the social institutions to achieve them. Therefore, justice is not just about the provision of water, for example, but how that provision leads to the maximization of an individual's potential (Bakker, 2010; Perreault, 2014).

As an example of this approach, in Central Mexico, the Coalition for the Defense of the Independence Aquifer, which is situated in the state of Guanajuato, filed a case against the government at the International Permanent Peoples' Tribunal. This coalition is attempting to hold the Mexican government accountable for enabling the overexploitation of the aquifer, which has led to toxic levels of arsenic and fluoride in an increasing number of community wells (Knappett et al., 2020). These wells are used for drinking water and thus communities face several public health threats related to prolonged exposure to arsenic and fluoride such as dental fluorosis, loss of memory, reduce mental capacity in children and kidney failure amongst others (Hoogesteger, 2018). These public health problems and the threat of future issues due to persistent aquifer overexploitation compromise the quality of life and the current and future potential of the inhabitants of the affected rural communities.

These theories of justice all carry challenges. Some of these theories create greater injustice in the distribution of water (*distributive justice*), processes and procedures through which decisions are made (*procedural justice*), and/or the recognition of certain actors and not others (*recognitional justice*). Other theories require a fundamental reorientation of power and politics, which has not occurred. What is critical, however, is to explicitly identify the underlying concept of justice that is being advanced in any reform and towards what end.

21.3 Defining water justice

With these theories of justice in mind, we adopt a relational theory of justice in which we examine the embedded context of particular concepts of justice and how a concept of justice is actualized through social practice (Boelens et al., 2018). By adopting a relational theory of justice, we gain a historical and comparative perspective that enables us to understand how diverse people define and actualize their definitions of justice within a particular context (ibid). Reflecting our relational theory, we define water justice as: (Boelens, 2015)

> *The interactive societal and academic endeavor to critically explore water knowledge production, allocation and governance and to combine struggles against water-based forms of material dispossession, cultural discrimination, political exclusion and ecological destruction, as rooted in particular contexts.*
> **Source**: *Adapted from: Boelens, R. 2015. Water Justice in Latin America: The Politics of Difference, Equality, and Indifference. Amsterdam: CEDLA and University of Amsterdam. 34)*

Water justice as a relational concept is multidimensional and is often associated with notions of equity. While universalist justice frames commonly deal with decontextualized definitions and generalized application of "rightness," the notion of equity deals with "fairness," as perceived real-life conditions and in accordance with people's and circumstances' diversity. Equity can be disaggregated into the following interrelated dimensions (Hoogesteger & Wester, 2015; McDermott et al., 2013): (1) *distributive equity* which looks at the distribution of resource access and its benefits. This relates to the material and economic dimensions of water justice (who gets what?); (2) *procedural equity* concerned with decision-making processes and procedures. This includes the cultural and political dimensions and include issues of procedure, recognition, and participation (Zwarteveen & Boelens, 2014) (who decides and through which procedures?); and (3) *contextual equity* concerned with the pre-existing conditions and power relations that shape the first two dimensions (what is the historically shaped and present socionatural context?). By focusing on these different dimensions, the questions about who, why and how, shape the content of equity and justice in water governance interventions can be addressed in struggles for and scholarship of water justice. In this context, climate change is and will impact the different dimensions of water justice by bringing about changes in the natural context (quality, accessibility, reliability, and availability of water) as well as in the interrelated social context (distribution of resources, decision making, power relations, and the discourses that advance and naturalize these). To explore how climate change affects the context in which water injustices are created, we discuss the close interrelations between water injustices and climate change vulnerability in the following section.

21.4 Water justice and climate change vulnerability

The relative impact of climate change on people, places, and systems depends on the vulnerability of those elements. Vulnerability is defined by the IPCC as, "the propensity of exposed elements such as human beings, their livelihoods, and assets to suffer adverse effects when impacted by hazard events" (Cardona et al., 2012; p. 69). Vulnerability to climate change depends on "social, geographic, demographic, cultural, institutional, governance, and environmental factors," meaning that vulnerability between and among individuals and communities depends on their class, race/ethnicity, age, disability, gender, health, and other axes of social difference (Cardona et al., 2012; p. 69). This type of vulnerability is often referred to as "contextual vulnerability," which is a process oriented and multidimensional view into climate and society linkages, where climate change vulnerability occurs in the context of political, economic, institutional and social structures and changes that interact dynamically (O'Brien et al., 2007; Thomas et al., 2019). Contextual vulnerability contrasts with "outcome vulnerability" in which vulnerability is a linear result of the projected impacts of climate change on an "exposure unit" which may be social or environmental and then mitigated through adaptation efforts (ibid). Overtime, contextual vulnerability and the related importance of nonclimatic factors in shaping vulnerability has grown more prevalent in the literature (O'Brien et al., 2007; Thomas et al., 2019).

Fundamentally, water injustices make communities more vulnerable to climate change. Many of the nonclimatic factors that shape climate change vulnerability are the same factors associated with water injustice and insecurity (Harris et al., 2018). As an example, gender is linked to both unequal water access and climate change vulnerability due to the gendered labor burdens placed on women (Harris et al., 2018; Wallace & Coles, 2005). Globally, women spend a disproportionate amount of time traveling to water sources particularly in areas experiencing water scarcity and/or contamination, often experiencing negative health, economic, and social outcomes as a result (Sultana, 2011; Wutich, 2009). Relatedly, women experience greater morbidity and mortality following climate-related disasters and droughts (Neumayer & Plümper, 2007). Water injustices create gender specific barriers to adapting to and coping with climate change (Denton, 2002; Harris et al., 2018; Terry, 2009).

The intersectionality of income, poverty, race, gender, Indigeneity, and other factors shapes climate change vulnerability and water justice (Crenshaw, 1991; Nightingale, 2011). These factors do not exist in isolation, but rather intersect with and shape the differential water access and climate change vulnerability of communities (Thomas et al., 2019). For example, structural racism leads to disparities in income, education, wealth, and access to services that leads to greater climate change vulnerability and water insecurity (Harris et al., 2018; Thomas et al., 2019). The impoverishment of marginalized communities is a key factor contributing to climate change vulnerability (Hallegatte & Rozenberg, 2017).

Poor households have fewer resources to respond to extreme climate and weather events and tend to lose more than wealthier households in the wake of these events, depleting savings and further driving households into poverty (Heltberg et al., 2013; Moser, 2008). The same can be observed between poorer and wealthier countries: while Bangladesh and the Netherlands are both low-lying delta countries that will be deeply impacted by sea level rise and flooding triggered by climate change, the Dutch have far more capital and means to avoid suffering and minimize climate change impacts. Repeated nonextreme and extreme weather and climate events erodes the adaptive capacity of communities over time and heightens their vulnerability to future events by wiping out assets and resources (Cardona et al., 2012). To understand and address the root cause of differential climate change vulnerability and water access, we must adopt an intersectional lens that enables us to see the interlocking nature of gender, race/ethnicity, class/caste, religion, and other types of social difference.

It is important to note that climate change did not create most water injustices and inequities, but it is poised to exacerbate and entrench them. As an example of this dynamic, in irrigation systems smallholders whose livelihoods depend on irrigated production tend to be more vulnerable to droughts than larger producers. This uneven vulnerability relates to smallholder farmers' vulnerable economic position as well as to the unequal allocation of irrigation water especially in periods of drought. In northern Mexico, for example, several irrigation districts allocate water to the most productive irrigated areas, which are mostly in the hands of commercial farmers, leaving smallholders with little alternative but to stop farming (Ahlers, 2010; Wolfe, 2017). Additionally, commercial farmers are often able to access additional groundwater to overcome periods of water scarcity; an option smallholders do not have (Hoogesteger, 2018). Similarly, rural communities in peri-urban zones lose their water to growing urban demands and the demands of politically connected urban/rural populations; processes that are accelerated during periods of drought (Díaz-Caravantes & Wilder, 2014; Hommes et al., 2019). These examples point to ways that land and water dispossession increase vulnerability by limiting the capacity of individuals and communities to adapt to climate change impacts.

Not all water injustices are driven by or directly relate to climate change, but all water injustices do create vulnerability to climate change. Water injustice impoverishes individuals and communities, undermining their wellbeing and producing greater vulnerability. Therefore addressing water injustices will also reduce climate change vulnerability.

21.5 *Water justice and climate change's proposed solutions*

Climate change solutions entail both efforts to reduce greenhouse gases (i.e., mitigation) and respond to climate change impacts (i.e., adaptation). Both mitigation and adaptation initiatives have implications for water justice. In some cases, mitigation and adaptation solutions will create or deepen existing water injustices while other solutions may represent a space for positive action. In this section, we present six examples of how responses to climate change are poised to affect water justice: lithium mining, REDD+/Payment of Ecosystem Services, hydropower dams, rural to urban water transfers, desalination, and adaptive management.

21.5.1 Mitigation

Transport emissions account for 23% of global carbon dioxide emissions (Sims et al., 2014). To promote "clean" or less fossil fuel intensive forms of transport, there has been a boom in electric cars, which has led to an eightfold increase in demand for lithium-ion batteries over the past decade (Agusdinata et al., 2018). 75% of known lithium reserves are in the "lithium triangle" in Chile, Bolivia, and Argentina (Jaskula, 2012). Lithium mining, like other extractive industries, affects local ecosystems and communities, but there has been little research on its local socioenvironmental impacts (Agusdinata et al., 2018). Initial studies point to environmental degradation (Liu et al., 2019), falling water tables due to high levels of groundwater pumping (Liu & Agusdinata, 2020; Marazuela et al., 2019), and a large influx of labor migrants with little local employment in the mines (Liu & Agusdinata, 2020). Very little is known about the impact of lithium mining on air quality and human health (Agusdinata et al., 2018). Research on the impact of lithium mining on water justice from the Altiplano to Nevada is urgently needed. If this sector continues to go unexamined by scholars, the water injustices experienced by communities near lithium extraction sites will remain absent from discussions about the transition to a "green economy."

Reducing Emissions from Deforestation and forest Degradation (REDD+) is a central strategy for conserving forest systems while safeguarding carbon stocks (the +). Through REDD+, carbon offsets are purchased by governments, businesses, or individuals in the Global North and used to fund good forest management practices in the Global South. Within the REDD+ framework, Payment for Ecosystem Services (PES) programs are a common strategy for providing direct compensation to people engaged in land stewardship. PES programs extend beyond simply considering carbon sequestration in forests to also encompassing the preservation of headwaters and riparian areas to preserve downstream water quality and availability (de Francisco & Boelens, 2015; Kolinjivadi et al., 2017). Advocates of PES promote these programs as providing needed income to local communities and emphasize that participation in PES is voluntary.

Indigenous communities and others, however, have raised concerns that PES and REDD+ programs endanger their access to traditional forest-based livelihoods, inadequately recognize the ancestral lands of Indigenous peoples, and endanger land tenure (De La Fuente & Hajjar, 2013). In practice, PES scholars have found that benefits have disproportionately flowed to large land holders and male-headed households (von Hedemann & Osborne, 2016). Procedurally, scholars have pointed to the fundamental power imbalances in PES negotiations and agreements, which leads to unequal benefits (de Francisco & Boelens, 2015, 2016; von Hedemann, 2020). A highly problematic facet of PES as a climate change mitigation and adaptation strategy is that it fails to recognize the cultural and cohesive function of local communities' land and water governance institutions. PES programs often neglect the historical rootedness and existence of complex rights systems and functional water management relationships. By ignoring these norms, universalistic PES programs, "impose commensurable values and a commodity-biased worldview" undermining "existing noncommodity exchange and collaboration mechanisms that are crucial to local peasant economies and sustainable land and water management" (Boelens et al., 2014; p. 86). These local governance arrangements link people, place, and production while shaping local water governance territories, organizational forms, institutions, identities, and cultures. In response to PES programs, Indigenous communities from Indonesia to Mexico have protested current PES models, which they feel curtail their territorial rights and reinforce unequal power relationships (Osborne et al., 2014). In Guatemala, Indigenous communities advocated for changes in the national PES program to better meet their needs and reflect their forest ontologies, which enabled them to benefit from these programs more fully (von Hedemann, 2020). When considering PES as a climate change solution, scholars and practitioners need to center local livelihoods and needs or risk further exacerbating longstanding water and land injustices.

The global boom in hydropower dam construction has been promoted as an alternative to fossil fuel-based energy production and as a form of adaptation by buffering intensifying flooding and drought (Crow-Miller et al., 2017; Mills-Novoa & Hermoza, 2017). Currently, 3700 major dams are either planned or under construction globally (Zarfl et al., 2015). As mitigation and adaptation strategies, however, dams have documented negative social-environmental impacts. Dams disrupt ecosystem connectivity (Anderson et al., 2018), displace communities (McDonald-Wilmsen & Webber, 2020; Tilt & Gerkey, 2016), and create irreversible changes in local livelihoods (Obour et al., 2016; Beck et al., 2012; Hausermann, 2018). Unlike earlier periods of dam development, this newest dam boom is being funded by private companies and private bilateral development banks (Gerlak et al., 2020). The Chinese Export-Import Bank and other Chinese financial institutions and state-owned companies are involved in a least 93 major dam projects globally, but have not accepted environmental and social guards that were established by the international community after substantial advocacy by civil society in earlier eras of dam development (Gerlak et al., 2020). Social movements continue to contest this latest dam boom, but the trend toward renewed mega dam development threatens rural livelihoods and territory (Borgias & Braun, 2017). It is also important to note that scientists have questioned the efficacy of hydropower dams as a mitigation strategy since dams generate methane and carbon dioxide emissions (Barros et al., 2011).

21.5.2 Adaptation

In response to growing urban water demand and declining freshwater availability, rural to urban water transfers are being proposed as a climate change adaptation strategy. These water transfers are not new. The mega-hydraulic projects that transfer water from rural to urban areas are often justified through the discourses of progress and necessity, but these projects also reflect a fundamental devaluing of rural livelihoods, ecosystems, and other nonurban water uses (Blackbourn, 2006; Hidalgo Bastidas et al., 2018; Hommes et al., 2019). Social movements and scholars are increasingly calling into question the rural-urban inequality and territorial transformation caused by rural to urban water transfers (Hoogendam, 2016; Punjabi & Johnson, 2019). Additionally, the focus on rural to water transfers to meet urban water needs sidesteps inefficiency in urban water infrastructure and use (Torio et al., 2019). As the climate crisis deepens, the call for rural to urban water transfers will likely intensify (Shi et al., 2021). It is crucial that scholars and activists continue to question the underlying values behind these transfers and the inequalities that they cement in the landscape.

Unlike rural to urban water transfers, desalination of marine or brackish water has the potential of providing a local approach for coastal areas to meet growing water demands (McEvoy, 2014). As a technological fix, desalination has been used for potable water, irrigation, and mining in arid coastal areas around the world (Bernabé-Crespo et al., 2019; Campero & Harris, 2019; Fragkou & Budds, 2020). Desalination, however, has major implications for water and climate justice. First, desalination technologies are energy intensive. Therefore a dependence on desalination technologies could also entrench carbon-intensive energy production unless desalination plant development is paired with renewable energy production (Nassrullah et al., 2020). Second, desalination processes create air, marine, and land pollution, which can be challenging and expensive to mitigate (Elsaid et al., 2020). Third, these environmental impacts affect nearby communities. Many of the existing desalination plants have been developed in ambiguous regulatory contexts where local stakeholders were often excluded or tokenized in regulatory processes meaning that the concerns and needs of the local community were not integrated into design or operation (Campero & Harris, 2019; Campero et al., 2021; McEvoy, 2014). Fourth, desalination has been funded by a diverse set of actors including private companies, public institutions, or public-private partnerships with major implications for who benefits. Scholars warn that in cases where private companies play a central role in desalination plant development and operation, the resulting privatization of water has led to municipalities and households being unable to afford the desalinated water (Loftus & March, 2016; McEvoy, 2014; Williams, 2018). Desalination may produce nonclimate dependent freshwater, but it does so at a high environmental, economic, and social cost, which needs to be explicitly addressed.

In the water governance literature, there has been a pivot toward adaptive management as an alternative management paradigm that better copes with the challenges of climate change and water insecurity. Adaptive management emphasizes the importance of social learning within management systems to cope with uncertainty. A core tenet of adaptive management is that managers can integrate new knowledge that is learned over time (Pahl-Wostl, 2007). Adaptive management sounds objectively positive, but has faced challenges in implementation

(Engle, 2011; Medema et al., 2008). In practice, attempts to undertake adaptive management have been stymied by the high costs for information gathering, fear of failure among managers, unstable funding, political risk associated with uncertain outcomes, and resistance from managers around increased transparency in management decisions (Medema et al., 2008).

Despite these implementation challenges, adaptive management has served as a powerful discursive tool for water governance reform. By evoking the global water and climate crisis, the World Bank in its 2010 World Development Report (World Bank, 2010; p. 137), stated that local communities will not be able to respond to climate change (Lynch, 2012). Therefore communities should conform to state authority, expert standards, and market rules. The report proposes transferrable water rights, full-value-pricing, and markets that are to be sustained by universalist expert information. These drastic interventions in water governance strips local communities of water governance authority (Lynch, 2012) and such simplified rules reduce their capacity to creatively respond through collective water control arrangements (Mills-Novoa et al., 2020). Fundamentally, adaptive management depoliticizes natural resource governance by emphasizing learning, flexibility, and uncertainty while sidestepping the root causes of water injustices (Medema et al., 2008). When reform is undertaken to create more "adaptive" water governance institutions, it is critical that the politics and power relations embedded in these new institutional arrangements be explicitly addressed otherwise adaptive management risks being used to consolidate state power and control over water resources.

21.6 Naturalizing climate change

There are material interactions between climate change and water justice, but there is also a pitfall in assigning climate change as the driver of water injustice. Climate change, as a discourse, can naturalize water scarcity and obscure the power and politics that drive water injustices. Scholars have begun to highlight cases where this is occurring. Perreault found that government officials blamed the drying of Lake Poopó in the Bolivian Altiplano on climate change to deflect attention away from the large-scale water withdrawals from mining, agriculture, and urban areas (Perreault, 2020). In the arid Peruvian coast, Mills-Novoa, (2020) found that the state water managers used the specter of climate change to justify the redistribution of water away from smallholder farmers and toward agroexport crops. In both cases, Perreault and Mills-Novoa do not deny that climate change is a factor in changing water availability, but both authors retain a wider analytical lens to understand the broader sociopolitical context driving water scarcity. These two cases highlight why scholars and practitioners need to be careful in assigning the cause of water injustices to climate change alone, obscuring historical and nonclimatic forces that produce water injustice.

Concerns over naturalizing the climate crisis echo earlier warnings over naturalizing the water crisis. The 2006 United Nations Development Program's Human Development report explicitly stated that the root of the global water crisis was not in physical scarcity but in power, poverty, and inequality (United Nations Development UNDP Program, 2006). We argue that the climate crisis is also not the cause of water scarcity. Any argument that asserts that water scarcity is driven solely by climate change needs to be interrogated. Attribution studies and critical social science are necessary for examining the additionality of climate change, historicizing water injustices, and accounting of power and politics.

21.7 Struggle(s) for water and climate justice

Struggles for climate justice have largely remained siloed from struggles for water justice, but this is changing. Initially, the climate justice movement was focused on prevention and mitigation, focusing on holding historical emitters accountable for greenhouse gas emissions and cutting global emissions. Over time, however, the climate justice movement has broadened to encompass equity and adaptation, reflecting core ideals of the environmental justice movement (Schlosberg & Lisette, 2014). The growing focus of the climate justice movement on the uneven impacts of climate change and equity of proposed climate change solutions, present an opportunity for unity between climate justice and water justice struggles.

In general, water justice struggles have tended to be local in nature with communities fighting threats such as hydropower dam development, agricultural water appropriation, extractivism or others (Boelens et al., 2018). Climate justice movements, alternatively, have tended to be more global in scale, focusing on international climate negotiations and north-south disparities in greenhouse gas emissions, climate change impacts, and development (Schlosberg & Lisette, 2014). The scalar differences between climate and water justice struggles are not irreconcilable but do reflect different strategies within these movements. Social movement scholars have argued for the necessity of movements that "look both ways," considering the global processes and linkages that create particular local injustices (Di Chiro, 2011; Schlosberg & Lisette, 2014). By building unity (not uniformity) between water justice and climate justice struggles, these movements could gain better insight into the local-global connections that exist between water injustice and climate injustice (Schlosberg & Lisette, 2014).

More research is needed into the increasing linkages between water and climate justice struggles to identify potential opportunities for forging greater solidarity. In building unity between these movements, climate justice struggles would gain a fuller view of the nonclimate and historical forces that have driven water dispossession and accumulation. Water justice struggles would benefit from climate justice movement's push for restorative justice.

21.8 Conclusions

Water justice and climate change are intertwined in three critical ways. First, we argue that water injustice produces climate change vulnerability and climate change entrenches water injustices. The intersectional nonclimatic factors such as gender, race, ethnicity, and class that shape climate change vulnerability are the same factors associated with water injustice and insecurity. Conversely, water and land dispossession because of water grabbing, hydropower development, rural to urban transfers, extractive industries and other drivers of water injustice impoverish individuals and communities, undermining their wellbeing and creating greater vulnerability to climate change. Simply put, not all water injustices are driven by or directly relate to climate change, but all water injustices do create vulnerability to climate change. Therefore addressing water injustices will also reduce climate change vulnerability.

Second, we argue that the proposed solutions to climate change can and will have implications for water justice. Mitigation solutions (e.g., lithium mining for electrification, REDD+/PES initiatives for carbon sequestration, and hydropower development for "green" energy) and adaptation solutions (e.g., desalination of salt or brackish water, rural to urban water transfers, and adaptive water management regimes) will reshape power relationships, water flows, and resource use. The implications of these solutions need to be studied and addressed or climate change solutions risk cementing and exacerbating water injustices.

Third, water justice and climate justice struggles can and should build greater unity. By building unity (not uniformity) between water justice and climate justice struggles, these movements could gain better insight into the local-global connections that exist between water injustice and climate injustice. Water justice movements would offer climate justice struggles a fuller view of the nonclimatic and historical forces that have driven water dispossession and accumulation. Conversely, water justice struggles would benefit from climate justice movement's emphasis on restorative justice.

While emphasizing the critical intersections between climate change and water justice, we also want to include an important caveat: there is risk in viewing climate change as *the* driver of water injustice. Climate change, as a discourse, can naturalize water scarcity and obscure the power and politics that drive water injustice. Therefore scholars and activists need to judiciously historize water injustices, critically analyze the additionality of climate change, and carefully account for power and politics in their analysis.

References

Agusdinata, D. A., Liu, W., Eakin, H., & Romero, H. (2018). Socio-environmental impacts of lithium mineral extraction: Towards a research agenda. *Environmental Research Letters*, *13*(12), 3001. Available from https://doi.org/10.1088/1748-9326/aae9b1.

Ahlers, R. (2010). Fixing and nixing: The politics of water privatization. *Review of Radical Political Economics*, *42*(2), 213–230. Available from https://doi.org/10.1177/0486613410368497.

Anderson, E., Jenkins, C., Heilpern, S., Maldonado-Ocampo, J., Carvajal-Vallejos, F. M., Encalada, A. C., Rivadeneira, J. F., et al. (2018). Fragmentation of Andes-to-Amazon connectivity by hydropower dams. *Science Advances*, *4*(1), eaao1642. Available from https://doi.org/10.1126/sciadv.aao1642.

Bakker, K. (2010). The limits of 'Neolibera Natures': Debatign Green Neoliberalism. *Progress in Human Geography*, *34*(6), 715–735. Available from https://doi.org/10.1177/0309132510376849.

Barros, N., Jonathan, J. C., Tranvik, L. J., Prairie, Y. T., Bastviken, D., Huszar, V. L. M., del Giorgio, P., & Roland, F. (2011). Carbon emission from hydroelectric reservoirs linked to reservoir age and latitude. *Nature Geoscience*, *4*(9), 593–596. Available from https://doi.org/10.1038/ngeo1211.

Bauer, C. J. (2015). Water conflicts and entrenched governance problems in Chile's Market Model. *Water Alternatives*, *8*(2), 147–172.

Beck, M. W., Claassen, A. H., & Hundt, P. J. (2012). Environmental and livelihood impacts of dams: common lessons across development gradients that challenge sustainability. *null*, *10*, 73–92. Available from https://doi.org/10.1080/15715124.2012.656133.

Bentham, J. (1988). *The principles of morals and legislation*. Amherst, NY: Prometheus Books.

Bernabé-Crespo, M. B., Gil-Meseguer, E., & Gómez-Espín, J. M. (2019). Desalination and water security in Southeastern Spain. *Journal of Political Ecology*, *26*(1), 486–499.

Blackbourn, D. (2006). *The conquest of nature: Water, landscape, and the making of modern Germany*. New York: W.W. Norton & Company.

Boelens, R. (2009). The politics of disciplining water rights. *Development and Change*, *40*(2), 307–331.

Boelens, R. (2015). *Water justice in Latin America: The politics of difference, equality, and indifference*. Amsterdam: CEDLA and University of Amsterdam.

Boelens, R., Hoogesteger, J., & de Francisco, J. C. R. (2014). Commoditizing water territories: The clash between Andean water rights cultures and payment for environmental services policies. *Capitalism Nature Socialism (after Jan 1, 2004)*, *25*(3), 84–102. Available from https://doi.org/10.1080/10455752.2013.876867.

Boelens, R., Vos, J., & Perrault, T. (2018). Introduction: The multiple challengse and layers of water justice struggles. In R. Boelens, T. Perreault, & J. Vos (Eds.), *Water Justice* (1–32). Cambridge, UK: Cambridge University Press.

Borgias, S., & Braun, Y. (2017). From dams to democracy: Framing processes and political opportunities in Chile's Patagonia without dams movement. *Interface: A Jounral for and about Social Movements*.

Campero, C., & Harris, L. M. (2019). The legal geographies of water claims: Seawater desalination in mining regions in Chile. *Water*, *11*(5). Available from https://doi.org/10.3390/w11050886.

Campero, C., Harris, L. M., & Kunz, N. C. (2021). De-politicising seawater desalination: Environmental impact assessments in the Atacama mining region, Chile. *Environmental Science & Policy*, *120*, 187–194. Available from https://doi.org/10.1016/j.envsci.2021.03.004.

Cardona, O. D. M. K. van Aalst, J., Birkmann, M., Fordham, G., McGregor, R., Perez, R. S., Pulwarty, E. L. F., Schipper, & Sinh, B. T. (2012). Determinants of risk: Exposure and vulnerability. In Managing the risks of extreme events and disasters to advance climate change adaptation, edited by C. B. Field, V. Barros, T. F. Stocker, D. Qin, D. J. Dokken, K. L. Ebi, M. D. Mastrandrea, et al., 65–108. A Special Report of Working Groups I and II of the Intergovernmental Panel on Climate Change (IPCC). Cambridge, UK and New York, NY: Cambridge University Press.

Cardoso, J. V., & Pacheco-Pizarro, M. R. (2021). water rights, indigenous legal mobilization and the hybridization of legal pluralism in Southern Chile. *The Journal of Legal Pluralism and Unofficial Law*, 1–30, July. Available from https://doi.org/10.1080/07329113.2021.1951483.

Crenshaw, K. (1991). Mapping the margins: Intersectionality, identity politics, and violence against women of color. *Stanford Law Review*, *43*(6), 1241–1299. Available from https://doi.org/10.2307/1229039.

Crow-Miller, B., Webber, M., & Molle, F. (2017). The (Re)turn of infrastructure for water management?". *Water Alternatives*, *10*(2), 195–207.

de Francisco, J. C. R., & Boelens, R. (2015). Payment for environmental services: Mobilising an epistemic community to construct dominant policy. *Environmental Politics*, *24*(3), 481–500. Available from https://doi.org/10.1080/09644016.2015.1014658.

de Francisco, J. C. R., & Boelens, R. (2016). PES hydrosocial territories: De-territorialization and re-patterning of water control Arenas in the Andean Highlands. *Water International*, *41*(1), 140–156. Available from https://doi.org/10.1080/02508060.2016.1129686.

De La Fuente, T., & Hajjar, R. (2013). Do current forest carbon standards include adequate requirements to ensure indigenous peoples' rights in REDD projects? *The International Forestry Review*, *15*(4), 427–441.

Denton, F. (2002). Climate change vulnerability, impacts, and adaptation: Why does gender matter? *Gender & Development*, *10*(2), 10–20. Available from https://doi.org/10.1080/13552070215903.

Di Chiro, G. (2011). Acting globally: Cultivating a thousand community solutions for climate justice. *Development (Cambridge, England)*, *54*, 232–236.

Díaz-Caravantes, R., & Wilder, M. (2014). Water, cities and peri-urban communities: Geographies of power in the context of drought in Northwest Mexico. *Water Alternatives*, *7*, 499–417.

Elsaid, K., Kamil, M., Sayed, E. T., Abdelkareem, M. A., Wilberforce, T., & Olabi, A. (2020). Environmental impact of desalination technologies: A review. *Science of The Total Environment*, *748*, 141528. Available from https://doi.org/10.1016/j.scitotenv.2020.141528.

Engle, N. L. (2011). Adaptive capacity and its assessment. *Global Environmental Change*, *21*(2), 647–656. Available from https://doi.org/10.1016/j.gloenvcha.2011.01.019.

Fragkou, M. C., & Budds, J. (2020). Desalination and the disarticulation of water resources: Stabilising the neoliberal model in Chile. *Transactions of the Institute of British Geographers*, *45*(2), 448–463. Available from https://doi.org/10.1111/tran.12351.

Fraser, N. (2000). Rethinking recognition: Overcoming displacement and reification in cultural politics. *New Left Review*, *3*, 107–120.

Friedman, M. (1963). *Capitalism and freedom*. Chicago, IL: University of Chicago Press.

Garrick, D. E., Schlager, E., & Villamayor-Tomas, S. (2016). Governing an international transboundary river: Opportunism, safeguards, and drought adaptation in the Rio Grande. *Publius: The Journal of Federalism*, *46*(2), 170–198. Available from https://doi.org/10.1093/publius/pjw002.

Gerlak, A. K., Saguier, M., Mills-Novoa, M., Fearnside, P. M., & Albrecht, T. R. (2020). Dams, Chinese investments, and EIAs: A race to the bottom in South America? *Ambio*, *49*(1), 156–164. Available from https://doi.org/10.1007/s13280-018-01145-y.

Hallegatte, S., & Rozenberg, J. (2017). Climate change through a poverty lens. *Nature Climate Change*, *7*, 250–256. Available from https://doi.org/10.1038/nclimate3253.

Harris, L., McKenzie, S., Rodina, L., Shah, S. H., & Wilson, N. (2018). Water justice: Key concepts, debates and research agendas. In J. Ryan Holifield, Chakraborty, & G. Walker (Eds.), *The routledge handbook of environmental justice*. London and New York: Routledge.

Harris, L. M., Cecilia, M., & Roa-García. (2013). Recent waves of water governance: Constitutional reform and resistance to neoliberalization in Latin America (1990–2012). *Geoforum; Journal of Physical, Human, and Regional Geosciences*, *50*, 20–30. Available from https://doi.org/10.1016/j.geoforum.2013.07.009.

Hasselman, L., & Stoker, G. (2017). Market-based governance and water management: The limits to economic rationalism in public policy. *Policy Studies*, *38*(5), 502–517. Available from https://doi.org/10.1080/01442872.2017.1360437.

Heltberg, R., Oviedo, A. M., & Talukar, F. (2013). What are the sources of risk and how do people cope? Insights from household surveys in 16 countries. World Development Report Background Paper. The World Bank.

Hausermann, H. (2018). Ghana must Progress, but we are Really Suffering": Bui Dam, Antipolitics Development, and the Livelihood Implications for Rural People. *null*, *31*, 633–648. Available from https://doi.org/10.1080/08941920.2017.1422062.

Hidalgo Bastidas, J. P., Boelens, R., & Isch, E. (2018). Hydroterritorial configuration and confrontation: The Daule-Peripa multipurpose hydraulic scheme in Coastal Ecuador. *Latin American Research Review*, *53*, 517. Available from https://doi.org/10.25222/larr.362.

Höhl, J., Rodríguez, S., Siemon, J., & Videla, A. (2021). Governance of water in Southern Chile: An analysis of the process of indigenous consultation as a part of environmental impact assessment. *Society & Natural Resources*, *34*(6), 745–764. Available from https://doi.org/10.1080/08941920.2021.1892893.

Hommes, L., & Boelens, R. (2018). From natural flow to 'working River': Hydropower development, modernity and socio-territorial transformations in Lima's RImac Watershed. *Journal of Historical Geography*, *62*, 85–95. Available from https://doi.org/10.1016/j.jhg.2018.04.001.

Hommes, L., Boelens, R., Harris, L. M., & Veldwisch, G. J. (2019). Rural–Urban water struggles: Urbanizing hydrosocial territories and evolving connections, discourses and identities. *Water International*, *44*(2), 81–94. Available from https://doi.org/10.1080/02508060.2019.1583311.

Hoogendam, P. (2016). Hydrosocial territories in the context of diverse and changing ruralities: The Case of Cochabamba's drinking water provision over time. *Water International*, *41*, 91–106.

Hoogesteger, J. (2018). The Ostrich politics of groundwater development and neoliberal regulation in Mexico. *Water Alternatives*, *11*.

Hoogesteger, J., & Wester, P. (2015). Intensive groundwater use and (in)equity: Processes and governance challenges. *Environmental Science & Policy*, *51*, 117–124. Available from https://doi.org/10.1016/j.envsci.2015.04.004.

IPCC. (2021). Summary for policymakers. In Climate change 2021: The physical science basis. Contribution of working group I to the sixth assessment report of the intergovernmental panel on climate change, edited by V. Masson-Delmotte, P. Zhai, A. Pirani, S. L. Connors, C. Péan, S. Berger, N. Caud, et al., Cambridge. Cambridge, UK: Cambridge University Press. <https://www.ipcc.ch/report/ar6/wg1/downloads/report/IPCC_AR6_WGI_SPM.pdf>.

Jaskula, B. (2012). *2010 minerals year books: Lithium*. Reston, VA: United States Geological Survey.

Knappett, P. S. K., Li, Y., Loza, I., Hernandez, H., Avilés, M., Haaf, D., Majumder, S., et al. (2020). Rising arsenic concentrations from dewatering a geothermally influenced aquifer in Central Mexico. *Water Research*, *185*, 116257. Available from https://doi.org/10.1016/j.watres.2020.116257.

Kolinjivadi, V., Van Hecken, G., Vela Almeida, D., Dupras, J., & Kosoy, N. (2017). Neoliberal performativities and the 'making' of payments for ecosystem services (PES). *Progress in Human Geography*, *43*(1), 3–25.

Liu, W., & Agusdinata, D. B. (2020). Interdependencies of lithium mining and communities sustainability in Salar de Atacama, Chile. *Journal of Cleaner Production*, *260*, 120838. Available from https://doi.org/10.1016/j.jclepro.2020.120838.

Liu, W., Agusdinata, D. B., & Myint, S. W. (2019). Spatiotemporal patterns of lithium mining and environmental degradation in the Atacama Salt Flat, Chile. *International Journal of Applied Earth Observation and Geoinformation*, *80*, 145–156. Available from https://doi.org/10.1016/j.jag.2019.04.016.

Loftus, A., & March, H. (2016). Financializing Desalination: Rethinking the Returns of Big Infrastructure. *International Journal of Urban and Regional Research*, *40*, 46–61. Available from https://doi.org/10.1111/1468-2427.12342.

Lynch, B. D. (2012). Vulnerabilities, competition and rights in a context of climate change toward equitable water governance in Peru's Rio Santa Valley. *Global Environmental Change*, *22*, 364–373.

Marazuela, M. A., Vázquez-Suñé, E., Ayora, C., García-Gil, A., & Palma, T. (2019). The effect of brine pumping on the natural hydrodynamics of the Salar de Atacama: The damping capacity of salt flats. *Science of The Total Environment*, *654*, 1118–1131. Available from https://doi.org/10.1016/j.scitotenv.2018.11.196.

McDermott, M., Mahanty, S., & Schreckenberg, K. (2013). Examining equity: A multidimensional framework for assessing equity in payments for ecosystem services. *Environmental Science & Policy*, *33*, 416–427. Available from https://doi.org/10.1016/j.envsci.2012.10.006.

McDonald-Wilmsen, B., & Webber, M. (2020). Dams and displacement: Raising the standards and broadening the research agenda. *Water Alternatives*, *3*(2), 142–161.

McEvoy, J. (2014). Desalination and water security: The promise and perils of a technological fix to the water crisis in Baja California Sur, Mexico. *Water Alternatives*, *7*, 518–541.

Medema, W., McIntosh, B. S., & Jeffrey, P. J. (2008). From premise to practice: A critical assessment of integrated water resources management and adaptive management approaches in the water sector. *Ecology & Society*, *12*(2), 29.

Mills-Novoa, M. (2020). Making agro-export entrepreneurs out of Campesinos: The role of water policy reform, agricultural development initiatives, and the specter of climate change in reshaping agricultural systems in Piura, Peru. *Agriculture and Human Values*, *37*, 667–682. Available from https://doi.org/10.1007/s10460-019-10008-5.

Mills-Novoa, M., Boelens, R., Hoogesteger, J., & Vos, J. (2020). Governmentalities, hydrosocial territories & recognition politics: The making of objects and subjects for climate change adaptation in Ecuador. *Geoforum; Journal of Physical, Human, and Regional Geosciences*, *115*, 90–101.

Mills-Novoa, M., & Hermoza, R. T. (2017). Coexistence and conflict: IWRM and large-scale water infrastructure development in Piura, Peru. *Water Alternatives*, *10*(2), 370–394.

Moser, C. (2008). *Reducing global poverty: The case of asset accumulation*. Reducing global poverty (15–50, pp. 15–50). Brookings Institutions Press.

Nassrullah, H., Anis, S., Hashaikeh, R., & Hilal, N. (2020). Energy for desalination: A state-of-the-art review. *Desalination*, *491*, 114569. Available from https://doi.org/10.1016/j.desal.2020.114569.

Neumayer, E., & Plümper, T. (2007). The gendered nature of natural disasters: The impact of catastrophic events on the gender gap in life expectancy, 1981–2002. *Annals of the Association of American Geographers*, *97*(3), 551–566. Available from https://doi.org/10.1111/j.1467-8306.2007.00563.x.

Nightingale, A. J. (2011). Bounding difference: Intersectionality and the material production of gender, caste, class and environment in Nepal. *Themed issue: New feminist political ecologies*, *42*(2), 153–162. Available from https://doi.org/10.1016/j.geoforum.2010.03.004.

Nozick, R. (1974). *Anarchy, state, and Utopia*. New York: Basic Books.

Obour, P. B., Owusu, K., Agyeman, E. A., Ahenkan, A., & Madrid, À. N. (2016). The impacts of dams on local livelihoods: a study of the Bui Hydroelectric Project in Ghana. *null*, *32*, 286–300. Available from https://doi.org/10.1080/07900627.2015.1022892.

O'Brien, K., Eriksen, S., Nygaard, L. P., & Schjolden, A. (2007). Why different interpretations of vulnerability matter in climate change discourses. *Climate Policy*, *7*(1), 73–88. Available from https://doi.org/10.1080/14693062.2007.9685639.

Osborne, T., L. Bellante, & N. von Hedemann (2014). Indigenous peoples and REDD+: A critical perspective. In *Indigenous peoples' biocultural climate change assessment initative*. Tucson, Arizona: Public Political Ecology Lab. <https://repository.arizona.edu/bitstream/handle/10150/605561/Osborne_IPCCA_REDD_English_2014.pdf?sequence=3&isAllowed=y>.

Pahl-Wostl, C. (2007). Transitions toward adaptive management of water facing climate and global change. *Water Resource Management*, *21*, 49–62.

Perreault, T. (2014). What kind of governance for what kind of equity? Towards a theorization of justice in water governance. *Water International*, *39*(2), 233–245. Available from https://doi.org/10.1080/02508060.2014.886843.

Perreault, T. (2020). Climate change and climate politics: Parsing the causes and effects of the drying of Lake Poopó, Bolivia. *Journal of Latin American Geography*, *19*(3), 26–46. Available from https://doi.org/10.1353/lag.2020.0070.

Prieto, M. (2016). Practicing costumbres and the decommodification of nature: The Chilean water markets and the Atacameño People. *Geoforum; Journal of Physical, Human, and Regional Geosciences*, *77*, 28–39. Available from https://doi.org/10.1016/j.geoforum.2016.10.004.

Punjabi, B., & Johnson, C. A. (2019). The politics of rural–urban water conflict in India: Untapping the power of institutional reform. *World Development*, *120*, 182–192. Available from https://doi.org/10.1016/j.worlddev.2018.03.021.

Rawls, J. (1971). *A theory of justice*. Cambridge, MA: Harvard University Press.

Rawls, J. (2005). *Political liberalism*. New York: Columbia University Press.

Roy, J., Tschakert, P., Waisman, H., Abdul Halim, S., Antwi-Agyei, P., Dasgupta, P., & Hayward, B. (2018). Sustainable development, poverty eradication and reducing inequalities. In Global warming of 1.5°C. An IPCC Special Report on the Impacts of Global Warming of 1.5°C above Pre-Industrial Levels and Related Global Greenhouse Gas Emission Pathways, in the Context of Strengthening the Global Response to the Threat of Climate Change, Sustainable Development, and Efforts to Eradicate Poverty, edited by V. Masson-Delmotte, P. Zhai, H.-O. Portner, D. Roberts, J. Skea, P.R. Shukla, A. Pirani, et al. Cambridge, UK: Cambridge University Press.

Schlosberg, D. (2004). Reconceiving environmental justice: Global movements and political theories. *Environmental Politics*, *13*(3), 517–540. Available from https://doi.org/10.1080/0964401042000229025.

Schlosberg, D., & Lisette, B. C. (2014). From environmental to climate justice: Climate change and the discourse of environmental justice. *Wiley Interdisciplinary Reviews: Climate Change*, *5*(3), 359–374. Available from https://doi.org/10.1002/wcc.275.

Sen, A. (2009). *The idea of justice*. Cambridge, MA: Harvard University Press.

Shi, L., Ahmad, S., Shukla, P., & Yupho, S. (2021). Shared injustice, splintered solidarity: Water governance across urban-rural divides. *Global Environmental Change*, *70*, 102354. Available from https://doi.org/10.1016/j.gloenvcha.2021.102354.

Sims, R., Schaeffer, R., Creutzig, F., Cruz-Núñez, X., D'Agosto, M., Dimitriu, D., & Meza, M. J. F. (2014). Transport. In Climate *change* 2014: Mitigation of *climate change*. Contribution of *working* group III to *the fifth assessment report of the intergovernmental panel on climate change*, edited by O. Edenhofer, R. Pichs-Madruga, Y. Sokona, E. Farahani, S. Kadner, K. Seyboth, A. Adler, et al. Cambridge, UK and New York, NY: Cambridge University Press.

Sultana, F. (2011). Suffering for water, suffering from water: Emotional geographies of resource access, control and conflict. *Themed issue: New feminist political ecologies*, *42*(2), 163–172. Available from https://doi.org/10.1016/j.geoforum.2010.12.002.

Sultana, F. (2018). Water justice: Why it matters and how to achieve it. *Water International*, *43*(4), 1–11. Available from https://doi.org/10.1080/02508060.2018.1458272.

Sultana, F., & Loftus, A. (2015). A human right to water: Critiques and condition of possibility. *Wiley Interdisciplinary Reviews: Water*, *2*(2), 97–105.

Terry, G. (2009). No climate justice without gender justice: An overview of the issues. *Gender & Development*, *17*(1), 5–18. Available from https://doi.org/10.1080/13552070802696839.

Thomas, K., Hardy, R. D., Lazrus, H., Mendez, M., Orlove, B., Rivera-Collazo, I., Roberts, J. T., Rockman, M., Warner, B. P., & Winthrop, R. (2019). Explaining differential vulnerability to climate change: A social science review. *Wiley Interdisciplinary Reviews: Climate Change*, *10*(2), e565. Available from https://doi.org/10.1002/wcc.565.

Tilt, B., & Gerkey, D. (2016). Dams and population displacement on China's Upper Mekong River: Implications for social capital and social–ecological resilience. *Global Environmental Change*, *36*, 153–162. Available from https://doi.org/10.1016/j.gloenvcha.2015.11.008.

Torio, P. C., Harris, L. M., & Angeles, L. C. (2019). The rural–urban equity nexus of Metro Manila's water system. *Water International*, *44*(2), 115–128. Available from https://doi.org/10.1080/02508060.2019.1560559.

UNICEF & WHO. (2019). *Progress on drinking water, sanitation and hygiene 2000–2017: Special focus on inequalities*. New York: United Nations Children's Fund and World Health Organization.

United Nations Development (UNDP) Program. (2006). Human development report 2006 - Beyond scarcity: Power, poverty and the global water crisis. New York: United Nations Development Program. <http://hdr.undp.org/sites/default/files/reports/267/hdr06-complete.pdf>.

United Nations (UN). (2010). *Human Right to Water and Sanitation (No. A/RES/64/292)*. New York, NY: United Nations.

von Hedemann, N. (2020). Transitions in payments for ecosystem services in Guatemala: Embedding forestry incentives into rural development value systems. *Development and Change*, *51*(1), 117–143. Available from https://doi.org/10.1111/dech.12547.

von Hedemann, N., & Osborne, T. (2016). State forestry incentives and community stewardship: A political ecology of payments and compensation for ecosystem services in Guatemala's highlands. *Journal of Latin American Geography, 15*(1), 83–110.

Wallace, T., & Coles, A. (2005). *Water, gender and development: An introduction.* (1st ed.). *Gender, water and development,* (20). New York and London: Routledge.

Williams, J. (2018). Assembling the water factory: Seawater desalination and the techno-politics of water privatisation in the San Diego–Tijuana metropolitan region. *Geoforum, 93,* 32–39. Available from https://doi.org/10.1016/j.geoforum.2018.04.022.

Wolfe, M. (2017). *Watering the revolution: An environmental and technological history of agrarian reform in Mexico.* Durham and London: Duke University Press.

World Bank. (2010). *Development and climate change: World development report.* Washington D.C: World Bank.

Wutich, A. (2009). Intrahousehold disparities in women and men's experiences of water insecurity and emotional distress in Urban Bolivia. *Medical Anthropology Quarterly, 23*(4), 436–454.

Zarfl, C., Alexander, E. L., Berlekamp, J., Tydecks, L., & Tockner, K. (2015). A global boom in hydrpower dam construction. *Aquatic Sciences, 77,* 161–170. Available from https://doi.org/10.1007/s00027-014-0377-0.

Zwarteveen, M. Z., & Boelens, R. (2014). Defining, researching and struggling for water justice: Some conceptual building blocks for research and action. *Water International, 39*(2), 143–158. Available from https://doi.org/10.1080/02508060.2014.891168.

CHAPTER 22

Environmental ethics and sustainable freshwater resource management

Oghenekaro Nelson Odume and Chris de Wet

Unilever Centre for Environmental Water Quality, Institute for Water Research, Rhodes University, Makhanda, South Africa

Chapter Outline

22.1 Introduction 419
22.2 Key issues and the need for environmental ethical considerations in water resource management 421
22.3 Approaches to environmental ethics from a western perspective 423
    22.3.1 Value-oriented environmental ethics 423
    22.3.2 Relationship-based environmental ethics 427
    22.3.3 African environmental ethics 430
22.4 Relating environmental ethics to water resource management in the context of complex socio-ecological systems 432
22.5 Conclusions 433
Acknowledgments 434
References 434

22.1 Introduction

Despite developments in water resource policy, law, monitoring, regulation, management and research, the quality, quantity, and biodiversity of water resources continue to be impacted globally (CSIR, 2010; Reid et al., 2019; Rockstrom et al., 2009). At the same time, there is a growing recognition that humans are integral components of complex socio-ecological systems; as such, their beliefs, values, and actions have direct implications, whether intended or unintended, for the environment (Folke, 2006; Partelow, 2018).

In South Africa, for example, considerable advances have been made in the inclusion of human values in the strategic adaptive management of natural resources (Roux & Foxcroft, 2011), where values are a central aspect of the established VSTEEP (values, social, technical, environmental, economic, and political) analysis (Rogers & Luton, 2011). In this chapter, we have sought to link ethical thinking and practice to current and emerging practices in integrated water resource management (IWRM).

Ethics and values pervade all aspects of water management, including water use allocation, social ordering, pollution control and the notion of healthy ecosystems (Brown & Schmidt, 2010a; De Wet & Odume, 2019; Odume and de Wet, 2019). Therefore explicit consideration of different value systems and the distillation of sets of environmental ethical principles and criteria by which such values can be negotiated in relation to each other, can contribute to sustainable freshwater resource management (Doorn, 2013; Grunwald, 2016). Ethical criteria can help to clarify the implications of different claims and claimants and courses of actions, and by addressing the concerns of claimants within what are effectively pluralistic and polycentric institutional frameworks of water governance and management (Brown & Schmidt, 2010a; Pradhan & Meinzen-Dick, 2010). These claims may range from calls for fair and equitable water allocation, to consensus around the ordering of social relationships and place-based values, such as cultural and spiritual beliefs around water (Schulz et al., 2017). Ethics, in seeking to establish broader principles to consider specific cases, plays an important balancing role between such claims and between claimants (Brown & Schmidt, 2010a).

Ethics is considered in this project as the systematic searching for and invocation of principles by which we reflectively seek to distinguish between right and wrong, good and bad, within ourselves and in our behavior and attitude towards other people, and towards nature—it is an inquiry into the nature and grounds of morality, value judgments, rules, norms and their implications (De Wet, 2009; De Wet & Odume, 2019; Okeja, 2018). Thus environmental ethics deals with the constitution of principles in relation to humans and nature and the value of it to nature (Minteer et al., 2004). Ethics is fundamentally different from morals/values because unlike ethics, morals/values are what specific individuals or groups, mostly at an unreflective level, believe and consider to be good or bad, right or wrong—without deep reflection of the implications of such claims.

In a highly plural society, the needs, desires, and values of people will differ widely. Each choice and action that is made tends to implicate other choices and actions, so that actions tend to open out and close down specific options in respect of access to the benefits of water use and/or protection and will often be contentious. Trade-offs between values (e.g., between equality and liberty) or a compromise in standards for reasons of affordability (e.g., resulting in water that is not always safe to drink) are not always meaningful or suitable, and there often will be winners and losers.

Thus the aim of this chapter is to review, analyze, and recommend ways by which environmental ethics can constructively contribute to water resource management in the context of socio-ecological systems. We draw our insights mainly from South Africa, as this is where we are most experienced. We begin by introducing environmental ethics and argue for its importance in water resource management. Following the analytical framework of Kronlid and

Öhman (2013), we present a brief review of recent (western) approaches to environmental ethics. A brief account of African environmental ethics is then provided. We then reflect on the significance of environmental ethics and water resource management in the context of socio-ecological systems. In the end, we argue for a relationship-based ethical approach as appropriate for sustainable water resource development, utilization and management.

22.2 Key issues and the need for environmental ethical considerations in water resource management

In as much as human beings ascribe value to nature, and do so in different kinds of ways, the discipline and practice of environmental ethics could provide a useful guide towards our thinking about, and attitude and behavior towards, the environment. A good ethical practice would reflect the way in which we value and relate to the environment and help to make explicit the values that guide us—and that should guide us.

Within the context of water resource management, environmental ethics are already implicit in water policies, regulations, and laws. For example, in South Africa, the National Department of Water and Sanitation slogan "Some for all, forever" speaks to the fact that it is seen as ethical to see water as a public good, not to be owned by a few wealthy individuals in the society, and to manage it sustainably. For example, the revolutionary National Water Act (Act No. 36 of 1998) of the Republic of South Africa, which shifted from the conventional "command-and-control" approach to an ecological basis for managing water resources, has an explicitly environmental ethical consideration, clearly captured in the notion of the Reserve, whereby rights accrue to the natural environment, as well as to people, for their basic needs for drinking, cooking, and cleanliness. By providing water for both ecological and basic human needs, the Reserve serves social and environmental justice. All other water is administratively allocated for other uses. It is, however, the decision-making around the amount and quality of water secured for resource protection, and the amount and quality of water allocated to various users with various levels of security, which makes real the notions of social and environmental justice.

The ecological Reserve concept, though difficult to implement, has led to some successful cases of improving water resource management in South Africa, for example in the Sabie River (Pollard & du Toit, 2011a, 2011b) Consequently, the development, adoption and embedding of environmental ethics, at different spatial and temporal scales, among a plurality of water resource users, and in specific contexts, has the potential to encourage best water and ecosystem management practices.

This brings us to the consideration of whether the environment has an intrinsic value of its own, and whether this should be the basis of its respect and protection; or whether environmental protection should simply ensure sustainable access by humans to ecosystem services (i.e., the environment as an instrumental value). Discourses on environmental ethics in the context of water and ecosystem management could be in several domains including, but not limited to, the water user sector, the institutional-governance domain, and/or environmental ethics in relation to environmental law. These various domains should not be seen in isolation from each other, but as aspects of a complex interconnected system. They can be viewed through the lenses of the different approaches to environmental ethics, such as whether the environment is seen as having intrinsic or instrumental value. An anthropocentric instrumentalism places human beings in the center of the moral universe, viewing humans as intrinsically valuable, as the only moral agents, and seeing nature as having moral value in only an instrumental sense, that is, in as much as it serves human beings, their needs and purview (Kronlid & Öhman, 2013:24). Placing an intrinsic value on the whole socio-ecological system decenters human beings and sees ecosystems as such—of which humans are part—as having an intrinsic moral value (Kronlid & Öhman, 2013:25). Using these perspectives and related questions from environmental ethics, we can then interrogate what is meant by the role and application of values in relation to water resource protection and sustainability.

Water and aquatic ecosystems, goods, and services have several potentially competing (i.e., human) stakeholders/users, for example, domestic, industrial, and agricultural users. Perspectives from environmental ethics have the potential to make players in each of these user sectors more aware of the issue of values in relation to ecosystem protection and sustainable utilization. Therefore, within the water user domain, environmental ethics could provide the framework and insights for interrogating pertinent questions such as:

- What are the moral obligations of the various water user sectors to protect and use water resources sustainably?
- Why should people invest in protecting aquatic ecosystems, even when this is not in their immediate financial interest?
- In the context of an industrial discharge into a water resource, for example, why should an industry stop discharging into a water resource just because it is obvious that the resource has exceeded its receiving capacity—even when the legal discharge standards are not being violated?

Environmental ethics offers valuable additional perspectives for analyzing the relationships between the biophysical aquatic ecosystems and their many users, with potential for improving ecosystems management through the recognition that these systems are part of large complex socio-ecological systems, in which values and interests play a significant role. Socio-ecological system is that in which humans and the rest of nature are seen as interdependent integral, complementary components of a unitary system, characterized by multiple feedback loops, unpredictability, and cross-scale dynamics.

Within the institutional-governance domain, perspectives, and practices from within environmental ethics could potentially contribute to a framework for reforming water policies and governance, management practices, and existing institutions, through a systematic investigation of the value systems, beliefs, and moral affiliations of people living within a catchment, both in time and space. In this regard, a concern with environmental ethics may infuse greater awareness and transparency into water policy formulations—better enabling adaptive responses to changing conditions. A rigorous and systematic analysis of values underlying present water policies and institutions could help expose existing deficiencies and provide for opportunities to guide a system that further embraces the vision of the biophysical world as having rights, rather than limiting moral and ethical concerns to human beings only (Groendfeldt & Schmidt, 2013; Hoffman, 1991).

This takes us into the heart of IWRM and the issues facing contemporary water managers—and the ways that values enter centrally into these matters. It needs to be noted that are potential tensions between the component (intrinsic or instrumental) and relationship-based approaches to value, with the former approach seeking to ascribe value to specific components of the socio-ecological system, and the latter approach seeking to decenter components and to locate value in relationships between the components of the socio-ecological system.

22.3 Approaches to environmental ethics from a western perspective

22.3.1 Value-oriented environmental ethics

Based on the Kronlid and Öhman's (2013) framework, environmental ethics, as generated by various western thinkers, can be divided into value-oriented and relation-oriented (relationship-based) environmental ethics. The primary focus of western environmental ethics is the notion of intrinsic value debate, that is, what/who in terms of humanity and (Jakobsen, 2017) nature has intrinsic value and who/what should therefore be considered as part of the "moral circle" (e.g., Goodpaster, 2003; O'Neill, 2003). Two distinct lines of thoughts are prominent in this regard—anthropocentrism and nonanthropocentrism.

22.3.1.1 Anthropocentric environmental ethics

Anthropocentric environmental ethics views only humans as the holders of intrinsic value and thus worthy of moral consideration (Heath, 2021; Light & Rolston, 2003; Thompson, 2016). According to anthropocentrism nonhuman life forms are only of instrumental value, in terms of the value which humans derive from such nonhuman entities. Analytically, the concept of anthropocentrism can be probed further within human relationships. For example, between humans, an important question, which may shape water resource policies and programs and the notion of environmental justice, is: who among humans/humanity has intrinsic value and thus moral standing? Values and moral standing could be ascribed to (1) only living humans (intra-generational); (2) living and future generations of humans (intergenerational); (3) dead, living and future generations of humans (trans-generational); (4) humans that are closely connected in some ways for example, socially, mentally, culturally, geographically etc. (local anthropocentrism), or (5) all humans, irrespective of their relative closeness (global anthropocentrism) (Kronlid & Öhman, 2013).

From an anthropocentric perspective, an explicit consideration of the question raised above (who should be part of the moral circle?) is therefore important in water resource management. For example, the value of equity, which refers to fairness to present generations, is underpinned by an intra-generational value judgment, whereas the value which is often emphasized in global water discourse on protecting water resources for future generations is underpinned by intergeneration anthropocentrism (United Nation, 2010). For an effective implementation of intergenerational environmental equity, government and relevant institutions must be seen as stewards of the present generation and of resources for future generations (McIntyre-Mills, 2013). This requires solidarity with present environmental concerns and the will to act to prevent unnecessary exploitation of natural resources. Intergenerational consideration in relation to water resources is addressed, for example, in the South African National Water Act through the value of sustainability, that is, balancing the use and protection of ecosystem for future generation.

The role of trans-generational solidarity in promoting intergenerational environmental equity is discussed later in the context of African environmental ethics (Behrens, 2012; Verharen et al., 2021). This involves taking intergenerational responsibility seriously in all aspects of water resources, underpinned by collective solidarity with the "concerns" of, and respect for, the well-being of, future generations.

An important aspect within anthropocentrism is the nature of the relationship between humans/humanity and the environment. Although anthropocentrism has traditionally been considered as human-centered and thus potentially un-eco-friendly, several authors (e.g., Hargrove, 2003; Norton, 2003; Sinha, 2021) have argued that, depending on the conceptualization of the relationship between humans and the environment, anthropocentrism can provide a strong basis upon which people can act to protect the environment. For example, the human−water relationship can be conceptualized based on the kind of value people obtain from water resources, as well as based on geography, history, and constructed social discourses. According to Kronlid and Öhman (2013), instrumental valuation of the environment can be categorized into demand values, transformative values, constitutive values and need values. In most cases, both demand values that satisfy human preference as well as need values, which are necessary for human survival and well-being, are economically and socially driven. Transformative values relate to our experience of water that can transform our relationship with it, whereas constitutive value shape our experience of water as an integral part of what it means to be human.

From the above, in relation to how people relate to and derive value from the natural environment, an important question becomes: can resource managers and policy makers identify values concerning water developed by local people within a specific context, which can enable them to protect the environment? For example, in some rural African communities, certain rivers/parts of certain rivers are considered sacred because they are regarded as places where the ancestors manifest themselves, that is, such communities see these sacred places as what is mean to be human (constitute value) (Ibanga, 2016; Kelbessa, 2014). In most such communities, people act to protect such rivers and to keep them clean—hence ecologically healthy. Such indigenous practices that lead to the protection of water resources are worth considering and exploring in formulating programs aimed at protecting water resources within the specific context

Place-centered aspects of the human—environment relationship are important to anthropocentrism. It has been argued that people with strong emotional, physical, spiritual, and social connections with their environment, often have a stronger sense of obligation towards the protection of the environment (e.g., Murove, 2009). Strong connection with the environment could arise out of culture, history, and identity. Societies with strong connections to their places in terms of dwelling and ecological understanding, usually develop ecological consciousness that tends to lead to resource protection and conservation (e.g., Gaard, 2010; Stamova et al., 2021). An important aspect of such place-based human—environment relationships is consideration of sacred sites, features and attributes (Raine, 2001). In many African cultures, the attribution of the value of sacredness to the environment, which is usually underpinned by ancestral and historical practices, has resulted in the development of a strong sense of environmental stewardship and guardianship (Benny et al., 2021; Bernard, 2010). On the other hand, a sense of "placelessness," conceived here as a lack of rootedness and an inability to connect with the society and the natural environment in which one dwells, can lead to a sense of lesser obligation towards the environment, and thus to its exploitation.

Biological, evolutionary, and ecological connections between people and the environment provide another dimension to viewing the human—environmental relationship (Norton, 1987). All species, humans, and nonhumans, depend on the external environment for survival. An understanding of these interconnections provides the potential for an attitude of care and stewardship towards the natural environment, with the realization that the survival of our own species is inextricably linked with the functionality and health of the environment with which we are biologically, evolutionarily, and ecologically connected.

All these different aspects of the way we understand the human—environmental relationship, and the ethics appropriate to this relationship, relate to and are reflected in our conception of the natural environment, and our narratives of our experience of nature and of what it means to be human within the broader context of socio-ecological systems (Kronlid & Öhman, 2013).

22.3.1.2 Nonanthropocentric environmental ethics

Critics of anthropocentrism have argued that restricting intrinsic value to only humans could lead to uncontrollable and unnecessary exploitation of natural resources (Kallhoff, 2016). Environmental crises confronting humanity and the planet, such as global warming, freshwater pollution, land degradation, species extinction etc., have been attributed to an anthropocentric moral outlook concerning the environment (Callicott, 2016; Stone, 2003). Nonanthropocentrism holds that humans and nonhuman beings have intrinsic value (Weckler, 2020). Individualism and holism are the two primary lines of thoughts regarding the matter of intrinsic value within nonanthropocentric thinking.

An individualistic moral outlook presupposes those individual nonhuman beings fulfilling certain criteria, for example sentience, communicative capacity, society and socialization, or having a sense of self etc., should be considered as having intrinsic value (Vandeveer & Pierce, 2003) and thus, as having a good/moral worth of their own. On the other hand, within a holistic approach, biocentrism regards all living organisms as holders of intrinsic value, because their flourishing and existence is a good that must be maintained (Taylor, 2003). An eco-centric outlook, as a line of thinking in holism, views the organic whole, that is, ecosystems—living and nonliving things—as holders of intrinsic value (Batavia et al., 2020; Heath, 2021; Taylor, 2003). Accepting the notion of intrinsic value as per ecocentrism implies that nonliving systems ought to be protected to the extent that they are life-sustaining. In this regard, Taylor (2003) invokes the attitude of respect for nature as a fundamental necessity needed to achieve the realization of ecosystems protection. Respect in this case implies explicit recognition of inherent worth of both humans and nonhumans within the ecosystem.

The attitude of intrinsic-type respect for nature may call into question the approach to ecological categories (i.e., Categories A-F) in South African water resource management, in which ecosystems considered heavily degraded in categories E and F may not be restored to pristine conditions (DWA—Department of Water Affairs, 2010; Kleynhans and Louw,2008). That is, the "good" of such ecosystems is no longer considered as the key value, but ecosystems are seen as instrumental to a water management system and its way of classifying ecosystems. The ethical question thus becomes: how are we, in the context of water resource management in South Africa for example, to respect the nonhuman-environment? How will water resource management ethics move beyond anthropocentric approaches, and, even so, what is the limit of respect for nature—especially from the perspective of resources and other practical constraints?

Another question that relates to the ethics of respect as constructed above, is that of potential trade-offs between, for example, human rights, and the rights of the environment, and between different evaluations of aquatic ecosystems. Trade-offs can exist between taxonomic diversity (e.g., rarity, vulnerability, richness, etc.), functional diversity, biotic welfare and environmental fidelity (e.g., water quality, sustained hydrology, riparian and geomorphological integrity), as well as ecosystem services and security (e.g., production, regulation, provisioning and cultural services) (Sarkar, 2013). In the face of limited resources, which of these environmental values deserve respect and attention, is an ethical question that must be deliberated upon, taking account of the specific socio-ecological context. Where the human right to water and environmental right to water are in conflict, how should such a trade-off be treated, and what ethical criteria should be used in the underlying value judgment of the final decision?

The attitude of respect would also require conferring legal or other rights onto the natural environment and respecting these rights. Stone (2003) suggests that rights on the environment involves two dimensions: (1) the legal-operational aspects and (2) the psychic and socio-psychic aspects. To promote the virtue of environmental inherent worth and dignity through operationalizing environmental rights, (1) the environment should be able to institute a legal action, (2) the injury or degradation to the environment must be considered in granting of the legal relief, and (3) the relief must run to the benefit of the environment, and not humans who are associated with the environment (Stone, 2003).

Since rights are also about relationships, and they become important only when respected, the question that arises is: what happens when the environmental right impedes on human rights? This is one of the reasons why, for example, the ecological Reserve has proved difficult to implement in South Africa (King & Pienaar, 2011). Although rights conferred on the environment through a legal system could help reduce environmental degradation, a reconceptualization of the human−environment relationship, involving a social re-ordering in which humans become more conscious of our inextricable link with nature, is needed to enable the attitude of respect for environmental rights. The implication here is that legal rights alone are not sufficient as the basis by which people would act voluntarily to protect the natural environment. Institutional change, that actively seeks to encourage the attitude of respect for nature by practically demonstrating various ways by which people are inextricably connected to the natural environment, and spelling out procedures that respect these connections, is also required.

22.3.2 Relationship-based environmental ethics

A key criticism of value-oriented environmental ethics is its ascription of values in a dichotomous manner (Gruen & Gaard, 2003). In a way it fails to recognize the complex relationships that exist between humans and the rest of nature (Berkes & Folke, 1998; Folke, 2006; Foy, 2016; Partelow, 2018). By emphasizing a dichotomy between humans and the rest of nature, value-oriented ethics fails to appreciate the complexity inherent in the context in which human−environment relationships are forged. (Heath, 2021; Minteer et al., 2004). The top-down application of a single principle or set of principles in the defense/or otherwise of the intrinsic value of nature may not help resolve environmental issues without taking account of the specific social context, or without asking whether all practical environmental decisions are in fact equally well-served by a single ethical principle (Minteer, 2016).

Therefore relationship-based environmental ethics holds the view that value relates to and derives from the notion of a system and the relationships between its component parts and processes, and that accordingly, the justification of behavior, and the prescription of moral principles must consider the specific systemic context (Gaard, 2010; Minteer et al., 2004). Pragmatist environmental ethics (Minteer, 2016), deep ecology, ecofeminism (e.g., Salleh, 2017), and African environmental ethics (e.g., Clarke, 2019; Kelbessa, 2014; Murove, 2009), are examples of relationship-based environmental ethics.

22.3.2.1 Pragmatic environmental ethics

Environmental pragmatists argue that environmental ethics should be more concerned with practical everyday situations in which environmental and societal needs come into conflict, and with how such conflicts could/should be resolved (Minteer & Manning, 2003). The specific context needs to be understood and reflected upon; in so doing, one evolves ethical principles which can be revisited and revised on an on-going basis, reflecting the specific environmental context (Minteer, 2016).

Applying preconceived sets of ethical principles to all environmental situations compromises the possibility of a relevant response to socio-ecological problems (Minteer et al., 2004). Socio-ecological systems are complex and are characterized by unpredictability, which factors require contextual and adaptive engagement (Odume and de Wet, 2019). Thus the justification of an ethical principle depends on whether it contributes meaningfully to resolving specific environmental issues. Emergence is a property of complex socio-ecological systems, which may give rise to new experience and insights (Folke, 2006, 2007; Partelow, 2018). Ethical approaches which focus on intrinsic value may not be flexible enough to respond to the dynamics and unpredictability associated with socio-ecological systems. Thus an adaptive ethical system, reflecting actual circumstances, new insights, and experiences, needs to be mainstreamed to manage water resources in the context of complex socio-ecological systems. The practical implication of this pragmatist viewpoint is that norms, rules, principles and standards would continuously undergo some form of revision and re-interpretation to reflect the existing socio-ecological context, while taking account of pluralism of environmental values and valuations in a democratic deliberative process.

This is, however, likely to raise problems in relation to the issue of power, as different actors with different ethical approaches are likely to be differently situated in terms of power in relation to each other, and the deliberation process around establishing the dynamics of the adaptive ethical system as it progresses is not necessarily going to be "democratic." There also needs to be some measure of continuity and understanding in institutions, and the dynamics of the water management system's adaptive ethical system cannot be too flexible in practice.

22.3.2.2 Ecofeminist environmental ethics (ecofeminism)

Ecofeminist environmental ethics holds the view that the exploitation of nature is a feminist concern because it relates to the twin domination of women and nature by a patriarchal society underpinned by value systems that promote such discrimination (Gaard, 2010; Salleh, 2017). Value dualism (where values associated with the "self" are considered superior to those associated with "others," usually women and nature), and the cultural evolutionary separation of men from women and nature (in which muscularity is seen as an indicator of superiority and as a means to dominate both women and nature), are issues that ecofeminists have advanced as to why environmental problems are feminist concerns (Gaard, 2010; Warren & Cheney, 2003). The implication of the ecofeminist analysis is that it surfaces a global world view in which (it is claimed) arrogance, domination, and conquest have replaced an ethics of harmony, reverence, and respect for all components of the socio-ecological system. The twin domination of nature and women could be seen as an outcome of value systems that seek to ignore the interconnectedness an interdependence between components of a socio-ecological system. Thus the ecofeminists' position is that environmental decision-making processes must take account of the interests of the most vulnerable in human society, and of the interests of nonhuman communities (Salleh, 2017).

Ecofeminism therefore advocates and seeks to promote an attitude, belief system and moral values that enhance respect for both the most vulnerable human constituencies and the environment. Ecofeminism emphasizes the importance of contextualizing environmental issues, taking account of history, the present, as well as the future, and avoiding a reductionist view to resolving environmental problems (Klemmer & McNamara, 2020). That is, it emphasizes the recognition of differences as well as commonalities and promotes a systemic approach to addressing the twin domination of women and nature.

The principle of inclusivity—that is, beyond only gender inclusivity—is an important aspect of ecofeminism. To avoid perpetuating inequalities and historical injustices in environmental issues and policies, ecofeminism emphasizes inclusivity, paying particular attention to the perspectives, insights and views of the marginalized (Stephens et al., 2010). It avoids the notion of a universal context-free practice and seeks to evolve practices that are best suited for resolving specific contextual environmental issues. For example, ecofeminists insist that the perspectives of indigenous people, women and other marginalized groups must be considered on an equal footing with other approaches, to avoid prejudice—unless (like any other approach) proven through rigorous debates, engagements and scientifically defensible methods and approaches to be environmentally damaging (Salleh, 2017; Stephens et al., 2010).

22.3.2.3 Deep ecology

Central to the deep ecological approach, is the view that environmental crises arise because of social-economic, cultural and other value systems that elevate humans above nonhumans. Deep ecology subscribes to the intrinsic value of nature and to the equal moral worth of both human and nonhuman entities. It advocates an ethic of respect for nature that translates into harmony rather than rivalries between human and nonhumans (Klemmer & McNamara, 2020). Devall and Sessions (2003) argue that self-realization and biocentric egalitarianism are the two ultimate norms of deep ecology. The first norm, self-realization, emphasizes the embedding of humans within the ecological system and thus discourages the narrow view of "self" and of the social-cultural values that separate humans from the rest of nature. Self-realization is therefore the mature experience of oneness in diversity—diversity that includes both other humans as well as nonhuman biological communities (Cisek & Jaglarz, 2021). The second norm, biocentric egalitarianism, emphasizes the intrinsic value of all living things and associated life-supporting systems (e.g., water, land, air etc.), and the principle of the equal moral worth of all living things (Devall & Sessions, 2003).

From a practical perspective, deep ecology seeks to promote environmental health through deep ecological consciousness based on self-reflection on the role and position of humans within the organic socio-ecological whole (Akamani, 2020). Deep ecology can contribute to water resource management via the development of policies that see humans as part of the socio-ecological system, rather than the notion of human superiority and dominance. However, quite how deep ecology would handle the problem of trade-offs and compromises being central to the domain of policy and its application, and how this would be accomplished when all components of the socio-ecological system are to be seen as having intrinsic (i.e., noncompromisable) value, is not clear.

22.3.3 African environmental ethics

African environmental ethics emphasizes relationships, that is, the notion that life is relationship-based, as opposed to most western environmental ethical movements, which are mostly concerned with issues of intrinsic value and moral extensionism—that is, whether nonhumans have moral standing (Bujo, 2009; Okeja, 2018). Central to the relational view of African environmental ethics, is the notion of mutual respect between all components of the socio-ecological system, because it is believed that destroying perceived lower components of the (hierarchical) system may alter the unity and intactness of the whole (Bujo, 2009). The notion of hierarchy seems to suggest the perception that some members are less important than others in socio-ecological systems. According to Bujo (2009), whenever the system is altered, reconciliation is sought between its components—for example, through religious means, to maintain the unity or intactness of the whole.

Consequently, the rationale for environmental conservation from an African environmental ethical perspective, relates to the maintenance of this unity/intactness of the organic whole.

Ubuntu and Ukama are two well-known African concepts that emphasize the importance of relationships, not just between humans, but between humans and the environment (Murove, 2009; Okeja, 2018; Ramose, 2009). Insights from these two ethical concepts can contribute towards achieving social and environmental sustainability (Okeja, 2018). While *Ubuntu/Botho* implies shared humanity or humanness and presupposes that meaningful existence in one's life depends on mutual and caring relationships with other members of humanity—poor and rich alike (Okeja, 2018), *Ukama* on the other hand, extends this concept to the natural environment (Murove, 2009; Ramose, 2009). Although on the surface, *Ubuntu* seems anthropocentric, its affirmation of care for other humans indirectly leads to care for the environment on which other humans depend (Ramose, 2009).

The practice of *Ukama* implies recognizing the interrelatedness and interdependency between humans, and between humans and the rest of nature (Murove, 2009). *Ubuntu and Ukama* provide a strong basis for relationship-based ethics from an African perspective through their promotion of ecological consciousness and moral obligation towards future generations (Behrens, 2012; Murove, 2009). The two ethical concepts recognize the intrinsic value of all components of the socio-ecological system through personal and cultural experience, as well as their instrumental value through interrelatedness, in which each component of the socio-ecological system is understood to serve the entire system, and each component is served by the entire system (De Wet & Odume, 2019).

The environment is viewed as a common resource, and a trans-generational value system becomes a key motivation for preserving its integrity (Behrens, 2012; Murove, 2009). The environment as a common resource belongs to the past (ancestors), present and future generations and, as such, it ought to be treated with utmost respect and care. This invokes moral obligation on the present generation to preserve it for future generations as a way of showing gratitude, solidarity and respect to the ancestors who have left behind an environment capable of supporting the needs of the present generation (Behrens, 2012).

Thus, in the African context, the object of moral consideration is viewed with the lenses of a trans-generational value system in which the dead, the living and future generations are all included in the moral circle (Wiredu, 1994). Many Africans, for whom their indigenous culture is still important, hold the view that the environment is an inherited commons from past generations, which must be taken care of for future generations, as a way of showing respect and gratitude to their predecessors (Terblanché-Greeff, 2019). Maintaining such value systems underpinned by trans-generational solidarity, may contribute to the sustainable utilization of freshwater resources because, even if we do not know the future generations, as a way of showing respect to the ancestors, we can preserve the environment and maintain it the way we found it. An important question becomes whether such a value system might restrict development. In a way, showing gratitude to the ancestors does necessarily implies social-economic development within safe, planetary boundaries (Rockstrom et al., 2009) that safeguards the rights of future generations to environment that is not harmful.

This review has sought to outline the basic aspects of the major approaches to environmental ethics, in terms of the main issues that they raise. These include:

- The role (central or otherwise) of human beings and its ethical implications in the human—natural environment relationship.
- The usefulness of the idea of intrinsic value in considering the ethical status of, and ethical behavior towards, components of the environment.
- That the socio-ecological environment may be seen as an integrated unit, in which the various components parts all have inherent value, and in which human beings do not have primary status—but in which all aspects are interrelated and support each other.
- That water and other components of the aquatic ecosystem may thus be seen as having intrinsic value, as well as instrumental value. This requires a considered ethical—and managerial—balancing act.

22.4 Relating environmental ethics to water resource management in the context of complex socio-ecological systems

There is a growing recognition that humans are integral components of complex socio-ecological systems, sharing the biophysical environment with nonhuman communities (Virapongse et al., 2016). The concept of socio-ecological system recognizes the tightly coupled and integrated nature of social and ecological systems, and therefore represents a departure from the notion of two separate independent systems (De Wet & Odume, 2019; Folke, 2006). Socio-ecological systems are complex, and are characterized by unpredictability, nonlinearity, cross-scale dynamic interactions and multiple feedback mechanisms (Folke, 2007). Managing water resources within the framework of complex socio-ecological systems would require a careful consideration of management decisions on both the social (human societal aspects) and the ecological (nonhuman aspects) levels, with an explicit recognition of their interdependence on multiple spatial and temporal scales.

Folke (2007) argues that interconnection and coevolution, nonlinearity, and cross-scale dynamic interactions are the three most prominent characteristics of socio-ecological systems. Complexity analysis arose in part because, in natural systems, it is limiting to use conventional, reductionist, single cause—effect relationship-based analysis as a basis for managing systems that are inherently complex, with many linkages and interactions (Kliskey et al., 2021). For example, deteriorating water quality of a water resource which is receiving effluent from a treatment plant could be attributed to failure of the plant from a reductionist perspective. In reality however, other factors, including demography, politics, governance failure and socio-economics etc. could also be contributing to the water quality problem. Thus complexity challenges the notion of linearity in complex systems, and instead proposes that systems be viewed holistically (Kliskey et al., 2021). Accepting socio-ecological systems as being fundamentally complex would lead to a rethink about resource management (Berkes et al., 2003). For example, Berkes et al. (2003) argue that qualitative analysis must complement quantitative approaches, as the latter alone is inadequate in dealing with complex systems (Berkes et al., 2003), and multiple perspectives must be considered as sources of data and analysis in managing resources within complex systems.

Managing water resources in the context of socio-ecological systems would thus require a fundamental shift in the way the relationship between humans and the rest of nature is conceived and interpreted. Western approaches promoting a utilitarian ethic have dominated water resource management in most parts of the world (McGee, 2010). Since utilitarianism is defined with reference to only the human component of socio-ecological systems, a new kind of ethic is required that re-assesses the relationship between humans and the rest of nature. Even one of the most widely cited definitions of IWRM, which also appears in the South African National Water Resource Strategy (NWRS), is utilitarian in emphasis, where the envisaged outcome is to maximize economic and social welfare, while ensuring environmental integrity (DWA—Department of Water Affairs, 2013).

Brown and Schmidt (2010b), in arguing for a new water resource management ethic, which they refer to as the "*ethic of compassionate retreat*," suggest that human relationship with water is not only underpinned by science and technology, but also by factors such as customs, beliefs, values, emotions, morals, and religion. An appropriate ethic should therefore recognize the multiple values influencing our relationship with water and strive to incorporate them.

The ethic of compassionate retreat requires the humility to acknowledge our incomplete understanding of complex, dynamic social-ecological systems, and to accept that this acknowledgment must guide water resource management. It also emphasizes the redefinition of the place of humans within the socio-ecological system, appealing for the decentering of humans regarding water resource management, as well as for an incorporation of wider practical wisdom, rather than using only science and technology in water resource management.

22.5 Conclusions

Taking the idea of socio-ecological system seriously, we argue for a relationship-based approach to environmental ethics (De Wet & Odume, 2019; Odume and de Wet, 2019). The relationship-based approach posit that values reside in the socio-ecological system as a whole and in the relationships between the components. The relationship-based approach decenters humans and emphasized respect for all components of the socio-ecological system, and in so doing call to action the need to carefully analyze the implications of water resource development and utilization for all components of the socio-ecological system. Relationships and interdependence between the components of the socio-ecological systems are emphasized and respected, and the socio-ecological system seen as the holder of both instrumental and intrinsic value. We favor the relationship-based approach because it is ethically preferable in respecting all components of the socio-ecological systems, and it holds more promise for improved water management. This is because several implications flow from the relationship-based approach. First, the pursuit of sustainability from a relationship-based perspective is all encompassing in as much as the intrinsic and instrumental values of the entire socio-ecological systems are upheld and brought to constructive balance. Second, locating the primary value at the system level implies that rights are accrued to all components of the socio-ecological systems, implying the onerous need for analyzing the implications of moral and practical decisions for (1) the relationships between the components of the socio-ecological system, (2) the entire socio-ecological systems, and (3) the contributions of each of the component to the functionality of the socio-ecological system. Third, the relationship-based approach advances the central argument that to minimize further degradation of aquatic ecosystems, a fundamental shift in our relationship with other people and with water is required, and an ecologically genuinely interactive process is needed, in which government and associated institutions are not only accountable to people, but also to the rest of nature. Such interactive practices would, in addition to recognizing and respecting human rights, promote the rights of the biophysical components of aquatic ecosystems and place responsibility on all people, as morally responsible agents, to respect those rights.

Acknowledgments

We thank the Water Research Commission of South Africa for funding the research project (Project No. K5/2342), on which this chapter is based.

References

Akamani, K. (2020). Integrating deep ecology and adaptive governance for sustainable development: Implications for protected areas management. *Sustainability, 12*(14), 5757. Available from https://doi.org/10.3390/su12145757.

Batavia, C., Bruskotter, J. T., Jones, J. A., & Nelson, M. P. (2020). Exploring the ins and out of biodiversity in the moral community. *Biological Conservation, 245,* 108580.

Behrens, K. G. (2012). Moral obligations towards future generations in African thought. *Journal of Global Ethics, 8*(2–3), 178–191.

Benny, F. F., Martoredjo, N. T., & Arivia, G. (2021). Environmental ethics of Belu society in Indonesia and Bobonaro society in Timor Leste: An ecological reflection. *11th International Conference on Future Environment and Energy, 801,* 012010.

Berkes, F., Colding, J., & Folke, C. (2003). *Navigating socio-ecological systems*. Cambridge, UK: Cambridge University Press.

Berkes, F., & Folke, C. (1998). *Linking social and ecological systems: Management practices and social mechanisms for building resilience.* Cambridge, UK: Cambridge University Press.

Bernard, P.S. (2010). Messages from the deep — water divinities, dreams and diviners in Southern Africa. Rhodes University, PhD thesis.

Brown, P. G., & Schmidt, J. J. (Eds.), (2010a). *Water ethics—Foundational readings for students and professionals.* Washington: Island Press.

Brown, P. G., & Schmidt, J. J. (2010b). An ethic of compassionate retreat. In P. G. Brown, & J. J. Schmidt (Eds.), *Water ethics—Foundational readings for students and professionals* (pp. 265–286). Washington: Island Press.

Bujo, B. (2009). Ecology and ethical responsibility from an African perspective. In M. F. Murove (Ed.), *African ethics—An anthology of comparative and applied ethics* (pp. 281–297). Scottsville: University of KwaZulu-Natal Press.

Callicott, J. B. (2016). How ecological collectives are morally considerable. In S. M. Gardiner, & A. Thompson (Eds.), *The Oxford handbook of environmental ethics* (pp. 113–124). New York: Oxford University Press.

Cisek, E., & Jaglarz, A. (2021). Architectural education in the current of deep ecology and sustainability. *Buildings, 11*(8), 358. Available from https://doi.org/10.3390/buildings11080358.

Clarke, M. L. (2019). New waves: African environmental ethics and ocean ecosystems. In M. Chemhuru (Ed.), *African environmental ethics. The International Library of Environmental, Agricultural and Food Ethics* (vol 29). Cham: Springer. Available from https://0-doi.org.wam.seals.ac.za/10.1007/978-3-030-18807-8_11.

CSIR (2010). A CSIR perspective on water in South Africa. CSIR Report No: CSIR /NRE /PW/IR/0012/A. Council for Scientific and Industrial Research, Pretoria, South Africa.

De Wet, C., & Odume, O. N. (2019). Developing a systemic relationship-based approach to environmental ethics in water resource management. *Journal of Environmental Science and Policy, 93*, 139–145.

De Wet, C. J. (2009). Does development displace ethics? The challenge of forced resettlement. In A. Oliver-Smith (Ed.), *Development and dispossession: The crisis of forced displacement and resettlement* (pp. 77–96). Santa Fe: School for Advanced Research Press.

Devall, B., & Sessions, G. (2003). Deep ecology. In D. Vandeveer, & C. Pierce (Eds.), *The environmental ethics and policy book* (3rd edition, pp. 263–268). Wadsworth/Thompson Learning, Inc.

Doorn, N. (2013). Water and justice: Towards an ethics if water governance. *Public Reasons, 5*(1), 97–114.

DWA—(Department of Water Affairs) (2010). Regulation for the establishment of a water resources classification system. No. R. 810. Department of Water Affairs, Pretoria, South Africa.

DWA—(Department of Water Affairs). (2013). *National water resource strategy* (Second edition). Pretoria, South Africa: Department of Water Affairs.

Folke, C. (2006). Resilience: The emergence of a perspective for socio-ecological systems analyses. Global Environmental Change 16: 253–267. *E3S Web Conferences, 258*, 07015.

Folke, C. (2007). Socio-ecological systems and adaptive governance of the commons. *Ecological Research, 22*, 14–15.

Foy, K. C. (2016). Balancing multiple goals at the local level: Water quality, water equity, and water conservation. *Duke Environmental Law and Policy Forum, 26*, 241–272.

Gaard, G. (2010). Women, water and energy: an ecofeminist approach. In P. G. Brown, & J. J. Schmidt (Eds.), *Water ethics—Foundational readings for students and professionals* (pp. 59–75). Washington: Island Press.

Goodpaster, K. E. (2003). On being morally considerable. In D. Vandeveer, & C. Pierce (Eds.), *The environmental ethics and policy book* (3rd edition, pp. 183–189). Wadsworth/Thompson Learning, Inc.

Groendfeldt, D., & Schmidt. (2013). Ethics and water governance. *Ecology and Society, 18*(1), 14. Available from https://doi.org/10.5751/ES-04629-180114.

Gruen, L., & Gaard, G. (2003). Ecofeminism: Towards global justice and planetary health. In A. Light, & H. Rolston, III (Eds.), *Environmental ethics: An anthology* (pp. 276–294). Malden, USA: Blackwell Publishing Ltd.

Grunwald, A. (2016). Water ethics—Orientation for water conflicts as part of inter-and transdisciplinary deliberation. In R. Huttl, O. Bens, C. Bismuth, & S. Hoechstetter (Eds.), *Society-water-technology—A critical appraisal of major water engineering projects (water resources development and management)*. Springer Open, DOI 10.1007/978-3-319-18971-0.

Hargrove, E. (2003). Weak anthropocentric intrinsic value. In A. Light, & I. I. I. H. Rolston (Eds.), *Environmental ethics: An anthology* (pp. 163–190). Malden, USA: Blackwell Publishing Ltd.

Heath, J. (2021). The failure of traditional environmental philosophy. *Res Publica (Liverpool, England)*. Available from https://doi.org/10.1007/s11158-021-09520-5.

Hoffman, W. M. (1991). Business and environmental ethics. *Business Ethics Quarterly*, *1*(2), 169–184.

Ibanga, D. (2016). Logical and theoretical foundations of African environmental ethics. *Africology: The Journal of Pan African Studies*, *9*(9), 3–24.

Jakobsen, T. G. (2017). Environmental ethics: Anthropocentrism and non-anthropocentrism revised in the light of critical realism. *Journal of Critical Realism*, *16*(2), 184–199.

Kallhoff, A. (2016). Water ethics: Toward ecological cooperation. In S. M. Gardiner, & A. Thompson (Eds.), *The Oxford handbook of environmental ethics* (pp. 416–426). New York: Oxford University Press.

Kelbessa, W. (2014). Can African environmental ethics contribute to environmental policy in Africa? *Environmental Ethics*, *36*(1), 31–61.

King, J., & Pienaar, H. (2011). Sustainable use of South Africa's inland waters. WRC Report No. TT 491/11. Water Research Commission, Pretoria, South Africa.

Klemmer, C. L., & McNamara, K. A. (2020). Deep ecology and ecofeminism: social work to address global environmental crisis. *Affilia: Feminist Inquiry in Social Work*, *35*(4), 503–515.

Kleynhans, C.J., & Louw, M.D. (2008). River ecoclassification: Manual for ecostatus determination (version 2)—Module A: EcoClassification and EcoStatus determination. Joint Water Research Commission and Department of Water Affairs and Forestry Report. WRC Report No TT 329/08. Water Research Commission, Pretoria, South Africa.

Kliskey, A., Alessa, L., Griffith, D., Olsen, S., Williams, P., Matsaw, S., Cenek, M., Gosz, J., & Dengler, S. (2021). Transforming sustainability science for practice: A social–ecological systems framework for training sustainability professionals. *Sustain Science*, *16*, 283–294. Available from https://doi.org/10.1007/s11625-020-00846-2.

Kronlid, D. O., & Öhman, J. (2013). An environmental ethical conceptual framework for research on sustainability and environmental education. *Environmental Education Research; a Journal of Science and its Applications*, *19*(1), 21–44.

Light, A., & Rolston, H., III (Eds.), (2003). *Environmental ethics: An anthology*. Malden, USA: Blackwell Publishing Ltd.

McGee, W. J. (2010). Water as a resource. In P. G. Brown, & J. J. Schmidt (Eds.), *Water ethics—Foundational readings for students and professionals* (pp. 87–90). Washington: Island Press.

McIntyre-Mills, J. J. (2013). Anthropocentrism and well-being: A way out of the lobster pot? Systems research and behavioural. *Science (New York, N.Y.)*, *30*, 136–155.

Minteer, B. A. (2016). Environmental ethics, sustainability science and the recovery of pragmatism. In S. M. Gardiner, & A. Thompson (Eds.), *The Oxford handbook of environmental ethics* (pp. 528–540). New York: Oxford University Press.

Minteer, B. A., Corley, E. A., & Manning, R. E. (2004). Environmental ethics beyond principle? The case for a pragmatic contextualism. *Journal of Agricultural and Environmental Ethics*, *17*, 131–156.

Minteer, B. A., & Manning, R. E. (2003). Pragmatism in environmental ethics: Democracy, pluralism, and the management of nature. In A. Light, & H. Rolston, III (Eds.), *Environmental ethics: An anthology* (pp. 319–330). Malden, USA: Blackwell Publishing Ltd.

Murove, M. F. (2009). An African environmental ethic based on the concepts of Ukama and Ubuntu. In M. F. Murove (Ed.), *African ethics—An anthology of comparative and applied ethics* (pp. 315–331). Scottsville: University of KwaZulu-Natal Press.

Norton, B. G. (1987). *Why preserve natural variety?* Princeton, NJ: Princeton University Press.

Norton, B. G. (2003). Environmental ethics and weak anthropocentrism. In A. Light, & H. Rolston, III (Eds.), *Environmental ethics: An anthology* (pp. 163–174). Malden, USA: Blackwell Publishing Ltd.

O'Neill, J. (2003). The varieties of intrinsic value. In A. Light, & H. Rolston, III (Eds.), *Environmental ethics: An anthology* (pp. 131–142). Malden, USA: Blackwell Publishing Ltd.

Odume, O. N., & De Wet, C. (2019). A systemic-relationship-based ethical framework for aquatic ecosystem health research and management in socio-ecological systems. *Journal of Sustainability*, *11*, 5261, 2019.

Okeja, U. (2018). Justification of moral norms in African philosophy. In E. Etieyibo (Ed.), *Method, substance and the future of African philosophy* (pp. 209−228). Basingstoke: Palgrave Macmilan.

Partelow, S. (2018). A review of the socio-ecological systems framework: Applications, methods, modifications, and challenges. *Ecology and Society*, *23*(4), 36.

Pollard, S., & du Toit, D. (2011a). Towards the sustainability of freshwater ecosystems in South Africa: An exploration of factors that enable or contain meeting the ecological Reserve in the context of integrated water resource management in the catchments of the lowveld. Water Research Commission, Pretoria. WRC Report No K5/1711.

Pollard, S., & du Toit, D. (2011b). Towards adaptive integrated water resources management in southern Africa: The role of self-organisation and multi-scale feedbacks for learning and responsiveness in the Letaba and Crocodile catchments. *Water Resources Management*, *25*(15), 4019−4035.

Pradhan, R., & Meinzen-Dick, R. (2010). Which rights are rights? Water rights, culture, and underlying values. In P. G. Brown, & J. J. Schmidt (Eds.), *Water ethics—Foundational readings for students and professionals* (pp. 39−58). Washington: Island Press.

Raine, P. (2001). The ethics of place and environmental change: Some examples from the South Pacific. *New Zealand Geographer*, *57*(2), 41−47.

Ramose, M. B. (2009). Ecology through Ubuntu. In M. F. Murove (Ed.), *African ethics—An anthology of comparative and applied ethics* (pp. 308−313). Scottsville: University of KwaZulu-Natal Press.

Reid, A. J., Carlson, A. K., Creed, I. F., Eliason, E. J., Gell, P. A., Johnson, P. T. J., Kidd, K. A., MacCormack, T. J., Olden, J. D., Ormerod, S. J., Smol, J. P., Taylor, W. W., Tockner, K., Vermaire, J. C., Dudgeon, D., & Cooke, S. J. (2019). Emerging threats and persistent conversation challenges for freshwater biodiversity. *Biological Reviews*, *94*, 849−873.

Rockstrom, J., Steffen, W., Noone, K., Persson, A., Chapin, F. S., Lambin, E., Lenton, T. M., Scheffer, M., Folke, C., Schellnhuber, H. J., Nykvist, B., de Wit, C. A., Hughes, T., van der Leeu, S., Rhode, H., Sorlin, S., Snder, P. K., Constanza, R., Svedin, U., ... Foley, J. (2009). Planetary boundaries: Exploring the safe operating space for humanity. *Ecology and Society*, *14*(2), 32.

Rogers, K.H., & Luton, R. (2011). Strategic adaptive management as a framework for implementing integrated water resource management in South Africa. Water Research Commission, Pretoria. WRC Report No KV 245/10.

Roux, D. J., & Foxcroft, L. C. (2011). The development and application of strategic adaptive management within South African National Parks. *Koedoe*, *53*, 5.

Salleh, A. (2017). *Ecofeminism as politics: Nature, max and the postmodern*. London, UK: Zed Books Ltd, the Foundry.

Sarkar, S. (2013). Multiple criteria and trade-offs in environmental ethics—Comment on "Ethics of species research and preservation" by Rob Irvine. *Bioethical Inquiry*, *10*, 533−537.

Schulz, C., Martin-Ortega, J., Glenk, K., & Loris, A. A. R. (2017). The value base of water governance: a multidisciplinary perspective. *Ecological Economics*, *131*, 241−249.

Sinha, R. C. (2021). Anthropocentric teleological environmental ethics. *Journal of Indian Council of Philosophical Research*, *38*, 125−136.

Stamova, R., Akmataliev, A., Yrazakov, D., Kambarova, N., & Salimov, R. (2021). Environmental ethics and the role of spiritual and moral values in crisis procedure. *E3S Web of Conferences*, *258*, 07015.

Stephens, A., Jacobson, C., & King, C. (2010). Towards a feminist-system theory. *Systemic Practice and Action Research*, *23*, 371−386.

Stone, C. D. (2003). Should trees have standing?—Towards legal rights for natural objects. In D. Vandeveer, & C. Pierce (Eds.), *The environmental ethics and policy book* (3rd edition, pp. 189−201). Wadsworth/Thompson Learning, Inc.

Taylor, P. W. (2003). The ethics of respect for nature. In D. Vandeveer, & C. Pierce (Eds.), *The environmental ethics and policy book* (3rd edition, pp. 189−201). Wadsworth/Thompson Learning, Inc.

Terblanché-Greeff, A. C. (2019). Ubuntu and environmental ethics: The west can learn from Africa when faced with climate change. In M. Chemhuru (Ed.), *African environmental ethics. The International Library of*

Environmental, Agricultural and Food Ethics (vol 29). Cham: Springer. Available from https://doi.org/10.1007/978-3-030-18807-8_7.

Thompson, A. (2016). Anthropocentrism: Humanity as peril and promise. In S. M. Gardiner, & A. Thompson (Eds.), *The Oxford handbook of environmental ethics* (pp. 77–90). New York: Oxford University Press.

United Nation (2010). General Assembly Resolution A/64/L63/Rev.1 on human right to water and sanitation. United Nations 26 July 2010.

Vandeveer, D., & Pierce, C. (Eds.), (2003). *The environmental ethics and policy book* (3rd edition). Wadsworth/Thompson Learning, Inc.

Verharen, C., Bugarin, F., Tharakan, J., Wensing, E., Gutema, B., Fortunak, J., & Midddendorf, G. (2021). African environmental ethics: Keys to sustainable development through agroecological villages. *Journal of Agricultural and Environmental Ethics*, *34*, 18.

Virapongse, A., Brooks, S., Metcalf, E. C., Zedalis, M., Gosz, J., Kliskey, A., & Alessa, L. (2016). A Socio-ecological systems approach for environmental management. *Journal of Environmental Management*, *178*, 83–91.

Warren, K. J., & Cheney, J. (2003). Ecological feminism and ecosystem ecology. In A. Light, & H. Rolston, III (Eds.), *Environmental ethics: An anthology* (pp. 294–305). Malden, USA: Blackwell Publishing Ltd.

Weckler I.I. (2020). In defense of non-anthropocentrism—A relationship-based account of value and how it can be integrated. Master of Art Thesis, University of Montana.

Wiredu, K. (1994). Philosophy, humankind and the environment. In O. Oruka (Ed.), *Philosophy, humanity and ecology* (pp. 30–48). Nairobi, Acts Press.

Index

Note: Page numbers followed by "*f*" and "*t*" refer to figures and tables respectively.

A

Abiotic process, 56, 78–79
"Absolute" water scarcity, 390–392
Abstraction water, 99–100
Açores tropical anticyclone system, 347–348
Acrylon, 165
Activated carbon filtration, 155
Activated sludge systems, 373
Ad hoc process, 349–350
Adaptation, 96, 105, 406, 408–410
Adaptive ethical system, 428–429
Adaptive management process, 217–218
Adaptive tipping points (ATP), 308
Adsorption, 48
Advanced oxidation processes (AOP), 79
Advanced treatment, 78
Adverse outcome pathway (AOP), 49–52
Aerated lagoons, 373
Aerobic bacteria population, 71
Aerobic process, 83, 86
Aerobic reactors, 83–86
  engineered and constructed wetland system, 84*f*
African communities, 425
African environmental ethics, 420–421, 428, 430–432
Agricultural lands, 368–369
Agricultural nonpoint source pollution, 369
Agricultural production, 211–212, 367
Agricultural Submodel, 325

Agricultural water, 296
Agricultural water pollution, 365–368
  agriculture and sustainable future, 377
  land use, 367–368
    phosphorus, nitrogen, and insect metrics compares CAFO impacted stream, 368*f*
  pollutants, 366–367
  pollution problems generated from agricultural practices, 368–371
  shifting practices and climate change, 372–374
    climate change, 372–373
    EC, 373–374
    globalization and trade, 372
    land management, 373
  water consumption and contamination tied to agriculture, 365–366
  water pollution control, 374–377
Agriculture, 317–318, 366, 372–373, 375
  management, 375–376
  and sustainable future, 377
  water consumption and contamination tied to, 365–366
Agroforestry, 263–264
Air stripping system, 89, 89*f*
Algae, 44–45, 56, 300–301
Algal toxins, 390–392
Allerheiligen Flood, 346–347
Aluminum foils, 149

"Ames *Salmonella*/microsome mutagenicity assay", 335–336
Ammonia (NH_3), 87
Ammonia nitrogen, 79
Anaerobic digestion, 87
Anaerobic processes, 83
Anaerobic reactors, 86–88
  air stripping system, 89*f*
Analytical functions, 217–218
ANEMI models, 321
ANEMI-2 (second-generation model), 321
Anomalous density of frozen water, 5–6
Anthropocentric environmental ethics, 423–425
Anthropocentric instrumentalism, 421–422
Anthropocentrism, 423–424
Anthropogenic CO_2 pollution, 59
Anthropogenic influence on climate change, 385
Anthropogenic origin, greenhouse gases of, 16–18
Anthropogenic pressure on global resources, 390–392
Anthropogenic sources of GHGs, 49
Antibiotics, 366
Antimicrobial resistant bacteria (AMR), 163
AOP. *See* Advanced oxidation processes (AOP); Adverse outcome pathway (AOP)
Apollo 13 mission, 295
Application programming interface (API), 225–226

Aquaculture, 263–264
Aquatic ecosystems, 274, 422
Aquifer, 296, 384
　recharge, 263–264
Arduino suite of products, 221–222
Array of contaminants, 373–374
Arsenic, 121–122, 403
　removal methods, 131
Art, 354, 360
　art-based intervention, 358
　art-driven process, 351
Art and science
　integration, 349
　local representation process, 349–356
　process, 358
Art-and-anthropology
　collaboration, 354–355
　cooperation, 356
Artificial intelligence, 216, 221
Artistic media, 353
Artistic process, 352–354
Asian rivers, 190–192
Assessment of food-energy-water nexus, 318–319
Atlantic Ocean, 347–348
Atmospheric circulation of air, 9
Atmospheric water harvesting, 101
ATP. *See* Adaptive tipping points (ATP)
"Attached growth" process, 76
Automatic workflow (AW), 227
Automobile tire wear on road surfaces, 182
Azadirachta indica. *See* Neem (*Azadirachta indica*)

B

Bacterial contamination, 370–371, 374–375
　heavy sediment contamination in agricultural dominated river system, 371*f*
Basic Human Need Reserve (BHNR), 277–278
Beas-Sutlej river basin, 32–33
Bed upflow reactor, 87–88
Benthamian rationale, 401
Benthic fauna, 345–346

Bergen, Norway
　art and science local representation processes by site and related challenges, 349
　local conditions of changing climate, 344–345
Best management practices (BMP), 305, 375
Big data, 216, 221
Bio-gas tanks, crop for, 348
Biocentric egalitarianism, 430
Biocentrism, 426
Biochemical oxygen demand (BOD), 66–67
　BOD5 value, 67
Biochemodynamics, 49–55
　adverse environmental outcome, 51*f*
　adverse outcome pathway, 53*f*
　AOP, consisting of sequence of KEs and KERs, 53*f*
　containment of microbe, 52*f*
Biodegradation efficiency, 72
Biodiversity, 389
　of water resources, 419
Biofilms, 82
Biogas, 87–88
Biogeochemical cycling, climate change on
　biochemodynamics, 49–55
　biogeochemistry of oxygen in water, 55–57
　　linked AOPs, 54*f*
　　normal temperature tolerances of aquatic organisms, 56*t*
　　relationship, 55*t*
　carbon cycle and greenhouse effect, 57–59
　　tropospheric carbon dioxide, 58*f*
　partitioning, 45–49
　thermodynamics of cycling, 42–45
Biological agents, 42
Biological pollutants, 147
Biological treatment system, 72, 88
Biophysical aquatic ecosystems, 422

Bioreactions, 86
Bioretention system, 73, 306
Biosand filtration, 130
Biosolids, 368–369
Biostimulation, 82–83
Biotic process, 78
Biotic systems, 56
Biotreatment systems, 72–73
Black Sea, 182
Blockchain technology, 221
Blood, 6
Blue Nile River Basin, 325
BMP. *See* Best management practices (BMP)
Boiling, water disinfection by, 150–152
Bottled water, 124
Brackish groundwater, 197–198
Brackish groundwater reverse osmosis (BWRO), 326
Brackish water desalination, 205–206
Brest, Kerourien, France
　art and science local representation processes by site and related challenges, 349–351
　local conditions of changing climate, 345–346
Brine disposal technologies, 200, 205
British Columbiás extreme wildfire, 387
"Byfjorden", 344–345

C

C-based GHGs, 59
CA. *See* Conservation agriculture (CA)
CAFO. *See* Concentrated animal feeding operations (CAFO)
Calcium carbonate, 71
Calcium hydroxide, 69
Calcium oxide, 69, 71
Caldor Fire in California, 399–400
Canadian Groundwater Information Network, 223–224
Cancer, 335–336
　colorectal, 129

Candidate contaminant list (CCL), 332
Capabilities approach, 403
Capillary action, 8
Carbon, 68, 155
Carbon capture, utilization, and storage (CCUS), 19
Carbon cycle, 57–59
Carbon dioxide (CO_2), 14–16, 71, 96
  emissions, 408
Carbon monoxide (CO), 203–205
Carman-Kozeny model, 157
Cartography, 356–358
Cash crop industry, 366
Cassia alata, 149
Cataravane, 352–353
"Catchment area", 163–164
Catchment visioning process, 274–275, 287–290
Cation exchange capacity, 48
Cavitation-based water hand pumps, 160–163, 161*t*
  hydrodynamic cavitation for Rankala Lake water treatment, 162*f*
CDC. *See* US Centers for Disease Control and Prevention (CDC)
Central nervous system (CNS), 332
Centralized systems, 101, 124–126
Centralized water management systems, 133
Centralized water supplies, 101–102
Centralized water treatment, 125–126
  and distribution systems, 134
Ceramic filters, 155–156
Ceramic filtration, 155–156
Ceramic water filters (CWFs), 155–156
CFU. *See* Colony forming units (CFU)
Channelization, 295
Chemical agents, 42, 45
Chemical cycling, 42
Chemical oxygen demand (COD), 67–68

Chemical treatment, 152–154
  chlorine for water disinfection, 153*t*
Chikungunya fever, 390–392
Chinese Export-Import Bank, 408
Chlorination, 129, 153–154, 164
Chlorine disinfection, 152
Chlorofluorocarbons (CFCs), 59
Chronic toxicity, 336
Chronotopes, 353
Citizen science website, 221
City-watershed system, 326–327
Clausius Clapeyron equation, 8–9
Clayey sediments, 80
Clean water, 104, 117–118
Clean water in developing countries, 115–118
  population using safely managed drinking water services, 117*f*
Climate, Land use, Energy-Water Strategies (CLEWS), 319–320
Climate adaptation, 349
Climate change, 14, 22–23, 27–30, 57, 95–97, 100, 118–119, 197–198, 294, 300, 302, 318–319, 349, 359, 372–373, 390, 404. *See also* Greenhouse gas (GHG)
  adaptation strategy, 407–409
  central premise impacting water resources, 294–295
  challenges and urban waters, 102–103
    service gap, 104*f*
    water supplies and demand, 103
  design of rule curves and hedging policy, 33–35
  effect on reservoir performance in Indus Basin, 32–37
  feedback mechanisms and tipping points, 18–19
  impacts, 400
    on country's hydrology, 384–385
    on poverty, 385–387, 386*f*
    on urban water, 104

    on water, 399–400
    on water resources, 383
  impacts from, 295–299
    changes in hydraulic retention time, 298
    changes in pollutant loading and processing, 298–299
    changes in precipitation patterns, 297–298
    changes in water storage, 296–297
    changes to evapotranspiration rates, 298
    changes to water cycle, 295–296
  main greenhouse gases of anthropogenic origin, 16–18
  managing worsening concerns, 299–304
    dead zones, 300
    drought, 302
    eutrophication, 300
    flooding, 301–302
    pollutants of emerging concern, 302–304
    red tides, 300–301
  mitigation, 407–408
  models, 294
  naturalizing, 410–411
  proposed solutions, 406–410
  reservoir performance indices, 35–37
  shifting practices and, 372–374
  solutions, 22–25, 406
    cost of producing electricity, 22*t*
  solving problem of global heating, 19–22
  struggle(s) for water and climate justice, 411
  system description, 32–33
    Beas-Sutlej river basin, 33*f*
    salient features, 33*t*
  and water, 385
  water and, 8–9
    CO_2 and greenhouse effect, 14–16
Climate change vulnerability, 404–406

Climate crisis, 410
Climate justice
   movement, 411
   struggle(s) for, 411
Climate knowledge coproduction, 359–360
"Climate Plan: Energy & Territory 2014–2019", 345–346
Climate related disasters, 118–119
"Climate Risk Management", 358
Climate science, 354–355
Climate service coproduction, 350–351
Climate-related disasters, 405
Climate-related droughts, 405
Climate-related pressure, 273–274
Climate–carbon cycle, 321
Closer water, 65
Cloud computing, 216, 221
Clouds, 351–352
CNS. *See* Central nervous system (CNS)
Coagulation, 125
Coalition for Defense of Independence Aquifer, 403
Coastal regions, 99–100
CoCliServ (codeveloping place-based climate services), 344–346, 356–359
COD. *See* Chemical oxygen demand (COD)
Collective process, 353
Colony forming units (CFU), 147
Colorado River basin, 399–400
Commercial farmers, 406
Commercial scenario for data usage, 229
Commons-based peer management, 233
Commons-based peer production, 233
Community acceptance, 309
Community management, 134
Comparison of purification techniques, 170–173
Complex social-ecological systems, 433
Complex socioecological systems, 419, 432
Complexity analysis, 432–433

Compound Parabolic Collector (CPC), 149
Concentrate disposal technologies, 205
Concentrated animal feeding operations (CAFO), 367–368
Conceptual maps, 325
Conferences of Parties (COPs), 24
Conservation agriculture (CA), 373
Consortium of Universities for the Advancement of Hydrologic Science (CUAHSI), 223
Constitutive values, 424
Consumption of agrochemicals, 366
Contaminant Candidate List, 332
Contaminated water, 117–118
Contamination, 368–369
"Contextual vulnerability", 404–405
Continental polar anticyclone systems, 347–348
Control volume, 46
Conventional "command-and-control" approach, 421
Conventional methods, 147
Conventional treatment processes, 124–125
Conventional urban water supplies, 99–100
Conventional water management, 197–198
Conveyance system, 163–164
Coproduction process, 349–350, 360
COPs. *See* Conferences of Parties (COPs)
Coronavirus pandemic, 21
Cost
   of conventional SODIS, 150
   effectiveness, 245
   of granulate activated carbon, 155
   of producing electricity, $22t$
   of recreational waterborne illness, 371
   of renewables, 22
   of water, 97–99, 102

Coupling climate models, 383
Covalent reaction, 48
Covid 19 virus (SARS-CoV-2), 21, 168, 353–354
   impact on water sector in developing countries, 168–170, $169f$
CPC. *See* Compound Parabolic Collector (CPC)
Crisis management theory, 295
Crop management, 373
Crop production, 366
Crude oil, 20–21
Cryptosporidium, 129, 152, 154–155, 298–299
   C. parvum, 149
CUAHSI. *See* Consortium of Universities for the Advancement of Hydrologic Science (CUAHSI)
CWFs. *See* Ceramic water filters (CWFs)
Cyber-physical systems (CPS), 216, 221
"Cycleclean", 166
Cylindrospermopsin, 390–392

D

Dam removal, 307–308
Dangast, 355
Data compilation process, 327
Data model, 224
Data processing, workflow for, 225–228
Data usage, scenarios for, 228–230
Data-driven machine learning models, 217–218
DBPs. *See* Disinfection byproducts (DBPs)
DDT, 370
Dead zones, 300
Decentralized systems, 124–125
Decentralized water supply, 126
Decision-making process, 274–275
Decomposition process, 69–71
Deep ecology, 428, 430

Deep groundwater sources, 197–198
Deep well injection, 205
Degradation, 367, 389
Deliberation process, 428–429
Demand values, 424
Dengue, 390–392
Dental fluorosis, 403
Desalination, 197–198, 409
　in agriculture, 326
　desalination centered food-energy-water systems, 211–212
　description of, 198–203
　　energy efficiency of thermal desalination, 200
　　MED, 199
　　membrane distillation, 201–203
　　MSF desalination, 199
　　RO, 200
　　solar stills, 198–199
　　vapor compression, 199–200
　future of sustainable desalination, 210–212
　industry, 199, 212
　plants, 203–205
　process, 206–207
　　criteria for selecting desalination technology, 209f
　　selection process of, 209
　renewable energy integration, 207–209
　of saline waters, 198
　and sustainability, 203–207
　　economics, 205–206
　　environmental footprint, 203–205
　　social aspects, 206–207, 207t
　technologies, 197–198, 207–208
Deteriorating water quality of water resource, 432–433
Detroit Water and Sewerage Department (DWSD), 327
Developing countries
　clean water in, 115–118
　Covid 19 pandemic impact on water sector in, 168–170
　environmental challenges to water purity in, 118–120
　water supply sources in, 123–124
　water treatment systems in, 124–133
Dichotomy, 372–373, 427–428
Diffusion, 73, 87
Digital groundwater, 220–221
Digital twin, 221, 232
Digital Water Program, 220–221
Dinitrogen monoxide, 59
Disasters, 387
Discarded plastic fishing gears, 181
Disease protection, 389
Disinfection, 125, 129
　chlorine, 152
　solar, 165
　water, 150–152
Disinfection byproducts (DBPs), 153
Dissolution process, 45–47
Dissolved organic carbon (DOC), 158
Dissolved organic matter (DOM), 298–299
Dissolved oxygen (DO), 42, 66–67, 69, 78
　sag curve, 68–69
Distributive equity, 404
Distributive justice, 403
Diverse workflows, 227
Do-it-yourself (DIY), 221–222
Dordrecht, the Netherlands
　art and science local representation processes by site and related challenges, 351–352
　local conditions of changing climate, 346–347
Dose-response
　curves, 332–336
　data, 334–335
　relationships, 332–337
　　examples of dose-response curves for emerging contaminants, 335f
　　U-shaped curve with dose regions of deficiency and toxicity, 336f

Dracunculus medinensis. See Guinea worm (*Dracunculus medinensis*)
Drainage ditches, 376
"Drinkable book", 166, 173
Drinking water, 101
　quality, 124
　sources, 124
　standards, 120
　treatment, 132
Drought, 102–103, 215–216, 302, 387
Drumstick seeds (*Moringa oleifera*), 159–160
Dry climates, 299
DWSD. See Detroit Water and Sewerage Department (DWSD)
Dynamic mapping perspectives for local representations, 356–358
Dynamic social-ecological systems, 433
Dynamics of water management system's adaptive ethical system, 428–429

E

E-Research, 221
E-Science, 221
Earth, 14, 41
　IR radiation, 15
　origin of water on, 9–10
　systems, 383
　temperature, 14
EarthCube, 222–223
　initiative for geosciences, 220–221
Ebro River, 187
Ecocentrism, 426
Ecofeminism, 429
Ecofeminist analysis, 429
Ecofeminist environmental ethics, 429
Ecological categories, 426
Ecological Focus Areas (EFA), 310
"Ecological integrity", 385
Ecological reserve (ER), 274, 421, 427

Ecological systems, 366
Ecological water requirements (EWR), 275
Economic deprivation, 388
Economic net water productivity trends, 254−256
Economic optimization models, 255
Economics, 245
 production and scarcity in, 248−249
Economy, 321
Ecosystem, 295, 400
 management, 422
 services, 263−264, 389
 stress, 79−80
Educational deprivation, 389
Effectiveness
 cost, 170, 173, 245
 of disaster risk management, 389
 of disinfection, 122
 of hedging, 32
 of system, 75−76
Electric field, 198
Electricity, 207−208
Electrodialysis (ED), 198
Electromagnetic radiation, 14
Electromechanical systems, 73−75
Elevated atmospheric CO_2 (eCO_2), 294
Emerging concerns (EC), 373−374
Emerging contaminants (EC), 303−304
Emotions, 351−352
Empowerment deprivation, 389
Energetic model, 326
Energy, 317−318
 cycling, 42
 energy-economy element, 321
 energy-intensive systems, 73
 energy-water nexus, 318
 system, 327
 wastewater as energy source, 268−269
Energy Submodel, 325
Enterococcus faecalis, 166
Enthalpy of vaporization of water, 4−5
Environmental contaminants, 45

Environmental decision-making process, 429
Environmental ethics
 approaches to environmental ethics from western perspective, 423−432
 African environmental ethics, 430−432
 relationship-based environmental ethics, 427−430
 value-oriented environmental ethics, 423−427
 key issues and need for considerations in water resource management, 421−423
 related to water resource management in context of complex socio-ecological systems, 432−433
Environmental ethics, 422
Environmental footprint, 203−205
 potential ecological/ environmental impacts of RO and MSF desalination process concentrates, 204t
Environmental impact assessments, 287
Environmental impacts on national/regional water resources, 384−385
Environmental pragmatists, 428
Environmental Protection Agency, 66
Environmental system, 42, 47
EPA. *See* United States Environmental Protection Agency (EPA)
Equal Priority, 325
Equity, 404
Erosion, 295
Errors, 339
Escherichia coli, 120−121, 148−149, 166, 298−299, 370
"Ethic of compassionate retreat", 433
Ethical criteria, 420
Ethics, 420

EU's nature conservation law, 385
European Union Water Framework Directive (EU-WFD), 215−216
Eutrophication, 300, 369−370, 390−392
 management, 308
Evaporation
 evaporation/condensation, 198
 ponds, 205
Evaporative cooling, 4−5
Evapotranspiration (ET), 298
 changes to evapotranspiration rates, 298
EWR. *See* Ecological water requirements (EWR)
Extended terminal subfluidization wash (ETSW), 157

F

FAIR Data Principles, 223−224
False positives, 339
Fecal coliform, 156, 279−280
Fecal contamination, 120
 of drinking water, 120−121
Feed seawater, 199−200
Feed water, 201
Feedback mechanisms of climate change, 18−19
Fengshan River in Taiwan, 187
Fenton oxidation, 82−83
Ferric oxide, 69
Fertilizers, 366, 369−370
Filtered water, 156−157
Filtration, 125, 130, 154−159, 165
Filtration process, 73
Financial management plans, 134
"First Flush Device", 163−164, 173
Fish population metrics, 374−375
"Fixed film" process, 76
Flash floods, 103
Flexibility, 356−358
Floating marine plastics, 182−183
Flocculation, 125
Flooding, 102−103, 295, 301−302
FloPy Python library, 227−228
Fluid dynamics, 49
Fluoride, 403

Index

Food, 317–318
  components, 317–318
  production, 372, 392
  security, 392
  system, 327
Food and Agricultural
    Organization (FAO), 107,
    320, 390–392
Food and Drug Administration, 66
Food-energy nexus, 318
Food-Energy-Water nexus models
    (F-E-W nexus models),
    317–318
  applications of, 322–328
  assessment of, 318–319
  comparison of different models,
    322
    summary of comparative
      elements of different F-E-
      W models, 323t
  development, 319–322
    CLEWS, 320
    Foreseer, 320–321
    MuSIASEM, 320
    WEAP-LEAP, 321–322
    WEF Nexus Rapid Appraisal
      Tool, 320
    WEF Nexus Tool 2.0, 319
  elements of, 318f
  in Phoenix, 325
  resilience, 325
Food-water nexus, 318
Foreseer Tool, 320–321
Forest ecosystems, 367
Forested watersheds, 367
Fossil fuel, 20
  fossil fuel-derived electricity, 20
4IR. *See* Fourth industrial
    revolution
Fourth industrial revolution, 216
FREEWAT platform, 227–228
Freshwater, 205–206
  allocations, 263
  pollution, 426
  production, 198
  resources, 273–274
  scarcity, 197–198
Frozen water, anomalous density
    of, 5–6
Fungal toxins, 390–392

G

Gain output ratio (GOR), 200
Ganga Water Information System
    (GangaWIS), 231–232
GDDL-CM3 model, 36
Geest, 348
Gender inclusivity, 429
General authorization (GA), 274
Genetic algorithm (GA), 33–34
Genetically modified crops
    (GMO), 366–367
Geo-referencing local
    representation process, 349
Geographers, 356–358
Geographic information systems
    (GIS), 256, 356–358
Geoscience domains, 221
Geothermal energy sources,
    207–208
GHG. *See* Greenhouse gas (GHG)
Giardia, 152, 154–155, 298–299
Giardia cysts, 151
Global access to safe drinking
    water, 123
Global carbon dioxide emissions,
    407
Global food system, 392
Global greenhouse gases
    emissions, 41, 96
Global heating, 14
  solving problem of, 19–22
Global initiatives, opportunities
    and, 104–106
  adaptation and mitigation, 105
  environmental impact modeling
    risk, 105–106
  understanding water baseline,
    105
Global marine plastic pollution,
    180–183
  distribution of microplastics in
    marine environment,
    182–183
  sources of marine plastic debris,
    180–182
Global warming, 8, 13–14, 18,
    426
  gases, 59
Global warming potential (GWP),
    208–209

Global water, 410
Global water cycle, 400
Global Water Partnership (GWP),
    107
Global Water Resources (GWR),
    95–96, 383–384, 390–392
  environmental impacts on
    exposure and vulnerability
    impacts, 387–389
  on national/regional water
    resources, 384–385
  slow-onset events, 389–393
Globalization and trade, 372
Granulated activated carbon
    (GAC), 89, 151
Graphical designers, 351
Grass swales, 306
"Gray infrastructure", 306
Green economy, 407
Green infrastructure, 306–307
Greenhouse
  effect of biogeochemical
    cycling, 57–59
  effect of climate change, 14–16
Greenhouse gas (GHG), 41, 203,
    385, 406. *See also* Climate
    change
  of anthropogenic origin, 16–18
    CO_2 produced from human
      activity, 16t
    greenhouse gas properties, 17t
    sources of methane, 17t
    worldwide source of CO_2
      emissions, 17t
  emissions, 400
  mitigation, 104
Greywater, 131–132
Groundwater (GW), 89, 99–100,
    215–216
  aquifers, 203
  systems, 105
  treating contaminated, 89–91
Groundwater sustainability,
    217–220
  components of groundwater
    sustainability evaluation,
    218f
  digital groundwater, 220–221
  inexact outline of four industrial
    revolutions, 217f

Groundwater sustainability (*Continued*)
  IoT, 221–222
  perspectives of web-based groundwater platforms, 230–233
  scenarios for data usage, 228–230
  web-based data sharing, 222–225
  workflow for data processing, 225–228
GroundWaterML2 (GWML2), 224
Guinea worm (*Dracunculus medinensis*), 121
Gulf of Morbihan, France, 347–348
  art and science local representation processes by site and related challenges, 352–354
  local conditions of changing climate, 347–348
Gypsum, 205

H

Haloacetic acids (HAAs), 153
Halocarbons, 59
Hanseatic League, 344–345
Harmful algal blooms (HAB), 300–301
Harvested rainwater, sustainable disinfection of, 163–166, 164f
Hazardous waste, 81–82, 85–86
Health and health services deprivation, 388
Health detriments, 390
Heat capacity of water, 5
Heat energy, 41, 207–208
Heatwaves, 387
Heavy metals, 131
Hedging policy, 33–35
Hepatotoxin, 332
Herbicide, 366–367
Herbs, 159–160
Hercynian western mountains, 347–348
Hibiscus esculentus. *See* Okra (*Hibiscus esculentus*)

"Hiring" process, 354
HLM tower project, 345–346
Hoedic islands, 347–348
Holism, 426
Holistic approach, 426
Homo sapiens, 338–339
Hormones, 6
Houat islands, 347–348
House deprivation, 388
Household scale evaluation, 327–328
Household water treatment (HWT), 150
Household water treatment and safe storage (HWTS), 126
Hub, 221
Human biological communities, 430
Human communities, 400
Human Development Index (HDI), 115–116, 190
Human health toxins, 300–301
Human Need Reserve (HNR), 274
Human Right to Water and Sanitation, 402
Human–environment relationship, 425
Human–environment relationship, reconceptualization of, 427
Hurricane Ida, 399–400
Hurricanes, 387
Hybrid desalination process, 198
Hybrid filtration methods, 158–159, 159t
Hybrid techniques, 165, 167
Hybrid unit, 167
Hybrid water treatment method, 163
Hydration, 69
Hydraulic retention time (HRT), 87
  changes in, 298
Hydrodynamic cavitation, 161–162, 173
Hydroeconomic model, 217–218, 253, 255–256
Hydrogen, 20
  atoms, 45
  bonds, 4, 6–7
  gas, 20

Hydrogen peroxide (H_2O_2), 82–83, 148–149
HydroLearn platform, 230
Hydrologic imbalances, 95–96
Hydrological cycle, 385
Hydrological models, 383
Hydrological systems, 385
Hydrology community, 224–225
Hydropower, 321
  dam
    construction, 408
    development, 411
  generation, 37
HydroShare, 223
Hydroxyl radicals, 148–149
Hypertrophic water's biotoxins, 390–392
Hypoxic water, 42–44

I

IBM Blockchain Platform, 232–233
Improved sanitation techniques, 124
Inclusive process, 353–354
Income deprivation, 388
Indigenous communities, 407–408
Individual nexuses, 318
Individualism, 426
Indus Basin
  climate change effect on reservoir performance in, 32–37
Industry 4.0. *See* Fourth industrial revolution
Inexpensive water purification methods, 146
Infectious diseases, 120–121, 392–393
Information and communications technology (ICT), 220–221
Infrastructure for Spatial Information in the European Community (INSPIRE), 220–221
  INSPIRE Directive, 220–221
  INSPIRE-Geology, 224
  INSPIRE-Hydrograph, 224
  INSPIRE-Landcover, 224
  INSPIRE-Landuse, 224

INSPIRE-Meteorology, 224
Innovation, 100, 106
Innovative hybrid technology, 167
"Insider perspective" of local communities, 351–352
Instrumental valuation of environment, 424
Integrated optimization model, 326
Integrated water resource management (IWRM), 27–28, 217–218, 226, 275, 419–420, 423, 433
Intelligent systems research, 221
Intelligent transportation systems, 221
Intelligent/smart systems, 221
Intergeneration anthropocentrism, 424
Intergovernmental Panel on Climate Change (IPCC), 14, 29, 96, 345–346, 385, 400, 404–405
    IPCC Technical Paper VI, 385
Interior solution, 254–255
International Agency for Research on Cancer (IARC), 153
International Decades for Action, 115–116
International Finance Corporation (IFC), 168–169
International Fund for Agricultural Development (IFAD), 107
'International Permanent Peoples', 403
International Renewable Energy Agency (IRENA), 22
Internet of Things (IoT), 216, 221–222
Interoperability, 224
Intraspecies uncertainty factors (UF_{intra}), 338–339
"Invisible or almost-visible" phenomenon of climate change, 360
Ion transfer, 198
Ionic compounds, 6
Ionic transfer, 198
Irrigated agriculture, 241–242, 368
Irrigation, 366, 406

J

Jade Bay, Germany
    art and science local representation processes by site and related challenges, 354–356
    local conditions of changing climate, 348
Joint Monitoring Program (JMP), 115–116, 123
Justice, 401–403
    environmental, 423–424
    justice, 401–404
        proposed solutions, 406–410
        struggle(s) for water and climate, 411
        vulnerability, 404–406
    theories of, 401

K

Kerourien reconstruction story, 345–346
Key event relationships (KERs), 51
Key events (KEs), 51

L

Labor productivity, 256–257
Lakes, 99–100, 123–124, 203
Land, 372
    application, 81–82
    degradation, 426
    disturbance, 374–375
    erosion, 389
    land-based plastic debris, 180–181
    land-based plastic wastes, 181–182
    land-farming facilities, 81–82
    management, 373
        fencing of cattle from streams, 374f
    productivity, 256–257
    surface, 29
    treatment, 81–82
    use, 321
Landfills, 81–82
Large scale arsenic poisoning, 121–122
Large scale desalination plants, 206–207
Lebreton model, 190
Legal-operational aspects (rights on environment dimensions), 427
Legionella, 165
Level of detection (LoD), 339
Libertarianism, 402
"Lifestraw", 147, 166–167, 170, 173
Lithium mining, 406–407
"Lithium triangle" in Chile, 407
Lithium-ion batteries, 407
Livestock
    density, 376
    grazing, 366
    production, 367
LOAEL, 338
Local representations of changing climate
    art and science local representation processes by site and related challenges, 349–356
        Bergen, Norway, 349
        Brest, Kerourien, France, 349–351
        Dordrecht, the Netherlands, 351–352
        Gulf of Morbihan, France, 352–354
        Jade Bay, Germany, 354–356
    lessons learned, 358–361
    local representations for codeveloping climate services, 359–360
    local conditions of changing climate, 344–348
        Bergen, Norway, 344–345
        Brest, Kerourien, France, 345–346
        Dordrecht, the Netherlands, 346–347
        Gulf of Morbihan, France, 347–348
        Jade Bay, Germany, 348
    metadata and dynamic mapping perspectives for local representations, 356–358

Long Range Alternatives Planning System (LEAP), 319, 321
Low impact development designs (LID designs), 306
Lower rule curve (LRC), 31–34
Lower Saxony Wadden Sea, 348, 355

M

Macro-algae biomass, 345–346
Malaria, 390–392
Management classes (MC), 275–276
Marginal physical product of water (MPP$_W$), 248
Marine biology, 345–346
Marine environment, microplastics distribution in, 182–183
Marine plastic debris, sources of, 180–182, 181f
Marine plastic pollution, 179–180
Marine pollution, 180–183
Marine sources, 180–181
Marine waste, 182
Mass fraction, 47
Material and energy flow analysis (MEFA), 327
Maximum contaminant levels (MCLs), 66
MDGs. See Millennium Development Goals (MDGs)
Mechanical energy, 198
Mechanical vapor compression (MVC), 198
MED. See Multi-effect evaporation/distillation (MED)
Mediterranean Sea, 182
Mega city, 97
Mega-hydraulic projects, 408–409
Membrane, 6
  filters, 130
  filtration, 130
  technologies, 100
Membrane bioreactors (MBR), 79
Membrane distillation (MD), 198, 201–203, 202t
Mercury, 57
Mesoscale convective systems, 384

MESS desalination process. See Multi-effect solar stills desalination process (MESS desalination process)
Metadata, 349
  model, 356
  perspectives for local representations, 356–358
  tool, 356
Metal compounds, 42–44
Metalloids, 131
Methane (CH$_4$), 17–18, 57, 96, 408
Methanogenic granular sludge structure, 72
Microbes, 72, 89–90
Microbial hazards, 121
Microbial mass, 78
Microbial optimization, 72
Microbial populations, 68, 76, 78, 83
Microbiological pathogens, 298–299
Microcystin, 390–392
Micronanotopography, 101
Microorganisms, 72, 121, 366
Microplastics (MPs), 179–180, 189–190
  to coastal oceans/seas, riverine transport of, 187–188, 187f, 188f
  concentrations, 183f
  in marine environment, 182–183
Mid-latitude Atlantic Ocean system, 347–348
Migrants, 388
Millennium Development Goals (MDGs), 95–96, 115–116
Mineralization, 44–45
Minimal tillage practices, 373
Mismanaged plastic waste (MPW), 190
Mitigation, 96, 105, 406–408
Mixed reactors, 82–83
Modeling
  environmental impact modeling risk, 105–106
  process, 325
  tools, 219, 318

water resource systems, 231–232
MODFLOW One-Water Hydrologic Flow Model version 2 (MF-OWHM2), 219
MODFLOW-based GW modeling framework, 227
Modular tool integrates models, 322
Molecular initiating event (MIE), 51–52
Molecule's polarity, 45
Molecules migrate, 48–49
Morbihan project, 353
Moringa oleifera, 128, 159–160
Moringa oleifera. See Drumstick seeds (*Moringa oleifera*)
MPs. See Microplastics (MPs)
MPW. See Mismanaged plastic waste (MPW)
Multi-effect distillation, 199
Multi-effect evaporation/distillation (MED), 198
Multi-effect solar stills desalination process (MESS desalination process), 198–199
Multi-stage flash desalination process (MSF), 198
  desalination, 199, 205–206
  distillation, 198
Multiatomic gases in atmosphere, 15
Multidimensional poverty, 383
Multiple land management problems, 375–376
Multiscale Integrated Analysis of Societal and Ecosystem Metabolism (MuSIASEM), 319–320, 326
Multisectoral Systems Analysis (MSA), 326–327
Municipal wastewater for resource recovery and reuse wastewater, 263–264
  as energy source, 268–269
  as nutrient source, 265–268
  as water source, 264–265
Mutagenicity studies, 335–336
MVC. See Mechanical vapor compression (MVC)

N

Nakdong River, 189–190
Nanofiltration, 73
Nanostructured surfaces, 101
National Aeronautics and Space Administration (NASA), 227
National Department of Water and Sanitation, 421
National Pollutant Discharge Elimination Permit (NPDES), 299
National Science Foundation (NSF), 220–221
National Water Act (NWA), 274, 421
  of South Africa, 215–216
National Water Initiative in Australia, 215–216
National Water Plan, 346–347
National/regional water resources, environmental impacts on, 384–385
Natural coagulants, 128
Natural environment, 427
Natural gas, 18, 200
Natural hydrological cycle, 321
Natural microbes, 81–82
Natural system, 73
Neem (*Azadirachta indica*), 159–160
Nephalometric turbidity units (NTUs), 122
Net water productivity, 252–253
Neurotoxin, 332
NexSym models, 321–322
Next-generation climate models, 384
Nexus Assessment framework, 320
Nexus context assessment, 320
Nexus intervention assessment, 320
Nitrate, 79
Nitric oxide (NO), 203–205
Nitrification, 79
Nitrogen, 87, 366
Nitrogen dioxide (NO_2), 203–205
Nitrous oxide (N_2O), 96
No observed adverse effect level (NOAEL), 338–339
NOAA Hydrometeorological Data, 305–306
No-renewable energy sources, 203
"No-threshold" model, 336
Non-conventional water
  management, 197–198
  sources, 100, 197–198
Non-phase change process, 198, 201
Nonanthropocentric environmental ethics, 426–427
Nonanthropocentrism, 423
Noncentralized systems, 102
Noncentralized water supplies, 101–102
Nonhuman biological communities, 430
Nonhuman entities, 430
Nonpoint source pollution, 373
Not In My Backyard (NIMBY), 206–207
NPDES. *See* National Pollutant Discharge Elimination Permit (NPDES)
NSF. *See* National Science Foundation (NSF)
NTUs. *See* Nephalometric turbidity units (NTUs)
Nutrient pollution, 369
Nutrients, 298–299, 366
  recovery from waste, 266
  wastewater as nutrient source, 265–268
NWA. *See* National Water Act (NWA)

O

Ocean water, 10, 100
Ocimum sanctum. *See* Tulsi (*Ocimum sanctum*)
O–H bonds, 4
Okra (*Hibiscus esculentus*), 159–160
Online banking, 221
Onsite treatment, 76
Onsite wastewater treatment systems (OWTSs), 75–76
Open depressions, 263–264
Open Geospatial Consortium (OGC), 224
Open Modeling Interface Standard (OpenMI), 226
Open Science Framework, 223
Openly Published Environmental Sensing project, 221–222
Operation and maintenance (O&M), 73
Optimization, 254–256
Organic matter, 81
Organisation for economic co-operation & development (OECD), 247
Origin of water on Earth, 9–10
  scarcity of water, 10
  water cycle, 9–10
Orthodox libertarianism, 402
Osmotic pressure, 200
"Outcome vulnerability", 404–405
Overgrazing, 376
Oxygen, 76–78, 82
  biogeochemistry of oxygen in water, 55–57
  gas, 6

P

Pacific Northwest National Laboratory, 322–325
Pacific Ocean, 182, 187
PACl. *See* Polyaluminum chloride (PACl)
Panel methods, 256–257
Particle size, 80
Partitioning, 45–49
  dissolution, 45–47
  environmental matrix, 46*f*
  sorption, 47–49
Pasteurization, 130
Pathogens, 298–299
Payment for Ecosystem Services (PES), 310, 407
Pearl River Delta (PRD), 186
Pearl River network, 186, 189–190
Peer-review validation process, 358
Per-and Polyfluoroalkyl substances (PFAS), 331–332
Perfluorooctanoate sulfonate (PFAS), 331–332

Perfluorooctanoic acid (PFOA), 331−332
Performance ratio (PR), 200
Pesticides, 366
  toxic contamination from, 370
"Petri dish" studies, 335−336
PFOA. *See* Perfluorooctanoic acid (PFOA)
"Phantoms of the Anthropocene", 354−355
Pharmacokinetic models (PK models), 331
Phase change desalination process, 198
Phase change process, 198, 201
Phosphate, 69
Phospholipids, 6
Phosphorus, 267, 366
Photocatalytic water treatment (PWT), 148−149
Photovoltaic (PV), 22, 167
PhreeqPy, 225−226
Phyllanthus niruri, 149
Physical agents, 42
Physical mediation, 351
Physical/Chemical Methods, 80
Phytoplankton, 300−301
  phenology, 345−346
PK models. *See* Pharmacokinetic models (PK models)
Place-based climate service coproduction, 350
Place-based coproduced climate services, 350
Place-based human−environment relationships, 425
Plantations, 368
Plastic(s), 179, 298−299, 302−303. *See also* Microplastics (MPs)
  concentrations, 182−183
  pollution, 302−303
    global marine plastic pollution, 180−183
    riverine plastic outflows, 188−192
    in rivers, 183−188
  products, 181
  wastes, 179, 181

Platform for Regional Integrated Modeling and Analysis (PRIMA), 322−325
Point of departure (POD), 334−335
Point of use method (PoU method), 126, 157, 160
Point sources, 99−100, 123
Point-of-entry systems (POE), 126
"Pointe de Quiberon", 347−348
Poiseuille's Law, 7
Polar nature of water, 4
Polarity, 45
Policy makers, 106−107
Pollutant(s), 44−45, 73−75, 198, 295, 298−299, 366−367
  changes in pollutant loading and processing, 298−299
  coffee plants in Costa Rica, 367f
  of emerging concern, 302−304
Pollution problems generated from agricultural practices, 368−371
  bacterial contamination, 370−371
  biosolids and contamination, 368−369
  fertilizers and eutrophication, 369−370
  toxic contamination from pesticides, 370
Polyaluminum chloride (PACl), 158
Polysemic concepts, 359
Polystyrene, 181
Polyvinyl chloride (PVC), 167
Ponds, 83, 306
  design and operation of, 85
  storage, 308
  waste stabilization, 73
Population, 321
  growth, 197−198
Portable water reuse, 101
Potassium, 267−268
Potassium chloride, 71
Potassium dichromate ($K_2Cr_2O_7$), 67−68
PoU method. *See* Point of use method (PoU method)

Poverty, 385−386
  environmental impacts on exposure and vulnerability impacts, 387−389, 388f
  impacts of climate change on, 385−387
  slow-onset events, 389−393
Power Priority, 325
PR. *See* Performance ratio (PR)
Practical intervention, 360
Pragmatist environmental ethics, 428−429
PRD. *See* Pearl River Delta (PRD)
Precipitation, 370
  changes in precipitation patterns, 297−298
Precipitation-Runoff Modeling System (PRMS-V), 219
Predictive models, 45
Present ecological state (PES), 275−276
PRIMA. *See* Platform for Regional Integrated Modeling and Analysis (PRIMA)
Primary MPs, 179−180
Primstav (ancient Norwegian stick), 349
Private contractors, 102
Procedural equity, 404
Procedural justice, 403
Production in economics, 248−249
Productive process, 351
Productivity, 366
  concepts as indicators, 243−248
  statistics, 242
Propositions, 353
Protect river systems, 105
Protists, 298−299
Protozoa, 151
Pseudomonas, 165
  P. aeruginosa, 163
Psychic and socio-psychic aspects (rights on environment dimensions), 427
"Pure Water Bottle", 167
Purification techniques, comparison of various, 170−173, 171t
PV. *See* Photovoltaic (PV)

PVC. *See* Polyvinyl chloride (PVC)
PWT. *See* Photocatalytic water treatment (PWT)

Q

Q-Nexus Model, 321–322
Quantitative structural activity relationships (QSAR), 337
Quite concerning, 376

R

Radial-flow constructed wetland system, 84–85
Rainwater, 102–103, 131
Rainwater harvesting system (RWHS), 164
Rapid sand filtration (RSF), 155
Raspberry Pi, 221–222
Rawls' theory of justice, 401–402
Rawlsian conceptualizations of justice, 401–402
Rawlsianism, 402
Reactive oxygen species (ROS), 148
Realistic environmental control volumes, 46
Recalcitrant organic compounds, 67–68
Recognitional justice, 403
Recommended ecological category (REC), 275–276
Red tides, 300–301
Reducing Emissions from Deforestation and forest Degradation (REDD +), 407
Reference concentration (RfC), 337–338
Reference doses (RfD), 337–338
Reflective solar boxes, 149
Regional analysis of wastewater production, 265
Regional level F-E-W nexus, 322–325
Regionalized Sensitivity Analysis, 326–327
Relational theory of justice, 403–404

Relationship-based environmental ethics, 427–430
  deep ecology, 430
  ecofeminist environmental ethics, 429
  pragmatic environmental ethics, 428–429
Reliability, 30–31
Renewable energy, 20–21, 207–208
  integration, 207–209
  integration in agriculture, 326
  renewable energy-powered desalination process, 209
  selection process of desalination process, 209
  sources, 207–208, 211–212
Reproduction cycle of poverty, 390
Reserve quality objectives, determination of, 277–278
Reservoir(s), 29–30, 132–133
  operation, 31–32
    water reservoir rule curves, 32*f*
  performance indices, 30–31, 35–37
    for hydropower generation, 37*f*
    in Indus Basin, 32–37
    for irrigation water demands, 36*f*
    optimized flood control reservoir levels, 37*f*
    reliability, 30–31
    resilience, 31
    vulnerability, 31
Resilience, 31, 325
Resource directed measures (RDM), 274–275
  and SDC instruments for sustainable freshwater resources management, 287–290
Resource quality objectives (RQO), 274
  determination of, 277–278
Resource recovery and reuse, 270
Restoration of floodplains, 305
Restoration of riparian buffers, 305
Return activated sludge (RAS), 78

Reuse of wastewater, 131–132
Reverse osmosis (RO), 198, 200, 203–206
Rhine-Meuse-Scheldt Delta, 346–347
Riparian buffer, 304
  improvement of, 376
  restoration of, 305
  sediment and water flow reduction for, 376
Riparian restoration, 305
River networks, 180
Riverine clay, 346–347
Riverine microplastic
  outflows, 188–190, 189*f*
  pollution, 183–186
Riverine plastic
  outflows, 188–192
    field measured riverine microplastic outflows, 188–190
    between measurements and model estimates, 190–192, 191*f*
  wastes, 182
Rivers, 188–189, 203
  distribution of plastics in global rivers, 183–186
    riverine microplastics collected by bulk water sampling, 185*t*
    riverine microplastics collected by Manta trawling nets, 184*t*
  plastic pollution in, 183–188
  riverine transport of microplastics to coastal oceans/seas, 187–188
Rooftop rainwater harvesting, 131
RSF. *See* Rapid sand filtration (RSF)
Rule curves design, 33–35
  integrated-hedging rule, 35*f*
  optimal zone-based rule curves, 34*f*
  performance measure for irrigation water supply, 35*t*
Run-off management system, 83
Run-on control system, 83
Rural water supply systems, 134

RWHS. *See* Rainwater harvesting system (RWHS)

S

Sabie River, 421
Safe drinking water, 115–116, 390–392
Safe Drinking Water Act (SDWA), 332
Safe water, 392–393
Saline water, 66
Salmonella, 335–336
Sanitation, 308–309, 392–393
Sankey diagrams, 320
SARS-CoV-2. *See* Covid 19 virus (SARS-CoV-2)
Scarcity, 242
  in economics, 248–249
Scenarios for data usage, 228–230
Schistosoma mansoni cercariae, 153
Schistosomiasis, 153
"Schmutzdecke" layer, 154–155
Science gateway. *See* Web-based platform
Scientific scenario, 229
SDC. *See* Source directed controls (SDC)
SDG. *See* Sustainable development goal (SDG)
Sea surface temperature (SST), 345–346
Sea-derived plastic wastes, 181
Seasonal adaptation, 349
Seawater, 197–198
Seawater desalination process (RO process), 210–211
Secondary MPs, 179–180
Secondary treatment, 76, 78
Secure shelter, 392–393
Sediment, 298–299, 366
Sedimentation, 125, 295
Self-realization, 430
Sense of place concept, 350
Separation process, 198
Sequent Peak Algorithm (SPA), 34
Service-oriented computing paradigm, 231–232
Sewer disposal, 205

SGMA. *See* Sustainable Groundwater Management Act in California (SGMA)
Shallow wells, 99–100
Sigmoidal-shaped curves, 335
Silica, 6
Silver-modified ceramic filters, 130
SIMILE (system dynamics modeling software), 327–328
Single factor productivity (SFP), 243–244
SLM. *See* Sustainable Land Management (SLM)
Slow sand filtration (SSF), 154–155
Sludge
  disposal of, 80–82
    particle size, solids content and extent, 81t
  system, 76–78
  treatment efficiency, 72
Slum, 97, 102
Slurry-phase lagoon, 85
Small scale desalination plants, 206–207
Small-scale systems (SSS), 126
Smart agriculture, 221
Smart buildings, 221
Smart city, 221
Smart contracts, 232–233
Smart devices, 221
Smart grids, 221
Smart GW management, 220–221
Smart mobility, 221
Smart street lighting, 221
Smart systems, 221
Social aspects of desalination, 206–207
Social cohesion, 389
Social justice, 325
Social stability, 392–393
Socio-ecological environment, 432
Socio-ecological system, 420–423, 425, 428, 432
Socio-hydrology, 217–218
Sodium chloride, 6
Soil
  hydrology, 29

  management, 369–370
  moisture, 29
  water, 29
Soil and Water Assessment Tool (SWAT), 219
"Solar ball" method, 166–167, 171t
Solar Collector Disinfection system (SOCODIS system), 149
Solar disinfection, 147–148, 165–167
Solar Disinfection (SODIS), 129, 147–148, 165
Solar energy, 150, 207–208
Solar irradiation, 197–198
Solar PV, 22
Solar radiation, 28
Solar stills method, 198–199
Solar treatment and intensification, 147–150, 149f
Solar Water Disinfection, 129
Solid ice caps of Earth influence climate, 8–9
Solid matrix, 48
Solid water, 6
Solids retention time (SRT), 87
Solubility, 56
Solution to global warming, 22–25
Solvent transfer, 198
SolWat (solar disinfection system), 167
Sorption
  isotherms, 48–49
  process, 47–49
Sorption combine physical and chemical phenomena, 48
Sound manure management, 369–370
Source directed controls (SDC), 274, 287
  and RDM instruments for sustainable freshwater management, 287–290
  tools, 288t
Source water, 99–100, 121
South African Department of Water, 280
South African Department of Water and Sanitation, 280–287

South African National Water Act, 424
South African National Water Resource Strategy (South African NWRS), 433
South African water resource management, 426
SPA. *See* Sequent Peak Algorithm (SPA)
Spatial data infrastructure (SDI), 220–223
Species extinction, 426
Springs, 384
SRT. *See* Solids retention time (SRT)
SSF. *See* Slow sand filtration (SSF)
SSS. *See* Small-scale systems (SSS)
SST. *See* Sea surface temperature (SST)
St. Elisabeth flood (1421), 346–347
Staphylococcus aureus, 163
State Water Code of Hawaii, 215–216
Stockholm Environment Institute, 321
Storage, 294, 296
Storm water management, 308
Stream
    degradation, 295
    restoration, 305–306
        destruction in wake of Tropical Storm Nate that hit Costa Rica, 301*f*
        plastics carried by water flow through ephemeral channel, 303*f*
        restored ephemeral channel receiving stormwater flow, 307*f*
        urbanized ephemeral channel, 306*f*
    water flow, 102–103
Structural racism, 405
Sub-basin management, 105
Sub-Saharan Africa (SSA), 116–117
Substance Flow Analysis, 326–327

Sulfur dioxide (SO_2), 203–205
Sulfuric acid (H_2SO_4), 67–68
Superhydrophobic surfaces, 101
Superhydrophobicity, 101
Surface discharge, 205
Surface impoundment, 81–82
Surface water (SW), 99–100, 215–216
    bodies, 390–392
    sources, 203
"Suspended growth processes", 76
Sustainability, 128–132, 294, 366
    and desalination, 203–207
        economics of desalination, 205–206
        environmental footprint, 203–205
        social aspects of desalination, 206–207, 207*t*
    in Yakima River Basin, 325
Sustainable desalination, 210–212
    at community or municipal scale, 211–212
    desalination centered food-energy-water nexus, 211*f*
    at household level, 210–211
    energy budget for desalination at domestic scale, 210*f*
Sustainable development goal (SDG), 27–28, 95–99, 115–116, 215–216, 245–247
    SDG 6, 215–216
    SDG 11, 215–216
    SDG 14, 215–216
    SDG 15, 215–216
    SGD 13, 215–216
Sustainable freshwater management
    classification system and classification of water resources, 275–276
    determination of reserve and resource quality objectives, 277–278
    linking RDM and SDC instruments for, 287–290, 289*f*
    reflections on resource directed measure

components using Vaal Barrage catchment, 278–287
    SDC, 287
resource management, 420
    environmental ethics in context of complex socio-ecological systems, 432–433
    key issues and need for environmental ethical considerations in, 421–423
Sustainable Groundwater Management Act in California (SGMA), 215–216
Sustainable Land Management (SLM), 29
Sustainable rainwater harvesting, 165–166
Sustainable water management, 294, 304–310
    challenges to effective management, 304–305
    dam removal and wetland creation, 307–308
    green infrastructure, 306–307
    managing worsening concerns, 299–304
        dead zones, 300
        drought, 302
        eutrophication, 300
        flooding, 301–302
        pollutants of emerging concern, 302–304
        red tides, 300–301
    mitigation, protection, and ecological services, 310
    sanitation, 308–309
    sociopolitics and economics, 309–310
    stormwater management, 308
    stream restoration, 305–306
    sustainable water future, 310–311
    systems, 133–135
        failed water pump, 135*f*
Sustainable water purification methods, 146
SW. *See* Surface water (SW)

SW-846 method, 80
SWAT. See Soil and Water Assessment Tool (SWAT)

T

Taaibosspruit Rives, 279–280
Tamarind (*Tamarindus indica*), 159–160
Tamarindus indica. See Tamarind (*Tamarindus indica*)
TDS. See Total dissolved solids (TDS)
Technologic scenario for data usage, 229
Telehealth, 221
Temperature, 298–299
　body, 5
　changes to evapotranspiration rates, 298
　Earth, 14
　normal temperature tolerances of aquatic organisms, 56t
　water, 42–44
Tertiary treatment, 78
Test Methods for Evaluating Solid Waste, 80
Tetrodotoxin, 390–392
Texas Water Development Board, GW database of, 223–224
TFP. See Total factor productivity (TFP)
The origin of water, 9–10
　scarcity of water, 10
　water cycle, 9–10
"The Safe Water Trust", 169–170
The water cycle, 9–10
Thermal desalination, energy efficiency of, 200
　desalination capacity, 201f
Thermal energy, 198, 200
Thermal vapor compression (TVC), 198
Thermocline development, 102–103
Thermodynamics of cycling, 42–45
　fate of matter leaving combustion chamber, 43f
　mass and energy flows in food chains, 44f

Thermotolerant coliforms (TTC), 120–121
THMs. See Trihalomethanes (THMs)
Threshold
　concentrations, 66–67
　of pollutant, 331–332
　criteria, 66
　"no effect", 338–339
　safety, 337–338
Time-reliability, 30–31
Tipping points of climate change, 18–19
Total dissolved solids (TDS), 47
Total factor productivity (TFP), 257
Total organic carbon (TOC), 81
Total suspended solids (TSS), 47
Tourism, 385
Toxic contamination from pesticides, 370
Toxins, 6, 298–299
Trade-offs, 317–318, 420, 426–427
Traditional coagulation systems, 128
Traditional filtration methods, 154–156, 173
　ceramic filtration, 156f
　recent advances/modifications in, 156–158
Traditional flocculation systems, 128
Traditional storage systems, 296
Trans-generational solidarity, 424, 431–432
Trans-generational value system, 431
Transboundary pollution, 102–103
Transformative values, 424
Transport emissions, 407
Treatment
　methods
　　hybrid water, 158–159, 163
　　natural, 159–160
　　techniques, 147
　　water supplies, 103
Trickling filter, 76, 82–83
Trigger for initiating global warming, 15

Triggering global warming, 16–17
Trihalomethanes (THMs), 153
Tropical storms, 97
TSS. See Total suspended solids (TSS)
TTC. See Thermotolerant coliforms (TTC)
Tulsi (*Ocimum sanctum*), 159–160
TVC. See Thermal vapor compression (TVC)

U

Ubuntu (African concepts), 431
Ubuntu/Botho, 431
Ukama, 431
Ultrafiltration (UF), 158
Ultraviolet (UV), 79
　rays, 147–148
　UV-A radiation, 148–149
　UV/vis radiation, 15
UN Economic Commission for Europe (UNECE), 107
UN Educational, Scientific and Cultural Organization (UNESCO), 107
UN Environment Program (UNEP), 107
UN General Assembly, 390–392
UN Multidimensional Poverty report, 386
UN Sustainable Development Goals, 392
UN University (UNU), 107
UN World Water Development Report, 365
Uncertainty
　analysis, 219
　assessment analysis, 327–328
　of water pollutants, 337–339
Uncertainty factors (UF), 338
Underground injection wells, 81–82
Uniform density function, 34
Unique chemical properties of water, 4–8
　acid–base property, 7
　anomalous density of frozen water, 5–6
　high enthalpy of vaporization of water, 4–5
　high heat capacity of water, 5

high surface tension, low
viscosity, and cohesive and
adhesive properties, 7–8
polar nature of water, 4
water, "universal" solvent, 6–7
Unitary system, 422
United Nation Development
Program's Human
Development, 410–411
United Nations (UN), 115–116
United Nations Children's Fund
(UNICEF), 107, 116–117,
123
United Nations Framework
Convention on Climate
Change (UNFCCC), 24
United Nations High
Commissioner for Refugees
(UNHCR), 392–393
United Nations Office for Disaster
Risk Reduction, 389
United States Environmental
Protection Agency (EPA),
299
University of Cambridge, 321
Unregulated waste disposal, 104
UN's Sendai Framework for
Disaster Risk Reduction,
387
Upflow anaerobic sludge blanket
process (UASB), 87–88
Upper rule curve (URC), 33–34
Urban development, 326–327
Urban flood resilience concept,
346–347
Urban metabolism, 326–327
Urban populations in developing
countries, 125–126
Urban water supplies, 97–102
centralized and noncentralized
water supplies, 101–102
climate change challenges and,
102–103
integrated approaches, 98f
opportunities and global
initiatives, 104–106
as resource, 101
source waters, 99–100
urbanization and extent in
developing countries, 98f

urbanization in developing
countries, 97
waste management and climate
change and impact on, 104
water security, 106–107
water sources innovative needs,
100–101
Urbanization, 263
in developing countries, 97
US Centers for Disease Control
and Prevention (CDC), 151
US Department of Agriculture
(USDA), 327
US Environmental Protection
Agency's Remediation
Guidance Document, 80
US Geological Survey (USGS),
327
US National Academy of Sciences,
369
US National Groundwater
Monitoring Network,
223–224
USDA Economics Research
Service, 327
Utilitarian approach, 401
Utilitarianism, 401–402, 433

V

Vaal Barrage catchment, resource
directed measure
components using,
278–287
reflection on MC and REC
for Suikerbosrand and Klip
River system, 281t
for Taaibosspruit and Vaal
River, 282t
reflections on water quality
component of proposed
ecological reserve
EWR, 283t, 284t
for Vaal River at De Neys,
285t
for Vaal River at Scandinavia,
286t
Vaal Barrage region, 280f
Vaal catchment system, 280–287
Vaal Dam, 279
Vaal River system, 278–279

Value added model, 254–256
Value dualism, 429
Value-oriented environmental
ethics, 423–428
anthropocentric environmental
ethics, 423–425
nonanthropocentric
environmental ethics,
426–427
Values, social, technical,
environmental, economic,
and political analysis
(VSTEEP analysis),
419–420
van Der Waals forces, 45
Vapor compression, 199–200
Vector-borne diseases, 390–392
Vector-transmitted diseases,
392–393
"Veil of ignorance", 401–402
Ventana Simulation Environment
software (Vensim
software), 321
Vertical evacuation concept,
346–347
Virtual community platform, 221
Virtual laboratory, 221
Virtual research environment, 221
Volumetric reliability, 30–31
Vulnerability, 31, 37, 105,
386–387, 404–405
water justice and climate
change, 404–406

W

Waste activated sludge (WAS), 78
Waste management and impact on
urban water, 104
Waste piles, 81–82
Wastewater, 78, 87–88, 131–132,
263–264
as energy source, 268–269
as nutrient source, 265–268
as water source, 264–265
Wastewater treatment, 66, 73–79
activated sludge treatment
system, 77f
flow of wastewater and
constituents, 75f
passive system, 74f

Wastewater treatment (*Continued*)
  process flow diagram, 79*f*
  trickling filter system, 76*f*
Wastewater treatment plant (WWTP), 72, 75–76, 269
Water, 3, 14–16, 27, 65, 95–99, 124, 145–146, 294, 309–310, 317–318, 372, 422
  "universal" solvent, 6–7
  biogeochemistry of oxygen in, 55–57
  consumption and contamination tied to agriculture, 365–366
  cycle, 9–10
    changes to, 295–296
  demand, 95–96, 321
  desalination, 100
  disinfection by boiling, 150–152
    microbes and temperature kills, 151*t*
  economics, 254
  extractions, 118
  molecule, 4
  origin of water on Earth, 9–10
  scarcity of, 10
  shortage, 31–32
  single oxygen, 45
  sources, 120, 123
  sources innovative needs, 100–101
    atmospheric water harvesting, 101
    water recycling, 101
  stress, 27
  sustainability, 97–99
  system, 327
  temperature, 42–44
  unique chemical properties of, 4–8
    acid–base property, 7
    anomalous density of frozen, 5–6
    high enthalpy of vaporization of, 4–5
    high heat capacity of, 5
    high surface tension, low viscosity, and cohesive and adhesive properties, 7–8
  polar nature of, 4
  as "universal" solvent, 6–7
  water and climate change, 8–9
Water baseline, understanding, 105
Water budget components, 28–29
Water energy food nexus Tool (WEF Nexus Tool), 319
Water Evaluation and Planning System (WEAP), 319
"Water for Life", 115–116
"Water for Sustainable Development", 115–116
Water governance, 401, 420
  context of, 401
Water Information Network System of Intergovernmental Hydrological Program, 223
Water Infrastructure, 220–221, 294
Water injustice, 400, 406
Water justice, 401–404
  proposed solutions, 406–410
    adaptation, 408–410
    mitigation, 407–408
  struggle(s) for water and climate justice, 411
  vulnerability, 404–406
Water management, 97, 420
  changes in, 242
  conventional, 197–198
  in developing countries, 121–122
  non-conventional, 197–198
  sustainable water management systems, 133–135
  systems, 294
Water pasteurization indicators (WAPIs), 151–152
Water policy formulations, 423
Water pollutants, 69
  dose-response relationships, 332–337
  drinking water contaminant candidate list 5-Draft, 333*t*
  uncertainty, 337–339
Water pollution, 370
  control, 374–377
    BMP, 375
    intensive management practices, 376–377
  land management practices, 375–376
  water quality, 374–375
Water Priority, 325
Water productivity (WP), 241–242
  examples of nomenclature for use with, 244*t*
  factors in estimating input and output in water productivity estimation, 250*t*
  production and scarcity in economics, 248–249
  productivity concepts as indicators, 243–248
  trends in, 253–257
    economic net water productivity trends, 254–256
    other economic approaches to water productivity and efficiency, 256–257
    physical and economic water productivity trends, 253–254
Water productivity indicators, 249–253
  economic considerations, 251–252
  net water productivity, 252–253
  physical considerations, 250–251
Water purification techniques, 147
  cavitation-based water hand pumps, 160–163
  chemical and microbial pollutants encountered in water, 146*t*
  chemical treatment, 152–154
  comparison of various purification techniques, 170–173
  Covid 19 pandemic impact on water sector in developing countries, 168–170
  filtration techniques, 154–159
  hybrid filtration methods, 158–159
  methodologies, 147
  natural treatment methods, 159–160

Index   457

recent advances/modifications in traditional filtration methods, 156–158
recent research on emerging methods, 166–167
  filtration, 166
  hybrid techniques, 167
  solar disinfection, 166–167
solar treatment and intensification, 147–150
sustainable disinfection of harvested rainwater, 163–166
traditional filtration methods, 154–156
water disinfection by boiling, 150–152
Water purity in developing countries
environmental challenges to, 118–120
  baseline global water stress, 119f
  WHO guidelines for water purity, 120–123
Water quality, 65, 120, 126–127, 263, 294, 374–375, 389
  engineering, 66–69
    aerobic reactors, 83–86
    anaerobic reactors, 86–88
    BOD curve, 68f
    chemical process, 69–73
    disposal of sludge, 80–82
    mixed reactors, 82–83
    treating contaminated ground water, 89–91
    wastewater treatment, 73–79
    water treatment to manage risks, 79–80
  indicators, 123
  metric evaluators, 368
  parameters, 298–299
  protection, 377
  system, 321
Water recycling, 101
Water reservoirs, 29–37
  climate change effect on reservoir performance in Indus Basin, 32–37

indices of reservoir performance, 30–31
reservoir operation, 31–32
Water resource, 27–28
  central premise of climate change impacting, 294–295
  classification system and classification of, 275–276
    ecological categories and description applicable to water resources, 276t
    relationship between ecological and user categories and management class, 277f
    water user requirement categories, 277t
  management, 220–221, 310–311, 420–421, 433
    context of, 421
    ethic, 426, 433
  planning
    climate change and water budget components, 28–29
    water reservoirs, 29–37
  policy, 419
Water resource classification system (WRCS), 274–275
Water scarcity, 118, 241, 372
Water sector in developing countries, Covid 19 pandemic impact on, 168–170
Water security, 96–99, 106–107
  policy makers, 106–107
Water storage
  changes in, 296–297
    hydropower dam representing water storage and electricity generation, 297f
  management, 28
  tanks, 132–133
Water supply, 66, 95–96
  and demand, 103
  and infrastructure, 97
  solutions in developing countries, 123
  sources in developing countries, 123–124
  and treatment technologies, 132

Water Sustainability Act in British Columbia, 215–216
Water treatment, 66, 124, 152
  to manage risks, 79–80
  model, 326
  process, 153, 326
  systems in developing countries, 124–133
    centralized water treatment systems, 124–126
    commonly used water treatment methods, 128–132
    decentralized water treatment systems, 126
    household water treatment system methods, 127f
    rooftop rainwater harvesting system, 132f
    SODIS steps and SODIS bottles, 129f
    water safety plans, 126–127
    water safety plan process and associated modules and tasks, 128f
    water storage, 132–133
Water use efficiency (WUE), 245–247
Water use licenses (WUL), 274, 287–290
Water-energy relationships, 52–55
Water-submodel, 325
Water–ecology–human system, 217–218
WaterShare, 223
Wave energy sources, 207–208
Weak land resource management, 99–100
WEAP (software models), 321, 327
WEAP-LEAP software models, 321–322
Weather conditions, 13–14
Web-based data sharing, 222–225
Web-based groundwater platforms, perspectives of, 230–233
Web-based GW platform, 229–232

Web-based platform, 221–222, 227
WEF Nexus Rapid Appraisal Tool, 320
WEF Nexus Tool 2.0, 319
Western environmental ethics, 423, 430–431
Wetlands, 84–85, 306
　creation, 307–308
Wildfires, 302, 387, 399–400
Wind energy sources, 207–208
Wind storms, 302
Wind turbines, crop for, 348
Wind-generated electricity, 326
Witwatersrand-Vaal Triangle region, 278–279
Wood's glass, 149–150
Workflow, 221
　for data processing, 225–228
World Bank, 388, 410
World Development Report, 410
World Health Organization (WHO), 107, 116–118, 123, 390–393
　guidelines for water purity, 120–123
　　acceptability, 122
　　chemical guidelines, 121–122
　　considerations, 122–123
　　microbial guidelines, 120–121
　　radiological guidelines, 122
World Meteorological Organization (WMO), 15, 107
World Wars, 345–346, 355
World Water Development Report (WWDR), 28
World-wide Hydrogeological Mapping and Assessment Program (WHYMAP), 223
WP4 project deliverables, 344

Y

Yangtze River, 186, 189–190
Yellow River, 186

Z

Zero liquid discharge systems (ZLD systems), 205
Zika virus fever, 390–392
Zoonotic diseases, 21

Printed in the United States
by Baker & Taylor Publisher Services